COASTAL DYNAMICS

ADVANCED SERIES ON OCEAN ENGINEERING

Series Editor-in-Chief
Philip L- F Liu (*Cornell University*)

Published

Vol. 20 The Theory and Practice of Hydrodynamics and Vibration
by Subrata K. Chakrabarti (Offshore Structure Analysis, Inc., Illinois, USA)

Vol. 21 Waves and Wave Forces on Coastal and Ocean Structures
by Robert T. Hudspeth (Oregon State Univ., USA)

Vol. 22 The Dynamics of Marine Craft: Maneuvering and Seakeeping
by Edward M. Lewandowski (Computer Sciences Corporation, USA)

Vol. 23 Theory and Applications of Ocean Surface Waves
Part 1: Linear Aspects
Part 2: Nonlinear Aspects
by Chiang C. Mei (Massachusetts Inst. of Technology, USA),
Michael Stiassnie (Technion–Israel Inst. of Technology, Israel) and
Dick K. P. Yue (Massachusetts Inst. of Technology, USA)

Vol. 24 Introduction to Nearshore Hydrodynamics
by Ib A. Svendsen (Univ. of Delaware, USA)

Vol. 25 Dynamics of Coastal Systems
by Job Dronkers (Rijkswaterstaat, The Netherlands)

Vol. 26 Hydrodynamics Around Cylindrical Structures (Revised Edition)
by B. Mutlu Sumer and Jørgen Fredsøe (Technical Univ. of Denmark, Denmark)

Vol. 27 Nonlinear Waves and Offshore Structures
by Cheung Hun Kim (Texas A&M Univ., USA)

Vol. 28 Coastal Processes: Concepts in Coastal Engineering and Their Applications to Multifarious Environments
by Tomoya Shibayama (Yokohama National Univ., Japan)

Vol. 29 Coastal and Estuarine Processes
by Peter Nielsen (The Univ. of Queensland, Australia)

Vol. 30 Introduction to Coastal Engineering and Management (2nd Edition)
by J. William Kamphuis (Queen's Univ., Canada)

Vol. 31 Japan's Beach Erosion: Reality and Future Measures
by Takaaki Uda (Public Works Research Center, Japan)

Vol. 32 Tsunami: To Survive from Tsunami
by Susumu Murata (Coastal Development Inst. of Technology, Japan),
Fumihiko Imamura (Tohoku Univ., Japan),
Kazumasa Katoh (Musashi Inst. of Technology, Japan),
Yoshiaki Kawata (Kyoto Univ., Japan),
Shigeo Takahashi (Port and Airport Research Inst., Japan) and
Tomotsuka Takayama (Kyoto Univ., Japan)

Vol. 33 Random Seas and Design of Maritime Structures, 3rd Edition
by Yoshimi Goda (Yokohama National University, Japan)

Vol. 34 Coastal Dynamics
by Willem T. Bakker (Delft Hydraulics, Netherlands)

*For the complete list of titles in this series, please write to the Publisher.

Advanced Series on Ocean Engineering — Volume 34

COASTAL DYNAMICS

Willem T Bakker
Delft Hydraulics, Netherlands

World Scientific

NEW JERSEY • LONDON • SINGAPORE • BEIJING • SHANGHAI • HONG KONG • TAIPEI • CHENNAI

Published by

World Scientific Publishing Co. Pte. Ltd.

5 Toh Tuck Link, Singapore 596224

USA office: 27 Warren Street, Suite 401-402, Hackensack, NJ 07601

UK office: 57 Shelton Street, Covent Garden, London WC2H 9HE

British Library Cataloguing-in-Publication Data
A catalogue record for this book is available from the British Library.

Advanced Series on Ocean Engineering — Vol. 34
COASTAL DYNAMICS

ISBN 978-981-270-373-6

Typeset by Stallion Press
Email: enquiries@stallionpress.com

Printed by FuIsland Offset Printing (S) Pte Ltd Singapore

Dear reader,

Many of the original ideas in this book have been published in series of papers in the past decades. Seldom these papers where purely my work alone. Others have contributed either in the underlying research or where co-author of these papers. This book is therefore really the product of

Willem T. Bakker and friends
THANK YOU, dear friends!

Agema
van der Biezen
Bilse
van Bochove
Dijkman
Grieve
Heemink
Hulsbergen
Jallah
Joustra
van der Kerk
Kersting
Klein Breteler
Knaapen
Niemeyer
Opdam
van Overeem
Pylarczyk
Roos
Smit
Speekenbrink
ten Hoopen
van Vechgel
de Vroeg
de Vriend

Forum

∗ ing. P. Roelse	**RWS RIKZ**
∗ dr. ir. J.A. Roelvink	**Delft Hydraulics**
∗ dr. ir. H.J. Steetzel	**Alkyon b.v.**
∗ ir. A. Bliek	**Svasek b.v.**
∗ ir. H.D. Rakhorst	**RWS dir. N-Holland**

Preface

When Lord created Man, he created two species:

A. the kind which listens to you;
B. the kind which does not listen to you.

Maybe your feeling is the same as mine, that species A is rather rare compared to B and your joy when you meet the A-kind can be compared with the one of an ornithologist, finding a rare species. Still, reviewing a human life time, I think I may be grateful for the number of findings.

Mostly, when they met, they said: "Listen, Wim, explaining something to someone is like showing your house to some people. You should take your guest by his hand, showing him the entrance first, and then...".

After that, muttering they moved to their desk and started to write; they wrote about what they thought I thought, or about what they thought I should think on things, if I had the time to think about those things. And, because they were intelligent people (by definition) what they wrote was always better than I could have thought. In this way species A and I wrote many papers together.

Given the opportunity[a] to organize a meeting for the A-species (mostly new findings; young people) and to tell them about the thoughts their predecessors and I had together, I decided to staple a number of former papers. I edited those in this way, that I tried to filter out duplications and that I made links.

In fact, all my friends together built a house for me, in which I felt myself very much at home. The aim of the present book is to show this house rather than their individual performances. Whether correct or not, whether outdated or not, it has one definite characteristic, in plain European: "Es ist aus einem Guss".

[a]May 29, June 3 and 5 1998.

I remember a frowning remark on the title of these notes: "Coastal Dynamics? Can't you be more specific?" The answer is: "This is Coastal Dynamics by Willem T.Bakker and friends". In the Introduction I shall try to show the limitations more verbalistic.

I should give the builders the credit they owe. At the end of this preface I made an alphabetic list of first-authors or co-authors of the stapled papers.

However, not everything is co-production and these notes contain as well information not published before.

Still, acknowledgment to the authors is not enough. Part of the A-species is A-species by appointment or by profession.

Writing on Coastal Engineering as an engineer of Rijkswaterstaat means interference with many people and your work could not be carried out without the support of many, mostly unnamed people. The same is the case at Universities. There is the direct support of the people, who make drawings for you or mark your lips.

I should name the whole staff of the former Department of Coastal Research of the former Directorate for Water Management and Hydraulic Research of the Rijkswaterstaat.

Many of the drawings in this booklet, Mr. Dofferhoff, originate from their work. I still remember the precious writing of my many formulas in my former reports by Ellen Brans; those reports are at the basis of the present lectures.

Part of this book consists of class notes made during my stay as Visiting Professor at the University of Hawaii. Thank you, Warren Bucher; thank you Gary Mocke, thank you Edith Katada and your staff for the work you did.

And how could I have delivered this work without the help of Edwin Elias (doing the Word processing) and Linda van Veen (translating some parts in English) and of Mark Voorendt, a super wizard spelling all the Word Wizards. Those Word Wizards put things on my laserprinter I never wanted to.

I wish to thank Christel Somers, who reviewed "Coastal Dynamics".

But I should like to honor as well the numerous skippers and observers on the measuring vessels and behind the water leveling apparatus. Names like Glim, Vonk, Blokland, Scheers, Goedegebuure[b] cross my mind.

[b]Apologies to the many people, which were in my mind when I wrote "Unnamed People" and which will be still remain unnamed after the publishing of this book.

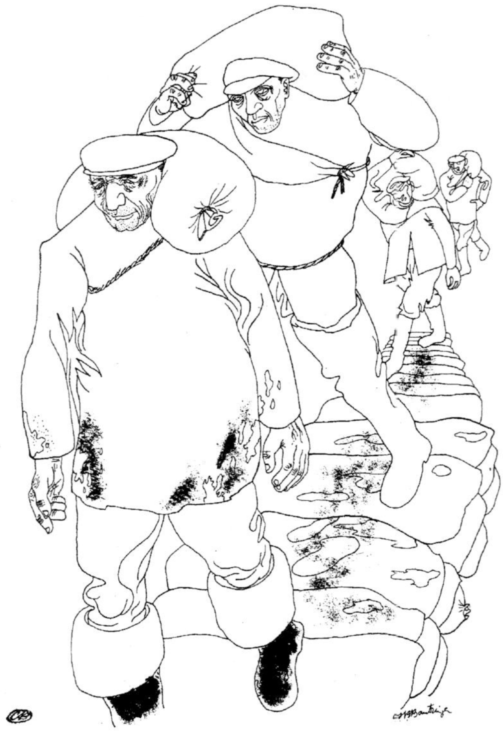

Fig. 0.1 This figure is to honor people like them: generally the brand of, what I would call "Rijkswaterstaat people": sturdy, reliable, no fuss (a drawing from Elsevier (by Bantzinger) from many years ago).

Chapter 2 is derived from three papers (references [37], [57] and [61]). Chapter 3 (translated by Gary Mocke) originates from the Dutch text of the course "De Dynamica van Kusten" on Delft Univ. of Technology, given by me, apart from Section 3.4 (refer to reference [42] in Chapter 3). Chapter 4 has been published in [37] of Chapter 4; (cf. references [20] and [36] of Chapter 4 as well). Section 5.1 is a report of Delft Hydraulics (special thanks to Kees Hulsbergen (reference [53] of Chapter 5). Refer to (reference [10] in Chapter 2) with respect to Section 5.2.2 and to (reference [35] in Chapter 5) concerning Subsections 5.2.3 and 5.2.4. Subsections 5.2.5 originates from ICCE'90 paper (reference [25] in Chapter 4) and 5.2.6 from the NOURTEC study (reference [54] in Chapter 5). The evaluation of the line theories (Chapter 6) is new. Chapter 7 deals on interaction between outer deltas and tidal basins.

Putting the co-authors of the various papers quoted above in alphabetic order, the result is: Agema; V.d.Biezen; Bilse; V Bochove; Dijkman; Grieve; Heemink; ten Hoopen; Hulsbergen; Jallah; Joustra; V.d.Kerk; Kersting; Klein Breteler; Knaapen; Niemeyer; v.Overeem; Pylarczyk; Roos; Smit; Speekenbrink; v.Vechgel; de Vroeg.; de Vriend, (last but not least).

Still, this list does unjustice to a number of people, who had either an essential part in underlying reports, or by papers, not directly cited here. I would like to mention specially Edelman (references [7], [31] and [48] in Chapter 2, references [6] and [9] in Chapter 3); Svasek (reference [9] in Chapter 2), (reference [8] in Chapter 4); V.d.Meulen [M918 model research), Casteleyn (reference [16] in Chapter 4), Swart (reference [9] in Chapter 4); Opdam (reference [34] in Chapter 2); Bult (reference [24] in Chapter 10); Dijkstra (reference [27] in Chapter 10); Andringa (reference [6] in Chapter 4) (Zeebrugge); the collaborators on the Thorsminde investigation (reference [55] in Chapter 5): Groenewoud, v.d. Graaff, Spanhoff, Laustrup.

I am grateful to Edelman and Bijker — my former bosses — for creating a biotope in which I could flourish; this product is one of the results. Bijker decided, we decided, that "Coastal Dynamics" should be finished in 1973. Sorry Eco, you had to do with a civil servant and at that time[c] work of civil servants sometimes, somehow, for some reason could have some delay. Here it is at last.

I want to thank Prof. Dette. Only mentioning the M. Sc. Thesis of Uwe Patzold (reference [15] in Chapter 5) (performed under Prof. Dette's guidance) does not express the years of close cooperation and friendship, which lead to the Sylt-study (Subsection 5.2.3).

May I mention my pin-ups? In a way, they were driving forces behind my work as well, eventually.

Prof. De Josselin de Jong (enflaming the greediness for understanding the essence of things and triggering the appreciation of "an elegant theory"); Prof. Thijsse (his audience got hilarious when he quoted his name and affiliation: so what should I say? He kept Lorentz from his work (relativity theory) by bothering him with the enclosure of the Zuyderzee; he made science an engineering tool); Tjebbe Visser (the Master of the Word-the Word which becomes steel and concrete; as a student he could silence thousand drunken students and after that he could evidently make a storm surge barrier).

---

[c]Not now, of course. Civil servants are always very quick and efficient.

I thank the Forum (Chapter 8) for their critical comments on the contents of this book. I did not want to leave all my biases to my audience, they succeeded to give the audience a more-dimensional view on the coast.

I thank the publishers (ASCE: Elsevier, the American Shore and Beach Preservation; SOGREAH; Delft Hydraulics; the Royal Institute of Engineers; Centre of Civil Engineering, Research, Codes and Specifications; Prof. Moutzouris; Mrs. Bantzinger) for giving permission to edit this book in the way it is and giving me a waiver for the copyright.

Finally, I would like to stress, that this report and this lecture series would not have been possible without the logistic and financial support of the Rijkswaterstaat (Prof. Dr. Ir. J. Dronkers) and Delft University of Technology (Prof. Ir. K. d'Angremond).

Contents

Note From the Publisher

Some of the figures in this book are of poor quality. However, the author has requested to publish the book as it is because the original figures are not available anymore. According to the author, the reader will be able to understand the meaning of the figures, because they are explained in the book.

We believe this book will still be of great value for anyone interested in Coastal Dynamics as it is. We hope you will agree!

Chapter 1

Introduction

The solving of coastal problems will remain for years rather an art than a knowledge, with the slogan: 'the art of engineering is to get sufficient conclusions from insufficient data". However also art needs skill, and a mathematical theory; which clarifies in an exact way the implications of several schematizations of reality, may provide a part of this skill. At the moment only several aspects of the processes, which occur in nature can be schematized and translated into a mathematical theory. In the transition of the real prototype to the mathematical model the danger is hidden, to concentrate on an insignificant feature, which gives pseudo exact results. This can be avoided by active knowledge of the prototype and carrying out measurements in prototype and hydraulic models, from which a "smoothing of data" can result. It is cheaper and faster to consider only one of the three aspects: prototype, theory and laboratory, however, often only a combination of all of these aspects reveals the truth. The theory indicates pertinent contradictions in the measurements, but measurements in prototype are necessary to keep the attention on invalid schematization.

This book deals on "Coastal Dynamics", which will be defined in a narrow sense as a mathematical theory, which starts from given equations of motion for the sediment, which leads with the continuity equation and given boundary conditions to a calculated (eventually schematized) coastal topography, which is generally a function of time. This is clearly analogous to aero- and hydrodynamics, thermo-dynamics, hydrology and other related fields.

The subject of this book, however, covers only a specific part of the Coastal Dynamics. It is based upon the notion of old masters, that for a manager the back-side of a cigar-box should be large enough to evaluate all the information and anti-information which is poured over him. For instance: statements, based upon high-tech number-crunching can find a sad end when those do not match large-scale continuity. Think for instance of tidal computations, in which time-varying boundary conditions (shoals which emerge above water level during a part of the tide) are not reproduced accurately enough.

The positive role of refined numerical techniques on solving problems in coastal dynamics should be stressed. However, one will not find much about it in this book. Emphasis is on physics, rather than on mathematics. It is meant for coastal managers, to inspire them, to enable them to put sensible questions and, if necessary, to say: "Oh, no sir".

The theory would be sterile without a consideration about the validity of the equation of motion, so much the more, while these equations of motion are not as evident, straight-forward and single-valued as for instance the Euler-equations in hydrodynamics.

The Newtonian laws are not quite sufficient for the computation of the sediment motion, because the motion of grains is subject to stochastic processes as the shape of the grains, the shape of the granular bed surface, turbulence, irregular wave motion, etc.

In modern sophisticated computermodels, a physical approach going far into details is possible, this includes the consideration of water and sand apart from each other, investigation of turbulent and viscous forces on the grains, and calculation of the sediment motion.

This is a difficult and tiresome way, which has to be followed and will give finally the best results. However, the building of structures cannot wait for that, as it has not done in the past and therefore a short-cut from theory to practice has to be made, which involves the use of experimental data. This includes entirely the loss of the paradise of exactness. As the accuracy of the experimental data is very poor, this is a serious draw-back. Instead of the real equations of motion substitutes have to be chosen and when it is not carefully examined, if and when these assumed equations of motion are valid, the mathematical theory is just an escape from reality.

In the mathematical theory, all constants and variables are indicated by symbols and it is suggested, that the numerical value of many of these constants can be found from prototype data. Without a discussion about at least the order of magnitude of these constants. the theory is merely a start for future research.

Under coastal dynamics in a broader sense also these discussions about the validity of the equations of motion and the order of magnitude of the assumed constants will be included.

In this book the sediment transport formulae, which are used, are not thoroughly discussed.

The coastal dynamics, as presented in this course, provides a far from complete theory.

The aim is threefold:

- It is a start for future research;
- it leads to qualitative conclusions about the effect of coastal structures;
- some of the results are already available for practical computations.

Although based upon universal physics, applications treated are more or less tailor made for Dutch circumstances. The same holds for schematizations used.

However, the book aims at a much broader public than only people of Dutch nationality.

Thus it is important to consider the physical characteristics of the considered examples rather than its location:

a. Nearly all considered (Dutch) beaches are sandy beaches ($D_{50} \approx 200\,\mu\text{m}$); rock never emerges at the surface; effect of peat layers (occurring seldomly, apart from in the Wadden area's and in the estuaries) may be neglected. Beaches like those emerge at many (most) of the places at the East- and Westcoast of the USA and in Japan. Grain size as given above has an ashtonishing university all over the world.
b. A difference between Dutch beaches and beaches elsewhere is the number of breaker bars, counted in cross-shore direction. Nearly nowhere this number exceeds two; however, in the Netherlands the occurrence of three successive breaker bars (Fig. 2.3) is not unusual. Up to now, the physics in detail of breaker bars is not well enough understood to give an explanation.
c. Effect of a rocky sublayer may occur on many places in the USA (for instance: Hawaii). However, the one- and two-coastline theory (displayed in this book) are based upon a solid sublayer; thus the application of the theory, the determination of the "closure depth" will be more straight forward at those kind of beaches than on beaches like the considered ones, where this closure depth depends on the wave height (cf. Chapters 4 and 5)

Thus, it may be concluded, that the theory developed has a much more general application than the choice of examples suggests. Theory is very much in line (qua topic and approach) with international literature, as presented in the bi-annual International Conferences on Coastal Engineering.

Chapter 2 gives an overview of the Dutch coast. The papers, on which it is based upon, are written for a broad public and thus sometimes might give more basics than necessary for most of the present readers.

Chapter 3 gives (at this moment more or less classical) line theory. Chapters 4–6 (determination coastal constants; experimental verification (laboratory and nature) and evaluation) try to expose, whether nature knows coastal dynamics.

Where the former chapters mainly deal with uninterrupted coasts (apart from single discontinuities like harbor moles or groins), in Chapter 7 some ideas are exposed concerning interaction of delta's and tidal basins. It is an important subject for the Netherlands, as a large part of the coast consists of tidal lagoons and estuaries (Chapter 2.1).

As mentioned in the preface, in this book coastal dynamics is considered from one viewpoint. Thus the attention should be drawn on Chapter 8, in which a number of Dutch coastal specialists give additional information and thus can deliver the three-dimensional view.

Chapter 2

The Dutch Coast: A Coastal Engineering View

2.1. Introduction

2.1.1. *Contents of this Chapter*

Coastal dynamics, as defined in the introduction (Chapter 1), is based on universal physics. However, as explained already in the introduction, the scientific language of this book will have a Dutch accent. General laws will be explained using examples regarding mainly to the Dutch coast. And as not all possible boundary conditions on our global sphere are considered, a loss of generality should be taken for granted.

Thus, this chapter (specifically Section 2.1) will start with an overview of the Dutch coast. An impression is given by the map of Fig. 2.1 and the aerial views are displayed in the photographs, Figs. 2.2–2.7. Those are derived from the site of the National Institute for Coastal and Marine Management (RIKZ).[a]

The country is unique in this way, that systematic measurements concerning its beach were already started (at least were saved) in 1850. A display and a phenomenological analysis of those antique measurements (up to 1968) are given in Section 2.2. Phenomenologically based conclusions about, for instance, the effect of groins can be drawn from those measurements. As could be expected, measurements were intensified in the course of time. Especially, the start of systematically profile measurements around 1960 (the so-called JARKUS-measurements[b] gave a more three-dimensional

[a]The author acknowledges gratefully the consent of RIKZ to publish those photographs.
[b]Abbreviation of: "Yearly coastal measurements" (in Dutch).

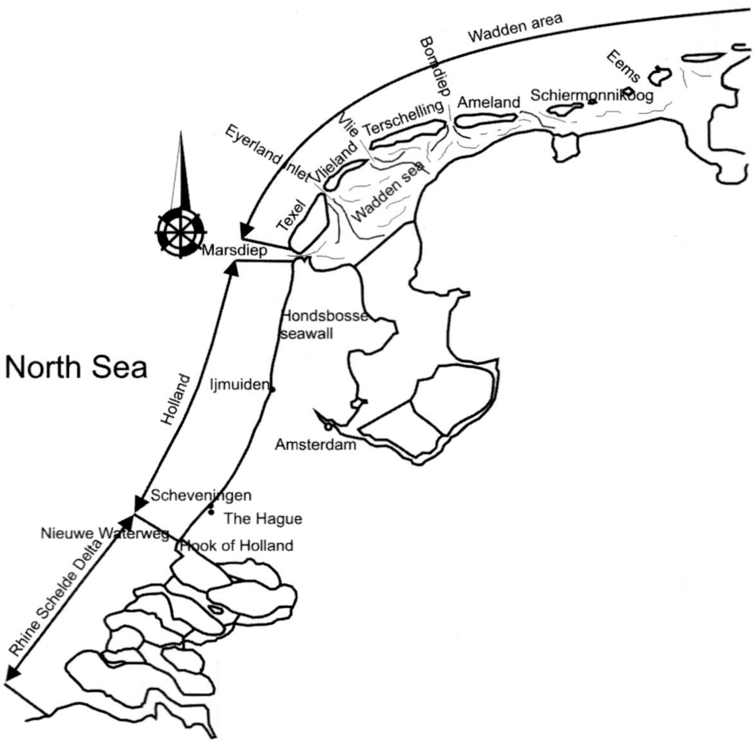

Fig. 2.1 The Dutch coast.

view. Description and analysis are given in Section 2.3. Behavior of breaker bars, effect of harbor moles and the overall coastal behavior, after depth-averaging over the breaker bar area (i.e. the coastal behavior on a larger scale) can be described.

Section 2.4 focuses on the sand transport along the Dutch coast,[c] as derived from the (external) effects of waves and tides. Starting from a simple (i.e.: simplified) transport relation, the CERC-formula, on a yearly basis probability distributions are derived for the longshore transport; this as function of the site along the coast.

However, as will be clarified in Subsection 2.1.2, tidal channels interrupt a large part of the Dutch coast, and as can be expected, morphology of those parts is dominated by those channels.

[c]That is, the part not interrupted by tidal channels.

Fig. 2.2 The Vliehors. Here, we will focus on Subsection 5.3.2. Figure 5.24 shows the site as well as its (historical) "structure": former dunes on the West side[d] of the island were washed away. So presently the island consists of a shoal of ca. 5 km (the Vliehors) on the West side and a remaining inhabited part of ca. 13 km in the East. Figure 2.2 shows the link between the two parts, where (on the landward side) the so-called "Kroon polders" show some result to cultivate a part of the Vliehors.

Section 2.5 deals with the aspect of coastal dynamics and classical theory is memorized concerning the morphodynamics of those areas. Those theories date already from the 1930s. Anticipating on the following chapters of this book, it may already be mentioned that in Chapter 8 other elements and more modern views will pass the review as well.

2.1.2. *Overview of the Dutch coast*

The Dutch coast can be divided into three parts (Fig. 2.1)

- the Wadden area;
- the uninterrupted coast of Holland;
- the Rhine-Scheldt Delta.

[d]Apart from a part (Eyerland) which [helped by the digging of a ditch (the Monniken Sloot, i.e. the Munk Ditch) across the island (by munks from Holland)] "deserted" (from Friesland to Holland) to the most Western Wadden island, Texel and was incorporated there in the sixteenth century. A rudimentary residual of a tidal channel (the Slufter) remained there on the seaside between the original island and its new accretion.

Fig. 2.3 "Hondsbossche Seawall". In the fifteenth and sixteenth centuries former "Wadden islands" (cf. Fig. 2.1) Southernly of the Marsdiep (Fig. 2.1) were linked. The coast straightened and (generally) eroded over a length of 6 km, so the coast was heavily protected. As witness of the fierce struggle of the Dutch against the sea, presently this strongest part protrudes 200 m into sea. Presently this "Hondsbossche Seawall' has been enheightened up to ca. NAP +12 m. (NAP; Normal Amsterdam Level; this is about Mean Sea Level). Variation in dike level (depending on slope, roughness and local wave attack): ca. 75 cm.

A visual impression of the coast, when flying from North to South, is given from the Figs. 2.2–2.7. Mind as well the photographs: Fig. 8.7b, d in Chapter 8.

The pictures are derived from the site of the Rijkswaterstaat, the National Institute for Coastal and Marine Management (RIKZ). The copyright belongs to RIKZ. The author acknowledges the consent of RIKZ to publish those photographs.

With respect to the Southern area (Zeeland), the construction of the Delta Works (ca. 1955–1985) may be remembered.

In this scope, Fig. 2.6 depicts the Storm Surge Barrier in the Eastern Scheldt. In this book, the Delta Works will not be given the attention those deserve, as many good publications are already given in the 3-monthly reports concerning those works. However, as an honor to the many, many people who gave their best forces to those works an impression: a picture of the Eastern Scheldt Storm Surge Barrier should not be omitted.

Fig. 2.4 The IJmuiden Harbor. The harbor moles of IJmuiden were built ca. 1870 and enlarged round 1970. They stretch into sea up to a length of 3 km. Clearly still the old harbor moles (which were shortened during the extension) can also be seen.

Fig. 2.5 The entrance of Rotterdam Harbor (Europort). The view shown is taken from the South and depicts Europort with its coal power plant on the foreground; the New Waterway with its outlet in the center and (vaguely) the South Holland coast ("Hook of Holland") in the middle, at a distance, on the right-hand side.

Fig. 2.6 "The Delta Works" depicts the Storm Surge Barrier in the Eastern Scheldt.

The coast of the Netherlands consists of sand ($D_m \approx 200\mu$). The tidal range varies between 4 m near the Belgian border, down to 1.5 m near Hook of Holland, to 1.3 m near Den Helder and up to 2.2 m near the German Island Borkum. The tidal channel systems of the Rhine-Scheldt delta and that of the inlets between the Wadden Islands in the North have a different shape, since the tidal basins are quite different.

The tidal channel systems in the Rhine-Scheldt delta and that of the inlets between the Wadden Islands in the North have a quite different shape. In the South, basins are rectangular with an open short side towards the sea; in the North these are rectangular with the long side parallel to the coastline; this side is bordered by the Wadden Islands with narrow inlets in between (cf. Section 2.5). Therefore, the Wadden coast can be compared with lagoon coasts, as those occur in many parts of the world. The Wadden Sea is a tidal flat, which is submerged during every high tide.

In order to enable international comparison, data is needed. Main purpose of this chapter is to supply these data, as far the Dutch coast is concerned. It gives a summary of many reports of regional study departments of Rijkswaterstaat, made over a period of about 50 years. Amongst the former papers, covering the same subject, special attention should be

Fig. 2.7 The *Zwin Area.* This series of pictures ends at the Dutch–Belgian border. Instead of reporting on the many dams, which were built, this book will report[e] on a dam, which was *not* built (Subsection 5.3.4).

given to Wentholt,[53] who investigated in particular the effect of groins, Van Veen,[48,49] who dealt with the origin of the Dutch coast and the shape of the tidal channels, Van Bendegom,[14] who investigated the hydraulic laws for the motion of the tidal channels, Van Straaten,[44] who considered the directional effects of winds, waves and currents, and concluded that the sand drift must be strong (west to east) along the northern barrier islands and small between Katwijk and Texel, Edelman and Eggink,[25–28] who drew morphological conclusions from the curvature of the coast. Per Bruun and Gerritsen[18] dealt with the cross-section of tidal channels and the stability of coastal inlets. Bijker and Svasek[19] gave a treatise about IJmuiden harbor. A good example of a recent in-depth morphodynamic study concerning tidal deltas is given by Elias.[29]

The present chapter gives more recent data than Wentholt. The conclusions of Van Veen, Van Straaten, Edelman and Eggink are reviewed in the light of "modern" theories about sand transport of CERC, Bijker and Svasek, making use of tidal computations.

[e]Among other things.

The eldest Dutch coastal data is data concerning the position of the coastline, measured since 1850. These measurements are discussed in Section 2.2. Especially it focuses onto the effect of harbor moles (Subsection 2.2.2), the change of the Wadden islands (Subsection 2.2.3) and on the effect of groins (Subsection 2.2.4).

Since ca. 1960 yearly the whole coastal topography is monitored, as described in Section 2.3. Apart from the morphological behavior of harbor moles, the behavior of breaker bars can be investigated in a similar way. Furthermore, the order of magnitude of the cross-shore transport can be estimated. The magnitude of the longshore transport along the Holland coast (also taking the effect of tides into consideration) is treated in Section 2.4. A quite different effect of tides may be expected near an inlet. The general phenomenology of tidal inlets will be exposed in Section 2.5; the Delta area in particular is highlighted in Section 2.6.

2.2. 150 Years of Beach Measurements

2.2.1. *Display of coastal changes*

Figures 2.8 and 2.9 show the erosion and accretion of the dune foot, the high-tide line and the low-tide line in intervals of 10 years. First the 10-year average of each of these lines was determined, for instance 1856–1865; the distance (in offshore direction) between two successive 10-year averaged "coast"-lines[f] is plotted in the above-mentioned figures. The positions of groins, seawalls or harbor moles are shown and their dates of construction are also indicated (cf. "explanation").

2.2.2. *The effect of harbor moles*

The area of the Holland coast (Fig. 2.8) has been highly influenced by the building of the harbor moles of Hook of Holland and IJmuiden in 1870. Apart from that, the low-tide line shows the influence of climatologic changes: Bakker and Joustra.[7] The overall picture is a large amount of erosion between 1860 and 1880 and accretion between 1880 and 1900. These periodical changes attenuate very strongly cross-shore, in landward direction, as shown by considering the line of the dune foot. Here one finds the general trend as indicated by Edelman and Eggink,[27,28] i.e. a general

[f]Dune foot, high-tide line or low-tide lines.

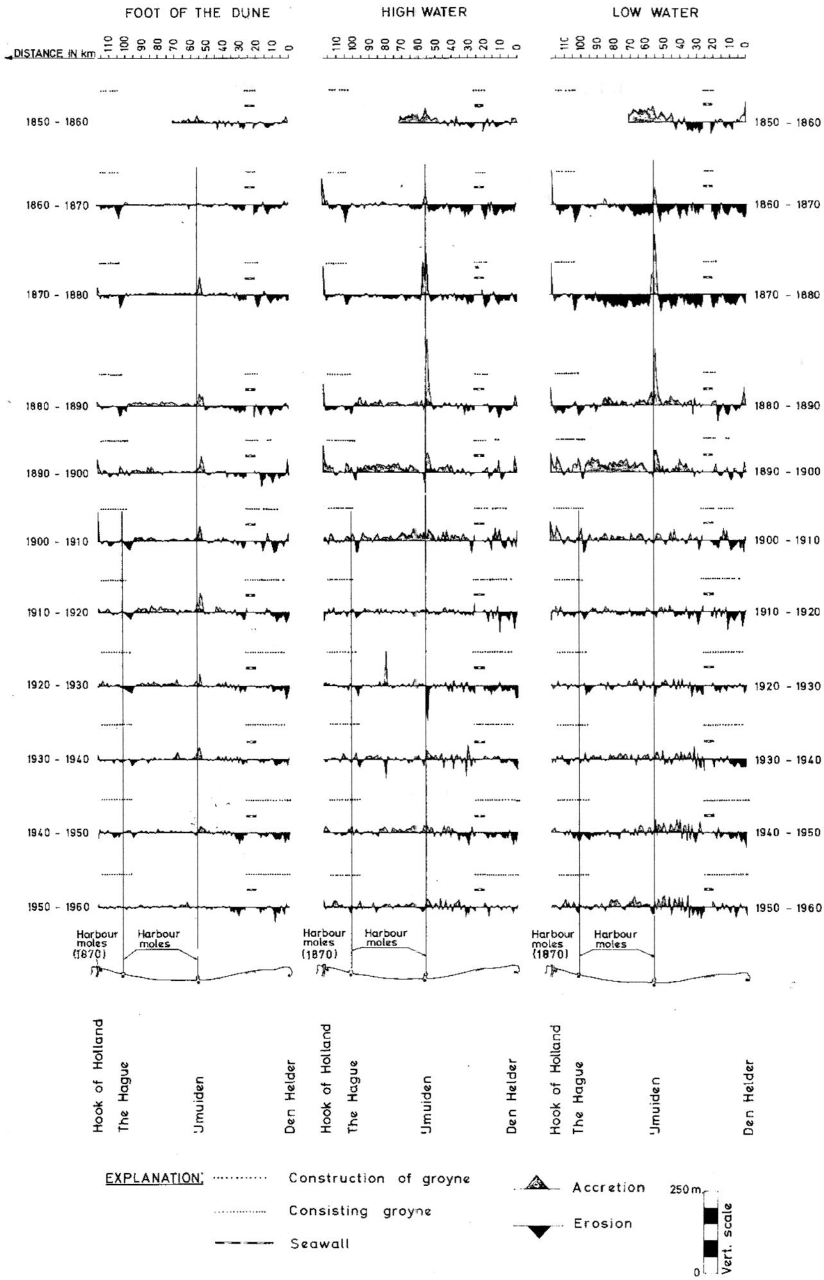

Fig. 2.8 Accretion and erosion of the coast of Holland from Bakker and Joustra.[7]

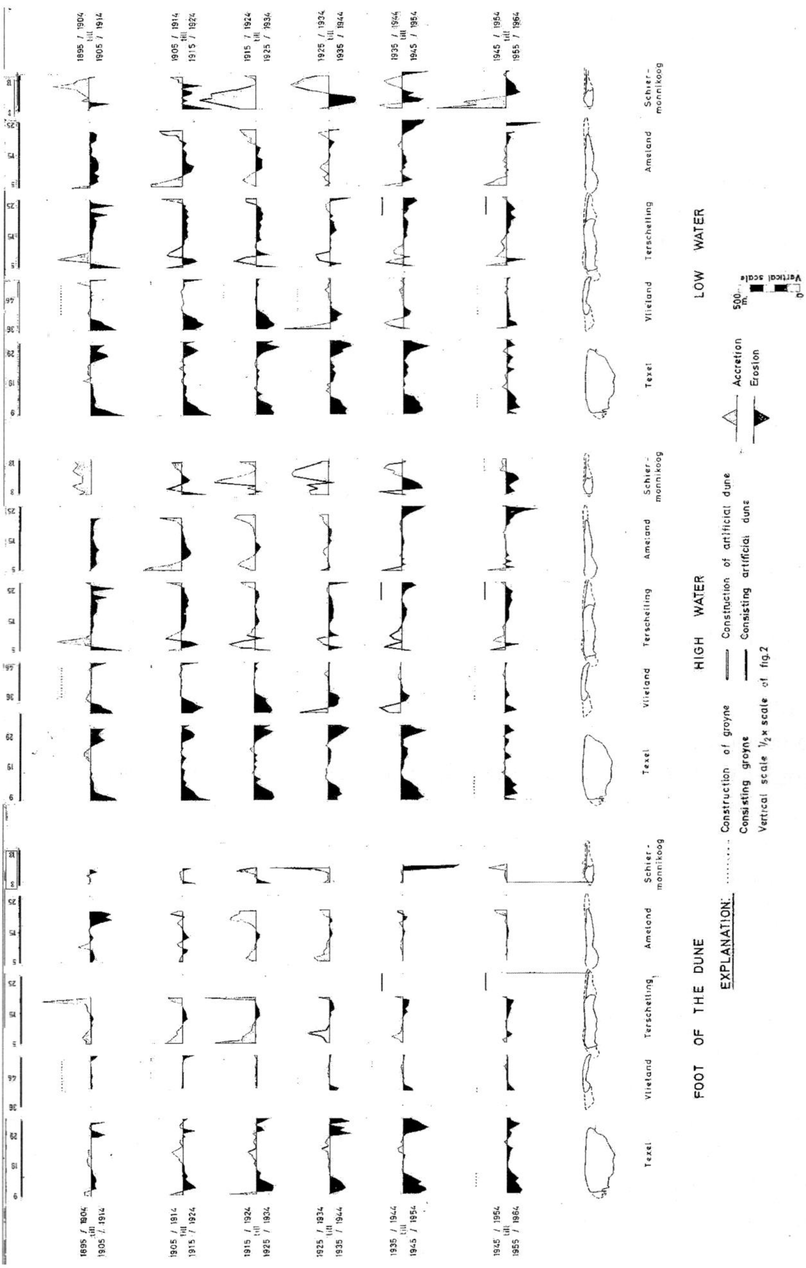

Fig. 2.9 Accretion and erosion of the Dutch Wadden Islands from Bakker and Joustra.[7]

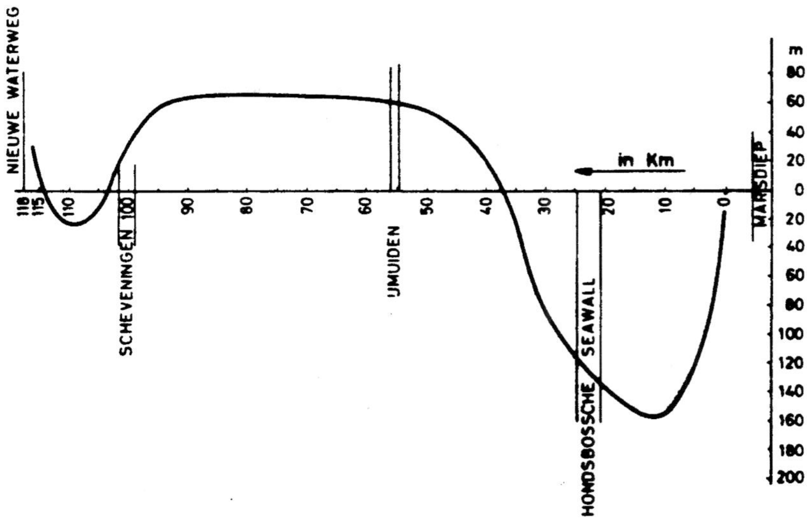

Fig. 2.10 Time-averaged dune foot motion from 1860 till 1960 between Nieuwe Waterweg and Marsdiep (cf. Fig. 2.1) (from [27]).

accretion of 0.5 m/yr, with erosion near Hook of Holland and Den Helder (Fig. 2.10). At present the Hondbossche Seawall (Figs. 2.3 and 2.8: km 20–26) protuberates in seaward direction relative to the adjacent dunes, due to the erosion of these dunes.

From Fig. 2.8, it is observed that the low-tide line and high-tide line near IJmuiden reach near stability about 1900, but that the dune foot changes up to 1930. The influence of the harbor moles (original length 1400 m) at IJmuiden, built in 1870 is shown in detail in Fig. 2.11. This figure gives a plan view of the 5-years average of the low-tide and high-tide lines (seaward scale exaggerated with respect to the longshore scale). The total gain of sand was $9 \times 10^6\,\mathrm{m}^3$ to the North and $6 \times 10^6\,\mathrm{m}^3$ to the South of the IJmuiden harbor moles.[19] In 1965, the harbor moles were extended to 3000 m. On the right hand side of Fig. 2.11, the change of the mean of the low-tide and high-tide lines since 1965 is shown. In the latter period, the accretion on the South side is more than on the North side. Although a changing wave climate may have had effect,[19,44] a more important influence may be, that at present an extensive area with only small currents prevents the entrainment of material from the surf zone around the head of the mole.

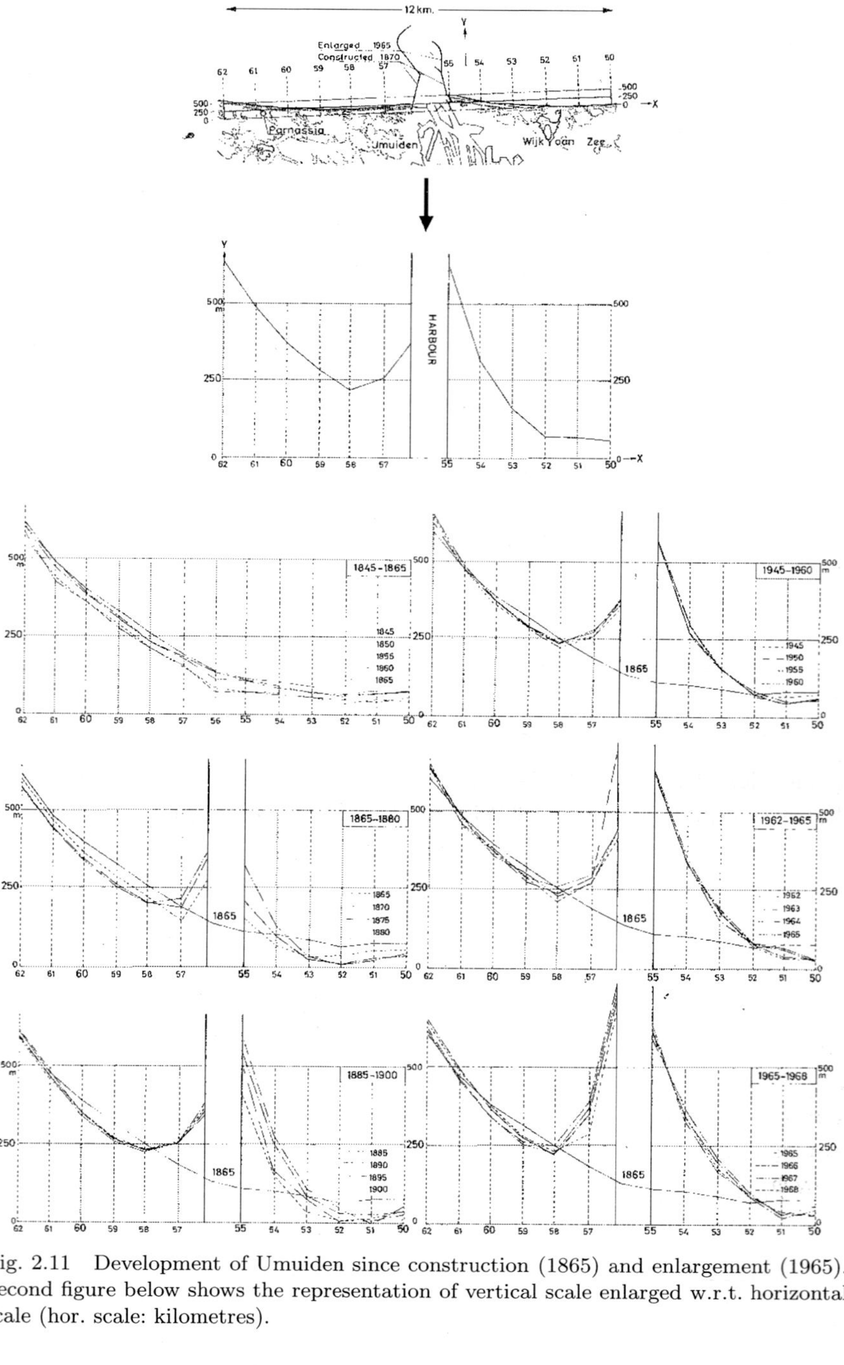

Fig. 2.11 Development of IJmuiden since construction (1865) and enlargement (1965). Second figure below shows the representation of vertical scale enlarged w.r.t. horizontal scale (hor. scale: kilometres).

2.2.3. *Change of the Wadden islands*

In Fig. 2.9, the changes of the Wadden Islands are shown. The change at head and tail of the islands are large compared to the changes in the middle, partly due to the changes in the channel-systems in the inlets. The full sedimentation and disappearance of a channel may sometimes have the effect that a shoal grows onto the end of an island. After that, a sandwave along the coast is generated (Subsection 5.3.2). Another obvious element is the relatively large erosion on both ends of Texel. We shall return to this subject (Subsection 6.3.2).

2.2.4. *The effect of groins*

The Dutch groins have a length of about 200 m. The distance between the successive groins is about 200 m as well.[g] In the considered area the groins are broad-crested stone structures, with its top at about mean sea level. In order to investigate the behavior of groins in practice three areas were considered, on which groins were constructed during the period of the measurements:

- South-Holland: 97–107 km (Fig. 2.12);
- North-Holland: 8–20 km (Fig. 2.13, derived from [36])
- Vlieland: 41–52 km (Fig. 2.14).

In the Figs. 2.12–2.14, the 5-year averages of the low water line have been shown (for instance 1858–1862 for 1860). The scale perpendicular to the reference line is exaggerated with respect to the longshore direction by 10 times in Fig. 2.14 and by 20 times in Figs. 2.12 and 2.13. In each figure, two successive lines are plotted together; the black areas show the erosion in 10 years, the white areas the accretion. The groins built between 1853 and 1862 are plotted through the line of 1860, and so on. The hatched area gives the protected coastal area.

Figure 2.14 (Vlieland) seems to show a striking proof of the benefit of groins. The erosion near 47–51 km in 1860/1870 can hardly be ascribed to the groins 5 km away. The reduction of the erosion after the building of the groins is quite clear. However, this does not mean that groins are the most economic way of coastal protection. The same kind of effect can be seen in Fig. 2.12, (South Holland) 1860/1870, although less convincing. Here,

[g]This could be derived from Figs. 2.12–2.14 as well, which show the sites of the groins in a symbolic way.

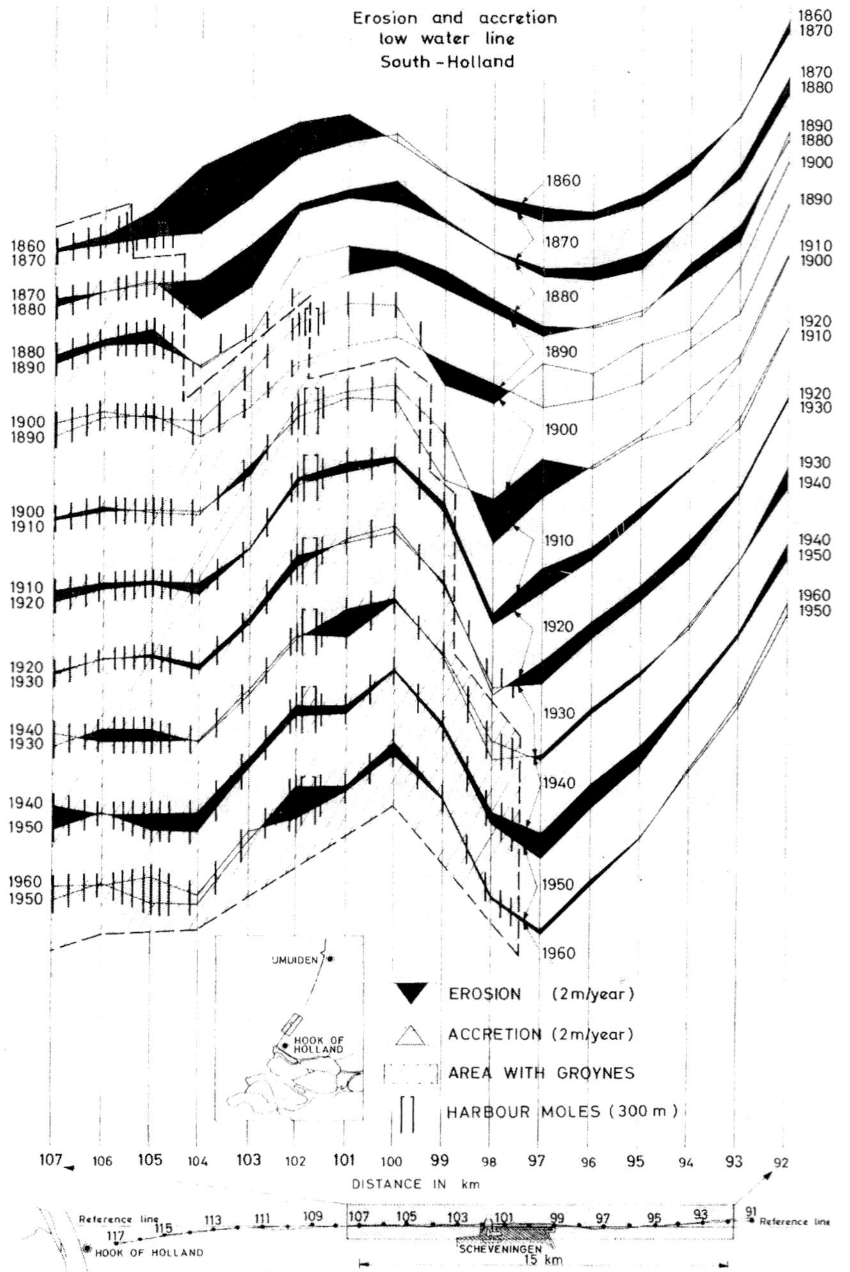

Fig. 2.12 Erosion and accretion (LW-line) of a part of the coast of South-Holland.

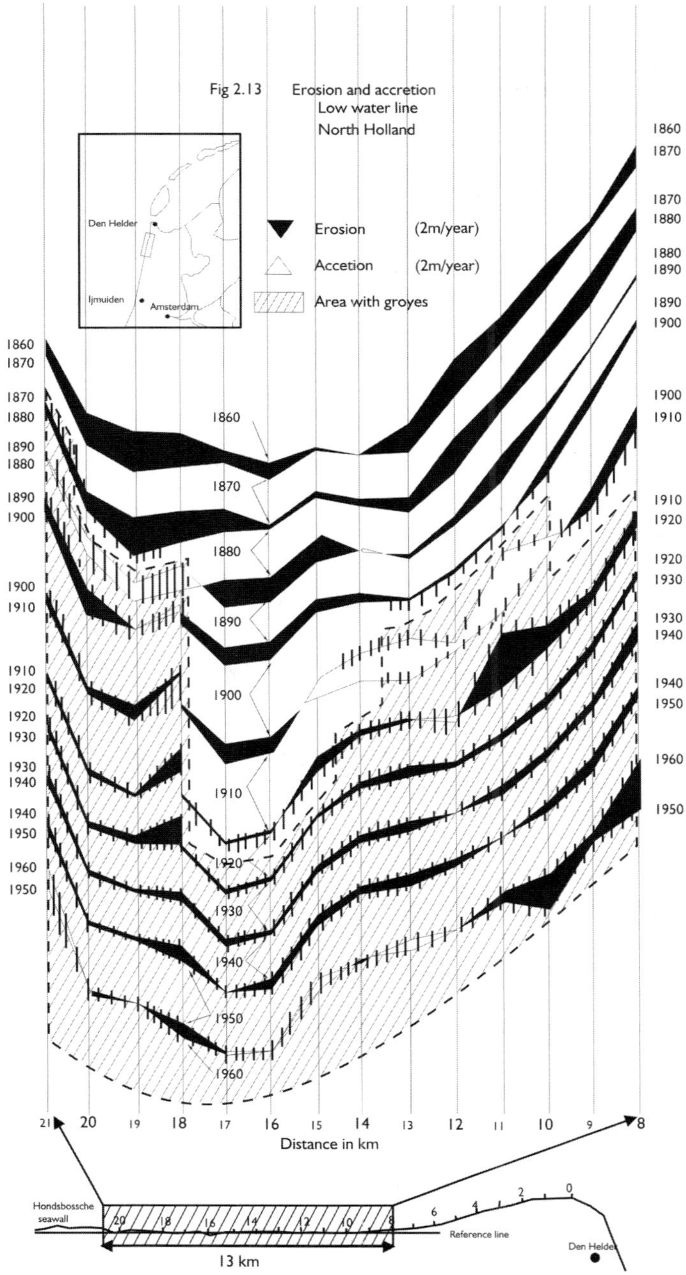

Fig. 2.13 Erosion and accretion (LW-line) of a part of the coast of North-Holland.

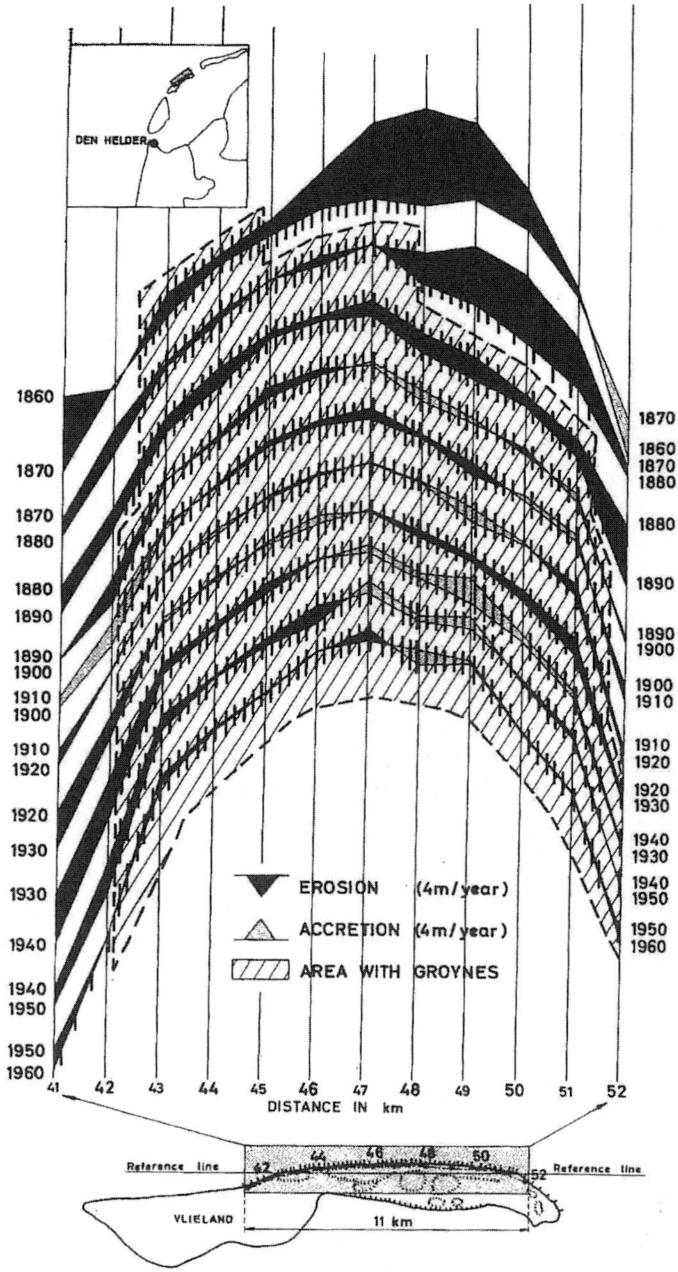

Fig. 2.14 Erosion and accretion of a part of the coast of Vlieland (LW-line). The vertical scale is 1/2 of the scale of Figs. 2.12 and 2.13.

the influence of the lee-side scour also plays a big role. A rough estimation of this lee-side scour (also giving a measure for the net littoral drift in this zone) is about 100,000 m^3/yr (i.e. erosion of about 2 m/yr over 3 km). Interpretation of Fig. 2.13 is not as self-evident. The Northern part of this area is subject to the movement of the Schulpengat, the Southern branch of the Texel inlet. The lee-side scour on the Northern side can be observed clearly (km 20, 21; 1860–1880, etc.). The inverse of this process can be found too: the accretion near km 13, 14 between 1900 and 1910 on the luff-side.

Analysis of the performance of groins is very difficult. Comparing protected areas with unprotected parts of the beach does not make sense, since groins are constructed only on eroding beaches. A before- and after-comparison could be obscured by climatologic changes (cf Figs. 2.8 and 2.9, low-waterline).[h] In the analysis, the areas of Figs. 2.12–2.14 have been chosen for comparison: all eroding beaches, gradually more and more protected with groins.

For each area and for each 10-year period, the mean erosion per year has been calculated, as well of the protected part as of the unprotected part (Fig. 2.15a,b,c). Thus the erosion on the same area could be compared for the time when this area was protected (later on), and for the time when it was not (in the beginning). In order to eliminate local influences all three areas were put together and again the mean erosion per year of the protected and of the unprotected areas was computed (Fig. 2.15d). An impression of the climatologic changes is given in Fig. 2.15e. Here the mean regression and progression per 10-year period of the low-tide line of the uninterrupted coast of Holland is shown. Only the relative changes are important. Finally in Fig. 2.15f the erosion of the protected part of Fig. 2.15d is plotted against the erosion of the unprotected part. From this plot, a considerable reduction of the erosion can be concluded. Although nearly all the considered unprotected areas were subject to lee-side scour, Fig. 2.14 (Vlieland) shows that this is not the main cause for (virtual?) reduction of erosion. In the area near 50 km, the erosion between 1860 and 1900 was more than in later times and here lee-side scour did not play a significant role. The erosion was just reduced by the building of groins.

[h] Cf. for instance Fig. 2.8: the differences between 1870–1880 and 1880–1890.

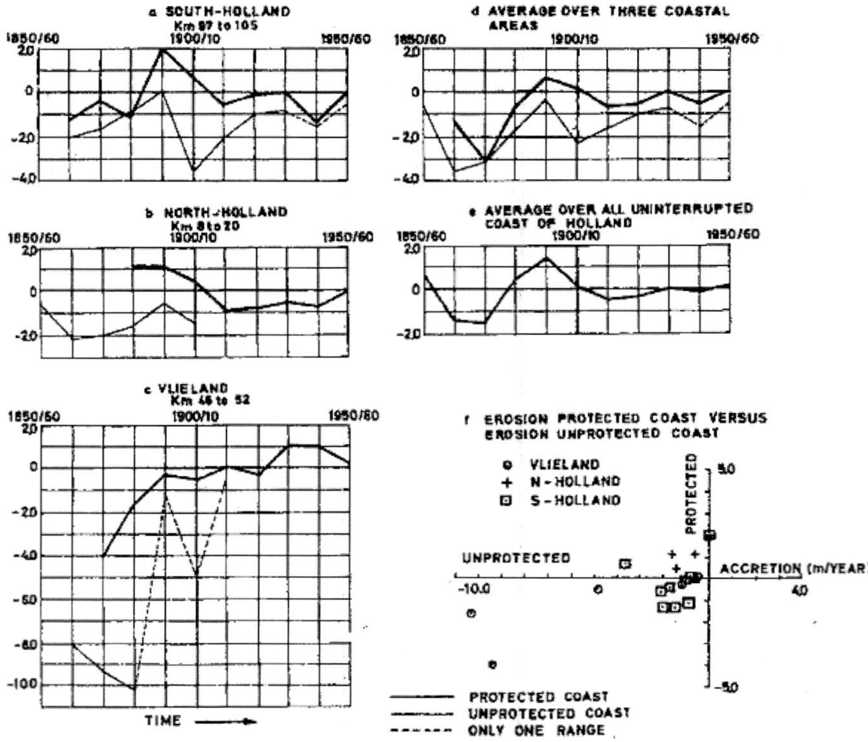

Fig. 2.15 Comparison of erosion (i.e. "negative accretion") in m/yr of protected and unprotected areas.

2.3. The Jarkus Measurements (Holland Coast)

2.3.1. *General remarks on the topography*

For the investigation of the bottom topography between den Helder and Hook of Holland use has been made of the yearly coastal measurements, which have been carried out by Rijkswaterstaat and the local regional boards ("waterschappen"). In seaward direction measurements extend to approx. 800 m from the waterline. The measured grid ranges have a mutual distance of 200–250 m longshore; measuring points being at a mutual distance of 10–20 m cross-shore.

In this subsection, the results of the measurements are displayed in two ways. According to the first kind of display (3D-space, Fig. 2.17a,b) the measurements of one year are displayed according to a three-dimensional pattern in oblique parallel projection; this results in a realistic picture of the bottom topography in the year considered. However, the coordinate reference line of the displayed picture is straight, where in reality it should be curvilinear. For this reason, the Southern part shows a distortion of the order of 45° with respect to the northern part.

The second type of display (3D-time) shows the cross-sections, measured in one range for all the successive years (cf. Fig. 2.19a,b), in which the site of the ranges in question are shown in an inset map. This results in a three-dimensional figure. If, in this kind of figure, the site of the top of a breaker bar keeps a constant distance to the time-axis, this indicates, that the breaker bar is stable in the course of time. Deviations (to the left or right) of the line through the top of the crests with respect to the time axis indicates seaward, c.q. landward motions. The principle is shown in the sketch of Fig. 2.16.

Figure 2.17a shows a 3D-space picture of the coast of North-Holland between Den Helder and IJmuiden, as measured in 1984. The landward boundary is near the dune foot (or somewhat more landward, as near Callantsoog); the area considered has a width of ca. 800 m. The distance Den Helder-IJmuiden is 55 km; so the length scale shown is reduced with respect to the width.

Near Den Helder the coastal protection is heavily under attack of the Nieuwe Schulpengat (depth ca. 25 m), the most southern branch of the channel system of the Texel inlet. For long an ongoing revetment has

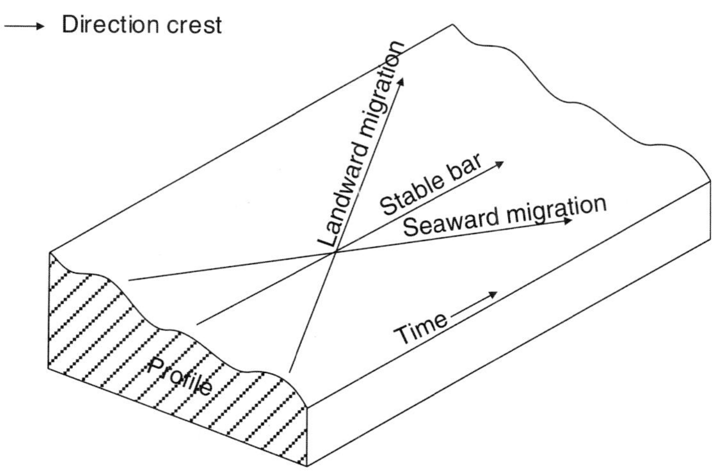

Fig. 2.16 Change of profile in the course of time. Lay-out of 3D-time figure (principle).

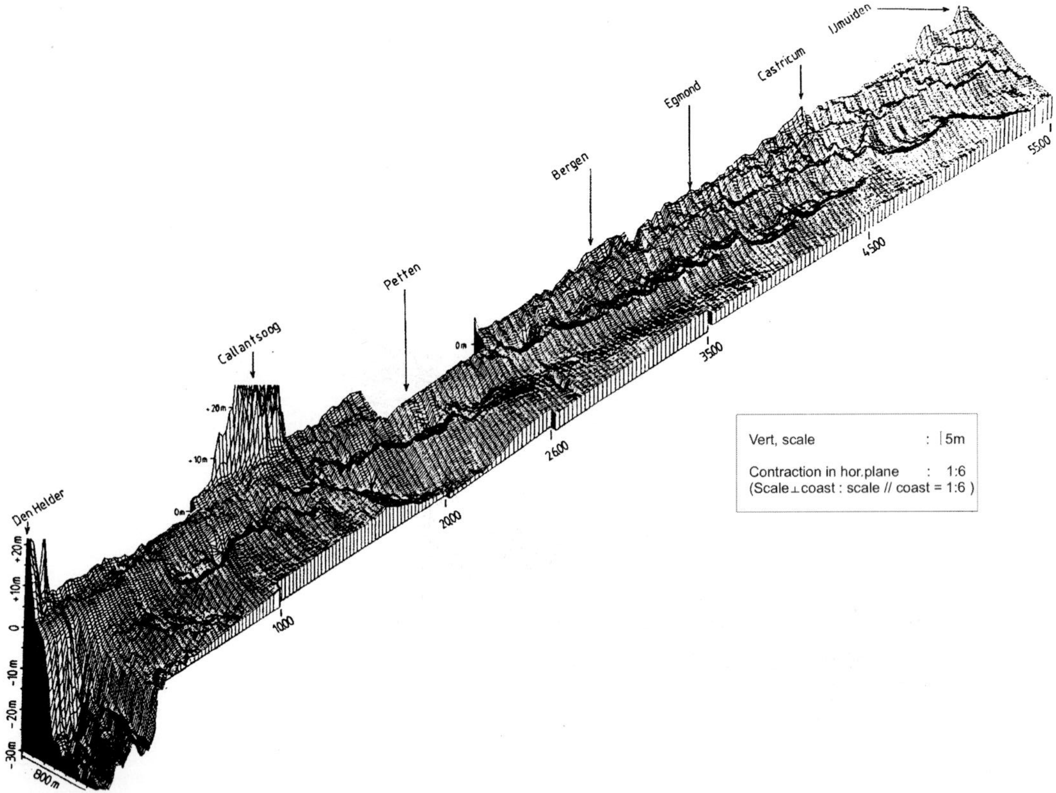

Fig. 2.17a Three-dimensional view of the coast of North-Holland (up to IJmuiden). The landward boundary of the area depends on local circumstances (it is the site of the main range, about parallel to the coast). Mind, that in the area of the Hondsbossche Seawall (20–26 km) the toe of the dike has been taken. At most places, the level is about NAP (Ordnance Level).

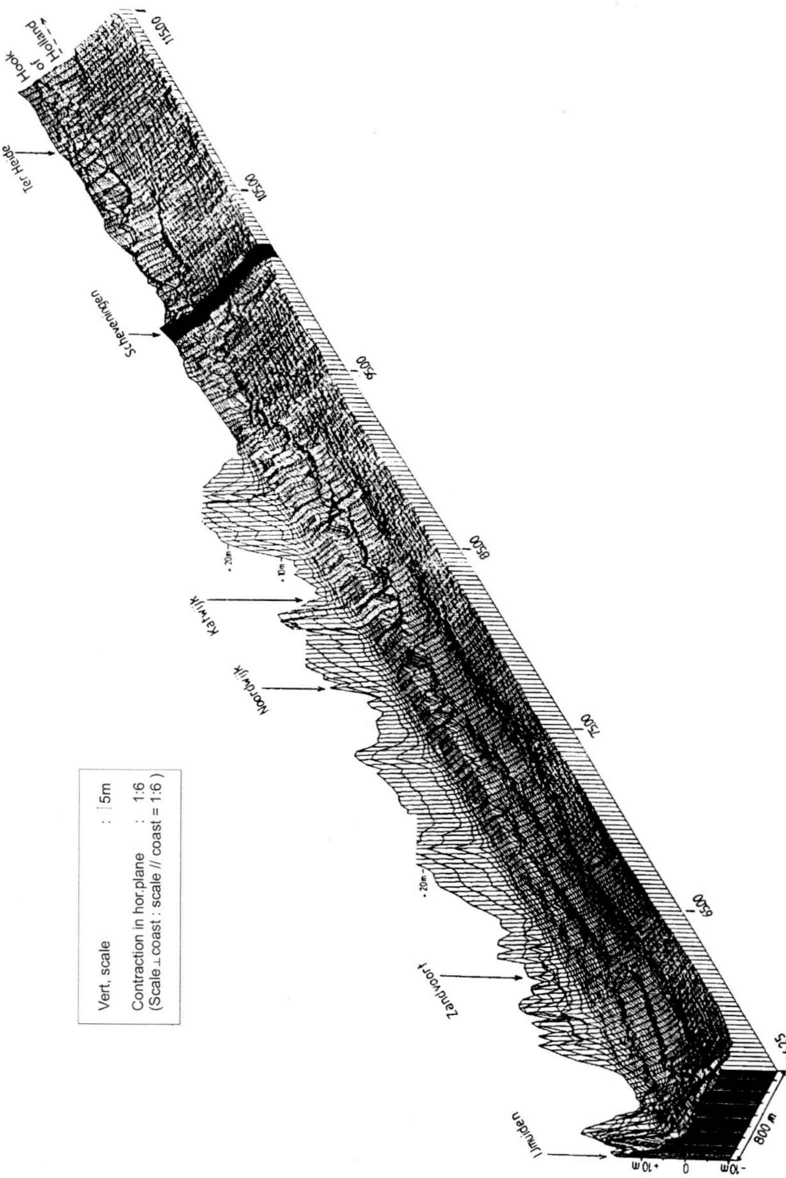

Fig. 2.17b 3D-display concerning the coast of (mainly) South Holland: IJmuiden to ca. Hook of Holland.

defended this coastal area positioned at the Northern point of the Holland coast, the so-called "Kaaphoofd". At least till RSP (governmental range system) km 10.00 the influence of the inlet can be felt. The seaward boundary depth of the area considered decreases in Southward direction until a minimum of NAP — 7.5 m near RSP km 16.88. One might presume a highly elevated boundary of an outer delta at this site, but its position would be extremely southward. Although this area defended by groins shows since long times an erosion of 1–1.5 m/yr, the shape of the coastal profile is surprisingly stable the last 20 years. Confer Fig. 2.19a where among others the 3D-time figures for the ranges RSP 9.94 and 20.15 km are shown: in these ranges the lines through the tops of the bars are parallel to the time-axis.

The three-dimensional figure (Fig. 2.17a) depicting the coastal area shows an omission at the landward side between RSP 20 and 26 km. Over here the Hondsbossche Seawall is situated, which protuberates into sea over ca. 200 m.

Considering the breaker bars (Figs. 2.17a and 2.19a): in the whole area between Callantsoog and IJmuiden these show — despite of the small-scale muddled patterns — a surprising regularity at a larger scale of space. Over a distance of 30 km the bars are "lined up" in a way a Dutch army sergeant with soldiers (being member of the trade union), would be proud of. Yet these features are no "fossilized" geological structures. In particular, in South of Petten the bars are clearly moving. This shows from the 3D-time presentations of the ranges 30.00, 40.00 and 50.00 km, given in Fig. 2.19a, from which clearly a seaward propagation can be observed. This does not imply *a priori* seaward transport of sand, averaged over the period of the bar motion: just oscillating transport is also a possibility (cf. Subsection 2.3.2).

For comparison of the behavior of bars and foot of the dunes, see Fig. 2.18, taken from Edelman,[25] in which the behavior of the dune foot, averaged over a period of 100 years is depicted. This shows that the dunes of, for example, Katwijk tend to accrete, while the breaker bars move seaward.

2.3.2. *Breaker bars*

In this subsection on breaker bars first (Ad "a") the attention will be drawn to the phenomenological aspects. After that (Ad "b"), an attempt will be

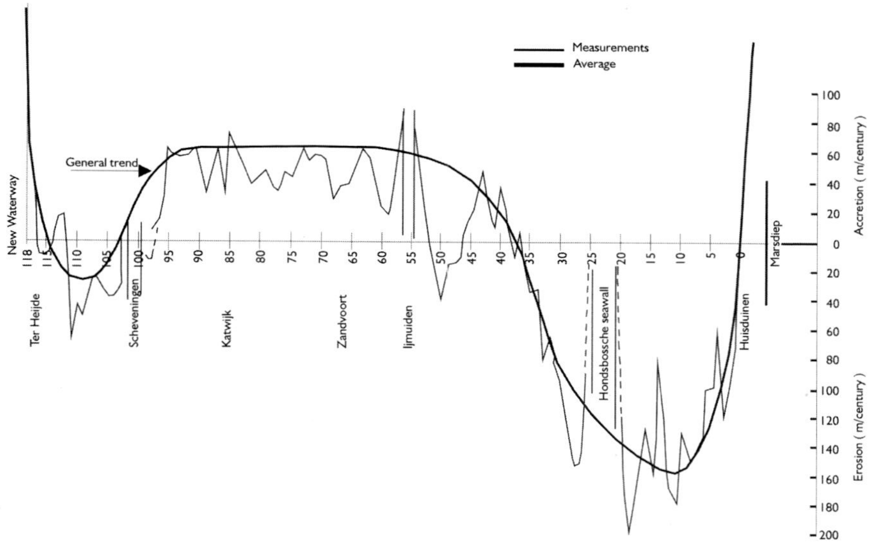

Fig. 2.18 Erosion and accretion of the Dutch Coast (dune foot) according to Edelman.[24] Calculated from measurements of the dune foot position between 1860 and 1960.

made to catch the shape in a mathematical formula (Eq. (2.1)) which gives the opportunity to schematize its shape by a few characteristics.

Practical questions involved are:

- Can the coastal behavior be predicted more accurately, if the pre-knowledge is implemented that breaker bars appear?
- How much cross-shore transport is involved with the motion of these bars?

Finally (Ad "c") in this subsection, the various elements of the cross-shore transport will be evaluated by comparison (cf. Figs. 2.21 and 2.22).

The problem, as formulated above, has been investigated by De Vroeg,[52] who schematized a breaker bar system to seaward propagating sandwaves with straight crests, which, according to the schematization, are oblique to the coast (Fig. 2.20). In seaward direction, the amplitude first increases and then decreases. The wavelength and the propagation velocity of these sandwaves over a coastal area are assumed to be constant. According to this schematization every point in the surf zone oscillates harmonically, but with its own amplitude and phase.

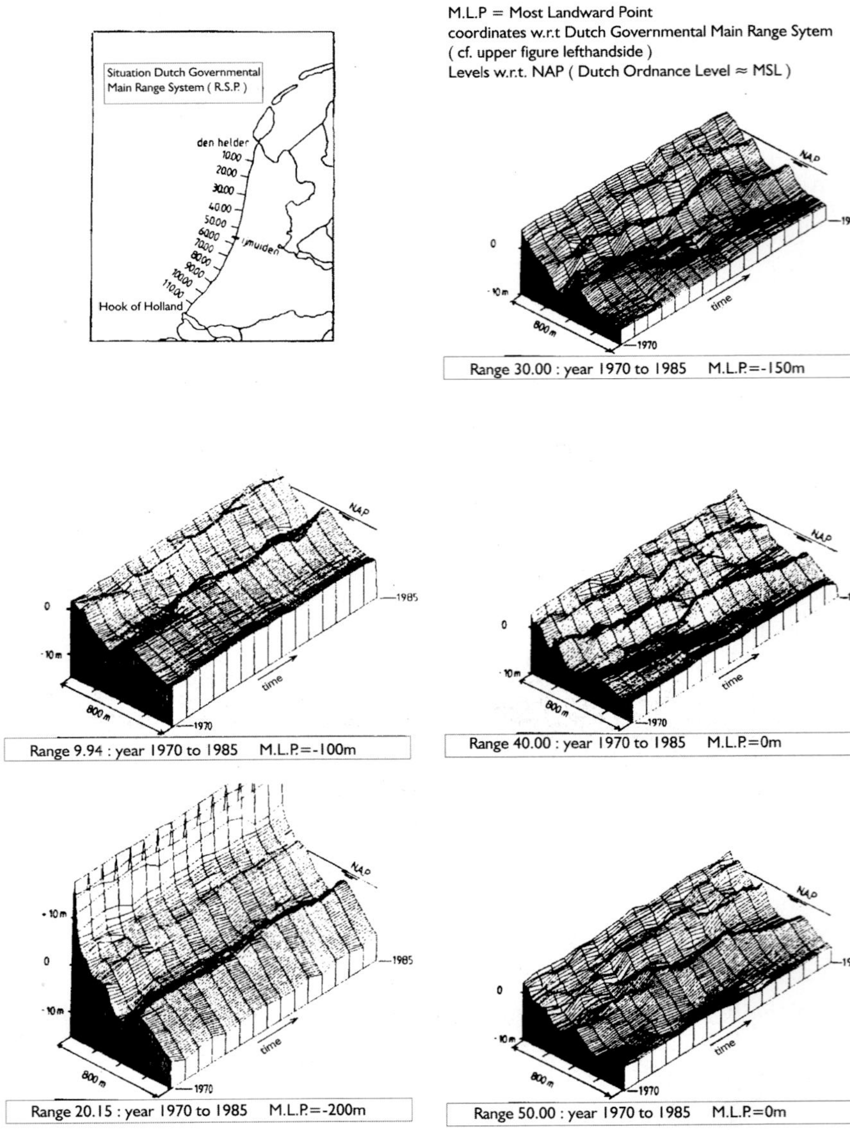

Fig. 2.19a 3D-time-diagrams of coastal profiles of the coast of North-Holland.

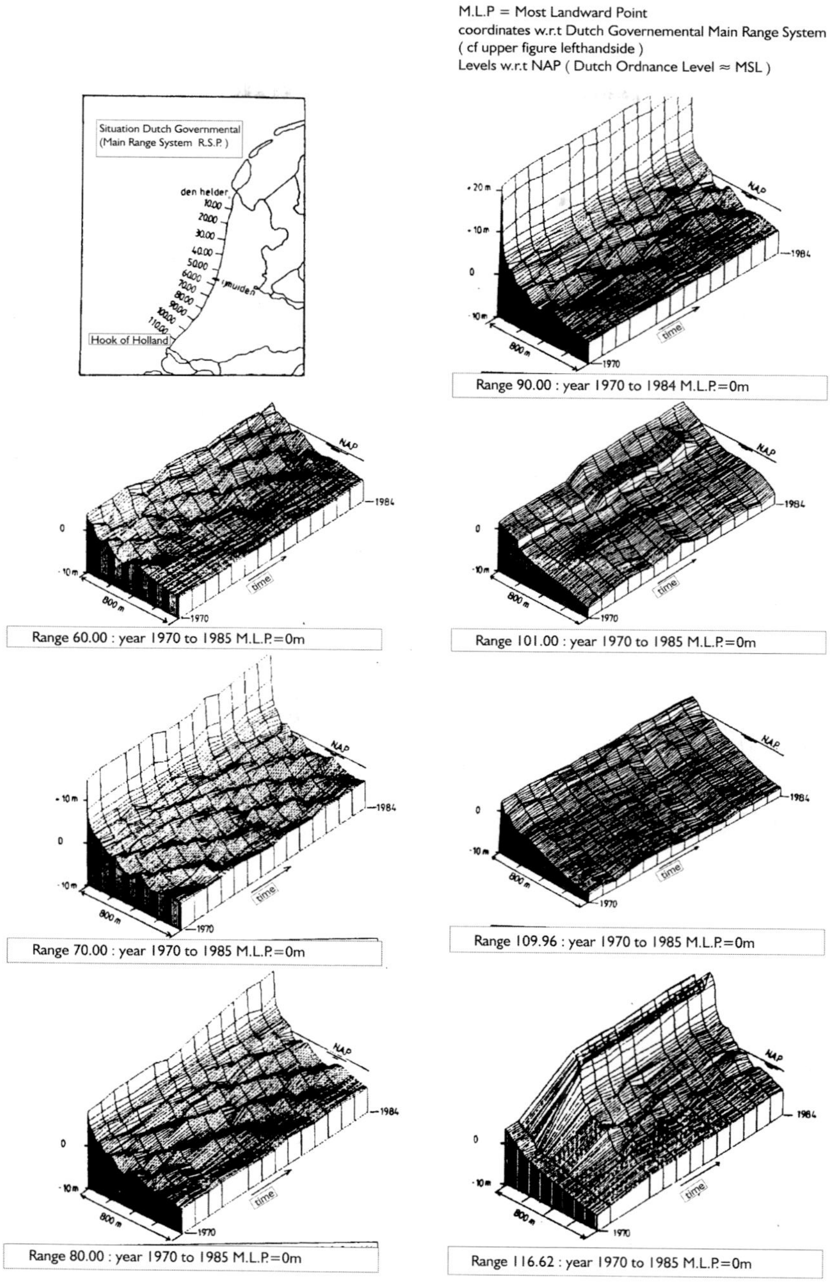

Fig. 2.19b 3D-time-diagrams of coastal profiles of the coast of South-Holland.

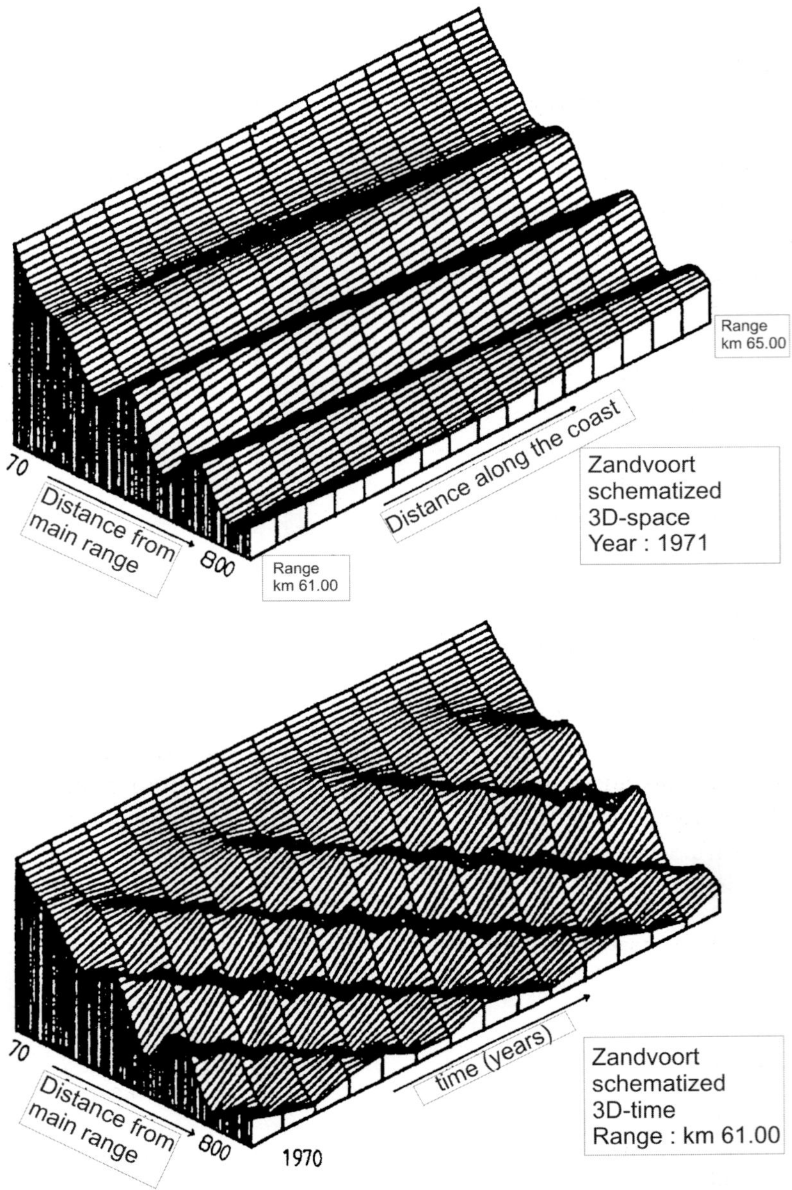

Fig. 2.20 Mathematical approximation of the coastal behavior of beaker bars.

Ad a: phenomenological description

Depending on the site of breaker bars, troughs may occur relatively near to the dunes or further away. This affects the amount of dune-erosion during storms. Because of this there are "safe" and "dangerous" years. This raises the question, whether from the breaker bar position in the past the "safe" and "dangerous" years in the future (every place having its own dangerous years) can be deduced.

The mathematical formulation, Eq. (2.1) (mentioned below, Ad "b") used is a compromise between, on one side, a reproduction as realistic as possible and, on the other side, a simple expression. Figure 2.20a,b gives an impression of this mathematical approximation of the behavior of the coast near RSP km 60.00 in the shape of one 3D-place and one 3D-time figure. The propagation of the wave in the figure should be superposed to a time-averaged profile.

For the sake of homogenuity, the coast first has been divided into pieces of 5 km. Next, with the aid of statistical techniques, it is possible to match the constants in the mathematical description as good as possible to reality. In this way, some characteristic values for breaker bars can be derived, which for three coastal areas are shown in Table 2.1. Details will be given in "Ad b".

Considering Table 2.1, mind, that from an engineering point of view a periodicity of 20 years is large enough to fine-tune the coastal management (for instance: the date of making revetments, etc.) where 5-year variations are a kind of "random noise" with which the coastal defense has to cope with.

Table 2.1 Average characteristic values for breaker bars for some coastal areas.

Coastal area	Max. ampl. (m)	Distance max. ampl. from coastline (m)	Period (year)	Av. distance between crests (m)	Propagation velocity. (m/yr)* angle
Bergen-IJmuiden	0.90	400	ca. 20	400	<28 m/y 0°
Zandvoort-Noordwijk	0.65	330	4	225	57 m/y − 3° à +2°†
Noordwijk-Scheveningen	0.50	250	5	300	62 m/y 0°

*Measured perpendicular to the crest, seaward is positive.
†Positive angle = opening to the South.

Ad b: The mathematical description of the breaker bars

Primary goal of the present investigation (carried out by De Vroeg[52] and Bakker & de Vroeg[10]) is to quantify the amount of cross-shore transport involved in the motion of breaker bars as a tool for the comparison, mentioned above under the heading: "Ad c".

Below the mathematical background of the way of treatment of this question is explained.

De Vroeg[52] answers these questions by approaching the shape and the behavior of a breaker bar system quantitatively. These bars are schematized to seaward propagating sandwaves with straight crests, which in principle are oblique to the coast. (cf. the introduction to this subsection).

A periodic transversal wave motion in an approximately horizontal plane can be simulated mathematically by a harmonic motion, A cos $(\omega t - k_x x - k_y y)$, where x and y represent the considered coordinates, parallel and perpendicular on the coast, ω is $2\pi/T$, T is the period of the motion, t is the time, (k_x, k_y) are the component of the wave number, i.e. the wavelength in the x- and y-positions are, respectively $2\pi/k_x$ and $2\pi/k_y$.

A is the amplitude of the wave motion; in the considered case of breaker bars it is assumed to be only a function of y. It first increases and later on decreases in seaward direction. In the investigation quoted above the magnitude of the amplitude has been approximated by the function:

$$A = A_1(y - y_1)\exp\left[-\left(\frac{y - y_0}{S}\right)^2\right] \tag{2.1}$$

where A_1, y_1, y_0 and S are coefficients. In this way, it is possible to simulate a wave, propagating in seaward direction, first increasing and later on decreasing. The used formulation is a compromise between, on the one hand, a simulation as realistic as possible and, on the other hand, an expression, which is simply to handle. The ratio k_x/k_y represents the tangens of the angle between the breaker bars and the coast. The magnitude of the different constants can be determined by curve-fitting. Figure 2.20 gives an impression of the mathematical approximation of the coastal behavior near RSP km 60.00 in the shape of a 3D-site and a 3D-time figure. The propagation of the wave in the figure is superponed on a time-averaged profile.

For application of this model to the Dutch coast the coast has been divided into parts of 5 km in order to warrant homogeneity (cf. Table 2.1).

After determination of the parameters, it is possible to deduce the values for:

- the maximum amplitude in a cross-section;
- the distance of this maximum related to the site of the coastline;
- the period T of the motion of the bars;
- the average distance $2\pi/k_y$ between the crests;
- the propagation speed and direction;
- the orientation of the bars w.r.t. the coastline (+ = opening to the south).

Table 2.1 gives an overview of the values for some coastal areas.

Purpose of the approach shown is that predictions can be made. For this goal the sand-volume (m^3/m') of the coastal zone between the contour lines OD + 1 m and OD − 3 m, seaward of the main range (parallel to the coast) as a function of time and location is compared to the "theoretical" function (Fig. 2.21). The drawn line gives for a number of ranges in a coastal stretch of 4 km in the surroundings of Zandvoort (Fig. 2.17b) the number of m^3/m sand according to the coastal measurements; the interrupted line shows this amount according to the simulation as described above. The fluctuation of volume caused by the variability of the site of the breaker bars is shown. According to the simulation this fluctuation should shift (because the bars are oblique to the coast) with a period of 4 years and a propagation velocity of approx. 1.7 km/yr in Northerly direction (from km 65 to km 61); The magnitude of this fluctuation is approx. $45\,m^3/m'$. With the mathematical formulation given, it should be possible to predict more precisely the dune erosion during storms. However, although the approximation is not too bad, for the purpose in question it appears to be too inaccurate, in that sense that the "hindcasting" including the periodical fluctuations is not better (in other words, it does not give a smaller standard deviation) than just taking the linear trend into account.

There are two causes for this inaccuracy:

- the position fixing of the bars in location and time by a cosine function is too simple. It suggests, that the bar length (top–top position) is constant over the whole profile, while in the nature the length of the bars increases in seaward direction[i]
- discontinuities, as occur in practice have a strongly negative effect on the results.

[i]Refer to Ref. [30] in Chapter 4 concerning this matter.

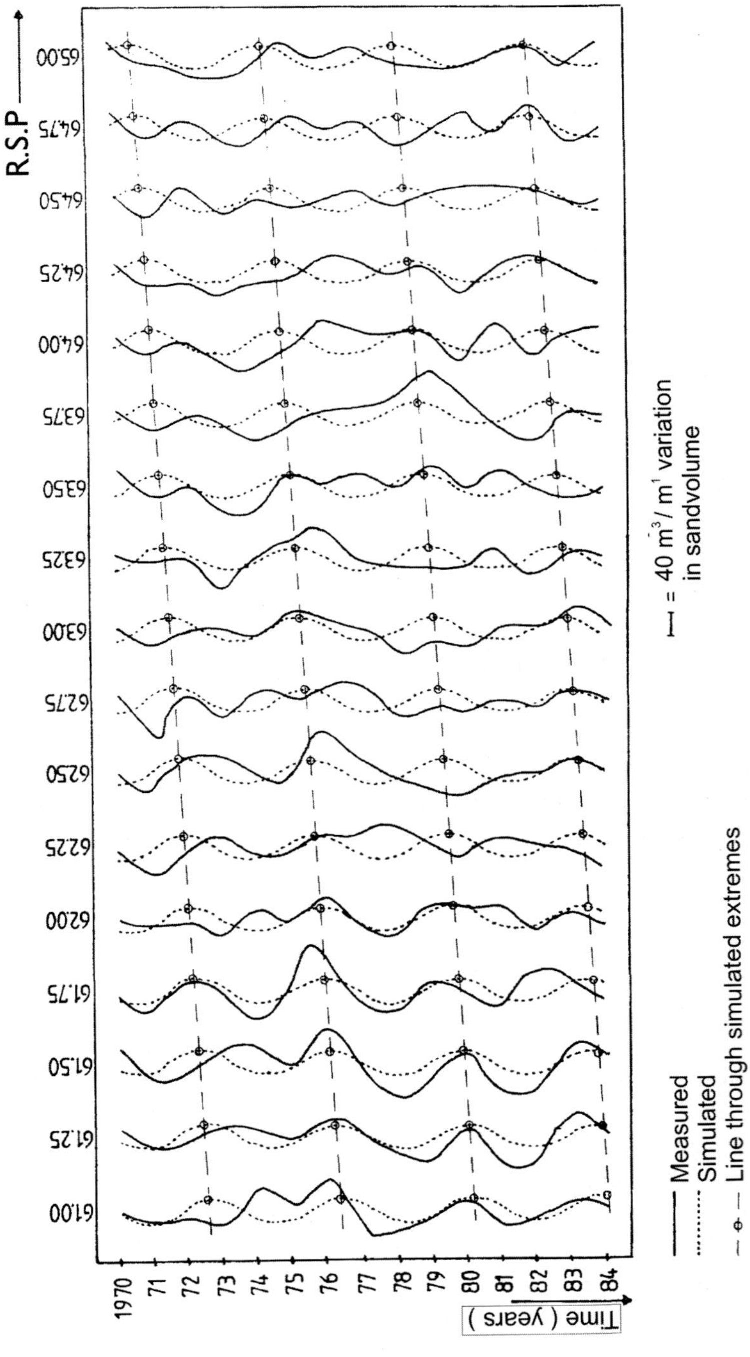

Fig. 2.21 Comparison variation measured and calculated sand volume.

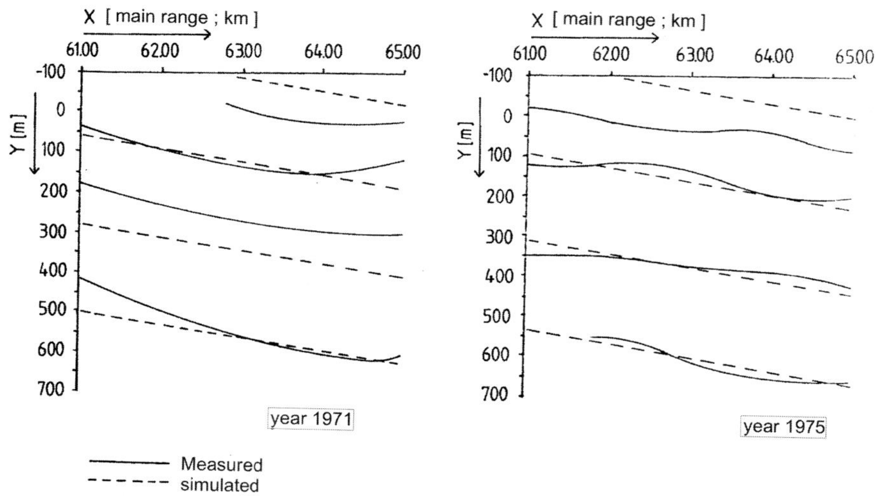

Fig. 2.22 Measured and simulated crests of the breaker bars.

To illustrate this, in Fig. 2.22 the measured and simulated crests of the breaker bars at RSP km 60.00 till 65.00 during the years 1971 to 1975 are displayed. Caused by both formerly mentioned symptoms phase differences of 180° may arise between the mathematical approximation and nature, causing a rapid growth of standard deviations between measurement and calculation. To achieve the aim, a more precise approach is necessary. The mathematical approach does give an impression of the amount of cross-shore transport, which the phenomenon "breaker bars" involves (Fig. 2.23).

For better understanding, consider some coast, which is stable at the long term; refer to "Ad c" below concerning amounts of sand transport linked with long term erosion and accretion.

The mathematical approach enables to express the amount of sand, positioned at a certain moment landward of a given coordinate y in the variables appearing in the formulae like V, y, t, etc. Differentiating this expression with respect to time yields the crossshore transport as a function of time and location. In this concept, the cross-shore transport landward of $y = y_1$ (the point where the amplitude of the sandwaves is zero) is supposed to be zero. As can be expected an oscillating transport with frequency ω is found (Figs. 2.23a,b and 2.24). This yields the location and magnitude of the maximum "amplitude of cross-shore transport". Starting from the

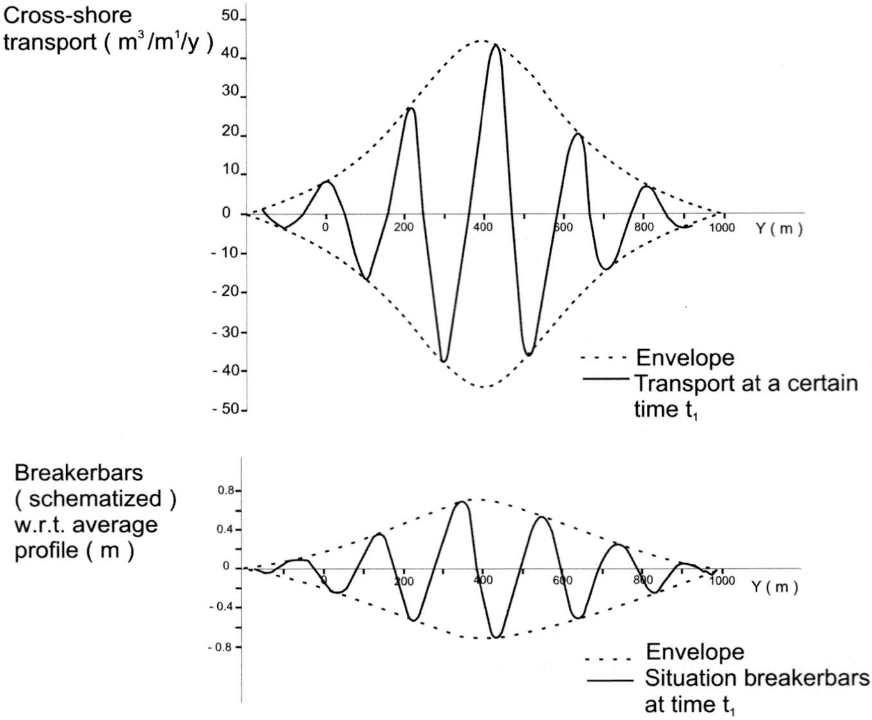

Fig. 2.23 Cross-shore transport due to breaker bars.

characteristics found for the breaker bars near Zandvoort (Fig. 2.22a,b), the result (per m in longshore direction) is approx. $45\,\text{m}^3/\text{yr}$.[52]

Ad c: the order of magnitude of the cross-shore transport

So far concerning the mathematical description of breaker bars. By integration and differentiation, the formulation (2.1); cf. "Ad b", allows to give an estimate concerning the amount of cross-shore transport, involved in the propagation of breaker bars as compared with the order of magnitude of the amounts as a consequence of the other morphological features. The result is depicted in Fig. 2.24, which will be explained below.

In order to be somewhat specific, focus has been placed upon the area between Zandvoort and Noordwijk (Fig. 2.17b), where the breaker bars are developed strongly and have a period of about 4 years; this implies a (semi)-periodical transport of $45\,\text{m}^3/\text{m}/\text{yr}$ (cf. "Ad a").

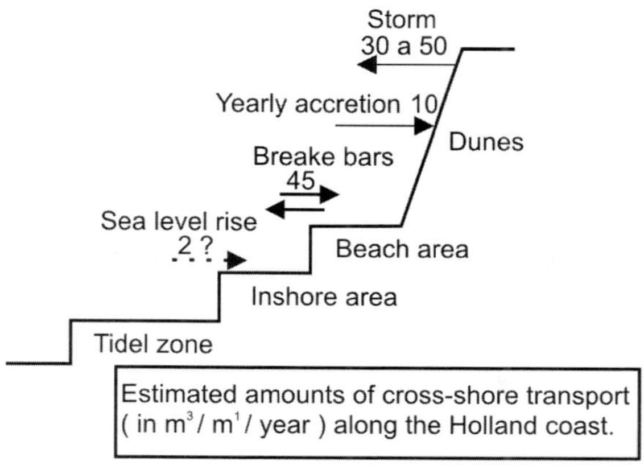

Fig. 2.24 Estimated cross shore transports near a coast as a result of waves, currents and tide.

However, the yearly coastal accretion of this coastal stretch has been kept out of consideration. If it may be assumed, that the considered coastal area extends from OD − 5 m to OD + 15 m, then the yearly accretion of 0.5 m/yr occurring at this site (Fig. 2.10) implies a landward transport of $10\,\text{m}^3/\text{m}/\text{yr}$.

One may compare this magnitude with the amount of dune erosion during a sturdy storm, which can be (in cross-shore direction) approx. 30 $\text{m}^3/\text{storm}/\text{m}$. During more quiet circumstances, often this amount is transported dune-wards again. How to compare this amount with the cross-shore transport, which is linked to the sea level rise and to the transport alongshore?

In this investigation, a coastal strip with a width of approx. 1 km is considered. During a sea level rise of 2 mm/yr, an amount of 2 m^3 per meter coastal length is necessary to maintain a constant local depth in this strip with respect to mean sea level.

Compared with the gross littoral drift along the coast this amount pales into insignificance.

Figure 2.25, made on the basis of calculations, mentioned by Bakker[4] (see as well Sections 2.4, 4.2 and 4.3; Fig. 2.25), gives the estimated transport along the coast as a result of the wave-driven longshore current and tide. Breaking waves, oblique to the coast, generate wave-driven longshore currents, which will dominate the tide (Section 2.4), even in the case with a small angle of wave incidence.

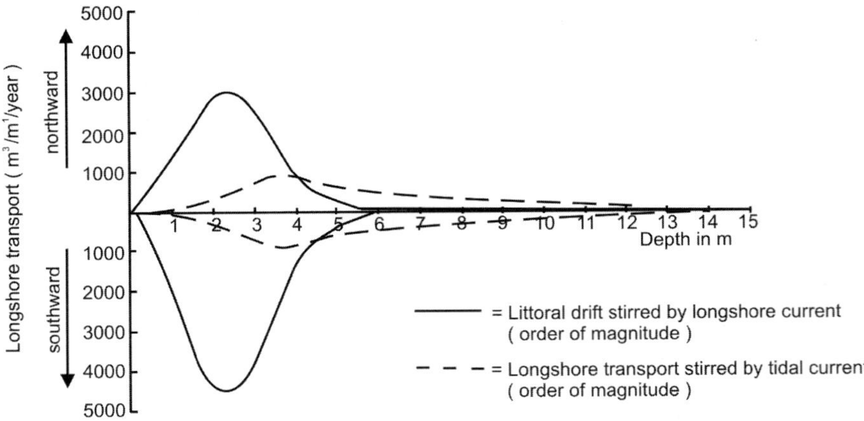

Fig. 2.25 Order of magnitude of the yearly transport *along* the coast as a result of the wave-driven longshore current and tide.

The drawn lines in Fig. 2.25 represent the year-averaged transports per meter coastal width, caused by this wave-driven current.

Herewith, the probability of wave conditions (significant height; period; main direction) has been taken into account.

Near Scheveningen the transport to the North and to the South appear to be of the same order of magnitude. The calculations are not accurate enough to make statements over the resulting transports; some more information will be given in Subsection 5.3.5.

Evidently, outside the surf zone the tide-driven transports dominate although the waves can still boost the transport by stirring effects. There, the direction of the waves is only of secondary importance. With an interrupted line in Fig. 2.25 is given, on basis of exploring calculation,[41] how this transport is distributed over the depth. Close to the coast the tidal transport decreases, as it is, at this location, mostly replaced by wave-driven transport. Far from the coast it also decreases, because the influence of the waves on the tide-driven transport becomes smaller. In the figure, the calculated transport per flood-(ebb-)period is shown, multiplied by the number of tides per year. Again, probability of wave occurrence has been taken into account. Also in this case inaccuracy of the calculations prohibits statements on the resulting transport.[j] From comparison of Figs. 2.24 and 2.25 it can be concluded that the longshore transport is at least a factor 100

[j]Refer to the RWS-Coastal Management Reports of 1990 and 1995 for more daring statements.

larger than the cross-shore transport. However, accretion and erosion only arise by local differences in transport. These differences ("gradients") can be of the same order of magnitude for both longshore- and cross-shore transport. In general, the longshore transport is large compared to the cross-shore transport. This means that the "to and fro moving slicer" that smoothens the breaker bars and models it in that way, is directed along the coast. But what "inspires" the coast to form breaker bars?[k]

One could have the following concept in mind: as soon as there is a small irregularity in the profile, this can cause wave breaking. The momentum of the breaking and broken waves will be large above the crest of the breaker bar with respect to the momentum below the crest, on the landward flank of the breaker bar. This could drive a circulation (Fig. 2.26) (an exact formula is given by Dyhr Nielsen and Sørensen[24]). Together with the longshore current in the trough, this results in a spiraling current.

The relative distance between the breaker bars is probably being ruled by a subtle mechanism, of which the essentials can only be guessed after.[17,20,30,46]

It is thought, that the relative distance between the breaker bars is regulated by resonance systems of long-periodic waves with wavelengths of hundreds of meters till 1 km (periods of some minutes, amplitudes max. some dm). The principle behind these wave patterns, the so-called "surf beat", can be described as follows.[46] Interference often causes grouping of alternately high and low waves. Breaking of those waves results in a shoreward directed pulse, which varies in strength, because of the variation of higher and lower waves in the wave groups. The water in the surf zone

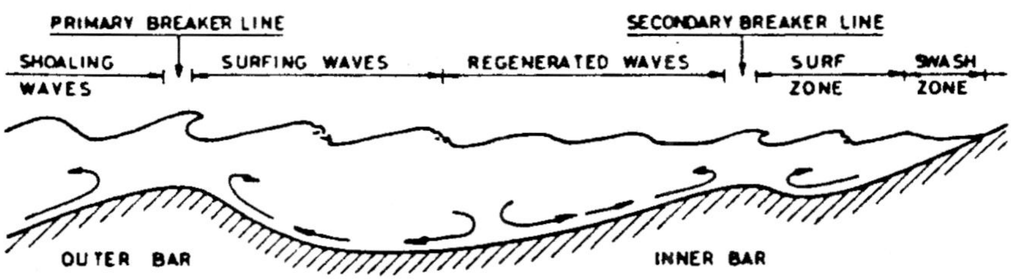

Fig. 2.26 Resultant current circulation near the bottom of a coast with two breaker bars. According to Dyhr Nielsen and Sørensen.[24]

---

[k]Since the original version of this text has been written (in Dutch) this topic has been subject of many excellent investigations, for instance by Roelvink, Ruessink, Wijnberg, Holman and Plant.

responds by means of an oscillation, not in the period of the separate waves, but in the period of the wave group. This process generates standing waves, of which the amplitude gradually increases in shore-ward direction. The amplitude of oscillation depends on whether the period of the surfbeat agrees with the "resonance frequency", to which the breaker bar system is "tuned": in the same way as a tuned radio can select from all waves in either only one or some frequencies to amplify. One could imagine that the resulting water motion above and between the breaker bars causes sand motion in this way, remoulding these breaker bars in a specific way. Because of the large scale of this phenomenon the moulding will be very slow, and thus the resonance frequency can adjust itself only very slowly (N.B.: a radio tuning itself!). Even if the incident waves differ from day to day, the same resonance frequency would be picked up and the remoulding of the breaker bars would occur in the same way. This could explain that breaker bars react more to a wave climate than to the instantaneous wave characteristics.

However, as usual the scholars do not agree on the process: this applies to the importance of surf beat as well[20] (Chapter 8) Evidently, a theory that can accurately predict the height and the propagation of the breaker bars, is not yet at disposal, although investigations of Roelvink [8.24] are encouraging.

A similar mechanism as being drawn up for the surf beat can be responsible for long-periodic waves, which propagate in longshore direction. Another driving force for this kind of waves can be propagating atmospheric disturbances. These so-called "edge waves", of which the amplitude again mostly decreases in seaward direction, could be responsible for the relative distance between the rip-currents perpendicular to the coast.[30]

2.3.3. *The effect of harbor moles and other effects of human interference*[1]

In the former subsection, the following questions were posed:

- Can the coastal behavior be predicted more accurately, if the pre-knowledge is implemented that breaker bars appear?
- How much cross-shore transport is involved with the motion of these bars?

[1]Cf. as well the comments of Rakhorst (Fig 8.5 etc.) in Chapter 8.

In the Netherlands, North of RSP 30 km the coast is defended by a groin system. South of this site, this is not the case. This fact does not seem to affect, the shape and site of the breaker bars very much; also the influence of the Hondsbossche Seawall on the breaker bars appears to be very small.

At the Northern side of the Hondsbossche Seawall (near Petten), at the seaward side, Fig. 2.17a shows a deep trough, of which the landward side is directed to Callantsoog. The coastal profile in this neighborhood is the steepest of the entire Dutch coast, apart from the vicinity of the Inlet of Texel. Figure 2.27 shows once more the situation on this spot, now in the shape of a map.

Do we see a relict of former centuries? Does the coast invite to sail the old Zijpe, leading towards the Zwanenwater? The latter nowadays is a nature reserve landward of the dunes, but in the thirteenth century the entrance of the Zijpe was situated here (see Fig. 2.27 after data from [43, 38]). An aerial photograph is given in Fig. 2.3.

At that time the coast extended 3 km seaward. Even now the old remains of the so-called "Pettemer Polder" can be found at a depth of 6 m and at

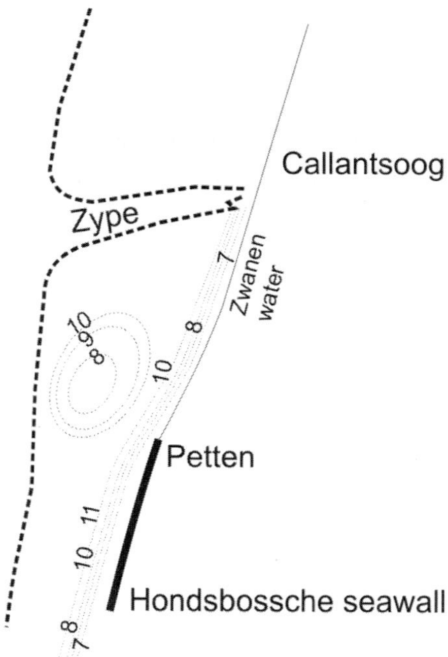

Fig. 2.27 The site of the Zijpe.

a distance of 2 km from the present coastline (Fig. 2.27). According to de Mulder[38] the Pleistocene is situated at the depth of NAP −8 m. The Pettemer Polder forms a protuberance on a peculiar horizontal plane at a depth of approx. OD — 10–12 m. It is located from RSP 10–20 km and from approx. 600 m seaward from the coastline upto 3 km from the coastline. Possibly a 4000-year-old current resistant clay and peat layer is located here. Bottom coring landward of the Hondsbossche Seawall[38] makes this probable and recent geological research could prove this. Although the old reconstruction[43] of the Zijpe channel locates this channel north of the Pettemer Polder, it seems hydraulically and morphologically feasible there is (possibly) a (second) Southern channel. In a clay bottom, this channel position could be preserved indeed.

However, one could also attribute the trough in the coastal profile at the North-Western side of the Hondsbossche Seawall to this coastal defense as a consequence of stream contraction. The fact that the depth does not adapt itself to that of a "normal" profile points at a hydraulically stable situation, or to a low amount of transport.

It shows that along the entire North-Holland coast these bars make a certain angle between Bergen and Egmond (Fig. 2.17a) one could, with some imagination, discover a terrace at the seaside, where the old clay and peat formations (lagunal depositions of more then 6000 years of age) could surface (see for instance [54]). However, from the 3D-time diagram of Fig. 2.19a it does not show clearly, that the bank at the seaside of the area in question is solid: at this site the bottom level varies somewhat.

Now take a look at the immediate surroundings North of IJmuiden (Fig. 2.17a).[m] Ever since the harbor moles of IJmuiden were built in 1870 sand has been accumulating here. The result is a gently sloping coastal profile. Along the Dutch coast a gentle sloping profile appears to coincide with many breaker bars; like here, where the number of breaker bars increases up to 4 or 5; for the Holland coast 2–3 breaker bars are usual with the coast. As appears from the 3D-time diagrams (Fig. 2.19a), the behavior of the IJmuiden coast is rather fussy. Possibly this is caused by the fact that short-term climatologic case) clearly are influences (accretion North of IJmuiden if the dominating waves come from the North, and erosion in the opposite manifest here, because the longshore transport gradients are large: the harbor moles completely interrupt the local longshore transport).

[m] Refer to Subsection 2.5.2 for a treatise on a mathematical–physical basis.

These harbor moles also have influenced the immediate surroundings Southward of IJmuiden. The extension (1970) of the Southern mole is positioned under an angle with respect to the original one: it is directed more Southward in comparison to the 1870 harbor moles, which were directed Westward (Fig. 2.28). Thus, at the landward side of it a sheltered area emerged, where sand, deposited by the sea during southwestern storms was able to stay during NW storms, when the area was in the leeside of the harbor mole. Because of this shelter effect, and possibly also by eddies (Fig. 2.28), after 1970 behind the harbor mole a rapid expansion of the beach area next to the Southern mole developed; this is also shown in Fig. 2.17b. The sloping terrace, which is present at the seaward side of the area considered, shows little variation (see Fig. 3b, and the 3D-time diagram of RSP km 60.00). The breaker bars, which at this spot clearly propagate in seaward direction, attenuate on this terrace. Do resistant layers from geological times show up here as well?

Further southward the 3D-time diagrams (Fig. 2.19b) show that the breaker bars between Zandvoort and Scheveningen also move in seaward direction; however, the amplitude decreases much slower as compared to RSP km 60.00, causing the bars to extend much further seaward than at RSP km 60.00 (Fig. 2.19b and Subsection 2.3.2). Here the bars propagate in a regular way, as the 3D-time diagrams depict. The regularity appears to arise above the relative chaos of the climatological circumstances: it is

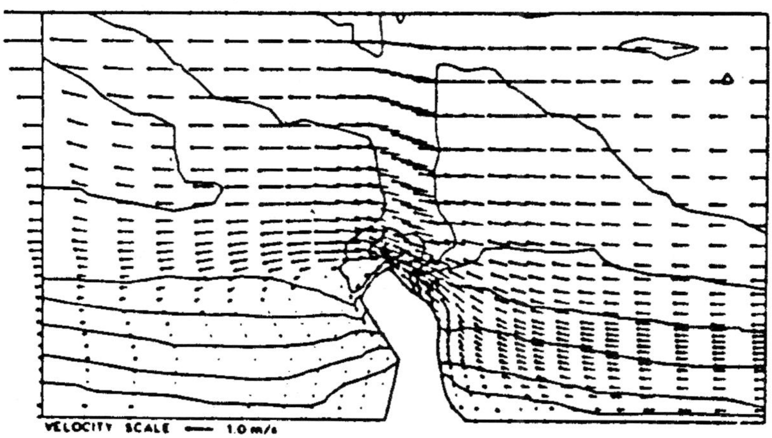

Fig. 2.28 Flow pattern during ebb near IJmuiden Harbor. The original harbor moles (before 1970) stretched only to the "crack" in the current Southern mole.

hard to discover the influence of relatively stormy years in these figures. In Subsection 2.3.2 this phenomenon has been treated more extensively.

More to the south the number of breaker bars decreases, while the coastal profile is steepening. At Scheveningen harbor monitoring is interrupted. North of the Scheveningen harbor a channel can be found parallel to the harbor head. South of Scheveningen, along the Delfland coast a steep foreshore is found together with an underdeveloped breaker bar at some places.

The movement of the coast near range RSP 101.00 km (Fig. 2.19b), just north of the harbor head of Scheveningen, is dominated at one side by the wave sheltering by the moles and the effect of the hampering of the littoral drift. However, the fact that in 1975, 0.7 million m^3 sand has been supplied should be taken into account as well.[n] Next to that, according to information of the "Hoogheemraadschap Delfland", periodically sand of the beach south of the harbor heads has been removed and dumped on the beach north of these moles, while the groin, directly north of the harbor moles has been enlarged in 1972 to affect the circulation of the tidal current in a positive way. Without thorough research it is hard to distinguish the differences between the various effects. This is even stronger at[o] RSP 116.62 km (Fig. 2.29b), where during the extension of the Hook of Holland harbor in 1971/1972 the coast had been brought also in a more forward position and 18,94 million m^3 sand was deposited in the scope of the Beach Project Hook of Holland (just North of the harbor mole).

The coast of Delfland (Scheveningen — Hook of Holland; Fig. 2.29b) is eroding already for centuries. This is illustrated by Fig. 2.18, after Edelman,[25] where for the entire coast of Holland the accretion and the erosion of the foot of the dunes between 1860 and 1960 is given. One could imagine that the sand initially would be withdrawn from deeper water, which would cause steepening. The shoreface in the immediate vicinity of Hook of Holland is at a slightly higher level. On one hand it is the effect of the Northern mole, constructed in 1870 and enlarged in 1970, being part of the works near the entrance of the Nieuwe Waterweg. On the other hand it is caused by the beach nourishment at Hook of Holland (this took place, at a site, somewhat South of the area, depicted in Fig. 2.29b).

Of the coastal area between IJmuiden and Hook van Holland groins only protect the Delfland coast: approximately from the north side of the

[n] This apart from other supplies in the 1980s and 1990s; cf. for instance, the CUR manual (Ref. [16] in Chapter 5).
[o] For location, see Fig. 2.17b.

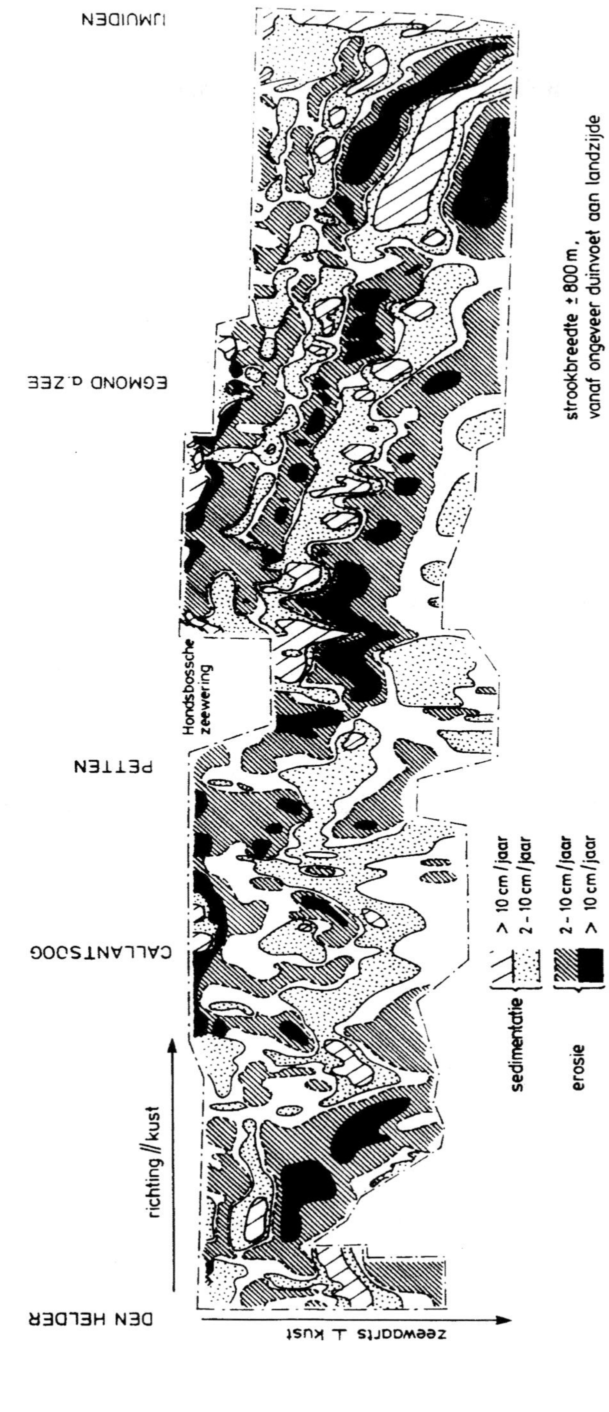

Fig. 2.29a Twenty-year averaged bottom trends North Holland. The area depicted has a width of ca. 800 m, starting from the foot of the dunes.

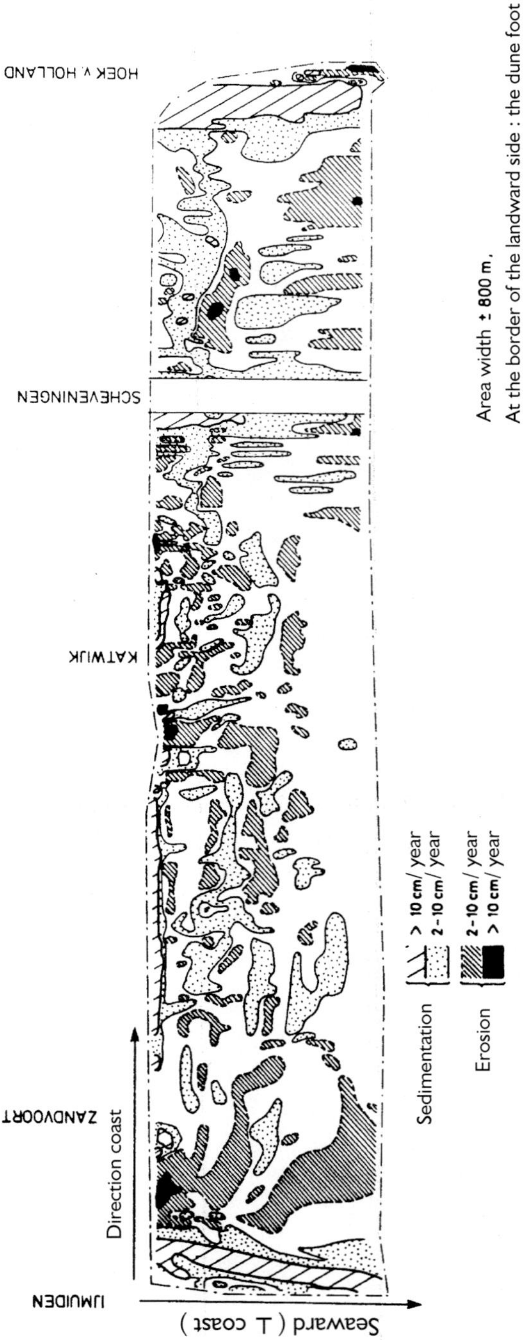

Fig. 2.29b Twenty-year averaged bottom trends South Holland. The area depicted has a width of ca. 800 m, starting from the foot of the dunes.

Scheveningen Boulevard (Fig. 2.12) to Hook of Holland. The effect of this coastal defense system appears to be modest. Investigation of this area is facilitated due to the fact that the Hoogheemraadschap Delfland made profile measurements for this part of the coast during more than 150 years.

2.3.4. *Coastal behavior after averaging of the breaker bars*

Knowledge of the period of motion of a bar system gives the opportunity to determine the accuracy, with which the coastal trend can be determined. This accuracy depends on the one hand upon the ratio between the period of the breaker bars, compared with the period of observation and on the other hand on the phase of the bar cycle at the start of the observation time. This statement will be elucidated in Subsection 2.3.5.

Figure 2.29a,b shows those trends over the period 1970–1984, however only for those measuring points, in which the series of observations does not show considerable gaps. With the least square method the linear trend of the bottom level rise of each point of the measuring grid has been determined. Depicted are lines of equal bottom level rise (which is expressed in cm/yr). These lines separate the area into sections, which are about stable (changes of 1 cm/yr or less), moderately accreting, c.q. eroding (2–10 cm/yr) and those changing strongly (10 cm/yr or more).

Although the pattern obtained is consistent, it is not easy to interpret. As Subsection 2.3.5 will elucidate, the "pseudo-trend", caused by the breaker bars, will play an important role. This pseudo-trend depends as well on the amplitude as on the period of the breaker bars. For instance (Table 2.1), the amplitude of the breaker bars is of the order of 1 m and the period is about 20 years for the coastal stretch Bergen-IJmuiden (Fig. 2.17a) and 5 years for IJmuiden-Katwijk (Fig. 2.17b). Thus, for a period of observation of 20 years one finds a pseudo-trend of ca. 1 dm/yr for Bergen-IJmuiden and ca. 2 cm/yr for IJmuiden-Katwijk.

As a matter of fact, in the picture of Fig. 2.29a one indeed observes a strong dominance by the motion of the breaker bars: shifting in parallel belts of alternately sedimentation and erosion. Indeed, this "trend" is of the order of 1–2 dm/yr. Topographically, those belts coincide rather well with the breaker bars. Visually, the macroscopic consistency of the features is striking (the same trends over a distance of 30 km!).

Near Callantsoog (Fig. 2.29a) sand was supplied for dune enforcement in 1976/1977 (350.000 m^3) and in 1979/1980 (470.000 m^3). This explains the sedimentation in Fig. 2.29a at the landward side. In the vicinity of Den Helder (Figs. 2.17a and 2.29a; Fig. 2.46) one observes sedimentation in the shallow part of the Schulpengat. Not enough data was available in the deeper part of the Schulpengat to make a statement concerning the trend at that site.

Consider now the part south of IJmuiden (Fig. 2.29b). Generally, here the trends vary between −9 to +9 cm/yr at the inshore; stronger trends occur in the dunes (order: ±20 cm/yr). Below OD − 6 m there are large areas (between 80 and 90 km) where the coast is practically stable; however, at some places, for instance between Scheveningen and Hook of Holland (Fig. 2.17b) variations of 5 cm/yr also do occur.

In Fig. 2.29a,b the direct or indirect outcomes of human interference are dominating. At Hook of Holland, the whole picture is dominated by the local Beach Project Hook of Holland; although less rigorous, the same holds for the coastal behavior North of Scheveningen (the effect of enlargement of the harbor moles; refer to Subsection 2.3.3 for both cases).

Indirect results of human interference can be observed from the sedimentation South of the harbor moles of IJmuiden; sheltering effect (Subsection 5.3.5) is probably dominant. In the direct vicinity of the Southern mole sedimentation has taken place at a rate of ca. 20 cm/yr at the inshore and even at a rate of 40 cm/yr. in the dune area. At a distance of 3 km Southerly of this mole (near 59 km; Figs. 2.17b and 2.29b) sedimentation rates are reduced to 1–5 cm/yr.

Another indirect outcome of human interference is the formation of a scour hole near the Scheveningen moles, caused by the contraction of the streamlines and local maxima of the current velocity.[p] In the local bottom trends (Fig. 2.29b) this reveals itself by the abrupt transition of some sedimentation into larger erosion (5–11 cm/yr) concentrating itself within a small area (RSP 100.50 km). In the 3D-picture (Fig. 2.17b) also the scour hole is visible.

Considering Fig. 2.29b, longshore as well as cross-shore features are showing off, i.e. longshore and profile changes. The erosion of the coast near Zandvoort clearly turns out. This seduces to ponder that the sand for the sedimentation near IJmuiden has to come from somewhere. Here (RSP 59.00–64.00 km) bottom level changes occur at a rate of −1 to −5 cm/yr (sedimentation being positive, evidently).

[p] An example of this kind of process, however valid for IJmuiden, is depicted in Fig. 2.22. (Subsection 2.3.3).

Longshore belts are visible in Fig. 2.29b. This leads to think of the effect of breaker bars. However, the bar-structure does not show up in the location and orientation of those belts. Furthermore, the feature is present as well near the coast of Delfland (i.e. Scheveningen — Hook of Holland), where (if bars are present anyway) bar activity is of minimal importance.

Excluding the area of the Beach Project Hook of Holland at the seaward end of the area under consideration some offshore erosion can be observed. Seaward of the area considered, near Terheide (114 km; Fig. 2.17b), one finds a long term dumping site of silt from the Nieuwe Waterweg.[q] Figure 2.29b does not show coastal accretion at that site, but rather coastal erosion.

Generally speaking, changes in the curvature of the coastal profile show up more explicitly than a general steepening or flattening of the profile. Adequate explanations are still lacking.

A 100-yr analysis probably would result in smaller trends than the 20 yrs considered. As one cannot assign the sky the limit to those trends and furthermore opposite trends in neighboring areas would cause reactive forces, a similar analysis as displayed in this chapter would be advisable at the turn of each millennium.

2.3.5. *Philosophy concerning "trend"*

Do 20-year averaged linear bottom trends (cf. Subsection 2.3.4 and Fig. 2.29a,b) make sense? This is the topic of this Subsection.

In the first place the question arises, how consistent the measured trends are. For instance, the pattern of Fig. 2.29b is rather mixed-up (in contrast with the one shown in Fig. 2.29a, where the 20-year motion of the breaker bars strongly dominates). The consistence appears to be better than what Fig. 2.29b suggests. This is illustrated in Fig. 2.31, where as a detail of Fig. 2.29b the trends in any of the respective measuring points in the surroundings of IJmuiden have been depicted. It shows, that de spatial gradients in the accretion velocity are small and over belts with a width of several hundreds of meters and over the length of 5–10 km the same tendencies are found.

Bakker[9] deals with the influence of accidental measurement errors on trends. From the calculations the following result is found. If the standard deviation of the accuracy of sounding in every separate measuring point

[q] "Loswal Noord" in Dutch.

amounts 20 cm, with annual surveys over a period of 20 years a standard deviation of the trend of approx. 1.4 cm/yr is found. This results if the deviations from the trend from year to year are not correlated.

But is this the case?

Suppose, a sinusoidal signal is averaged over a certain observation time, using the least square method (Fig. 2.30). This implies the assumption of a linear trend; every deviation from this straight line will be seen as a random fluctuation. With reason statisticians will object to this method; however, if one has to deal with periods of the signal of four years and sample frequencies of once a year with a relatively small observation period, it is hard to distinguish signal and noise.

At first, presume that the observation time is very short compared to the period of the sinus and that at the starting time of observation the signal has the value zero (Fig. 2.30). As a (pseudo)-trend one will find the tangent to the sine curve. Clearly, this is the maximum pseudo-trend that can be obtained. Figure 2.32 displays this pseudo-trend (as fraction of the earlier mentioned maximum) as a function of the observation period. As mentioned, the starting point was the assumption that the signal was zero at the beginning of the observation time and after that it became positive. The calculations are given in Ref. [9].

For the area between Zandvoort and Noordwijk (Table 2.1, Section 2.3.2) in the longshore belt with maximum breaker bar motion a maximum pseudo-trend (small observation period) of 1.14 m/yr is found, caused by the periodical motion of the breaker bars. Choosing an observation period of 4 times the period of the breaker bars, i.e. 16 years, still a

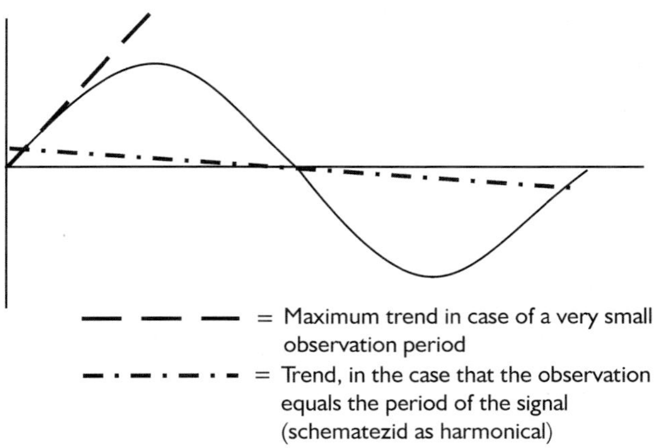

Fig. 2.30 Shortterm (- - - -) and long-term (-.-.-.) trends.

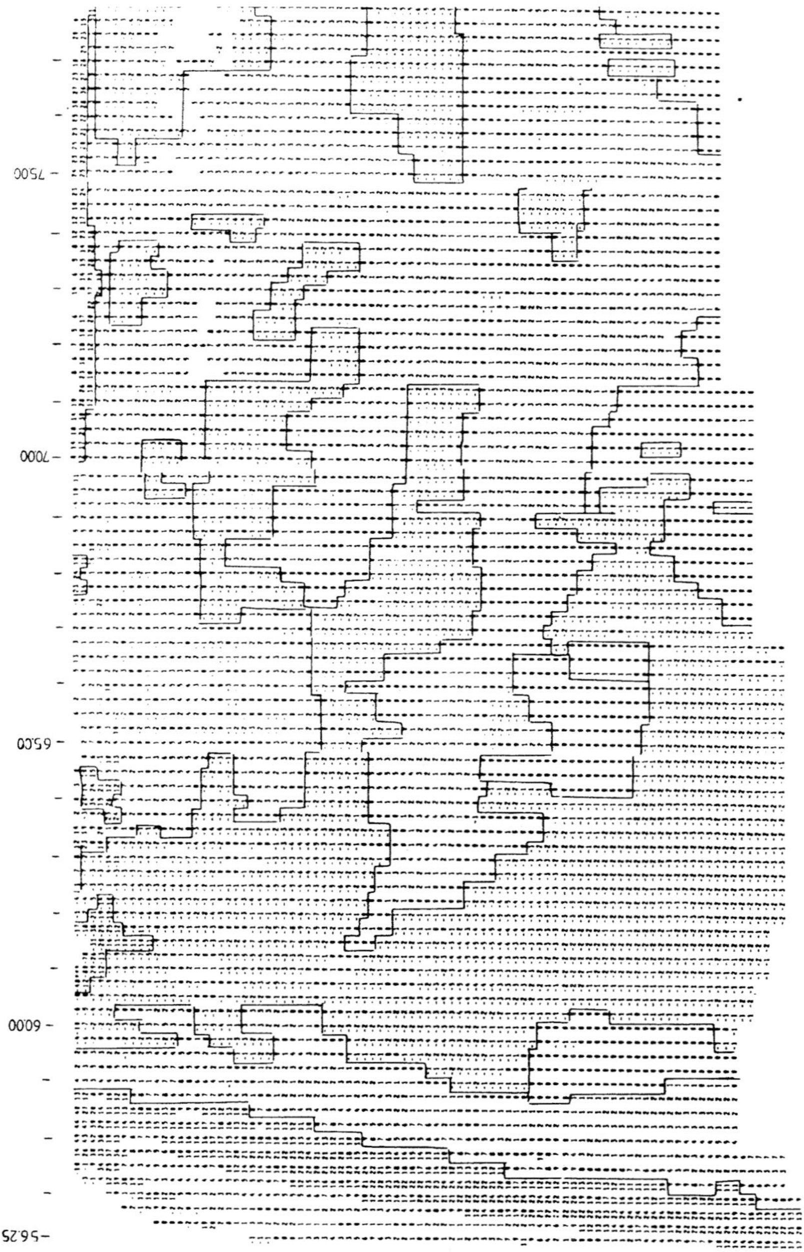

Fig. 2.31 Twenty-years average bottom trends; South Holland; 56.75–80.00 km (cf. Fig. 2.29b). Near the coast and IJmuiden, sand supplies took place. Compare with fig. 2.29b, of which the present figure shows a detail (IJmuiden to ca. Katwijk).

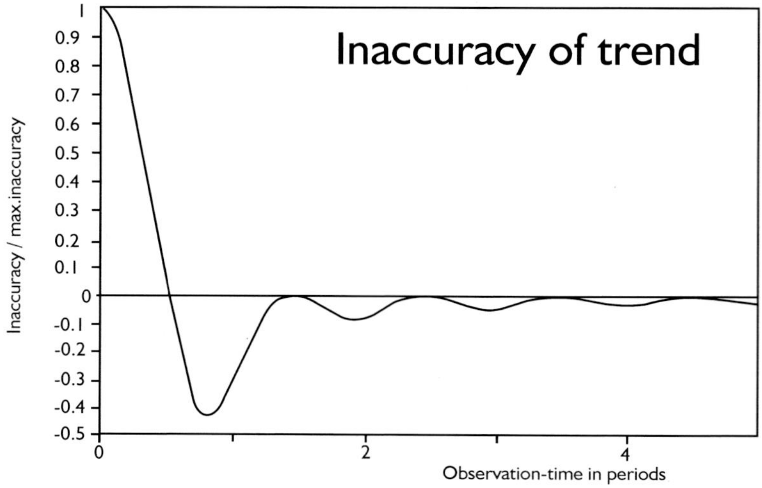

Fig. 2.32 Inaccuracy of a trend.

maximum pseudo trend of 1.9% (Fig. 2.32) remains: this means 2 cm/yr. In fact the pseudo trend will be somewhat smaller, as at the beginning of the observation period the phase of the oscillation will be different everywhere.

It is a matter of philosophy, to which rate one wants to involve the influence of the long period oscillations (like for instance with a period of 30 years) into a 20 years trend. In the first place it will be disputable, if one would be able to recognize those signals as periodic; however, even if this is the case, it is "allowed" that two succeeding 20-year trends are different. However, it makes sense to realize how long periodic motion affects the trend.

The breaker bars in the surroundings of Egmond (Fig. 2.17a; 35 km) for instance, have a period in the order of 20 years; Fig. 2.29a is not corrected for this. According to Fig. 2.32 this can cause "errors" in the order of the trend of 30–40% of the maximum pseudo trend, which has a value of 30 cm/yr at this location. In order to give an impression, how the breaker bars affect the accuracy of the trends, in Fig. 2.33 this has been indicated for the area between IJmuiden and Scheveningen. The way of presentation corresponds to Fig. 2.29; it is shown which pseudo trend would arise due to the breaker bars, if these are schematized according to description in Subsection 2.3.2 and calculated for the considered area. At the most an inaccuracy of 3–4 cm/yr is found. One has to compare this pattern with the data in Fig. 2.29b. Another illustration of the non-linearity of the trend is

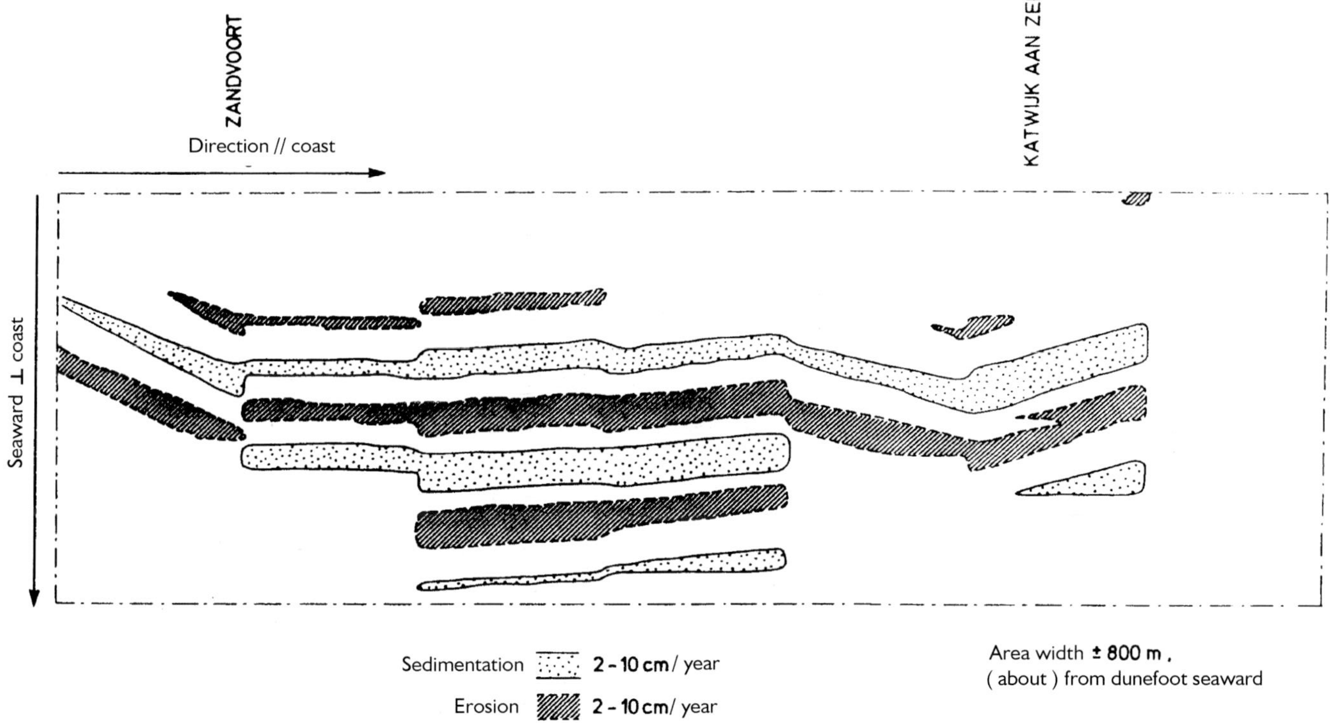

Fig. 2.33 Twenty-years pseudo trend, for explanation see text.

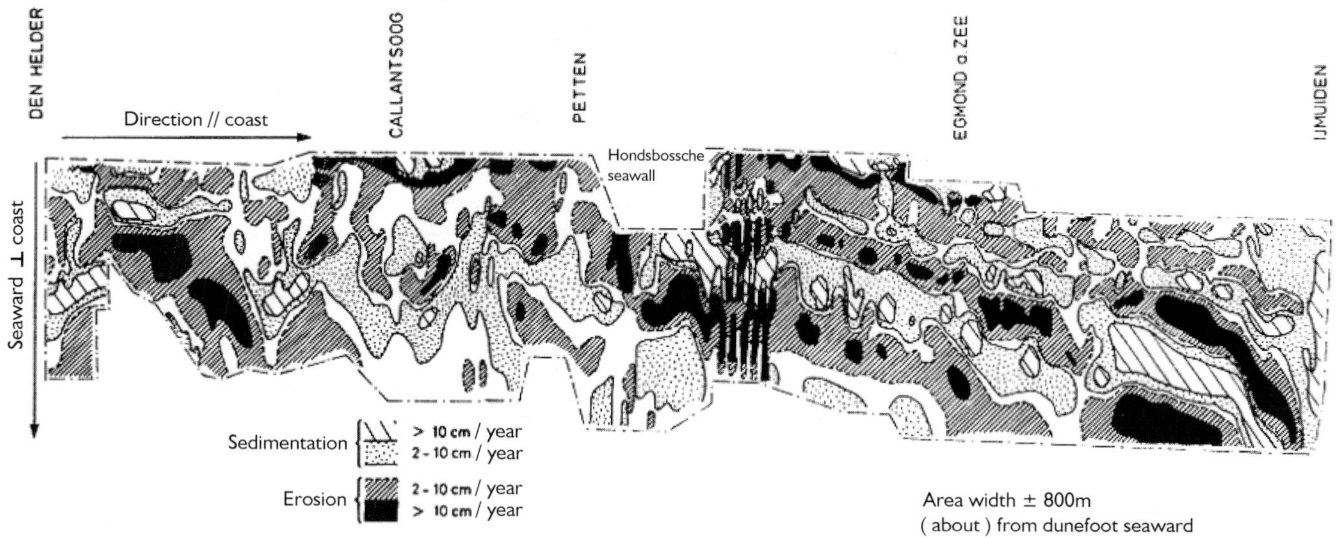

Fig. 2.34 A discontinuity in the starting time of the observation period (and thus of the observation time; cf. the area just South of the Hondsbossche seawall, where the measuring grid has been refined during the (general) observation period) gives a distinct discontinuity in the result of the analysis. Refer to the text: and compare Fig. 2.29a.

given in Fig. 2.34, an early version of Fig. 2.29a South of the Hondsbossche Seawall five extra ranges have been taken into account, however this data is only available for the last 7 years (from 1975 to 1985). As the measurement period differs, these ranges are found to have a completely different trend than in the surrounding area.

It is in any case useful to realize how time-dependent the calculated 20-year trends (still) are. A trend of 5 cm/yr is (mathematically) equal to 5 m/century. The area between the Delfland groins accretes with approx. 5 cm/yr. However, one has to be a very big optimist to expect this to go on for a century.

The 100-years trend of erosion and accretion of the coast of 1–2 m/yr (Fig. 2.18) should, at an average coastal slope of 1:100, be translated to vertical trends of 1–2 cm/yr. From the previous, it is found that the scatter of short-term phenomena is too big to monitor those trends. This is even more the case for a sea level rise at 2 mm/yr.

2.3.6. *General conclusions concerning Section 2.3.*

(1) The Dutch coast is far from uniform (Fig. 2.17a,b). The part near Den Helder is determined by the dynamics of the Texel inlet. South of the outer delta occasionally one breaker bar shows up. Farther South the Hondsbossche and Pettemer Seawall (Fig. 2.3) dominate the bottom topography. At the South of this (Egmond) a somewhat tangled pattern of 2 or 3 breaker bars appear more or less obliquely to the coast (the opening heading south). Continuing in Southward direction, the coastal profile becomes gentler; this is paired to more breaker bars.

For example, between Bergen and Egmond in the deeper regions bottom changes are restrained by the expose of old clay and peat formations.[54]

The harbor moles of IJmuiden strongly determine the local topography. Further South, away from this influence, a regular pattern of 2–3 breaker bars is found. Number and size of the breaker bars decrease in southward direction, as the coast tends to steepen. South of Scheveningen no breaker bars are found.

(2) South of IJmuiden the breaker bars propagate seaward with a velocity of 60–80 m/yr (Fig. 2.19b); its wavelength is of the order of 200–300 m and its period is 3–4 years. The influence of storms on the behavior of the breaker bars can hardly be recognized. North of IJmuiden the movements are much smaller and less regular (Fig. 2.19a).

(3) Probably, "surf beat" and "edge waves" play a big role in the dynamics of breaker bars. These are low-frequent waves with a period of the order of 5 min, emerging because the momentum of groups of alternately high and low waves forces resonance of the (total) water mass in the surf zone.

(4) The gross longshore transport along the Dutch coast is of the order of a 100 times larger than the cross-shore transport (compare Figs. 2.25 and 2.26). However, sedimentation and erosion appear by local differences in transport. These can become of the same size order for the longshore and cross-shore transport.

(5) The direct and the indirect consequences of human interference such as at one hand the extension of the harbor heads of IJmuiden, Scheveningen and Hook of Holland and on the other hand the large-scale sand supply of the Beach project Hook of Holland are very clear. These locally dominate the natural changes along the coast (Fig. 2.29a,b). As the breaker bars North of IJmuiden have a period of the same order of magnitude as the observation period, the propagation of these breaker bars is reflected in the linear trend. This results in trends, varying alternately in belts (300 m wide; 30 km long) of sedimentation and erosion in the order of 10–20 cm/yr. The site of those belts coincides with the site of the breaker bars.

South of IJmuiden, where the period of those breaker bars is considerably lower, the motion of these bars averages out (to a large extent) over the observation period, although the influence of the breaker bars still cannot be neglected. In Subsection 2.3.5 this averaging effect is investigated.

South of IJmuiden changes of the underwater topography, remaining after time-averaging the breaker bars, generally vary between +10 and −10 cm/yr. Spatial gradients in sedimentation velocity are small: over belts with a width of several hundreds of meters and lengths of 5–10 km the same tendency is found (Fig. 2.29b).

(6) Only locally, like between (about) Zandvoort and Scheveningen (Fig. 2.29b), the most seaward part of the measured coastal profile (until about 800 m seaward of the coastline) is stable. In front of Monster however an erosion rate between 5 and 10 cm/yr is found.

(7) Changes in the curvature of the coastal profile (after time-averaging the breaker bars) show off over the last 20 years more clearly than an overall steepening or leveling.

(8) Rather much for sure the 20-year trends are still large in comparison with the 100-year trends, in which even more phenomena are time-averaged.

Many of the causes of the registered coastal behavior are still unknown. It will be clear, that monitoring the topography of the coastal zone requires a great effort in measuring and processing capacity.

However, this is necessary to maintain the validity of the statement: "the coast is safe".

2.4. Sand Transport by Waves and Tides (Uninterrupted Coast)

Van Straaten[44] and Edelman and Eggink[27] give qualitative considerations about the sand transport by waves. Since their publications, so much data have been collected about the relation between the longshore component of the wave energy and the sand transport, that it is worthwhile to apply such formulae to the Dutch coast in order to obtain more quantitative conclusions. However these conclusions cannot be better than the available data; i.e. the visual wave observations made on the Dutch lightvessels (Table 2.2). These measurements from 1949 to 1957 have been statistically analyzed by Dorrestein,[22] giving the probability of occurrence for each wave condition (characterized by height period and direction of the waves). Van Harreveld,[33] correlated the visual wave height of the Goeree lightvessel (at 10 km from the coast) with the wave height observed on the step resistance wave gage "Triton" at 3 km from the coast, and concluded from this that the low waves were estimated too low and high waves too high, the (significant) waves of 1–2 m being estimated correctly. The regression coefficients between lightvessel and wave gage were different for each wave direction, which is clear for the offshore waves because these waves were generated with a different fetch. However, also for the onshore waves there appears to be some influence of site. From the visually measured wave period at the lightvessels an "equivalent wave period" T_{eq} has been derived, according to Battjes[12]

$$T_{\text{eq}} = 1.23\, T_{\text{m}} \tag{2.1}$$

in which T_{m} is the mean period of upward zero-crossings. The equivalent wave period is the period, for which the energy/rn^2 and the energy-flux/rn of the replacing sinusoidal wave are equal to those in the wave in nature.

Table 2.2 The visual wave observations of the Netherlands wave vessels.

Site	Orientation of the coast with respect to the North				
The size of P_l	Vlieland[1] 060°–240°	Texel[2] 030°–210°	IJmuiden[3] 15°–195°	Scheveningen[4] 045°–225°	
$-33 < P_l < -31$				0.13	North to South
$-31 < P_l < -29$			0.52		
$-25 < P_l < -23$			0.17		
$-23 < P_l < -21$		0.13	0.24		
$-19 < P_l < -17$			0.42	0.27	
$-17 < P_l < -15$		0.13	0.70		
$-15 < P_l < -13$	0.03	0.25		0.92	
$-13 < P_l < -11$	0.27	0.50	1.1	1.6	
$-11 < P_l < -9$	0.25	0.76	3.1	0.35	
$-9 < P_l < -7$	1.2	2.4	8.7	5.6	
$-7 < P_l < -5$	0.82	4.3	3.4	2.0	
$-5 < P_l < -3$	8.9	19	26	26	
$-3 < P_l < -1$	32	56	120	119	
$-1 < P_l < 0$	166	203	159	200	
$-\infty < P_l < 0$	210	287	323	355	
$P_l = 0$	529	401	299	316	
$0 < P_l < 1$	155	196	272	188	South to North
$1 < P_l < 3$	53	70	86	92	
$3 < P_l < 5$	29	31	15	28	
$5 < P_l < 7$	11	6.1	4.0	10	
$7 < P_l < 9$	5.4	4.2	0.88	6.0	
$9 < P_l < 11$	1.4	2.1	1.3	0.56	
$11 < P_l < 13$	1.0	0.61	0.15	2.9	
$13 < P_l < 15$	1.1	0.84			
$15 < P_l < 17$	0.84	0.47		0.56	
$17 < P_l < 19$	0.81	0.24		0.22	
$21 < P_l < 23$	0.34	0.23		0.44	
$23 < P_l < 25$	0.42				
$29 < P_l < 31$	0.12				
$0 < P_l < +\infty$	261	312	379	329	

[1]From data lightvessel.
[2]From data lightvessel Texel.
[3]Probability of wave characteristics: mean between lightvessels Texel and Goeree.
[4]From data lightvessel Goeree.

Van Harreveld[33] found the visual Goeree lightvessel period T_{vi} about equal to $1.5T_{\text{m}}$ from which $T_{\text{eq}} = 0.82T_{\text{vi}}$. The longshore component of the wave energy-flux, P_l has been computed from:

$$P_l = 1380 h_{\text{br}}^3 \frac{\sin\varphi_0}{c_0}, \quad (P_l \text{ in W/m}^1, h_{\text{br}} \text{ in m}) \tag{2.2}$$

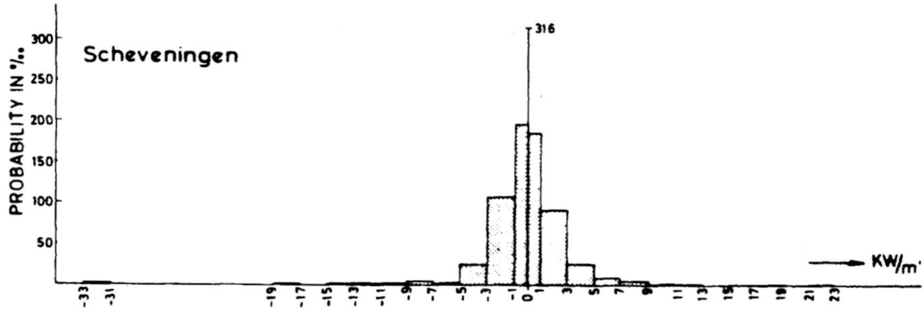

Fig. 2.35 The probability distribution of the longshore current of the wave-energy flux near Scheveningen.

in which φ_0 and C_0 were the deep-water wave direction and phase velocity, and the breakerdepth $h_{\rm br}$ was found from:

$$h_{\rm br} = \left[\frac{H_{\rm sign}^2 \, c_0 \cos\varphi_0}{2(0.4)^2\sqrt{1.4g}\cos\varphi_{\rm br}}\right]^{\frac{2}{5}} \tag{2.3}$$

The theoretical basis of these formulae is given by Bakker.[4] The way of calculation (however for a more general kind of topography) is explained in Chapter 4.3.2 as well: Eq. (2.2) is analogous to Eq. (4.5b). The results are given in Table 2.2.

Figure 2.35 gives one of the probability distributions. From this it is easy to compute the mean energy flux and the mean littoral drift S^* (integrated over the surf zone), using the CERC-formula:

$$S^* = 1.4 \times 10^{-2} \frac{P_1}{\frac{1}{16}\rho_{\rm w} g} \quad (\rho_{\rm w} g = \text{specific weight water}) \tag{2.4}$$

where $\rho_{\rm w}$ is the specific density of the fluid and g the acceleration of gravity.

$$S^* = 2300\, P_1 \quad (S^* \text{ in m}^3/\text{yr}, \ P_1 \text{ in W/m}') \tag{2.5}$$

However, the used data and formulae are not accurate enough to justify this computation since the probability distributions are about symmetrical. Only along the Vlieland coast the resultant drift is significant (about 1/2 mln. m^3/yr). The resultant net energy-flux was always less than 0.3 KN/m′. The gross littoral drift (sum of transport in both directions) can be computed from the summations of the products of the absolute value of

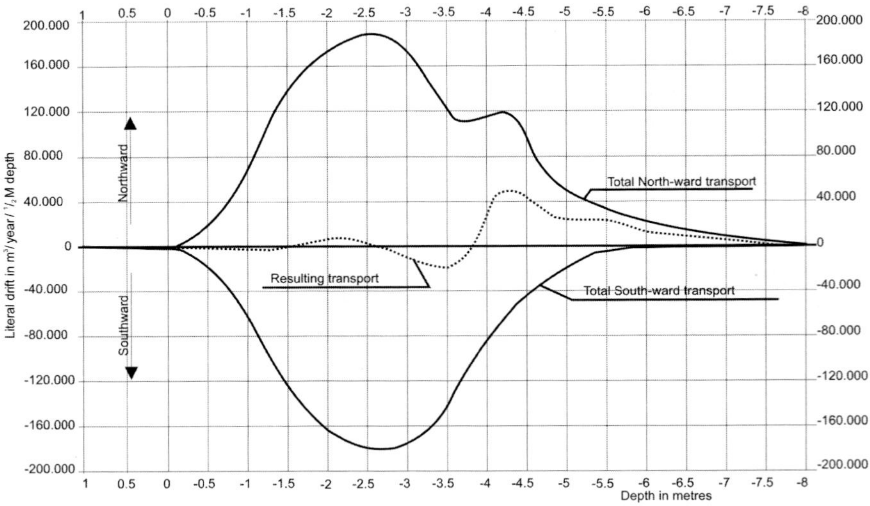

Fig. 2.36 Cross-shore distribution (as function of depth) of year-averaged littoral drift. It is expressed in $m^3/yr/(0.5\,m$ depth). Site: Scheveningen (cf. Fig. 2.17b).

P_l and the corresponding probability of occurrence (called pr (P_l)).

$$S^*_{\text{gross}} = 2300 \sum_{\text{all prob.}} \{|P_l| \times \text{pr}(P_l)\} \tag{2.6}$$

This gross littoral drift is found to be of the order of 1.5–2 rnln m^3/yr (thus about 1 mln m^3 in each direction).

Assuming for a moment no influence of tides, the distribution of the littoral drift over the surf zone has been computed by an adapted method proposed by Svasek[19]; the method of computation is illustrated in Bakker *et al.*[11] and is treated in more detail by Bakker[4]; In Section 4.2 a more general case is treated.[r] The result for Scheveningen is shown in Fig. 2.36.

Considering which influence is more important, the one of waves or the one of tides, Bakker and Opdam,[5,6] investigate the driving forces of waves and currents.

The total driving force of the *longshore current* in the surf zone is, Bowen[16]:

$$\frac{1}{2} E_{\text{br}} \sin 2\varphi_{\text{br}} = \frac{1}{16} \rho g \gamma^2 h^2_{\text{br}} \sin \varphi_{\text{br}'} \tag{2.7}$$

[r]If h_1 in Eq. (4.5a) is taken equal to zero, the result is Eq. (2.2). Equation (2.3) is a specific case of Eq. (4.18a): $n_0 = 0.5$; $\gamma = 0.4$; $F_r = \sqrt{}1.4$.

in which E_{br} is the wave energy per m^2 in the breaker zone, φ_{br} the angle of wave incidence, h_{br} the breaker depth and γ the ratio between breaker height H_{br} and h_{br}.

The driving *tidal force* over the breaker zone assuming a rigid bottom slope with gradient m is:

$$\frac{1}{2m}h^2 \times \rho g \frac{\partial h}{\partial x} \tag{2.8}$$

in which $\partial h/\partial x$ is the gradient of the water level in longshore direction, of which for a sinusoidal tide (a. $\cos(\omega t - kx)$) is taken the maximum value, i.e. *ak*. The ratio between the longshore-current force and the tidal force is therefore:

$$\frac{1}{8}\frac{m\gamma^2}{ak}\sin 2\varphi_{\text{br}} \tag{2.9}$$

which is mostly large except for very small angles of wave incidence.

This estimate of order of magnitudes can be illustrated with the aid of Fig. 2.37. This figure shows the result of current measurements in a wave basin in Delft Hydraulics Laboratory, in which the current was measured caused by obliquely incident waves, respectively (first alternative:) without a gradient of the water level (then only longshore current results), (second alternative:) with a surface current in the direction of the longshore current and (third:) with a surface gradient counteracting the longshore current. Figure 2.37, from Hulsbergen *et al.* Delft Hydraulics (reference [51] of Chapter 5) demonstrates, that the longshore current over rules the tidal current. Assuming that the waves stir the material and the currents transport it, it thus is reasonable that the transport in the surf zone, according to the former computation, is rather well reproduced, except for the case where zero transport is computed.

In order to get an impression on the influence of the tide a tidal computation has been carried out with the numerical tidal model of which Fig. 2.38 gives the scheme.[35] Assumed is one channel system parallel and another one perpendicular to the coast. At the boundaries (the ends of the coastal-parallel channels) the vertical tide was given; at the junctions the vertical tide was computed and in the channels the tide. The computer program was developed by Booy according to the horizontal explicit leap-frog method; non-linear terms were considered, but Coriolis was neglected.

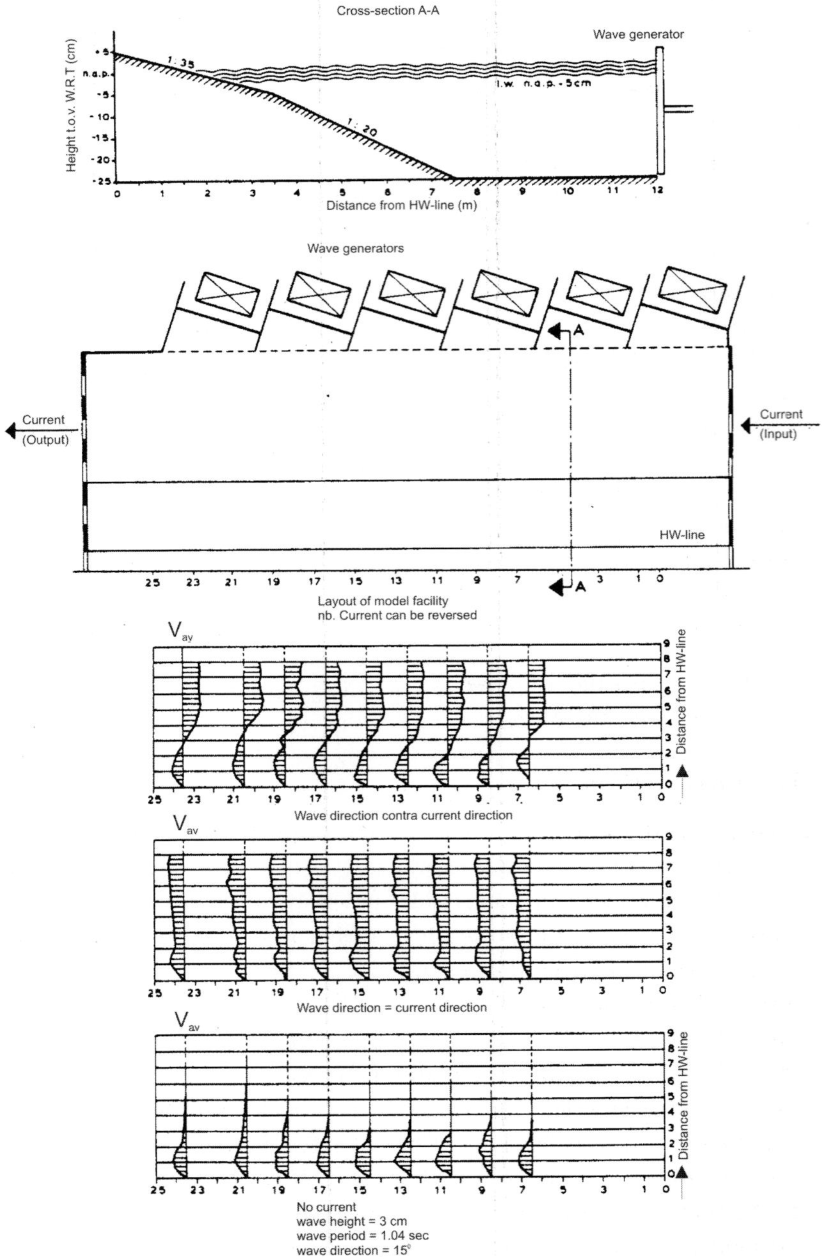

Fig. 2.37 Experimental data concerning currents, generated by a combination of waves and (tidal) water level gradients.

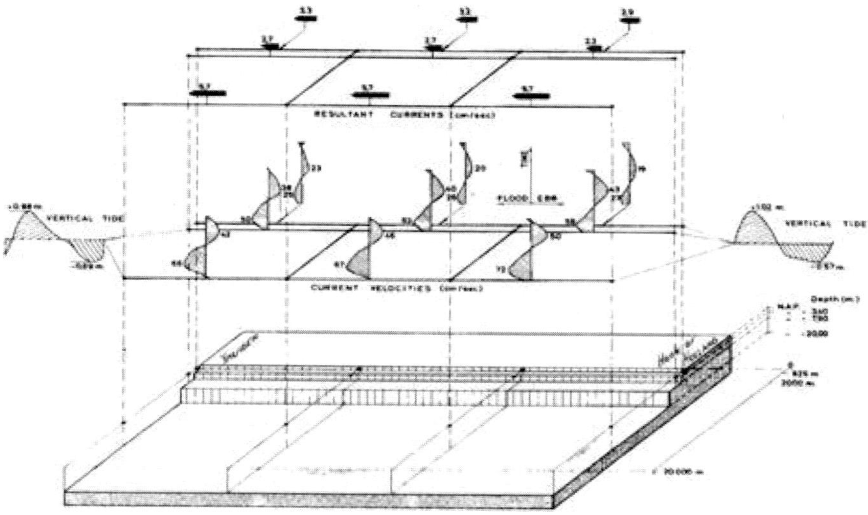

Fig. 2.38 Tidal module according to Klein Breteler.[34] Lowest part: schematization of the coastal area.

2.5. Tides, Shaping a Coast: Lessons from Van Veen and Other Old Masters

2.5.1. *Introduction*

Many coasts in the world are molded dominantly by the tides, especially at sites where tidal inlets or estuaries are present. The effect of waves and tides near those inlets will greatly differ from the one along an uninterrupted coast (as treated in Section 2.4). Evidently, current effects are very important in tidal inlets and estuaries. However, waves play a counterpart in adding diffusive effects.

The following Sections 2.5 and 2.6 should be seen in connection with Chapter 7 (Chapter 2: phenomenology; Chapter 7 emphasis on mathematics). Mathematical elements are treated furthermore in Subsection 3.2.5 (case e) in which an inlet (c.q. a tidal delta) is considered as a suction point (c.q. point source) of sand, which affects the shape of the coast. In Subsection 3.2.5 only the effect of waves is considered.

The modeling under consideration (Subsection 3.2.5 and Chapter 7) still has the character of pilot studies, as accuracy of sand transport computations is rather low.

Making "binary" selections on the topic "coastal dynamics; tides" instead of splitting in "phenomenology" (Chapter 2) or "mathematics" (Chapter 7) other possible "dimensions" are:

(a) "outer deltas" versus "tidal basins", assigned as "O"(uter) and "I"(nner)
(b) "linear effects" versus "quadratic effects".

This gives $2^3 = 8$ cases to be treated: four in this Section 2.5; the other Four in Chapter 3, c.q. 7.

Thus in this section and, in various subsections one or more of four sub-selections (O&linear, O&quadratic, etc.) can be treated. In the following paragraph, for each of the subsections those four sub-selections are put in an 2*2 array; a "dot" featuring in this array denotes, that the sub-selection is not treated in that subsection. Thus "O." denotes that linear as well as quadratic features will be considered.

The introduction of Chapter 8 (Mathematics: interaction between outer delta's and tidal basins) will give similar arrays for the sections of Chapter 7. "O" 's appear only in the first column of the arrays; "I" 's in the second one.

Subsection 2.5.2 (to be refered to below) is missing in this schedule, as it deals on terminology.

	2.5.3	2.5.4	2.5.5
linear	O.	.I	..
quadr	O.	.I	OI

Linear effects depend (mainly) on the tidal motion as such; it will show, that for instance the orientation and cross-section of the channels depend on linear effects.

"Quadratic effects" (for simplicity, also resultant flow, a zero-order effect, is comprised as well under this subject) are those effects of tide, which remain, even when the primary motion is cancelled by averaging.

Examples are: tide-averaging flow, which can have the shape of eddies in an estuary, with a horizontal or a vertical axis, but also meandering (or, more general: transversal motion) of tidal channels.

There remain features, which do not fit in directly with the rigid binary schedules, for instance: sand pressure. This subject is treated (as well) in Subsection 2.5.3.

Subsections 2.5.3 and 2.5.4 will give an abstract of a classical paper of Van Veen,[49] originally written in Dutch, but which deserves international common knowledge.[s]

A number of his original figures will be reproduced, which contain text in Dutch as well. Van Veen,[49] "solved" this problem, by adding a Dutch–English glossary. Even here, he demonstrates his wit, reason for the present author not to skip this glossary (mentioned before: Subsection 2.5.2).

Subsection 2.5.3. deals on the *outer deltas*. It starts by the display of the Van Veen theory (as far as the subject is: "the outer delta") and proceeds with an investigation of the actual behavior of the tidal channels in the Dutch Wadden deltas. In this subsection it is considered, whether this behavior is in line with the Van Veen theory.

Subsection 2.5.4 deals on the theory Van Veen developed concerning *ebb- and flood channels*. This subject is more important for the flood basin than for the outer delta. Often it is difficult to determine straight forward whether a channel is a flood channel or an ebb channel.

During the flood period sand is transported in flood direction; during the ebb-tide the inverse. Difference between ebb and flood transport is called the tidal transport; this is mostly small compared to ebb- and flood-transports. As its calculation involves the subtraction of two large and relatively inaccurate numbers, the result will be very inaccurate. In Subsection 2.5.5, a striking example will be given.

Apart from Van Veen the names of other "old masters" will emerge in this section (cf. for instance Subsection 2.5.3). For instance, Lorentz,[t] Thijsse (Chapter 7), and Van Bendegom[u] contributed in the 1940s and 1950s to a line of thoughts, which lead to the generation of papers like the ones of Van Veen.

With respect to practical examples, this section will focus on the Northern part of the Netherlands. The Southern part will be considered in Section 2.6. This is only for practical reasons. Essential difference between the Northern part and the Southern part is the shape of the inlets. Its

---

[s]Since the editing of this chapter this wish has been substantiated in 2005 by Stive *et al*; cf. p. 2–48.

[t]During his classes Thijsse told, that when attending the funeral of Lorentz he got complaints for keeping Lorentz from his work on relativity theory by trivial things as tidal computations.

[u]He described (for instance) in Ref. [22] in Chapter 3, the phenomenology of the groin behavior, which was in the present Chapter 3 translated in Mathematical terms and well-considered up to its consequences.

rectangular shapes are turned with a long, respectively, short side to the Northsea. As it should, the "lessons of old masters" will echo as well from the North and the South, although with different "accent".

2.5.2. *Terminology*

The following glossary is copied from the original Van Veen paper.[49] The links made by Van Veen between Dutch and English language are too interesting to be covered up.

Bochtwerking: centrifugal action, meander-action.

Ebschaar[v]*: ebb-channel, open to the ebb-stream, bar at the lower end. Compare "to scour"?*

Drempel: literally sill, i.e. bar; often meeting-place of the two opposite sandstreams.

Hoofdaanvoer- en afvoerweg: Main direction of water supply.

Kentering or kanteling: slack water or turnover. Compare "to cant" in the meaning of turn over.

Komvulling or komberging: flood-fill of a tidal basin between the high and low levels of slack water (Fill of basin, Dutch "kom" has primary function in shaping a tidal channel).

Neer or maalstroom: slow maelstrom, maximum velocity at the outer side.

Schilddelta: Semi-circular submarine delta in front of each inlet or river mouth.

Tijdlijnen: lines of equal phase velocity.

Vloedkuil: flood-depression, or area where the high tides or storm-floods do not reach the heights they attain closer to the sea.

Vloedschaar: flood-channel, open to the flood, a bar at its upper end.

Wadden: tidal flats, covered at high tide, largely dry at low tide; compare Engl. "to wade".

Wantij: meeting place of tidal waves behind an island; i.e. irregular, "sickly" tidal currents and a regular vertical movement of the water level.[w] *Compare Engl. Wan = sickly, Skand. vanskabt, Dutch wanstaltig, etc. Therefore wantij = wan-tide.*

---

[v]Note for Dutch readers: mind that Van Veen (as all real morphologists) uses "*het* schaar" and does not use "*de* schaar". Such a channel is "the result of the scouring" and it is *not* "a pair of scissors". Certainly "*het* schaar" owes a place in the Dutch dictionary.

[w]Apart from the one at the "Amelander wantij" where during spring tide conditions the only and original Dutch "bore" develops (height: 20 cm). Compare the (only) Dutch "mountain" in the utmost South-East of the country with a height of approx. 250 m. (W.B.).

Wervel: Vortex.; maximum velocity in the center.

Zandneer: Circulating sand stream (Fig. 2.59a and b), i.e. moving of sand up in a flood channel; down in the ebb-channel; therefore not a gyratory or circular movement but a slow, intermittent movement where a sand grain after many weeks may return to the same place where it was before.

Zandstroom: intermittent, or non-intermittent strem of sand; in tidal waters an alternating movement.

2.5.3. *Morphological behavior of tidal channels in outer delta's. The Van Veen theory on outer deltas and its application for the Dutch coast, especially the Wadden coast*

Outer delta's exist because of "rain in the drain". This is vertical water circulation, caused by non-linear water motion. The feature is explained in Section 7.4. It is counteracted by the effect of density differences: cf. Section 7.3. Evidently geological motion plays a role in configuring those tidal delta's as well.

Wise words from old masters like Van Veen[49] (honored, memorized and translated fortunately, by Van der Spek *et al.*[50]) will be checked and evaluated in the present subsection.[x]

After those general considerations, there will be focussed on the cross-section of those channels (subject of other old masters like Bruun and Gerritsen[18]) and on channel rotation.

The boundary conditions for the motion of the coast are given by the inlets. Therefore, it is important to consider the motion of the tidal channels. What is known about these deltas? Consider first the "linear aspects" (Subsection 2.5.1) start with the tides.

The Southern part of the North Sea (roughly between Den Helder–Cromer at the Northen side and the English Channel at the Southern side) has a length of about one quarter of the length of the tidal wave. Thus one would expect a system of nodes and antinodes with a large vertical tide near Calais and practical no vertical tide near Den Helder–Cromer. However, because of the effect of earth rotation during flood water is pressed to the English coast and during the ebb against the Dutch coast. This has the effect, that the tidal wave turns around in the North Sea: from North to South along the English coast and from South to North along the Belgian and Dutch coast (Fig. 2.39).

[x] As well on the following concerning ebb- and flood-channels.

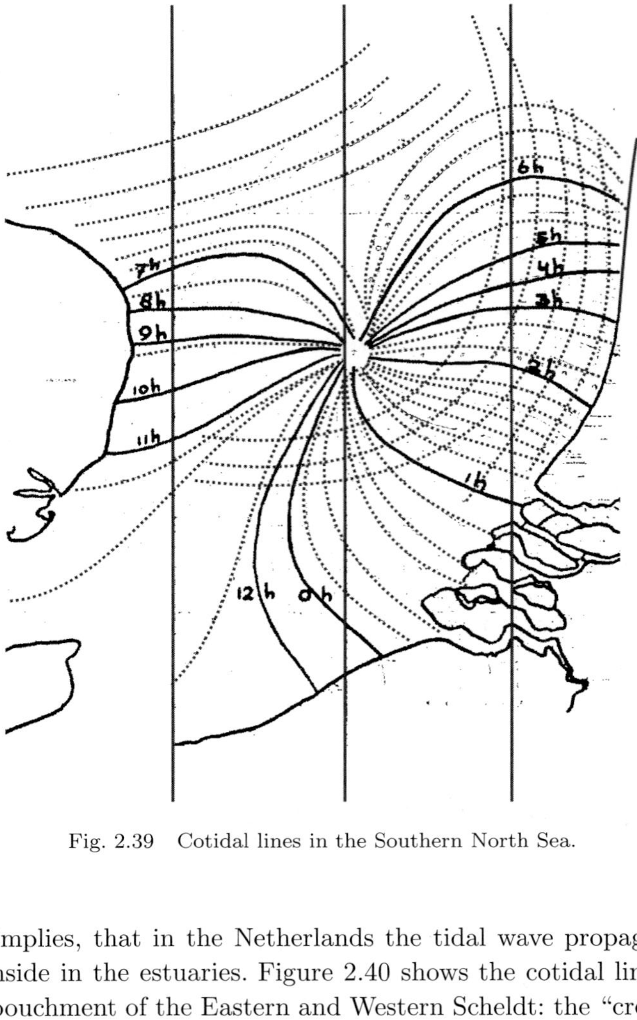

Fig. 2.39 Cotidal lines in the Southern North Sea.

This implies, that in the Netherlands the tidal wave propagates from the Southside in the estuaries. Figure 2.40 shows the cotidal lines of HW in de embouchment of the Eastern and Western Scheldt: the "crest" of the tidal wave intrudes from the Southern side in the estuary.

In the direction of the crest ("AC" in Fig. 2.41) the water levels at a certain time are equal (by approximation) and as, generally speaking, water directs itself to the lower level, current velocities in the direction of the crest are small. As the water surface gradient is perpendicular to the tidal wave crest, the current direction will be perpendicular to the crest as well. It may be clear that direction of channel scouring coincides with current direction, i.e. (again: generally speaking) perpendicular to the crest of the tidal wave, i.e. to the cotidal lines ("AB" in Fig. 2.41). This implies, that the main

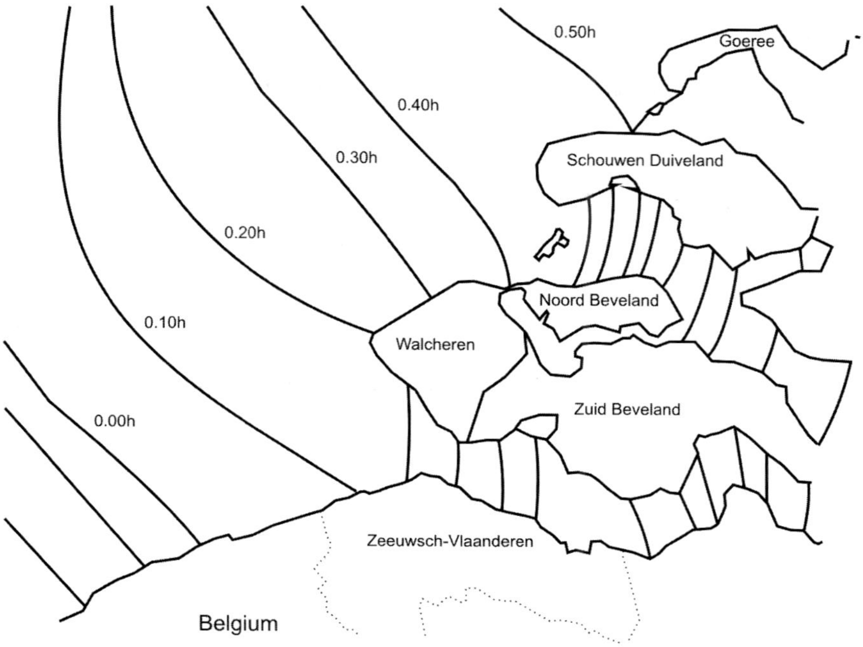

Fig. 2.40 Cotidal lines in the Zealand area (before enclosure of the Eastern Scheldt).

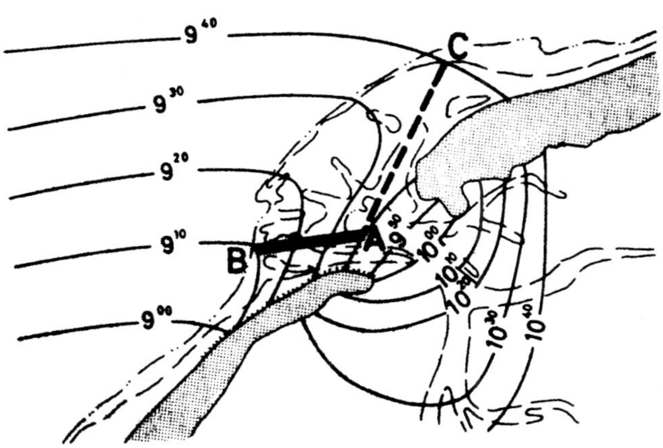

Fig. 2.41 Cotidal lines in an inlet and waterlevel gradients. From Van Veen.[49]

channels in the estuaries are mostly directed in this way, i.e. in Southward direction.[y]

Van Veen[49] tries to quantify this feature by defining a "motoric power", keeping the channel open, being the time-integrated slope[z] of the water level (= time integrated difference in water level between two points, divided by the distance between those points[aa] (Fig. 2.42).

The fact, that in spite of this there still exist channels with another orientation, is caused by the over-simplification of the explanation above. One of the effects neglected is the difference in behavior of ebb and flood[bb] (Fig. 2.46). The estuary sucks its flood-water from all sides in a funnel, but it spits its ebb-water as a thin jet into sea. More formally: mind the (unidirectional) momentum of the ebb-current. So the main channel (mentioned before) generally will be an ebb-channel, where at the borders of the basin "flood-channels" will emerge (Figs. 2.44 and 2.46). Thus results a sand transport circulation as depicted in Fig. 2.46.

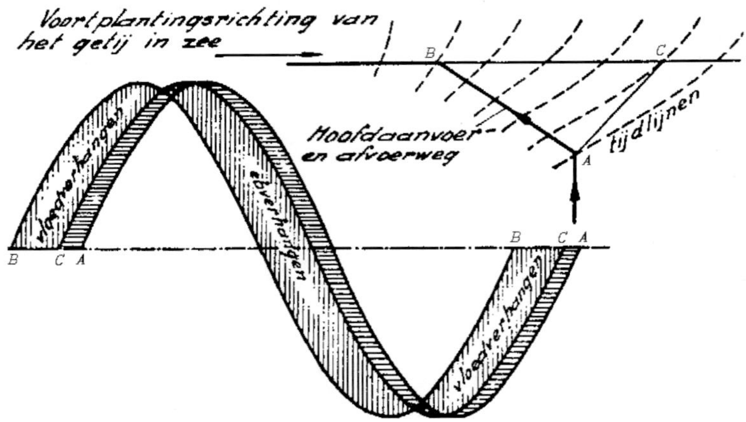

Fig. 2.42 Belonging tidal registration.; *(Voortplantings .... = propagation velocity of the tide (seaward of the delta); vloed-, c.q. ebverhangen = tidal gradients during flood, c.q. ebb).* "motorial area". From Van Veen.[49]

[y]In the terms of the introduction 2.5.1, this is essentially a linear effect. According to the harmonic tide theory (Lorentz, Chapter 8) vertical and horizontal tide will be equal but opposite at their ebb- and flood maximum, thus giving the same direction of water level gradients during maximum ebb and flood conditions.

[z]Ignoring the sign (absolute value).

[aa] Mind that this "power" is dimensionless (if the time axis is normalized by the tidal period). Cf. Section 8.2 in[47] calls the variable "motorial area"; cf. Fig. 2.42.

[bb] Quadratic effect.

If, for a minute, one takes[cc] a coastline in North-South direction in mind (inset of Fig. 2.45), interrupted by a number of East-West orientated estuaries, then generally the North-West corner will be cut off strongly, where the South-Western corner suffers less current attack (although evidently, the flood channel still will play its role). In this way the coastal shape will look like a kind of a stepping stone. See fig 2.41 from Van Veen, 1950[49] and Fig. 2.45 after Edelman[26] and Edelman and Eggink.[27]

Two reasons can be given for *the transversal motion of the tidal channels*: a meandering effect and longshore sand drift. Clearly, the resulting littoral sand drift perpendicular to the tidal channel cannot be derived from the velocity of the tidal channel, because of the meandering effect. The high amount of sand transport in the tidal channels can be due to high current velocities and to this meandering.

As the tide comes in from the South,[dd] (Figs. 2.39, 2.40), i.e. the highest water level gradients are in SW–NE direction, the current will bend off to W–E. As the velocity at the surface is higher than at the bottom[14] the centrifugal force on the particles at the surface will be higher than near the bottom, thus generating a spiraling current (to the outer bend at the surface, from the outer bend near the bottom) which results in scour at the outer bend and (given the orientation of the channels) to a clockwise rotation of the channels (Fig. 2.43). Mind that this mechanism works in the same direction during ebb and flood; clearly it is a *non-linear (quadratic)* feature.

Longshore sand drift[ee] encountering a tidal channel will result in sedimentation of this amount of sand on the windward side in this channel, as the stirring and driving power suddenly disappears. This under the condition, that the driving mechanisms which are characteristic for tidal channels (for instance, the spiraling current treated in the former paragraph) do not take over. This outer-bend scouring is a self-contained sand transport mechanism with its own dynamic equilibrium. Thus longshore sand drift "pressing" the channel aside and the meandering effect could (algebraically) add up.[ff]

[cc] Dutch tidal conditions.

[dd] This remark makes sense for the Delta area in the South. Read respectively "NW–SE" and "N–S" for the Wadden area. This means, one should turn the picture over 90° for the Wadden, but the essence remains the same.

[ee] More general: "wave driven sand motion".

[ff] Mind the investigation to which Roelse refers to (Bankje van Zoutelande; Chapter 8; Fig. 8.23).

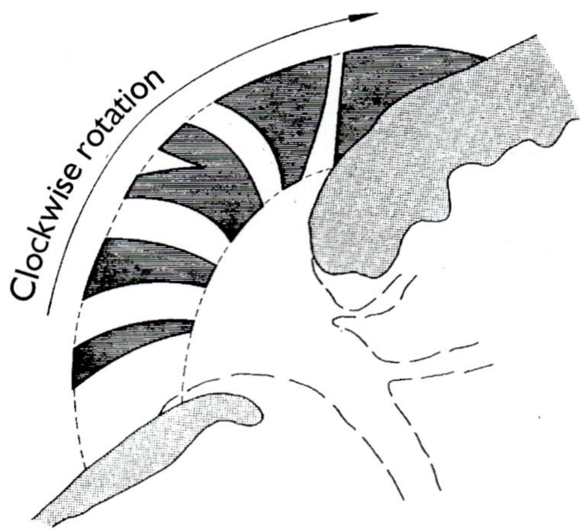

Fig. 2.43 Clockwise turning of channels because of centrifugal forces, acting on the flood current. N.B.: the ebb current concentrates much more in the shape of a jet stream (cf. Figs. 2.44, 2.46). After Van Veen.[49]

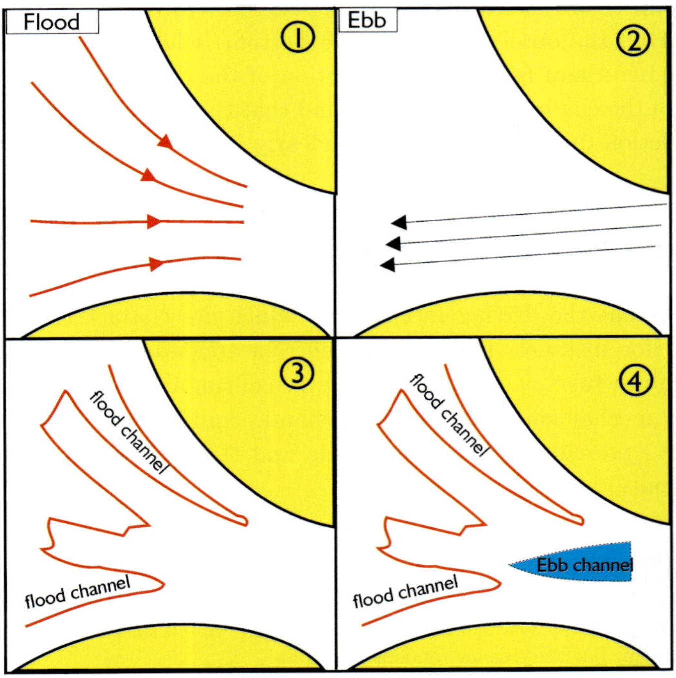

Fig. 2.44 Flood- and ebb-channels in a tidal estuary.

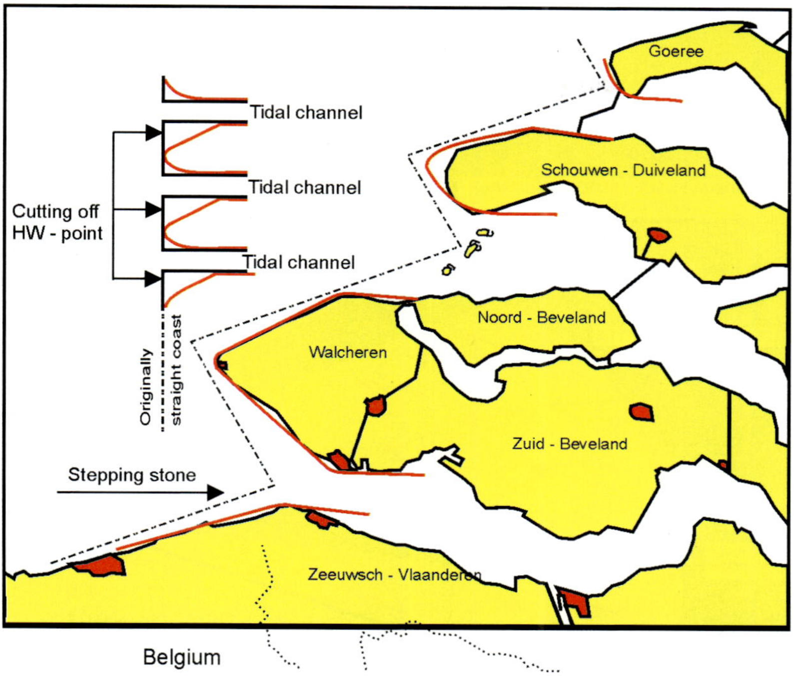

Fig. 2.45 "Stepping stone" shape of coastline near an inlet.

The spiraling current, Fig. 2.43 (and the tidal discharge, Fig. 2.46) could force the channel to keep its cross-section, and thus give scour on the leeward side of the channel. Thus the whole channel "drifts transversely in tidal wave direction". Last-mentioned driving force is called "sand pressure".

Another feature, to be kept in mind when forecasting the behavior of tidal delta's is the characteristic of tidal channels, pointed out by Bruun and Gerritsen[18,gg] and afterwards quoted and commented by many others. This deals on the proportionality between the cross-section (during spring tide) of a tidal channel and the maximum tidal flow. Dividing the last figure (m^3/s) by the first one (m^2) results in a mean (crosssection-averaged) top velocity of 1 m/s (cf. Fig. 2.47; derived from their paper).

[gg] Inspired by a paper of O' Brien (1931) concerning the relation between tidal prisms and entrance areas.

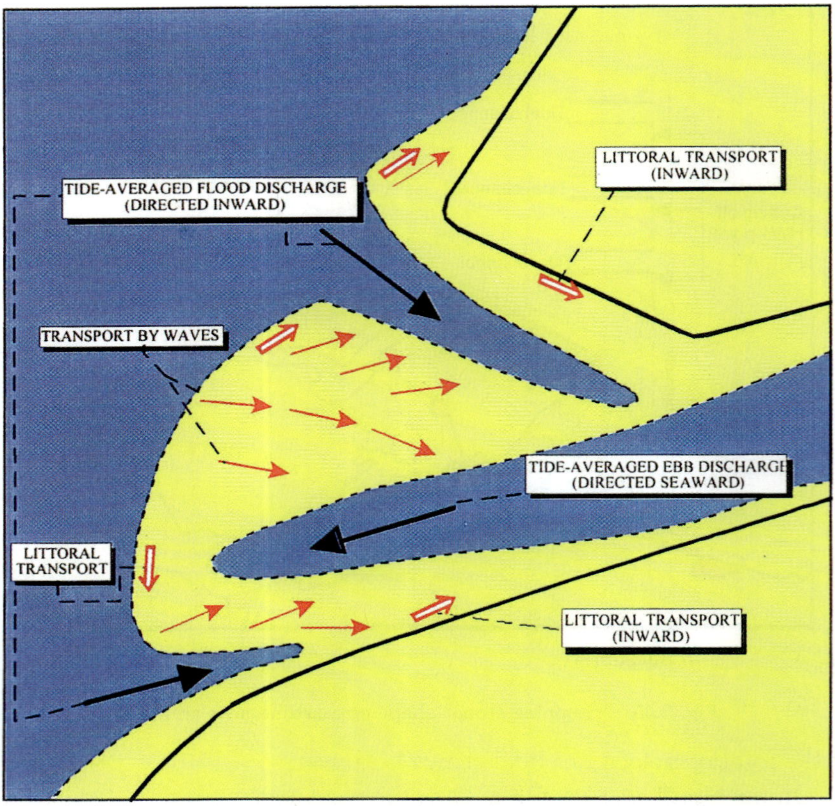

Fig. 2.46 Schematic display sand transport near an inlet.

The next part of this subsection (outer delta's) will focus especially on checking the afore-mentioned theories with data. For this, the channel systems in the outer deltas of the Wadden islands are considered. In this part of the subsection the morphodynamic behavior of tidal channels in the Wadden area (as depicted in the following Figs. 2.48–2.50 (data) and 2.51–2.53 (analysis)) is investigated.[hh]

Thus questions (investigated with the aid of the maps on the forecoming pages) are:

a. To evaluate the Van Veen theory: do the channels move clockwise?
b. More in general: does the way, channel systems develop, make sense?

[hh] On changes in the Zealand area is focussed on in Section 2.6.

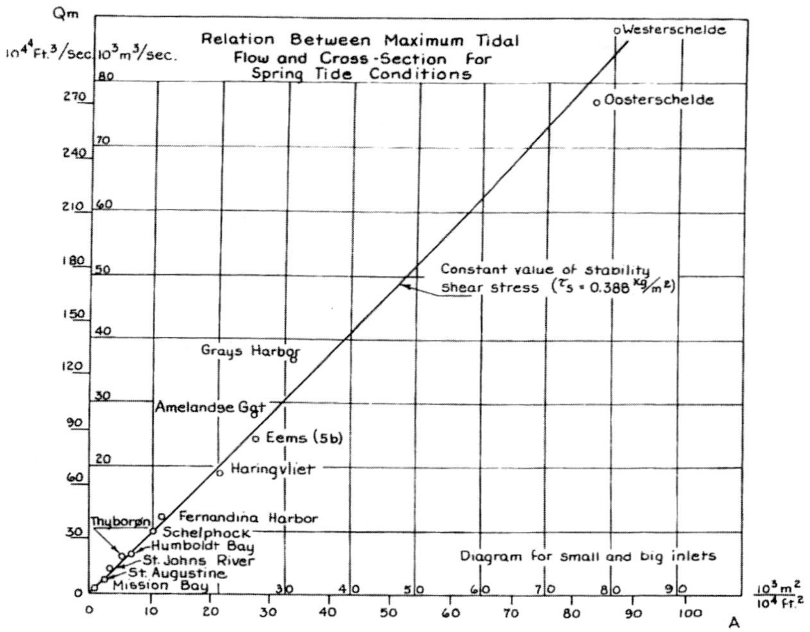

Fig. 2.47 The relation between maximum tidal flow and cross-section. From Ref. [18].

After this, this subsection will end with a further investigation (by Huys[34]) of the rotation velocity of tidal channels in inlets.

On the *righthand* side of Fig. 2.51, three channel systems[ii] are depicted. To each (present or ancient) tidal channel, a number has been assigned.

On the *lefthand* side of the figure, the centerline of the bars denotes the orientation of the channel as function of time. Clockwise turning channels result in rising lines. The thickness of the bar is proportional to the cross-section ("wet surface" in a representative section): of the channel in the year under consideration. As an example, consider channel III in the Vlie, which turns clockwise, until its fades away. This is the behavior, depicted in Fig. 2.43.

The graphs *in the middle* denote a kind of "popularity index" of the orientation of the channel: cf. Figs. 2.41 and 2.42. The length of those bars is proportional to the time-integrated absolute values of the water level gradients, i.e. the hatched areas in Fig. 2.42.

[ii] Eems, Vlie, Marsdiep (cf. Fig. 2.1a).

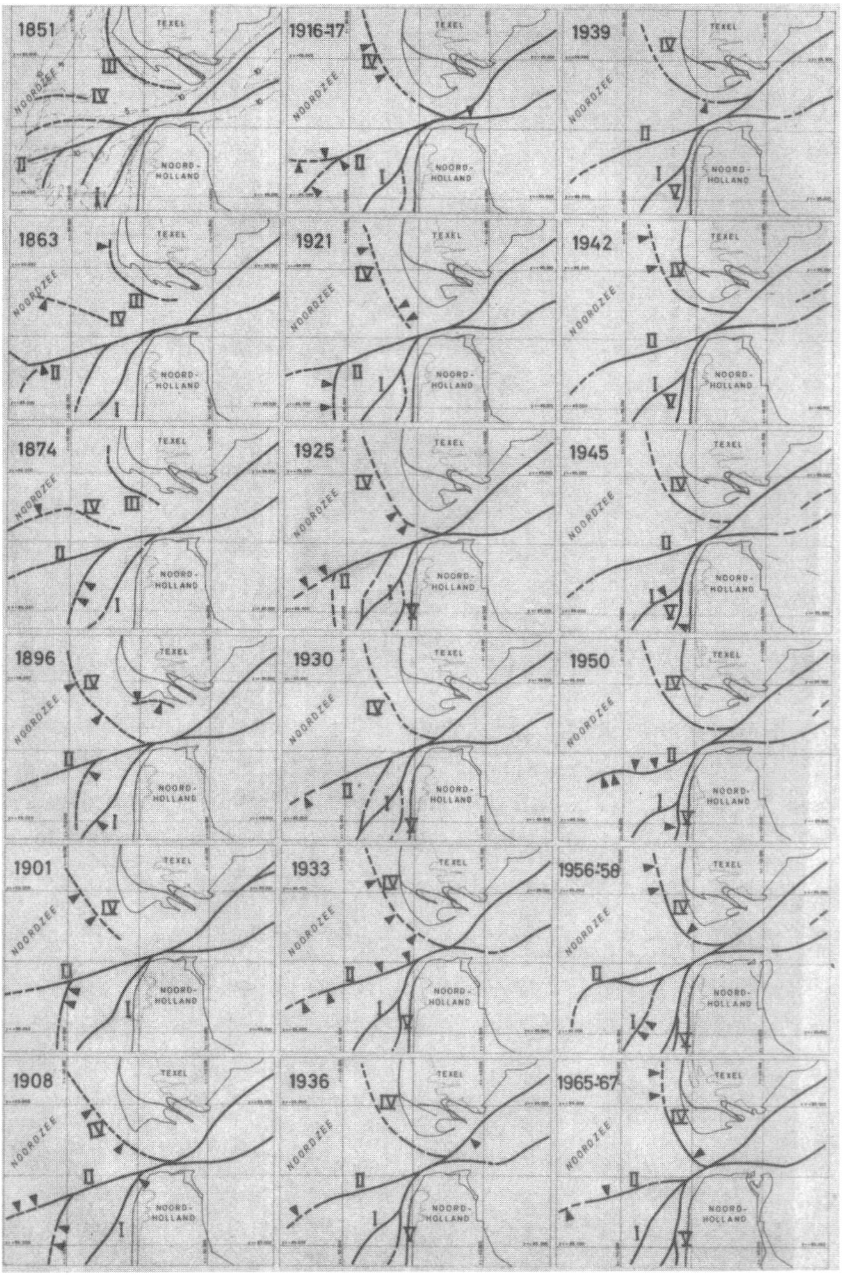

Fig. 2.48 Changes of the site of the thalway of the channels in the Marsdiep. Situation: cf. Fig. 2.1. The arrows show the motion since the last recording. From Ref. [7].

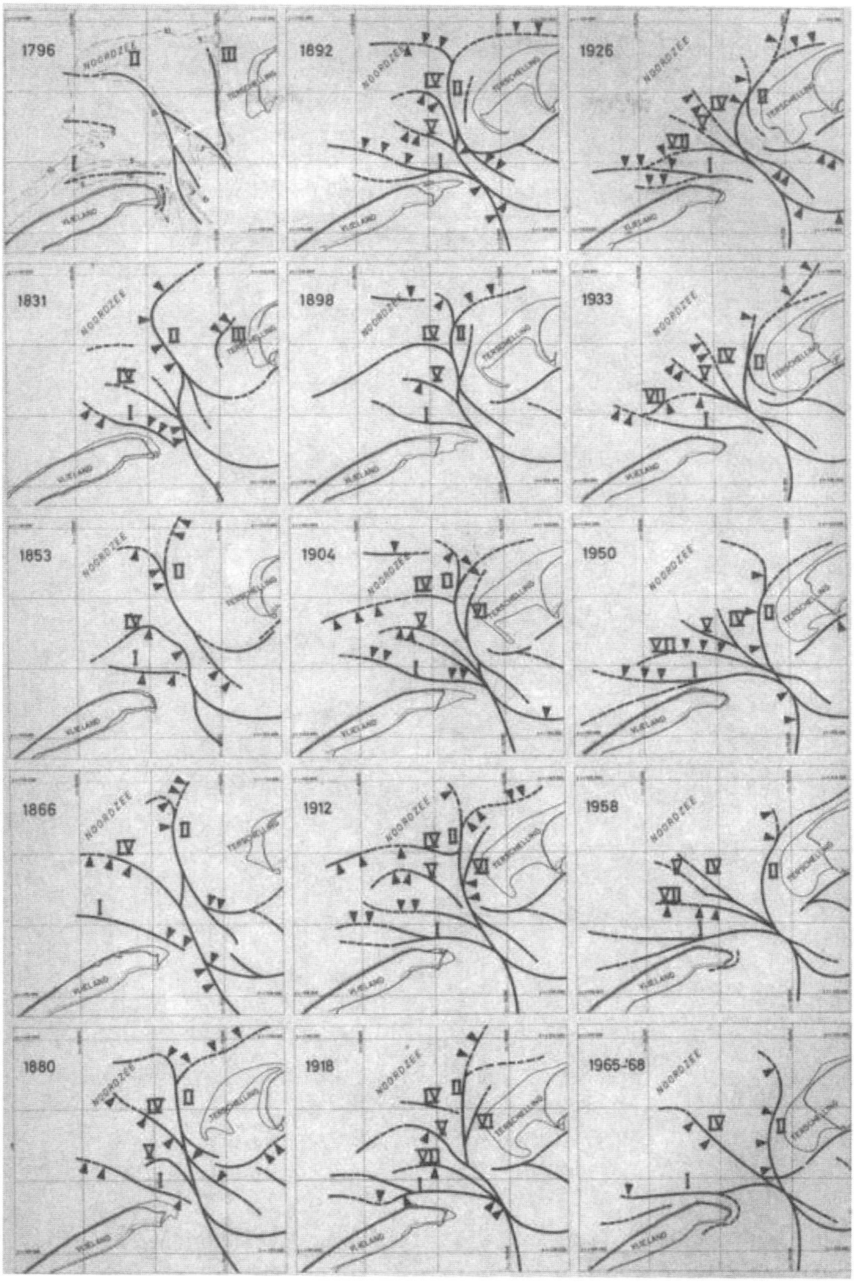

Fig. 2.49 Changes of the site of the thalway of the channels in the Vlie. Situation: cf. Fig. 2.1). The arrows show the motion since the last recording. From Ref. [7].

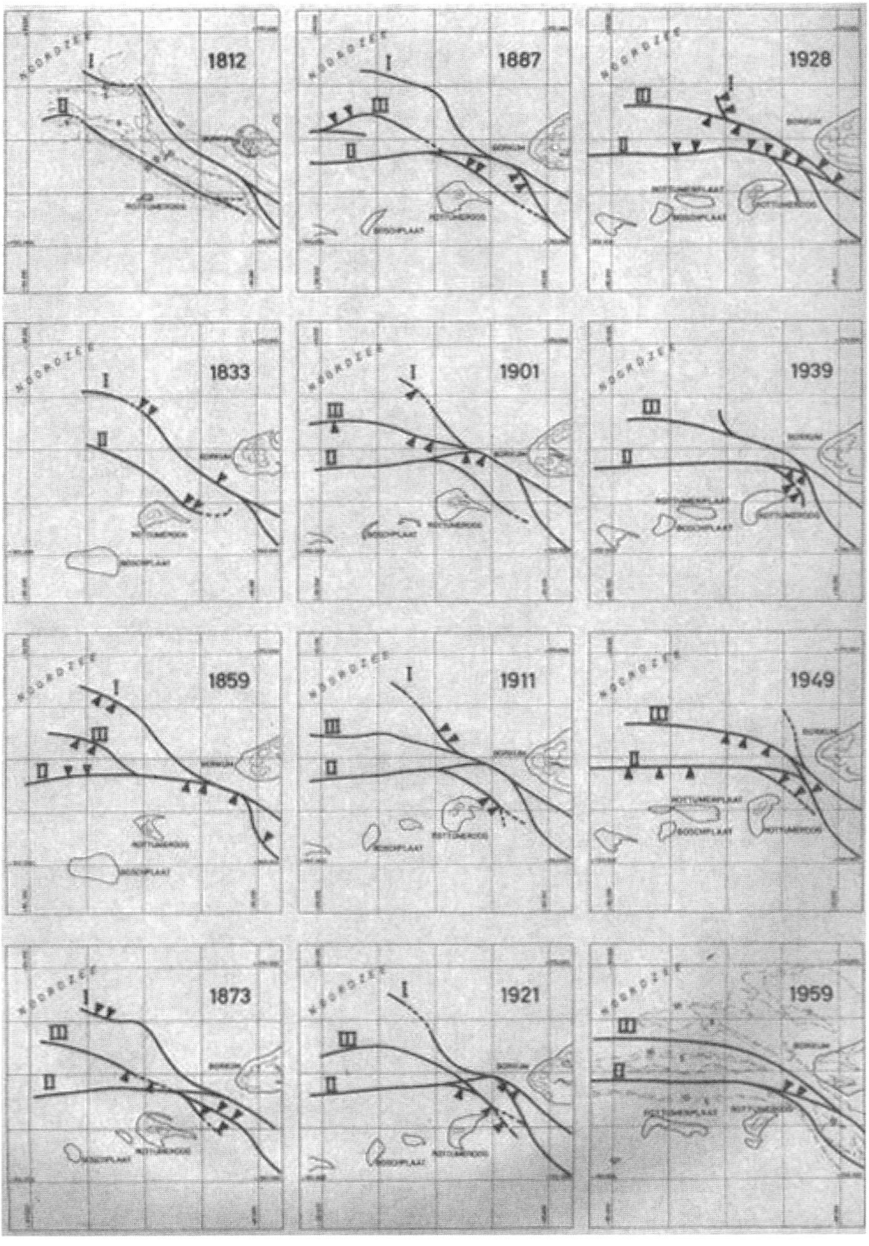

Fig. 2.50 Changes of the site of the thalway of the channels in the Eems. Situation: cf. Fig. 2.1.

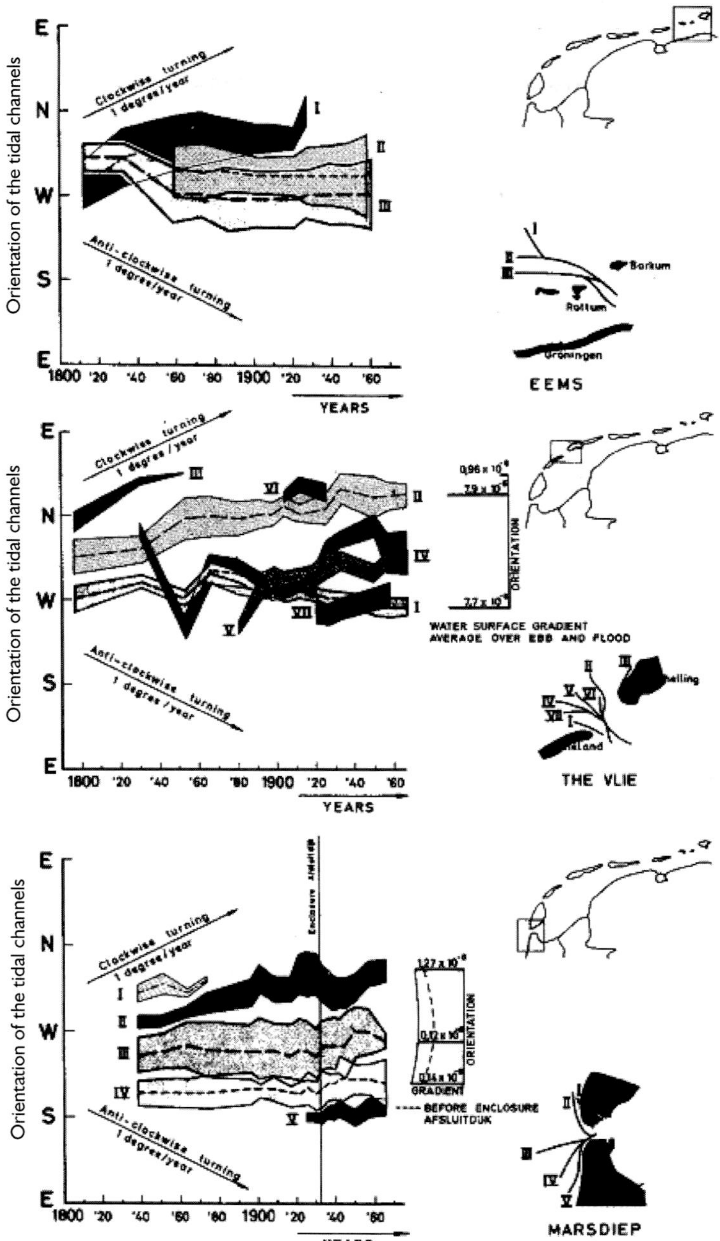

Fig. 2.51 The "rise and fall" of tidal channels. Cf. Figs. 2.48–2.50, in which the channel numbers are indicated. Direction and cross-section of tidal channels: check of the Van Veen theory (cf. pp. 2–49 and 2–50 and Figs. 2.41 and 2.42). From Ref. [7].

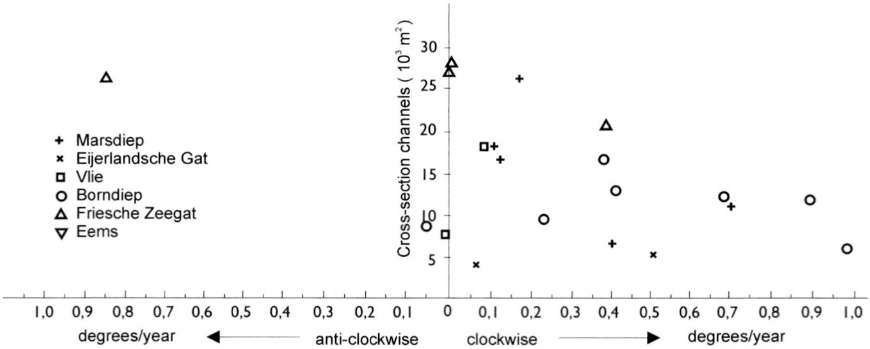

Fig. 2.52 Correlation cross-section and rotation velocity of tidal channels.

Referring to question (a): from Fig. 2.51 some tendency to clockwise turning (with exception of the Eems estuary) can be perceived.

Figure 2.52 depicts the correlation between cross-section and rotation velocity of tidal channels in the Dutch Wadden delta. Generally speaking, larger (deeper) channels move more slowly and (apart from the Friese Zeegat) the channels investigated moved in clockwise direction.

From those data, another rule of the thumb on the behavior of tidal channels in outer deltas, is derived: a formula which deals on the transverse velocity (i.e. rotation) of tidal channnels. It is based upon the data from Bakker and Joustra[7] and of Huys.[34] The latter investigated the Friese Zeegat and the Amelander Gat in particular. She considered the rotation of tidal channels (Fig. 2.53a,b). Mainly these turn clockwise, although very slowly, and there is a slight indication (correlation coefficient 0.24), that the large tidal channels turn slower than the small ones.

As a challenge to future researchers Bakker drafted from her data tentatively to the DC-formula (2.10),[jj] indicated above, stating that the rotation velocity of a tidal channel is inversely proportional to its depth and to its radius of curvature, The background idea is, that the spiral current will probably be inversely proportional to the longitudinal curvature and the cross-transport in the channel roughly proportional to this current (convection more important than stirring). The assumption of inverse proportionality with channel depth implies a cross-transport of sand, independent of depth.

jj "D.C." stands for Dutch Courage.

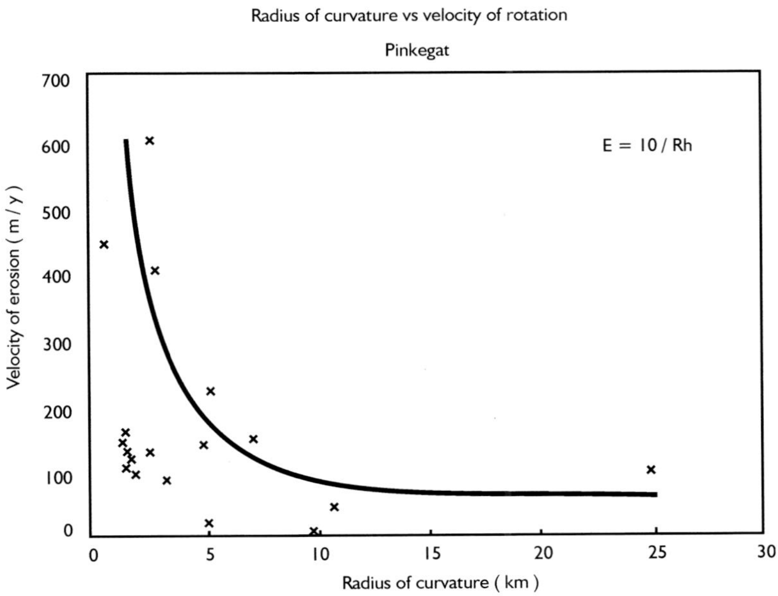

(a)

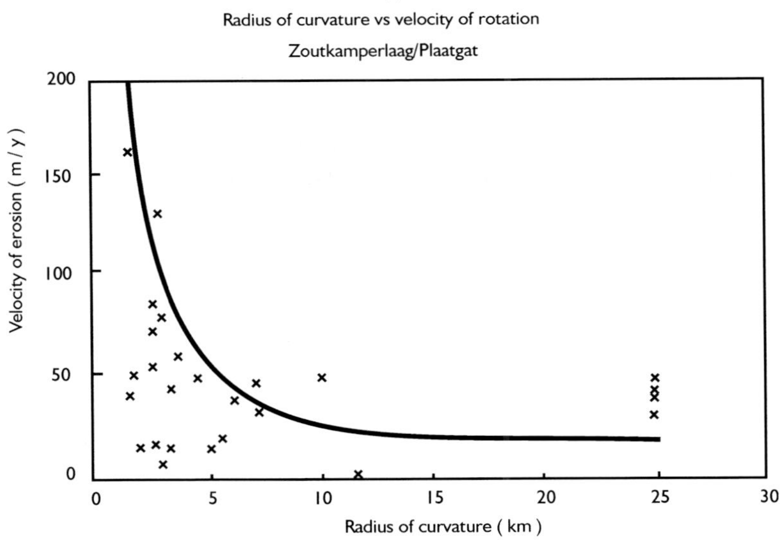

(b)

Fig. 2.53 (a, b) The DC-formula (cf. Eq. (2.10)).

The formula reads:

$$e_r = \frac{10}{R \times h} \tag{2.10}$$

where e_r is the transversal erosion velocity of the channel in km/yr; R is the radius of curvature in km and h is the channel depth in m.

Two reasons can be given (also; cf. p. 2–49) for *erosion of the coast near these inlets.*

— As has been pointed out by van Bendegom,[14] the submerging of the Wadden during flood-tide takes place with higher velocities than the retreating of the water over the shoals during ebb-tide. Thus, the water loses a part of its sediment here, which causes the Wadden shoals to reach equilibrium at about the mean water level. The relative sea-level rise in the Netherlands during the last 20 centuries was about 20 cm/century, which would result in a "sand hunger" of the Wadden shoals of about 1 mln m^3/yr (distributed over all inlets). However, this will mostly be confined to finer sediments (D_m about 100–150 μ). Enhanced sea level rise[kk] will result in larger sand hunger.

— The second reason is the one, depicted in Figs. 2.44 and 2.46, that the water during the flood tide gets an acceleration, entering the inlet, but that during the ebb-tide it gets a retardation and this will give a jet-stream with vortexes on its side. Therefore in the tidal channels near the beaches, there is surplus of flood discharge (Figs. 2.44 and 2.46) and in the center tidal channel a surplus of ebb discharge. The flood erodes the beaches and the ebb produces an outer delta, which can reach up to mean sealevel (Noorderhaaks in inlet of Texel). This delta gives a shelter against these waves, which would transport material away from the delta. As the waves come alternately from both sides, this process reinforces the erosion of the beaches near the inlets. It is the same kind of process as occurs near a harbor mole, gathering material around it (this will be explained in the text around Fig. 5.49 (Chapter 5)), however, in the present case the "gathered" material will flush through the inlet because of the local high (flood)-velocities. Thus the erosion of Texel could be rather well explained.[55] After some time equilibrium should be reached.[ll] An excellent, recent, comprehensive investigation

---

[kk] For instance: 20 cm/yr (last century) or 65 cm/yr (anticipated).

[ll] Evidently depending on interaction with other inlets.

of the sand transport mechanisms in the Western Wadden Sea is laid down in the Ph.D. thesis of Elias.[29,mm]

2.5.4. *Van Veen theory concerning ebb- and flood channels*

Section 2.5.3 dealt on outer deltas, the present subsection continues with the Van Veen theory,[nn] however focusing on the area's more inland. The subject is quite on the verge of the subject of this book, which deals on *coastal dynamics.* However, as the interaction between Wadden, estuaries and outer deltas will be considered in Chapter 7, it is thought that the interesting ideas of Van Veen concerning the phenomena ruling the tidal channels in the estuaries (which often determine the coastal erosion near the inlets) could not be missed.

In the North of the Netherlands ("Wadden area" Fig. 2.1) the coast has the character of a lagoon, where the dunes of the barrier islands are not far (ca. 10 km) away from the mainland. A tidal basin is confined by the wan-tides and by the coast (Fig. 2.54)

On the other hand, in the Zealand area ("Rhine-Scheldt delta", Figs. 2.1a and 2.6 and Chapter 7) the boundaries of the various basins are (apart from the mainland) islands, contrarily to tidal flats in the Northern part. Van Veen points out that the prevailing Westerly winds are perpendicular to the wan-tide in the North, and parallel to it in the South, which may be an explanation of the absence of those separating islands in the North. Strong wind-driven current during storms may prevent the generation of islands at this site in the Northern area.

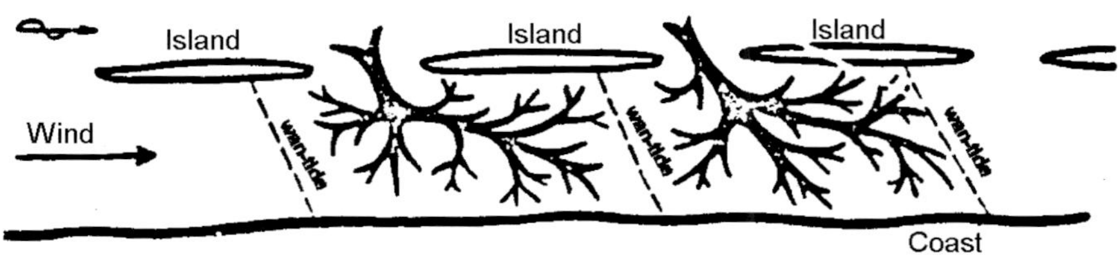

Fig. 2.54 Sketch of the Wadden tidal basins; short basins perpendicular to the prevailing wind (apple-tree shaped channel system). From Ref. [49].

[mm] Evidently involving modern techniques of sophisticated numerical modeling and data chrunching, underlining, however, qualitatively the indicated features.
[nn] Cf. as well the recent translation of Van Veens paper by Van Veen *et al.* (2005).

This is contrarily to the situation in the South (Fig. 2.56). Also because of the larger tidal difference in the South the tidal volumes per tidal basin in the South are very large and the tidal channels became wide and deep. Silting up of the wan-tides above high water in the South and emergence of the islands over there (Fig. 2.1) could be explained (Van Veen) by the dominating direction of the wind. The isles of Zealand were sheltered behind the barrier islands; in the North the wan-tides were unprotected.

Also the shape of the channel system in the North ("apple-tree shaped system") differs from the one in the South, where it is "poplar shaped" (cf. Fig. 2.54 with Fig. 2.56). Focus now on the poplar-shaped "trees" in the South (Figs. 2.54 and 2.56).

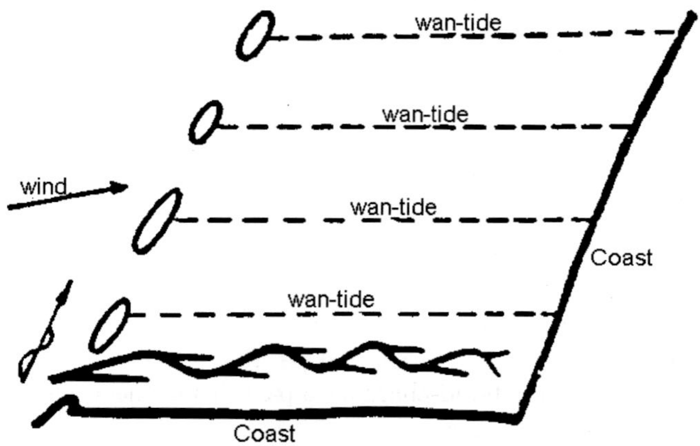

Fig. 2.55 Sketch of the tidal basins in Zealand. Short dune islands, long flood basins parallel to the prevailing wind (poplar shaped channel system). From Ref. [49].

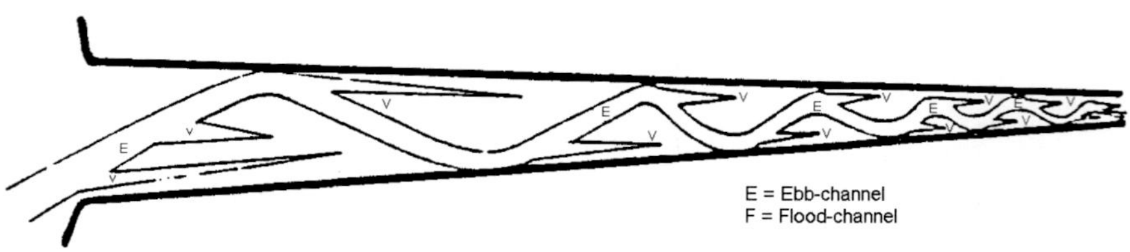

Fig. 2.56 Sketch of an ideal system of ebb- and flood-channels (Scheldt estuary). Sine-shaped main (ebb) channel, flood-channels starting in each bend. The latter have a double function, viz: (1) filling the tidal sand-flat in the inner bend of the main channel; (2) serving the cut-off currents of the bends. Notwithstanding considerable scour the depth of the bars at the upper end of the flood-channels are not permanently increased. From Ref. [49].

The asymmetric shape of the "apple-tree" in the North can be explained by looking at Fig. 2.54: at the sea-side of the barrier islands (where the depth is large) the propagation velocity of the tide is much larger than at the shore-side. However, as there is (still) phase lag between the tides at the successive inlets the wan-tide will not be in the middle between both inlets but approximately at three-fourth of the length of the island (measured in the propagation direction of the tide at the sea-side of the island). Focus now on the poplar-shaped "trees" in the South (Figs. 2.55 and 2.56).

Considering a "tree" as the one in (for instance) the Western Scheldt, then the "stem" meanders from left bank to right bank and the "branches" extend in every bend stream-upward[oo] along the banks. Those branches are flood channels, according to Van Veen's definition: a flood channel is a tidal channel accessible for the flood; it has a sill at the upstream side. Contrarily, an ebb-channel is a tidal channel accessible for the ebb; it has a sill at the downstream side. Remarkably, flood channels try to avoid ebb-channels and inversely (Figs. 2.56 and 2.57). Van Veen[49] announces a "battle of delta's", where the two antagonists (flood and ebb channel) approach each other by a flank attack. There are two reasons: "sand transport" and "bend action" (Figs. 2.57 and 2.58). For ordinarily in a flood channel flood sand transport will prevail and the inverse in the ebb-channel. "At the meeting point between flood- and ebb-channel sand is added from two sides and it forms a sill. Evidently the sand motion is more intricate then indicated in those few words".

In a set of ebb- and flood-channels apart from the oscillatory tidal motion a residual sand-drift will occur; mostly resulting in circulating[pp] sand currents.

Three more interesting story-telling pictures of Van Veen's paper[49] will be added here as well.

From Ref. [49]: Fig. 2.58a,b depicts the way scouring on one side or another of a Wadden isle can be explained by the shape of the tidal fill basin.

Concerning Figs. 2.57 and 2.60a,b, depicting the splicing of ebb- and flood-channels, still two relevant facts may be added to Van Veen's interesting and convincing story.

[oo] Considering the tidal-averaged current, evidently.

[pp] Circulation with a vertical axis (cf. Chapter 8; the interesting contribution of BLIEK); apart from that, there is a circulation with a horizontal axis, or even two, counteracting each other; the one by the "rain in the drain" (Chapter 7) and the one caused by longitudinal density differences (merging of fresh and salt water, Section 7.3).

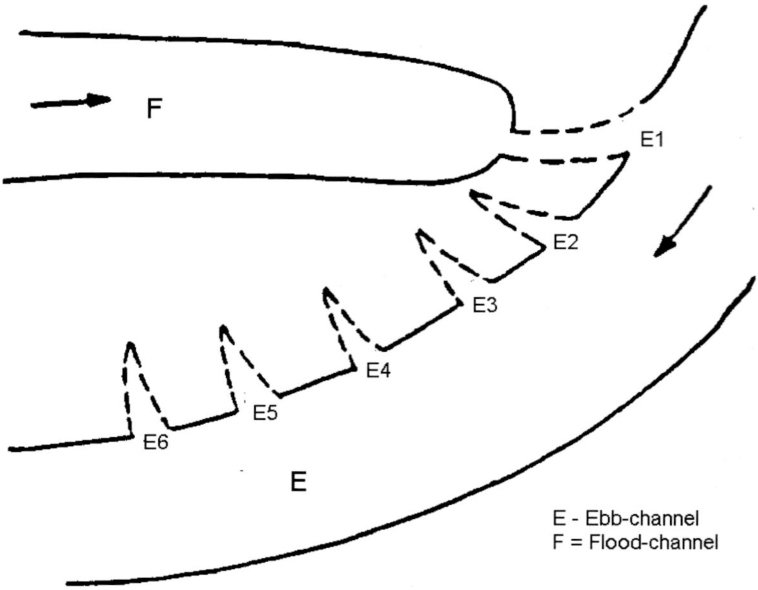

Fig. 2.57 Sketch of a wandering ebb-channel starting in the bar of a flood-channel. It is bend-action or meander-action again. From Ref. [49].

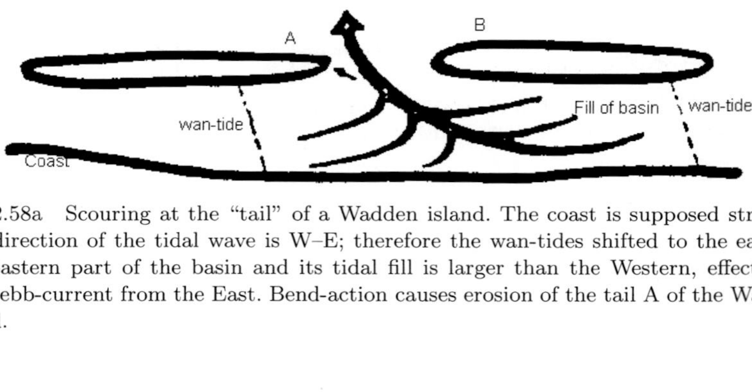

Fig. 2.58a Scouring at the "tail" of a Wadden island. The coast is supposed straight. The direction of the tidal wave is W–E; therefore the wan-tides shifted to the east; so the Eastern part of the basin and its tidal fill is larger than the Western, effecting a main ebb-current from the East. Bend-action causes erosion of the tail A of the Wadden island.

The current from the flood channel crossing the bar to the ebb-channel (Fig. 2.60) is submitted to the mechanism, physically formulated by Arthur.[2] Because of conservancy of vorticity, current, crossing a slope in downward direction, will converge to a jet, which (because of conservancy of mass) can reach very high velocities. Near Belgium, in the Eastern part of the Western Scheldt ("Zimmerman geul") during spring tide, velocities in this jet of 4 m/s were measured.[37]

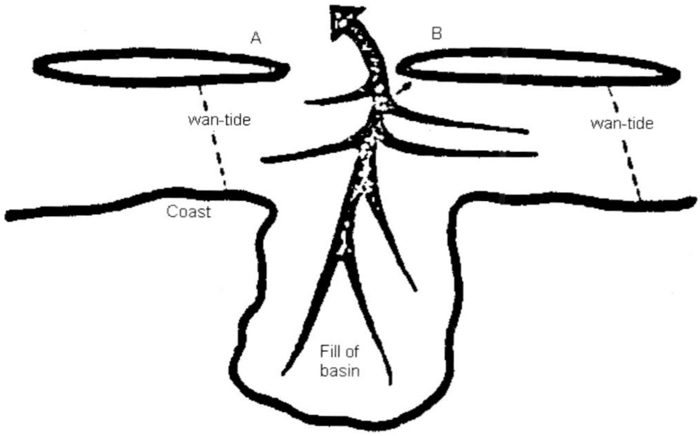

Fig. 2.58b Scouring at the "head" of a Wadden island. Here the coast is bent inward. In this case, the Western part of the total basin is the largest; so the main ebb channel comes from the West. Bend-action causes erosion of the head B of the Wadden island.

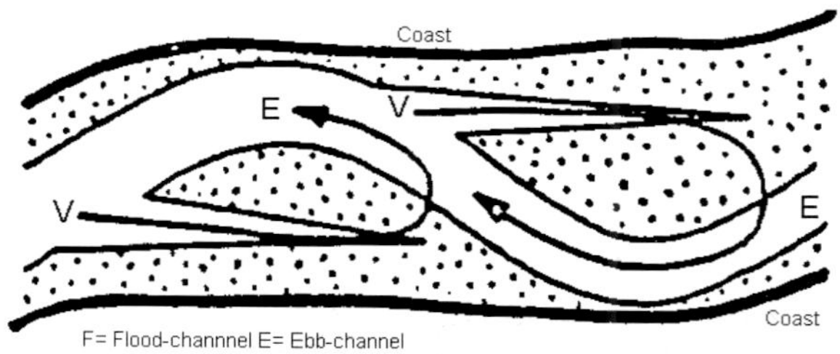

Fig. 2.59a Sketch of so-called circulating sand currents, the sand moving up-stream in the flood-channel, down-stream in the ebb-channel. From Ref. [49].

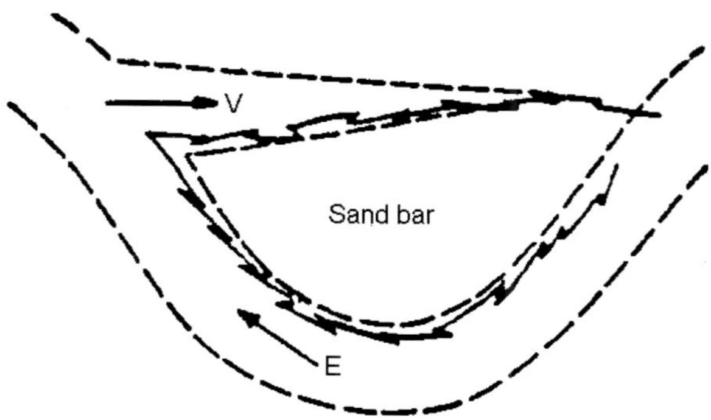

Fig. 2.59b Sketch of the true up- and downstream movements of the sand in so-called circulating sand currents. A grain of sand may come back to its original place; dredging may be of small avail. From Ref. [49].

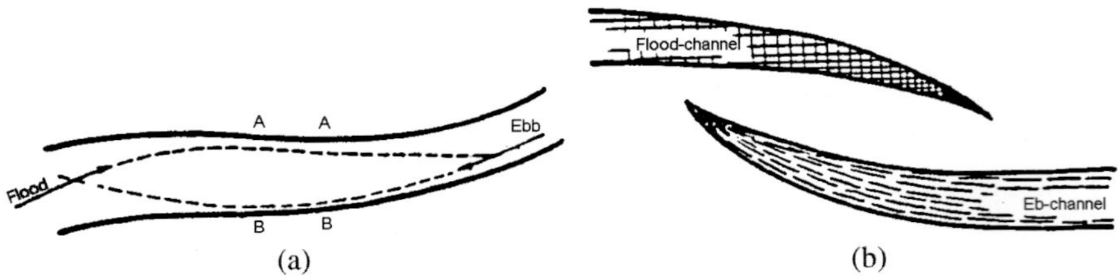

Fig. 2.60 (a, b) Bend-action explains to a large extent splicing of a channel, or the birth of ebb- and flood-channels. From Ref. [49].

The jet, crossing the ebb-channel has a width in the center of this channel of the order of several tens of meters and can provide a surprise party for draughtsmen of passing ships. First it hits at the bow; the turning of the rudder will (because of the inertia of the ship) only give a reaction on the ship when the jet reaches the tail; evidently not the reaction the draughtsman intended.

Secondly, because of high water velocities involved, the transverse motion of the ebb-channels "E" in Fig. 2.57 can be very large. The ebb-channels of the Zimmerman geul, mentioned before, with a depth of 6–7 m, used to move with a velocity of 0.5 m/day.

2.5.5. *Accuracy as limitation for the law of continuity*

This subsection gives an illustration of the remark in the introduction (Section 2.5.1):

"Consider the tide (Fig. 2.61). During the flood period sand is transported in flood direction; during the ebb-tide the inverse difference between ebb- and flood-transport is called the tidal transport; this is mostly small compared to ebb- and flood-transports. As its calculation involves the subtraction of two large and relatively inaccurate numbers, the result will be very inaccurate".

A striking example will be given.

Delft hydraulics[42] investigate the accuracy of the calculation of the erosion/sedimentation of the Valkenisse channel in the Western Scheldt (channel 14 in Fig. 2.62) and find an estimate of the sedimentation of −0.3 cm/yr with a standard deviation of 35 cm. Table 2.3a,b,c (referring to channel 14; Fig. 2.34) shows the way of calculation. Mind, that the calculated accuracy of the transport integrated over one flood-tide and one cross-section (i.e. integrated over a number of measuring positions

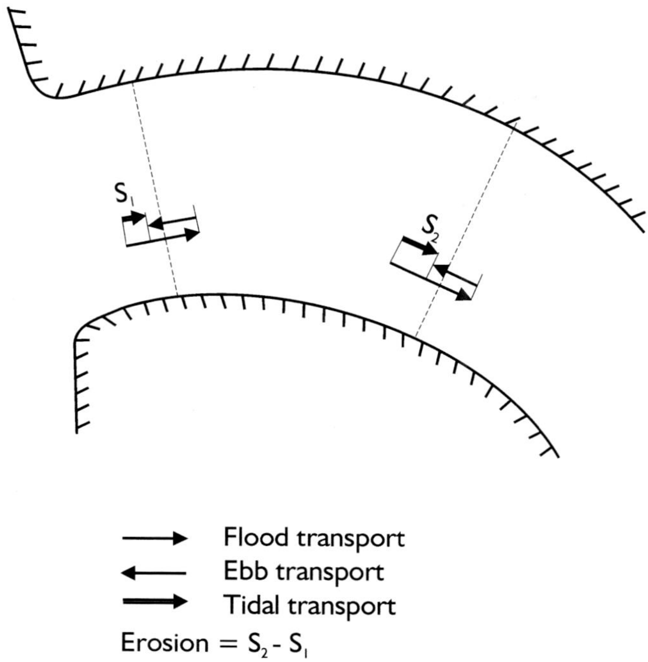

Fig. 2.61 Erosion difference between sand transport in two cross-sections.

as well) has "only" a standard deviation of the same order of magnitude of the transport itself. The sedimentation really measured in the channel was 20–30 cm/yr, which, as the report states: "is within the range of accuracy".

The investigation, quoted above illustrates also, that the existence of tidal transport does not imply, that the tidal channel is subject to changes. For if this transport is uniform, the channel is stable. The same statement holds for the adjacent coasts. Erosion results from a positive spatial transport gradient; sedimentation from the opposite case. As illustrated by the "channel 14"-story, determination of the rate of accretion, c.q. erosion is still more inaccurate than of the transport per tide, because again the difference between two large, inaccurate numbers has to be determined. The example given: "calculated rate of accretion/erosion orders of magnitude less than the standard deviation of the computation" is normative. Still, despite of all the reserves, qualitatively some statements on the mechanism of tidal transport can be made.

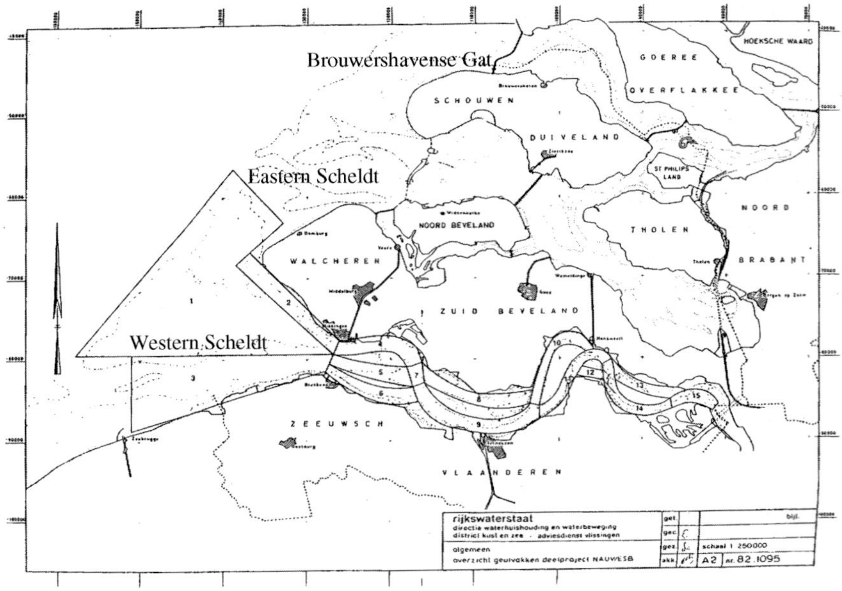

Fig. 2.62 Channel system Western Scheldt as considered by van Rijn and De Leeuw Ref. [42].

2.6. Changes of the Zealand coast by Man and Nature

In Zealand the mechanisms, treated in Section 2.5 cause three "turf's" of the various inlets, with specific influences, i.e. from North to South (Fig. 2.62):

- The one of the Brouwershavense Gat (Fig. 2.62)
- The one of the Eastern Scheldt (Fig. 2.6)
- The one of the Western Scheldt (Fig. 2.7)

Referring to Sections 2.4 and 2.5 concerning waves and tides, now those effects will be combined. Text of this section has been derived for a large part from Ref. [1].

With this general theory on the background the different inlets will be investigated more into detail:

- What has happened with the Brouwershavense Gat after the enclosure?
- What affects the outer delta of the Eastern Scheldt, in the past and in the future?
- What happened in the mouth of the Western Scheldt and which will be the influence of the harbor works have near Zeebrugge?

Table 2.3 Calculation of accretion/erosion and its accuracy. After Ref. [42].*

a. Sandtransport integrated over a tidal period. Channel 14: cf. Fig. 2.62.

Channel 14, 1981	S_{netto} (kg)	Standard deviation (kg)	Standard deviation %
Western boundary; flood	0.813×10^7	0.848×10^7	104.3
Western boundary, ebb	-0.806×10^7	-0.848×10^7	105.2
Eastern boundary; flood	0.482×10^7	0.470×10^7	97.5
Eastern boundary; ebb	-0.465×10^7	-0.470×10^7	101.1

b. Accretion, integrated over "half" a tidal period (ebb or flood) in a channel.

Channel 14, 1981	S_{netto}, ebb, c.q. flood (kg)	Variance (kg^2)	Standard deviation (kg)
Ebb	-0.341×10^7	0.798×10^{14}	0.894×10^7
Flood	0.331×10^7	0.798×10^{14}	0.894×10^7

c. Resulting accretion (tidal period).

Channel 14, 1981	S_{netto}, getijfase (kg)	Variance (kg^2)	Standard deviation (kg)
Ebb + flood	-0.10×10^6	0.160×10^{15}	0.126×108
Resulting bottom change		$\Delta z = -0.3 \pm 35$ cm/yr (1981)	

*The width of the channel varies from 1784 m at the Western side to 989 m at the Eastern side.

Figure 2.63 shows the mouth of the Brouwershavense Gat after the enclosure. The most important channel of the system is the Brouwershavense Gat itself, from which the name of the inlet is derived. Near Scharendijke (Figs. 2.65 and 2.66) the channel was "leaning" to the bank of Schouwen ("(in)scharen" in Dutch means: "to erode because of a meandering channel") and dug itself in; in this way a scour hole emerged with a depth of 47 m.

Explanation:

(1) Brouwersdam
(2) Eastern Scheldt Storm Surge Barrier (cf. Fig. 2.6)

After closing the inlet "Brouwershavense Gat"[qq] the current was "paralyzed".

[qq] Between Goeree and Schouwen (Figs. 2.63–2.66).

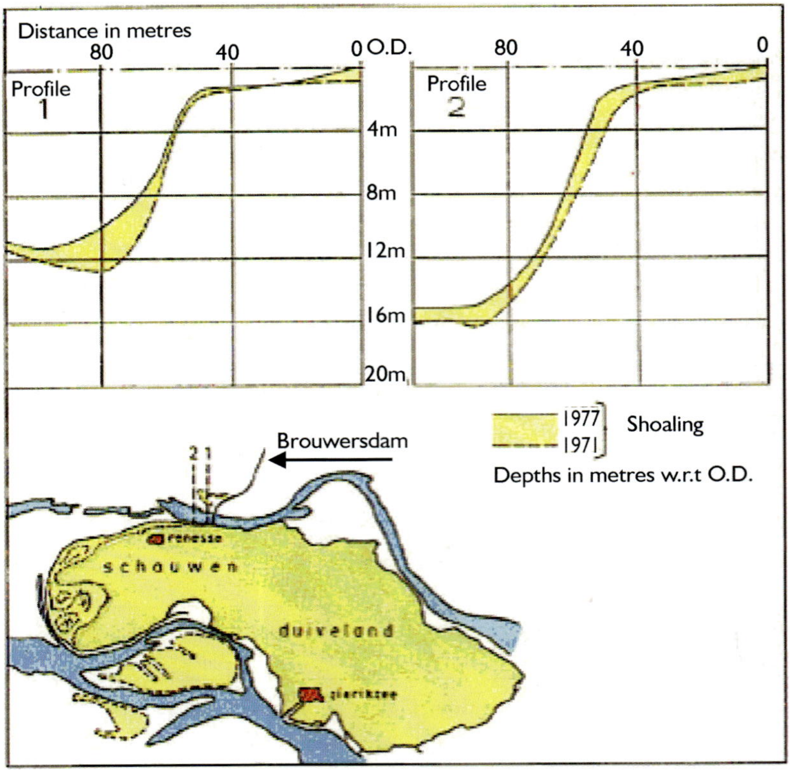

Fig. 2.63 Sedimentation in the Brouwershavense Gat after finishing the Brouwersdam.

Owing to this on the first hand the equilibrium of the sand balance between the outer delta and the estuary was disturbed, with the consequence that the shoreface of the outer delta is moving landward (Fig. 2.64).

Furthermore, because of this the old channels became potential areas of sedimentation. Figure 2.65 shows the accretion and erosion near the dam. The red numbers stand for erosion, the green ones for the shoals.

As far as the origin and the way of transport of the concerning sediment one has to remind a number of possibilities, which probably happen in combination.

Numbers indicated are expressed in 10^6 m^3(accretion positive).

Effects are:

— Settling of flood transport, which is not removed during ebb; think specially on fine material like silt.

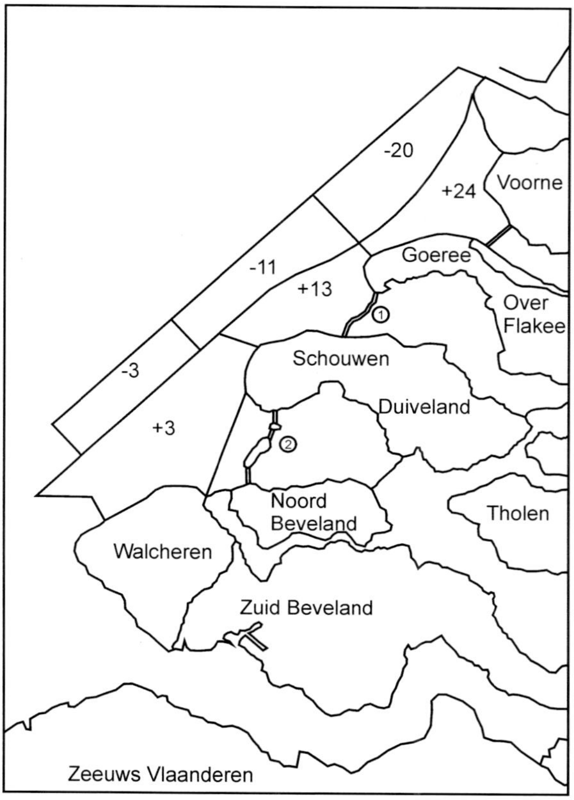

Fig. 2.64 Erosion and sedimentation in the Northern part of the Zealand area.

— Sand, which is removed from the shoals.
As a consequence of the reduction of the current velocity these shoals are not supplied with sand; the balance between current supply and erosion by wave action will be outweighted in favor of the last one. Consequence is lowering of the shoals. In this way, for example, the Middelplaat in front of the Brouwersdam lowered its level after the enclosure;
— Littoral drift (i.e. sand which originates from the heads of the islands and of the banks of the shoals) which sedimentates in the channel and is not carried away.

With respect to the Northern side of Schouwen the development is probably mainly defined by the littoral current; this is not much different since 1971, as Fig. 2.66 shows.

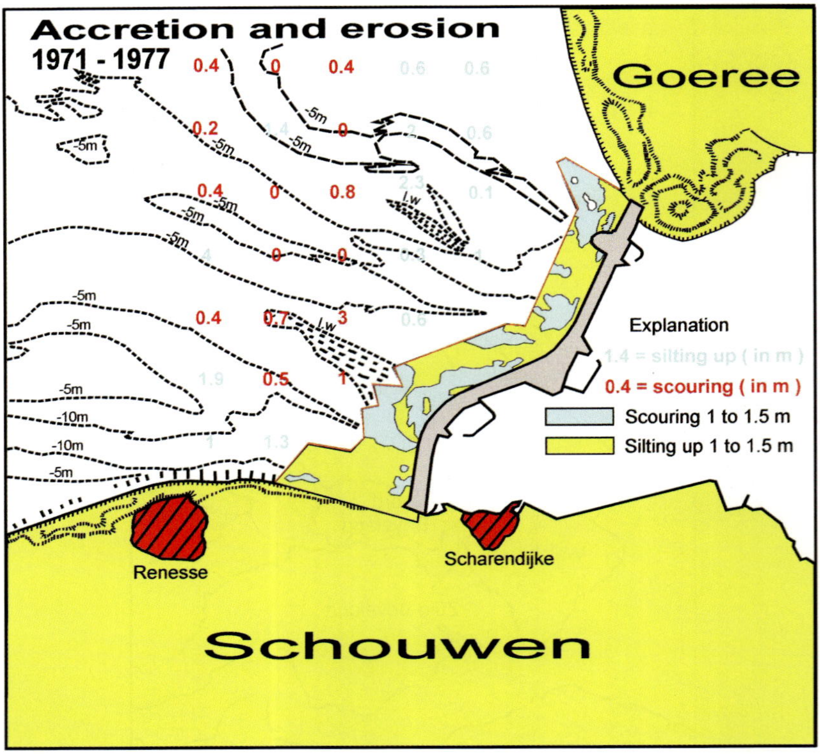

Fig. 2.65 Erosion and sedimentation near the Brouwersdam.

Only the immediate surroundings of the Brouwersdam showed a clearly stronger accretion. The accretion near the Zeepe dunes (cf. Fig. 2.66) is probably a consequence of the permeable groins, constructed between 1960 and 1975.

Can, on account of the present experience, a prognosis be made?

For *the short term prognosis* can be supposed that the changes are so slow, that it will last very long before the boundary conditions in the process change significantly, in other words, that for instance a beach will develop at the site of the Brouwershavense Gat. Likewise the shifting of the outer delta is so slow, that for long the wave boundary conditions will not be affected. There is no reason to presume, that the development the next years will be different from what is found so far. However, human behavior (large scale) can play a role; in the future as well!

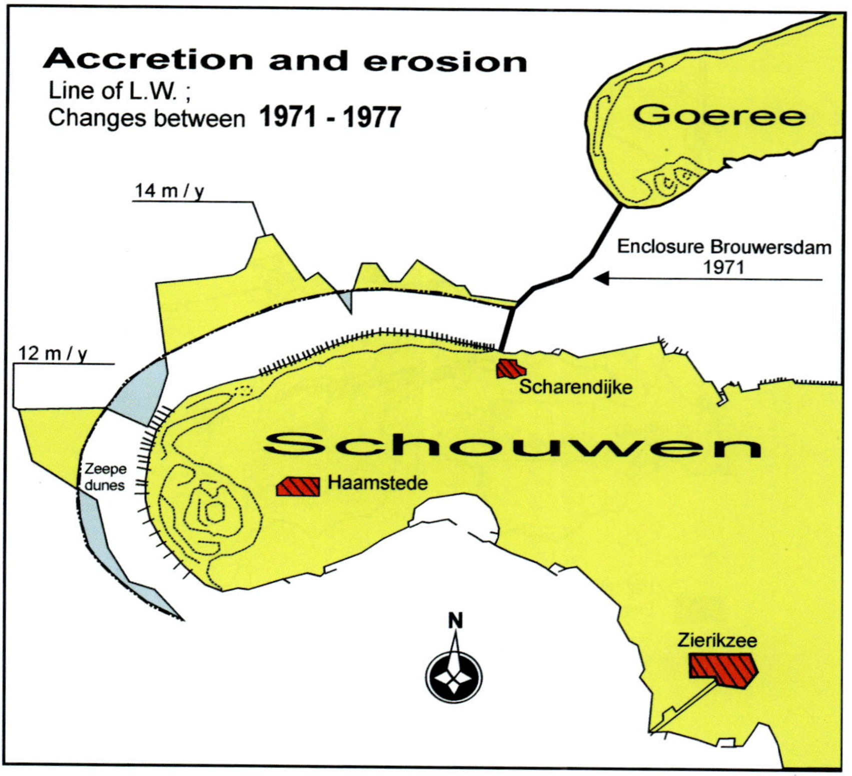

Fig. 2.66 Erosion and sedimentation along the Schouwen coast.

The *Northern tidal basin* stretching at the east side to Keeten-Mastgat-Zijpe and in the situation reached in 1980 till the Volkerakdam and the Hellegatsdam (Fig. 2.67).

The *Southern tidal basin*: the proper Eastern Scheldt.

Although some things mix up in the embouchure and although there are transverse connections, one could see, partly schematized, the Hammen as tidal gorge for the Northern tidal basin and the Roompot as tidal gorge for the Southern part.

Now specially the Northern part is artificially expanded, particularly by the closing of Grevelingen in 1964 and the Volkerak in 1969, but also before, since 1900 by several clearings and normalizations. Because the tidal capacity of the Eastern Scheldt increased, the current velocity increased as well with as consequence larger capacity of sand transportation. As profile

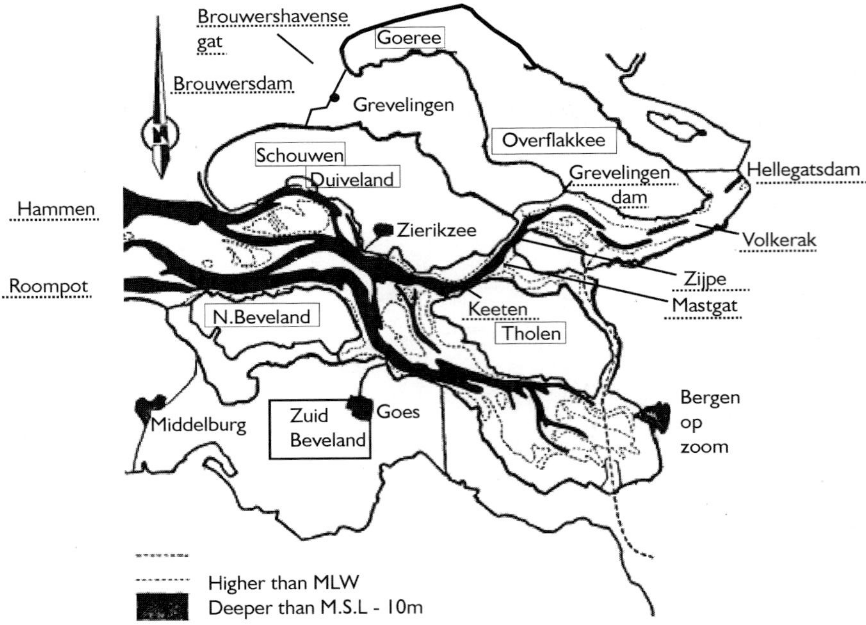

Fig. 2.67 The channel system of the Eastern Scheldt.

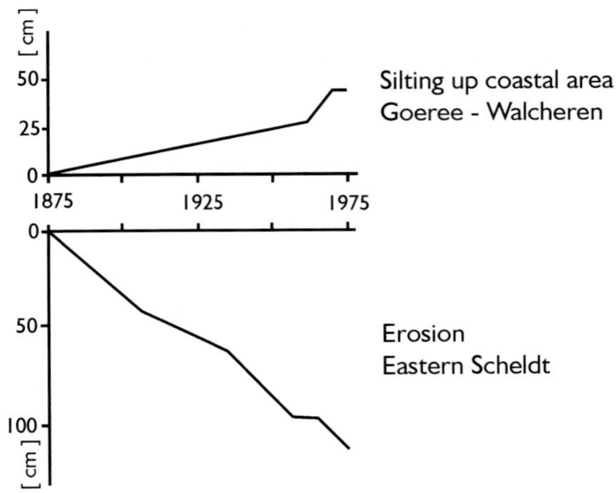

Fig. 2.68 This figure shows the average deepening of the basin between 1872 and 1972. Parallel to the scouring an increasing growth of the outer delta can be seen.

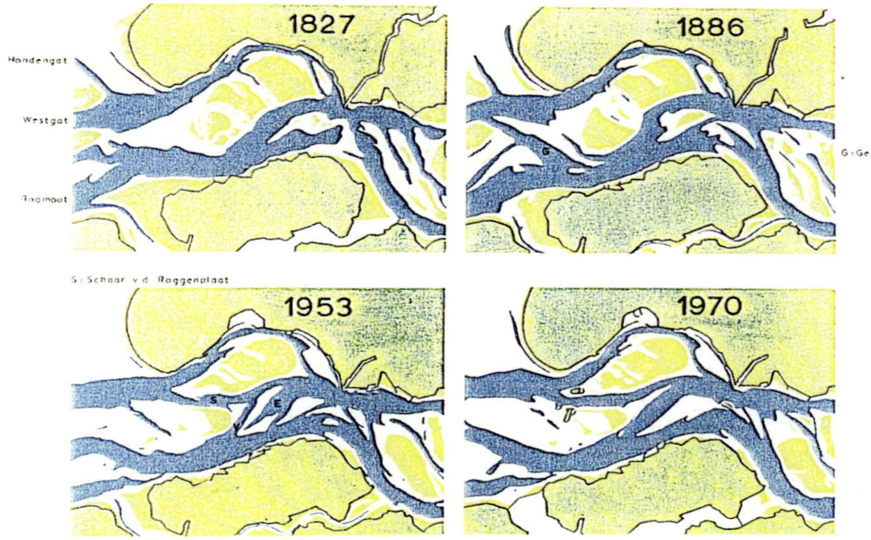

Fig. 2.69 Development of the tidal channels in the outer delta of the Eastern Scheldt.

expansion could not take place in horizontal sense, the basin deepened, linked with a seaward transportation of sand.

On the long term — order — order of half a century — there will be a straightening of the coast, presuming that man won't interfere any more.

Figure 2.69 shows the shifting of the channels in horizontal sense in the past century. A more specific review of the development of the Eastern Scheldt is displayed below (Fig. 2.69).

1827 The first figure one shows in the mouth of the Eastern Scheldt the two main channels the Hammen-Westgat and the Roompot, and in between a shoal without a channel connecting the two main channels. In origin the Schaar van de Roggenplaat can be observed, while the Hondengat, the channel along the Kop van Schouwen, in that time was a passable deep channel.

1886 This channel has become smaller in 1886. The most striking in this figure is that there exists a connection channel — de Geul — between de Hammen and the Roompot, which about this time has its largest capacity.

1953 After about 1900 the Geul is decreasing in capacity and a more inward laying connection between both the main channels the Westgat and the Roompot viz. the Schaar of the Roggeplaat appears. The Hondengat has disappeared. Also in this figure: the ebb-channel of the Vuilbaard and the flood-channel of the Roompot; here is clearly a difference between the ebb- and the flood-channel.

1970 In the situation of 1970 this system has almost become an indifferent channel. Also the shortcutting between the ebb-channel by the ebb-channel of the Roompot of the Westgat can be observed, to a certain extent. The Geul is almost disappeared while in 1971 it has been enclosed by the "damsection Geul".

The construction of the storm surge barrier created a situation in the outer delta, being between the complete blocking and opening of the Eastern Scheldt. The current velocities decreased. At a distance of approx. 12 km seaward of the storm surge barrier the reduction of the current velocity is negligible and, moving in the direction of the storm surge barrier, the reduction of the current velocity increased till approx. 26%.

Before closure, the influence has been discussed on the enlargement of the tidal basin and the tidal volume; one could expect that the reducing of the tidal volume means: "the way back"; in this case the reducing of the outer delta and the accretion of the Eastern Scheldt. Doubtlessly the channels will move, as a consequence of the changes in the tide propagation, but presumable this will have no effect on the seaward sides of the islands. For the littoral drift will hardly change. Naturally it is recommendable to follow the changes as scrupulously as possible. Even if the expectations would be disqualified, it will be technically possible to take shore-protecting measures, like beach nourishment to annihilate the unwanted consequences.

Shifting the attention to the embouchure of the Western Scheldt, (Fig. 2.62), one recognizes the channel system (Fig. 2.70) that fairly approaches the basic scheme, shown in Fig. 2.46.

At the North-side one finds the Oostgat, of which the Northern part surely has a resulting flood discharge. More Southward one finds the main channel the Wielingen, where theoretically a resulting ebb discharge may be expected. Indeed the fact that the water of the Wielingen has less salt content than in the Oostgat (as a consequence of the discharge of the Scheldt) points in that direction. The current and discharge measurements however give a very confused image. The Wielingen and the Oostgat enclose

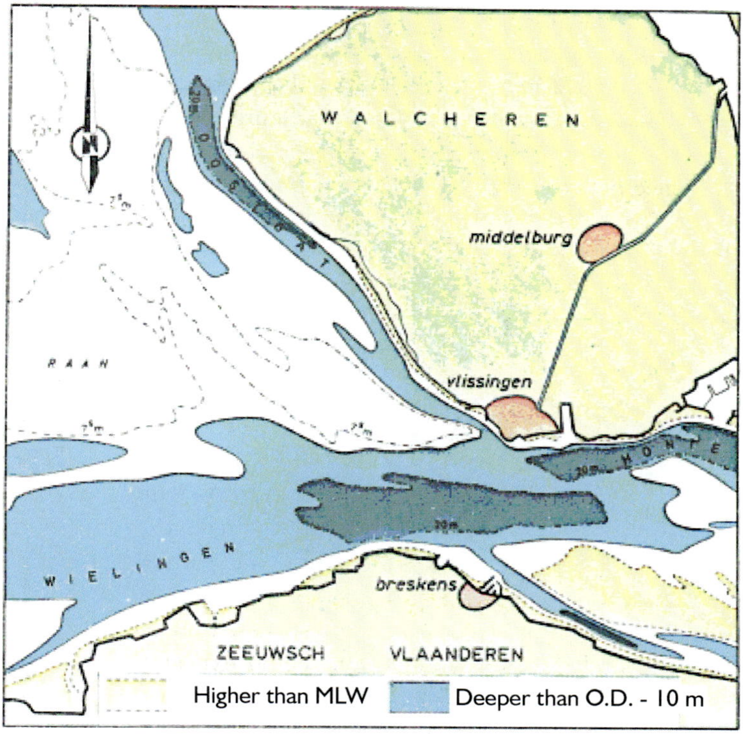

Fig. 2.70 The channel system in the estuary of the Western Scheldt.

the Vlakte of the Raam, on which, as a secondary channel, the Deurlo[rr] is situated. South of the Wielingen one finds the Appelzak,[ss] which, in the past, had a flood channel character. Although the Oostgat remains fairly stable, in the course of centuries considerable erosion took place. This is due to the fact that the littoral drift along the coast gets little natural nourishment at the Northside, (at the most-Western point of Walcheren one finds the "Westkappelse Seawall", a stronghold) neither from the outer delta because of the tidal channel "Oostgat", closely located to the coast. Because of this, near Westkapelle approx. 250 m dune disappeared into the sea since 1450 (Fig. 2.71).

[rr] Area below O.D. — 7.5 m, South–Westernly of the Oostgat/Deurlo, bordering the Walcheren coast (Fig. 2.70).
[ss] Reaching to the Belgian coast (Figs. 2.70 and 2.72).

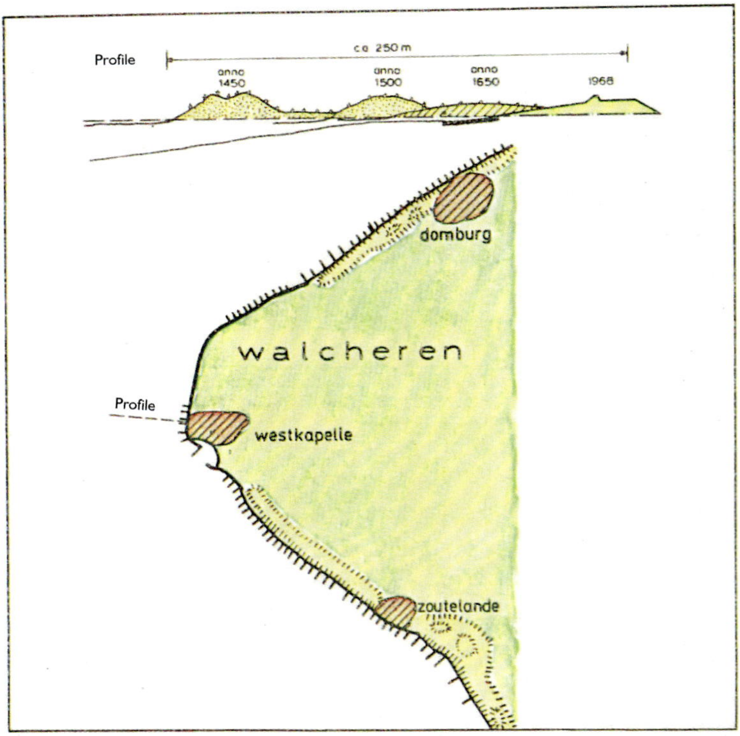

Fig. 2.71 Erosion of the Walcheren coast. Since 1450 the Westend eroded 450 m, cf. Fig. 2.64 for situation.

Focussing now on the Southern area, in front of the coast of Zeeuws-Vlaanderen (Figs. 2.70 and 2.72) it specially attracts the attention, that the Wielingen shifted somewhat between Cadzand (Fig. 2.72) and the Zwarte Polder[tt] to the South and West where it expanded in the deeper regions. The channel silted up from 25 m to 12 m near Groede (Fig. 2.72)[uu] and eroded in the Western part. The shallow part between Wielingen and Appelzak largely disappeared: the intermediate bar the Binnen-Paardenmarkt eroded, where the Appelzak itself became shallower, just like the southern part of the Wielingen on the Belgian territory. Only the eastern outlet of the Appelzak got more or less the character of an outlet of the Wielingen and eroded and shifted.

[tt] The discontinuity in the coastline halfway Cadzand and Groede (Fig. 2.72).
[uu] The crosses denote the Dutch–Belgian border.

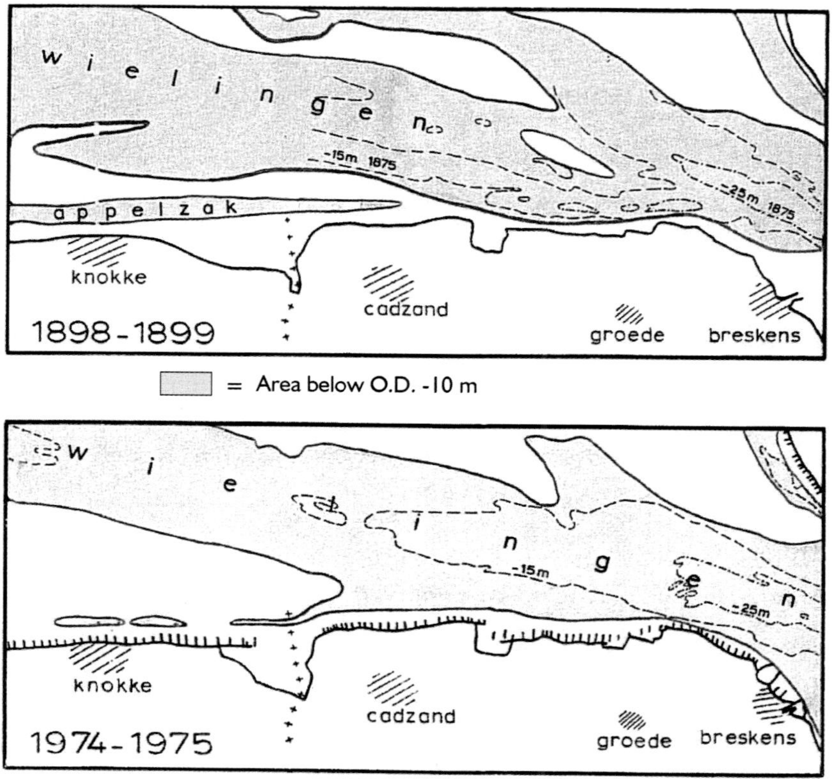

Fig. 2.72 Development of tidal channels in the outer delta of the Western Scheldt. Heavy lines: contour lines coastline at NAP −10 m

Here also man made changes are performed, as mentioned below:

Thus this end of Section 2.6 will be devoted to anticipated effects of a coastal engineering project round 1978: a groin with a length of 1350 m (Fig. 2.74) near the Belgian–Dutch nature reserve "the Zwin"[vv] (Fig. 2.73). Its projected global situation (Easternly of Knokke; Fig. 2.74, Westernly of the groin at "16.820 km"; will be recognized from Figs. 2.72 to 2.77c. Mind this project has not been carried out.

It would lack the conditions for implementation in this section, if not for the following reasons:

(a) it certainly would have affected the Zealand coast, and, as could be anticipated, in an unfavorable way;

[vv] The ancient shipping entrance (silted up) to the Belgian town of Brugge.

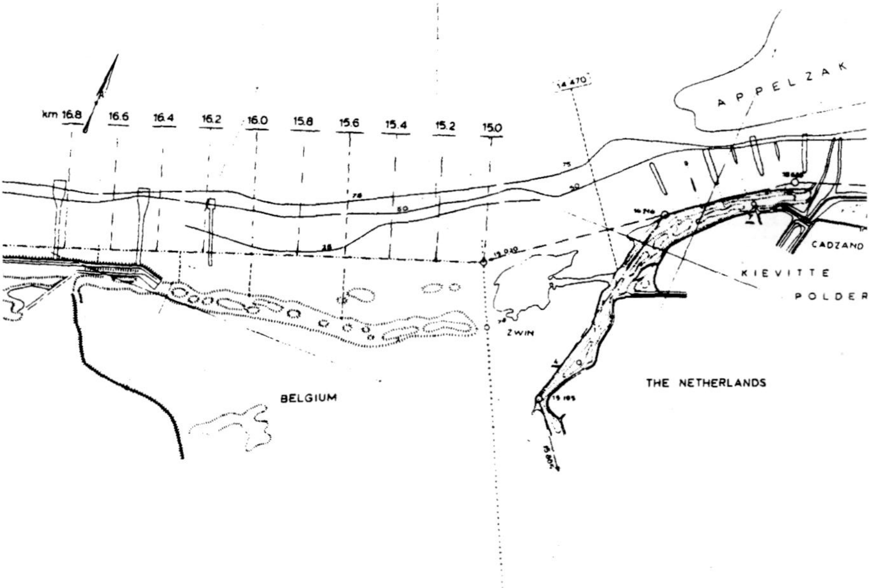

Fig. 2.73 The Zwin area at the Dutch–Belgian border (15.0 km); cf. areal view (Fig. 2.7). Investigated has been the construction of a long groin (Fig. 2.74) at 16.8 km.

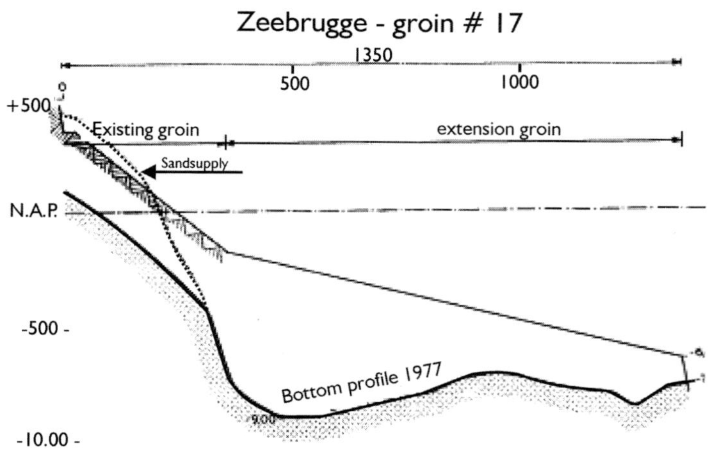

Fig. 2.74 Design of a long groin projected ca. 2 km Westernly of the Dutch–Belgian border (at 17 km; cf. Figs. 2.73 and 2.75).

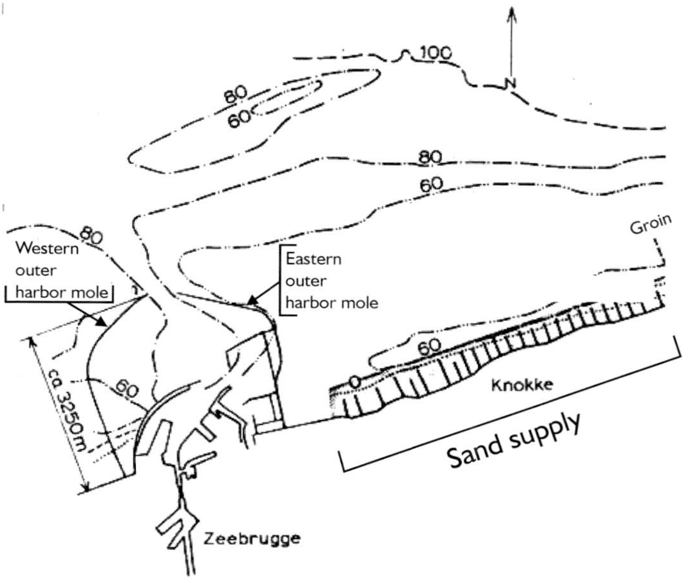

Fig. 2.75 Enlargement harbor moles of Zeebrugge; location of long groin (cf. Fig. 2.74).

(b) thus the project would have had potentional international consequences, with respect to conflict of interests between the Netherlands and Belgium;

(c) it has not been carried out because of calculations in the field of coastal dynamics, which showed, that the project would not have been feasible. In order to achieve the Belgian goals (protection of the Belgian coast Easternly of Zeebrugge) another solution (just sand supply, without building the dam) turned out to be more economic: a solution beneficial *as well* for Belgium as for the Netherlands!

Consensus (common interest) replaced a potential conflict of interests!

It thus demonstrates perfectly the practical importance of the theory, displayed in this book:

Knowledge pays off!

The background of the aforementioned project must be seen in connection with the plans, which the Belgian Government had, 1978, (and since then carried out) concerning the neighboring coastal area: expansion of the harbor of Zeebrugge.

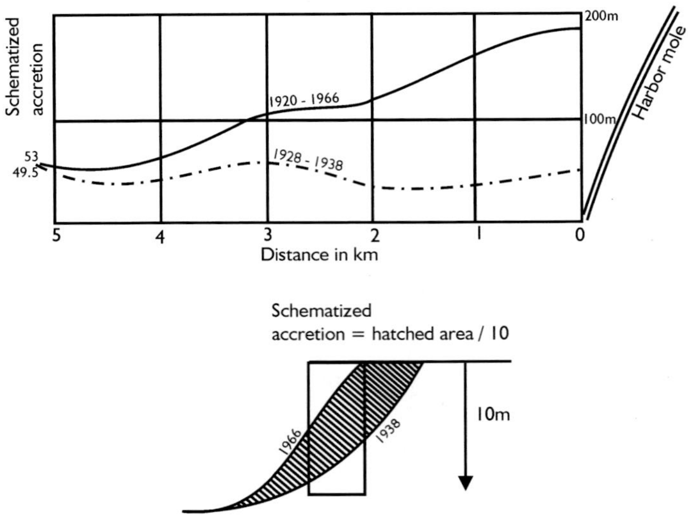

Fig. 2.76 Effect (accretion) in 1966 at the luffside (Westside) of the harbor moles of Zeebrugge as a consequence of its construction in 1928.

Those plans implied, Fig. 2.75, an extension of the harbor moles till approx. 3250 m of the coast and beach nourishment of 9 mln m^3 on the Belgian beach, Easternly of the moles (construction since then performed). Furthermore, the expansion of the aforementioned long groin specially through the deeper parts of the Appelzak. It should hamper the tidal and littoral transport of the supplied sand (Fig. 2.75). This longshore drift could be estimated as considerable, keeping in mind, that the original Western harbor mole (the strongly curved one on the innerside of the "Western outer harbor mole" (Fig. 2.75) lead to an accretion at the Westside of the harbor mole between 1928 and 1966 of ca. 150 m (Fig. 2.76). Figure 2.77 shows this local accretion at the mole by means of a cross-section.

Note:

It must be pointed out that because of thorough knowledge of coastal dynamics of the Belgian coastal engineers involved in this project consensus between the Dutch and Belgian coastal authorities could be reached on improvement of the plans with respect to the coast, to mutual benefit of Belgium as well as the Netherlands.

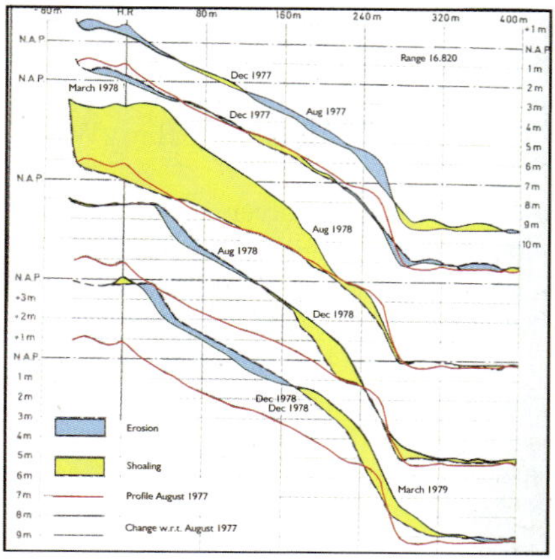

(a)

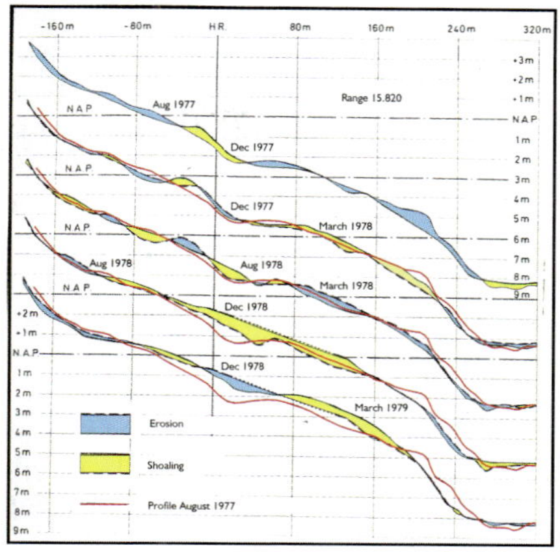

(b)

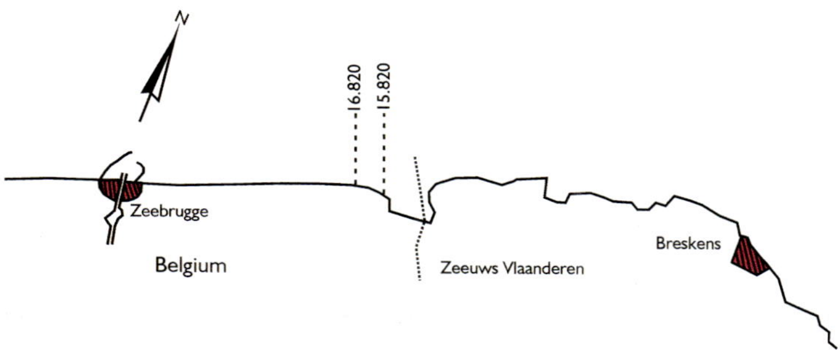

(c)

Fig. 2.77 (a–c) The coastal supply from August 1978 near the Dutch–Belgian border and its deformation. (a) Changes in range 16.820; (b) Changes in range 15.820; (c) Situation (. . . : Dutch–Belgian border).

References

1. Agema, J. F., W. T. Bakker, H. Speekenbrink, R. van Veghel, C. de Smit, Verandering van de Zeeuwse kusten door mens en natuur (Changes of the coast of Zealand by man and nature), Rijkswaterstaat, internal rep. WWKZ-80.V016, 1980.
2. Arthur, R. S., A note on the dynamics of rip currents. *J. Geoph. Res.* **67**(7), 2777–2779, 1962.
3. Bakker, W. T., *A Mathematical Theory about Sandwaves*. Shore and Beach, Oct. 1968.
4. Bakker, W. T., Berekening van het langstransport door golven met de methode van evenwijdige dieptelijnen (Computation of the wave-induced

littoral drift with the Svasek-method in case of parallel depth contours) Rijkswaterstaat, Dept. for Coastal Res., Rep. WWK 69-7, 1969.
5. Bakker, W. T. and H. J. Opdam, Influence of waves and tides on the littoral drift in the surf zone, Rijkswaterstaat, Dept. for Coastal Res., Rep. WWK 70-8, 1970.
6. Bakker, W. T. and H. J. Opdam, Over de invloed van golven en getij op het zandtransport in de brandingszone (About the influence of the waves and tide on the transportation of sand in the breaker zone), Rijkswaterstaat, Dept. for Coastal Res., Rep. WWK 70-9, 1970.
7. Bakker, W. T. and D. S. Joustra, The History of the Dutch Coast in the last century, Washington, Xllth Int. Conf. on Coastal Eng. (ICCE), 1970 (also; Rijkswaterstaat, Dept. for Coastal Res., Rep. WWK 70-12).
8. Bakker, W. T. and J. H. M. de Ruig, Opmerkingen over de gewenste nauwkeurigheid en informatiedichtheid, benodigd voor het kustmeetsysteem (Wanted information density and accuracy for coastal topographic monitoring), Rijkswaterstaat, Dept. Getijdewateren, Rep. GWAO-87.460, 1987.
9. Bakker, W. T., De nauwkeurigheid, waarmee een trend kan worden bepaald (The accuracy to define a trend), Rijkswaterstaat, Dept. Getijdewateren, Rep. GWAO-87.
10. Bakker, W. T. and H. de Vroeg, De Hollandse kust bekeken: analyse van 20 jaar kustgedrag (A review of the Hollandse coast: analysis of 20 years of coastal behavior), De Ingenieur, May 1988.
11. Bakker, W. T., E. H. J. Klein Breteler and A. Roos, *The Dynamics of a Coast with a Groyne System*, Washington, XXIInd Int. Conf. on Coastal Eng. (ICCE), 1990.
12. J. A. Battjes, Private communication to Ir. H.N. Svasek.
13. Van Bendegom, L., Beschouwingen over riviermorphologie (Considerations about river morphology), De Ingenieur no. 59, Jan. 24, 1947.
14. van Bendegom, L., *Beschouwingen over de Grondslagen der Kustverdediging (Principles of Coastal Protection)*, Rijkswaterstaat Hoorn, Ministry of Public Works, 1949.
15. Boer, S., H. J. de Vriend and H. G. Wind, *A System of Mathematical Models for the Simulation of Morphological Processes in the Coastal Area*, Houston, XXth Int. Conf. on Coastal Eng. (ICCE), 1984.
16. Bowen, A. J., The generation of longshore current on a plane beach, *J. Marine Res.* **27**(2), May 1969.
17. Bowen, A. J. and D. L. Inman, Edge waves and crescentic bars, *J. Geoph. Res.* **76**, 8662–8671, 1971.
18. Bruun, P. and F. Gerritsen, *Stability of Coastal Inlets*, Scheveningen, VIIIth Conf. on Coastal Engng, 1960.
19. Bijker, E. W. and J. N. Svasek, *Two Methods for Determination of Morphological Changes, Induced by Coastal Structures*, Int. Nav. Congr., 1969.
20. Dally, W. R., *Longshore Bar Formation — Surf Beat or Undertow?* New Orleans, Coastal Sediments, pp. 71–86, 1987.
21. Delft Hydraulics and Rijkswaterstaat, *Kustontwikkeling; verleden, heden, toekomst, (Development of the Coast, Past, Present, Future)*, Summary

colloquium May 1985 Also: paper series in Polytechnisch Tijdschrift, Oct./Nov. 1985.

22. R. Dorrestein, *Wind and Wave Data of Netherlands Lightvessels*, Med. en Verh, K.N.M.I. no. 90; 1976.
23. Dronkers, J. J., *Research for the Coastal Area of the Delta Region of the Netherlands*, Washington, XIIth Int. Conf. on Coastal Eng. (ICCE), 1970.
24. Dyhr-Nielsen, M. en T. Sorensen, *Some Sand Transport Phenomena on Coasts with Bars*, Washington, XIIth Int. Conf. on Coastal Eng. (ICCE), pp. 855–866, 1970.
25. Edelman, T., Erosie en aanwas van het kustvak Den Helder-Hoek van Holland (Erosion and accretion of the coastal area Den Helder-Hoek van Holland), Rijkswaterstaat, Dept. for Coastal Res., Rep. WWK 61-1, 1961.
26. Edelman, T., Verstoringen in de kustlijn t.g.v. de aanwezigheid van zeegaten (Disturbances in the coastline, caused by the presence of inlets), Rijkswaterstaat, Dept. for Coastal Res., Rep. WWK-61-3, 1961.
27. Edelman, T. and D. N. Eggink, *Some Characteristics of the Dutch Coast*, Scheveningen, VIIIth Int. Conf. on Coastal Eng. (ICCE), 1962.
28. Edelman, T., *Systematic Measurements Along the Dutch Coast*, Tokyo, Xth Int. Conf. on Coastal Eng. (ICCE), 1966.
29. Elias, E., Morphodynamics of Texel inlet, Ph.D. thesis; Delft University of Technology, 2006.
30. Gerritsen, F. and J. van Heteren, *Effect of Low-Frequency Oscillations on Coastal Morphology*, Houston, XIXth Int. Conf. on Coastal Eng. (ICCE), 1984.
31. Groenenboom, J., Onderzoek naar de kustontwikkeling bij de havendammen van IJmuiden (Investigation on the coastal development near the harbor moles of IJmuiden), Delft University of Technology, Hydraulic and Offshore Engineering Division, M.Sc. thesis.
32. Haring, J., Oppervlakte van het dwarsprofiel van de Nederlandse zeegaten als functie van het getij-volume (Area of the cross-section of the Dutch inlets as a function of the tidal volume), Rijkswaterstaat, Delta Works, Rep. K 251.
33. Van Harreveld K., Correlatie van golfgegevens Triton-1964/65 met overeenkomende gegevens van het lichtschip Goeree (Correlation of wave measurements Triton-1964/65 with corresponding data of the lightvessel Goeree), Rijkswaterstaat, Delta Works, Dept. of Hydraulics, Rep. K 288.
34. Huys S., *Geulmigratie op de Buitendelta. Het Friese zeegat en het Amelander Gat (Migration of The Tidal Channels in the Outer Delta. The Friese Inlet and the Amelander Gat)*, Rijksuniversiteit Utrecht, 1992.
35. Klein Breteler, E. H. J., *A Tidal Model for the Coast of Hook or Holland — IJmuiden*, Rijkswaterstaat, Dept. for Coastal Res., Rep. WWK 70-10, 1970.
36. L. Knop, Onderzoek Noordzeekust Petten-Huisduinen (Investigations Northsea coast Petten-Huisduinen), Study Department Hoorn Rijkswaterstaat, no. 584.
37. De Looff, D. and Van Malde, J., *Over de cyclus der zogenaamde drempelgeulen in de Westerschelde (On the Morphological Cycle of the So-called "Sill*

Channels" in the Western Scheldt), Rijkswaterstaat, Directorate of Zealand, Flushing, 1973.

38. de Mulder, E. F. J., *Geologische geschiedenis van de Hondsbossche zeewering (Geological History of the Hondsbossche Seawall)*, Derde uitgave Kring van Vrienden van de Hondsbossche, 1983.
39. Mulder, J. P. M., H. J. de Vriend and A. W. Heemink, Validation of large scale morphological models, Report on a workshop at Noordwijk (Netherlands) on 7 and 8 February 1994.
40. Opdam, H. J., A wave-tide model for the Dutch coast, Delft, M.Sc. thesis, 1970.
41. De Ras, M. A. M., Onderzoek naar de grootte van het langstransport langs de kust bij IJmuiden door de combinatie van getijstroom en golforbitaalbeweqing (Investigation on the magnitude of the currents transport along the IJmuiden coast by the combination of tidal streams and waveorbital motion), Rijkswaterstaat, Dept. for Coastal Res., Rep. WWK 71-11, 1971.
42. Van Rijn, L. and De Leeuw, *Zandtransportmodel Westerschelde, stochastisch onderzoek zandtransportformule (Sand Transportation Model Westerschelde, Stochastic Research on a Sand Transport Formula)*, Delft Hydraulics Laboratory, June 1985.
43. Schoorl, H., *Several Publications (1973, 1979, 1982 en 1983 compilation of De Mulder).*
44. Van Straaten, L. M. J. U., Directional Effects of winds, waves and currents, Geologie en Mijnbouw no. 23, 1961.
45. Van Straaten, L. M. J. U., Coastal barrier deposits in South and North Holland in particular in the areas around Scheveningen and IJmuiden, *Meded. Geol. Stichting Serie* **17**, 41–75, 1965.
46. Symonds, G., D. A. Huntley and A. J. Bowen, Two-dimensional Surf-beat: long-wave generation by a time-varying breakpoint, *Geoph. Res.* no. 87, C1, pp. 492–498, 1982.
47. Van Veen, J., *Zeegat van het Vlie (Inlet of the Vlie)*, Rijkswaterstaat, Dir. Benedenrivieren, 1934.
48. Van Veen, J., Onderzoekingen in de hoofden (Investigations In the Southern Northsea), Delft, Ph.D. thesis, 1936.
49. Van Veen, J., Eb- en vloedschaarsysteem in de Nederlandse getijwateren (Ebb- and flood channel systems in the Dutch tidal waters), *J. Kon. Ned. Aardr. Gen.* **67**, 45–65, 1950.
50. Van Veen, J., A. J. F. v. d. Spek, M. J. F. Stive and T. Zitman, Ebb- and flood channel systems in the Netherlands tidal waters, *J. Coastal Res.* **21**(6), 1107–1120, Nov. 2005.
51. Vellinga, P., Beach and dune erosion during Storm Surges, Delft University of Technology, Ph.D. thesis, 1965.
52. De Vroeg, H., Schematisering brandingsruggen m.b.v. jaarlijkse kustmetingen (Schematizing of breaker bars using annual measurements along the coast) Delft University of Technology, Hydraulic and Offshore Engineering Section, M.Sc. thesis, 1987.

53. Wentholt, L. R., Stranden en strandverdediging (Beaches and beach protection), Delft, Ph.D. thesis, 1912.
54. Wiersma, J., De kust in vier dimensies (The coast in four dimensions), part of Ref. [20].
55. Wijnant, Th. J. C., *Littoral Drift Near Texel*, Rijkswaterstaat, Dept. for Coastal Res. 1968.

Chapter 3

Line Theories in Coastal Dynamics

3.1. Introduction

The present chapter gives the (classical) theory concerning line theories in coastal dynamics. Using the Dutch coast as an example, Chapter 2 gave practical insight:

- in the way coasts can reshape in the course of time,
- in the agents that cause this coastal change;
- in orders of magnitude of those changes.

The present chapter deals with the line theories, of which the first one was published by Pelnard-Considère,[48] after an idea of Bossen.[a] The clever idea was to link coastal theory to classical theory, well known from heat conduction and electricity (cf. Table 3.1 page 115).

In this mathematical theory, all constants and variables are indicated by symbols and it is suggested, that the numerical value of many of these constants can be found from prototype data.

On the problem of finding those constants will be focused in Chapter 4, where Chapters 5 and 6 will deal on verification and evaluation, respectively.

First (Section 3.2) the classical one-line theory, as originally developed by Bossen and Pelnard Considère,[48] is reviewed. Various solutions (depending on boundary conditions) will be considered.

The coastal profile can be schematized with one part lying at a depth h which is deep enough to neglect the littoral transport at that site

[a]As Thijsse told during his classes, Bossen and Pelnard-Considère collaborated on Delft Hydraulics, where Pelnard Considere did the practical work and Bossen the theoretical work.

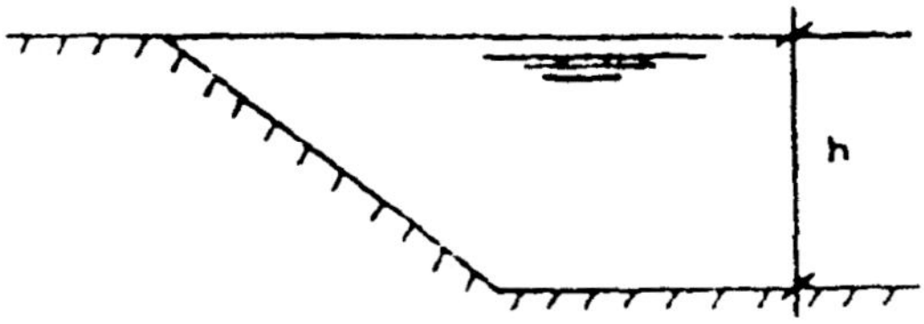

Fig. 3.1 Schematized coastal profile.

and another part where an equilibrium profile is formed. This part shifts seaward, respectively landward, as accretion and erosion occur, but its shape does not change (Fig. 3.1).

In Section 3.3, an extension by Bakker *et al.*,[3,5–7,9,11] is given, the two-line theory, in which the assumption of a uniform coastal profile alongshore is abandoned. This facilitates the simulation of coastal structures, which hamper only a part of the transport, for instance: the transport along the beach only, but not along the inshore.

Still more sophisticated is Section 3.4, in which (still) more-line theories are considered. More and more, during this exercise, the subject of investigation will be the cross-shore transport rather than the longshore transport. Furthermore, also the time-scale of the topics of investigation increases, up to geological scales (Subsection 3.4.3).

3.2. One Line Theory, No Diffraction[b]

3.2.1. *Assumptions*

The coastal profile can be schematized with one part lying at a depth h which is deep enough that the littoral transport can be neglected and another part where an equilibrium profile is formed. This part shifts to seaward and landward, as accretion and erosion occur, but its shape does not change (Fig. 3.1).

The dynamics of one "Coastline" will be considered using one of the depth contours (it is immaterial which one, as they are all considered to be parallel). Assume all the transport equations apply, then the total transport[c] S^* along the coast (only) is a function of the coastal direction;

[b]This chapter (up to Section 3.3) is translated from Bakker (int. rep;) by G. Mocke.
[c]The star indicates an integral transport over a zone (to distinguish between this variable and a transport per unit of width, as for instance S_y in Eq. (3.45), later on).

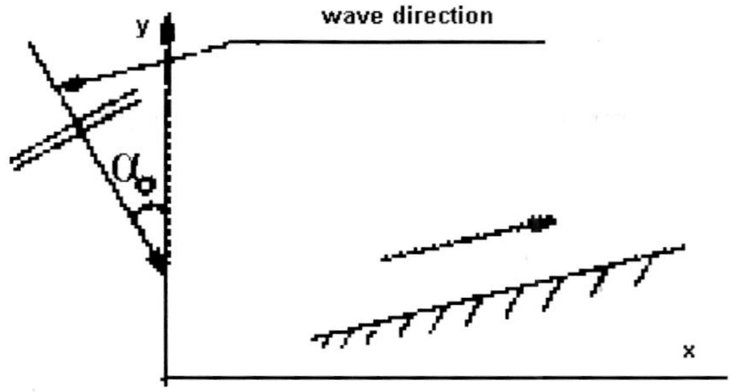

Fig. 3.2 Definition wave direction α_0.

this function is linearized:

$$S^* = S_0^* - s\frac{\partial y}{\partial x} \tag{3.1}$$

Denoting symbols:

x, an axis lying in the mean coastal direction; y, an axis perpendicular to x and positive in the seaward direction; S_0^*, the net transport when the coast is parallel to the x direction; s, a proportionality constant ("coastal constant").

From Fig. 3.2 one can observe that the angle of wave incidence decreases when $\partial y/\partial x$ increases and as long as the angle is not too big (which has been assumed), a decrease in the amount of transported material results. Assuming transport equation applicable means therefore that the transport due to wave action is considered the most significant.

In Chapter 4, the magnitude of S_0^* and s is investigated.

Initially we assume, small angle of wave incidence is equal to α_0, if the coastal direction is parallel to the x-axis:

$$S_0^* = s\ tg\,\alpha_0 \tag{3.2}$$

$$S^* = s\left(tg\,\alpha_0 - \frac{\partial y}{\partial x}\right) \tag{3.3}$$

3.2.2. *The coastal equation of Pelnard-Considère*

The continuity equation gives (Fig. 3.3a,b):

$$\frac{\partial S^*}{\partial x}\Delta x \Delta t + h\Delta x \Delta y = 0 \tag{3.4}$$

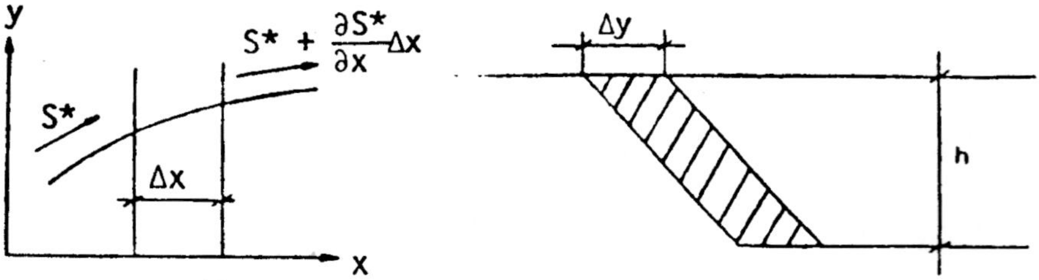

Fig. 3.3 (a) Plan view of coast. (b) Belonging cross-section.

where t is the time and:

$$\Delta y = \frac{\partial y}{\partial t}\Delta t$$

this gives:

$$\frac{\partial S^*}{\partial x} + h\frac{\partial y}{\partial t} = 0 \tag{3.5}$$

Substitution of Eq. (3.1) in Eq. (3.5) yields:

$$\frac{\partial y}{\partial t} = \frac{s}{h}\frac{\partial^2 y}{\partial x^2} \tag{3.6}$$

Equation (3.6) will be considered to be the "Coastal equation of Pelnard-Considère." Pelnard-Considère,[48] was the first to publish a paper on this theory as well as to check it against experiment. The original idea however stems from Bossen.

Grijm,[35,36] developed a similar theory with a more complicated transport formula. He assumed the transport to be proportional to sin 2α, in which α is the angle of wave incidence.

3.2.3. *Analogies*

Equation (3.6) is the "diffusivity equation" applicable to many disciplines, as depicted in Table 3.1. In this table the analogy with tidal theory (Lorentz method) is somewhat dubious. An analogy only exists when the inertial term is small with respect to the friction term (shallow water). For this case an analogy exists between water surface and coastline and current with sand transport.

Table 3.1 Analogies with Coastal Dynamics, after a similar table of De Josselin de Jong (1963).

	Hydrology	Thermodynamics	Electricity	Tidal theory Lorentz method	Coastal Dynamics
Current	v Water	q Heat flux	i Electronics	v Water	S^* Sand
Because of a gradient of	z Piezometric level	T Temperature	v Potential difference	z Water level	y Coastal direction
Dynamic equation	$v = -K \times \partial z/\partial s$ K = permeability	$q = -\lambda \times \partial T/\partial s$ λ = conductivity	$i = -1/R \times \partial v/\partial s$ R = resistance	$v + \frac{1}{K} \cdot \frac{\partial v}{\partial t} = -g \times \frac{\partial z}{\kappa \partial s}$ K = linear coefficient $= \frac{8g}{3C_h^2} \cdot \frac{V}{h}$ C_h = chezy coeff. h = depth	$S^* = S_0^* - s\partial y/\partial x$ S_0^* = stationary transport $s = [\partial S^*/\partial \alpha]_{\alpha=0}$ α = angle wave incidence
Continuity equation	$\frac{\partial z}{\partial t} = -\frac{KD}{\beta}\nabla^2 z$ β = porosity D = thickness soil layer	$\frac{\partial q}{\partial t} = \frac{\lambda}{c}\nabla^2 T$ c = heat capacity $\nabla^2 = \frac{\partial^2}{\partial x^2} + \frac{\partial^2}{\partial y^2}$	$\frac{\partial v}{\partial} = \frac{D}{R_c}\nabla^2 v$ c = capacity per unit of surface D = thickness material	$\frac{\partial z}{\partial t} + \frac{1}{K}\frac{\partial^2 z}{\partial y^2} = -\frac{gD}{K}\frac{\partial^2 z}{\partial s^2}$ D = depth	$\frac{\partial y}{\partial x} = \frac{s}{h}\frac{\partial^2 y}{\partial x^2}$ h = depth $\frac{s}{h}$ = coast constant

3.2.4. *Law of Scale*

N.B. The Coastal Eq. (3.6) is independent of S_0^*!

Theorem 1. *If $y_1(x,t)$ and $y_2(x,t)$ both satisfy the coastal Eq.* (3.6) *then this equation is also satisfied by:*

$$y_3 = Ay_1 + By_2$$

where A and B are arbitrary constants. The proof follows from the linearity of Eq. (3.6).

Theorem 2. *If $y(x,t)$ satisfies the coastal Eq.* (3.6), *then also $y(Ax, A^2t)$ satisfies this equation. The proof follows by substituting the latter expression in Eq.* (3.6).

Example of Theorem 2: Let the coastline after 1, 2 and 4 years be given by the curves A, B and C, respectively in Fig. 3.4a and let a certain coastline after 4 years be such that it can be found from curve A by multiplying the x-scale by 2 and leaving the y-scale constant (curve D). In that case, the coastlines E and F originating from D and corresponding to 8 and 16 years may be found by multiplying the x-scales of the curves B and C by a factor of 2.

Example of Theorems 1 and 2: let the coastline after 1, 2 and 4 years be depicted by the curves A, B and C in Fig. 3.5a and assume a certain coastline after 4 years to be such that it can be found from curve A by multiplying the x- and y-scale by a factor of 2 (curve D). Then the coastlines E and F for 8 and 16 years, respectively, may be found from D by multiplying the x- and y-scale of curve B and C by 2.

This is the case because according to Theorem 1 the y-scale may be multiplied by any factor. Here, $A = 2$, $B = 0$ in Theorem 1.

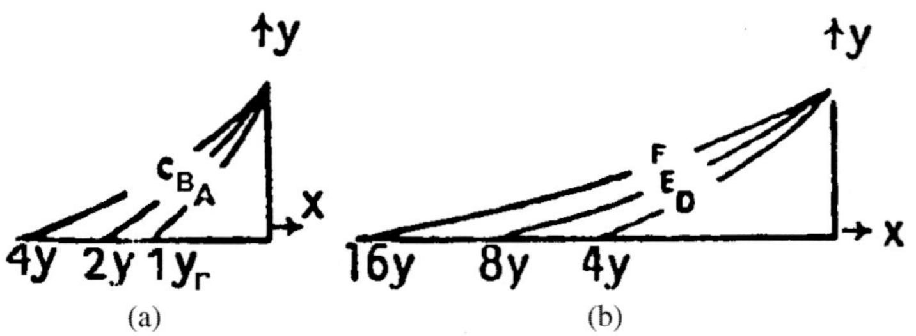

Fig. 3.4 (a) Successive coastlines A, B, C. (b) Successive coastlines D,E,F $y_D(2x,t) = y_A(x,t)$.

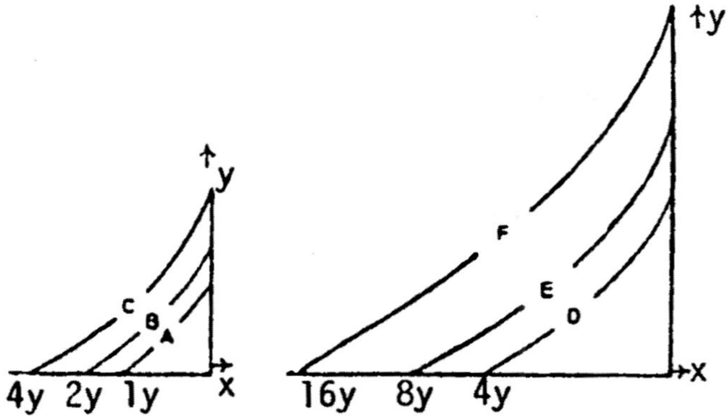

Fig. 3.5 (a) Successive coastlines A, B, C. (b) Succesive coastlines D,E,F $y_{\mathrm{D}}(2x,t) = 2y_{\mathrm{A}}(x,t)$.

Remark in connection with these examples: if it would be possible to choose curve A such that curve C coincides with curve D, then the whole coastline would be predictable with time. For instance curves G and H for 32 and 64 years may be determined by multiplying curves E and H by a factor of 2 and so on. In this case the x-scale is proportional to $\sqrt{t}$.

Indeed it shows to be possible to choose curve A in this way, as will be elaborated in the proof of the following Theorem 3.

Theorem 3. *There exists such curves y^*, only a function of x^*, such that from these solutions of the coastal equation can be derived by multiplying y^* by t^n and X^* by $\sqrt{t}$. In this n may be arbitrary.*

Proof.[d] Let $y = y^* \times t^n$ and $x = x^* \times t^m \to x^* = x \times t^{-m}$.

$$\frac{\partial y}{\partial x} = t^n \frac{dy^*}{dx^*} \times \frac{\partial x^*}{\partial x} = t^{n-m}\frac{dy^*}{dx^*}$$

$$\frac{\partial^2 y}{\partial x^2} = t^{n-m}\frac{d^2y^*}{dx^{*2}} \times \frac{\partial x^*}{\partial x} = t^{n-2m}\frac{d^2y^*}{dx^{*2}}$$

$$\frac{\partial y}{\partial t} = \frac{d(y \times t^n)}{dt} = nt^{n-1} \times y^* + t^n \frac{dy^*}{dx^*} \times \frac{dx^*}{dt}$$

In which:

$$\frac{dx^*}{dt} = \frac{\partial(xt^{-m})}{dt} = -mxt^{-m-1} = -m\frac{x^*}{t}$$

[d]The derivation is given by Edelman.[33,34]

By this:

$$\frac{dy}{dt} = t^{n-1}\left(ny^* - mx^*\frac{dy^*}{dx^*}\right)$$

The coastal equation thus gives, see Eq. (3.6)

$$t^{n-1}\left(ny^* - mx^*\frac{dy^*}{dx^*}\right) = \frac{s}{h}t^{n-2m}\frac{d^2y^*}{dx^{*2}}$$

$$ny^* - mx^*\frac{dy^*}{dx^*} = \frac{s}{h}t^{1-2m}\frac{d^2y^*}{dx^{*2}}$$

On the left hand side is a function, independent of t. Also the right hand side must be independent of t.

This can only happen if:

$$m = \frac{1}{2}$$

Furthermore y^* must be a solution of the (ordinary) differential equation

$$\frac{s}{h}\frac{d^2y^*}{dx^{*2}} + \frac{1}{2}x^*\frac{dy^*}{dx^*} - ny^* = 0 \tag{3.7}$$

As was shown to be reasonable from Theorem 2, it appears that m should be 1/2. It will be clear that n may take any arbitrary value; however, the shape of y^* is influenced by n. □

Theorem 4. *Let $y(x,t)$ satisfy the coastal Eq. (3.6). Then the solution of the equation:*

$$\frac{1}{A^2}\frac{s}{h}\frac{d^2y}{dx^2} = \frac{\partial y}{\partial t}$$

Is given by the following functions:

4a: $y(x/A, t)$

4b: $y(x, A^2t)$

This is found from substitution. In Fig. 3.6 these theorems are depicted.

Theorems 2 and 4 may be summarized by reducing Eq. (3.6) to basic form:

$$\frac{d^2y}{dx'^2} = \frac{\partial y}{\partial t'} \tag{3.8}$$

this can be done by introducing $x^1 = x/L_1$ and $t' = t/T_1$ in which should be valid:

$$\frac{L_1^2}{T_1} = \frac{s}{h} \tag{3.9}$$

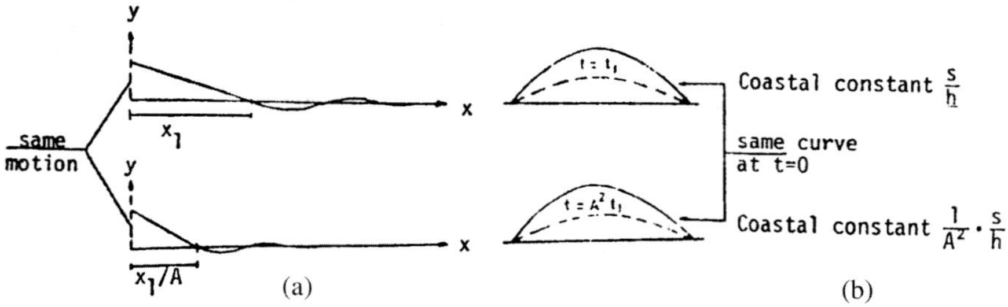

Fig. 3.6 (a) Depicting Theorem 4a. (b) Depicting Theorem 4b.

3.2.5. *Applications of the laws of scale in some simple cases*

The solution of the coastal equation depends largely on the boundary conditions. According to Theorem 1, a case with difficult boundary conditions can be decomposed in a number of cases with simple boundary conditions, after which these solutions may be superimposed. In the course of this section, a number of these simple cases will be dealt with; the superposition is left to the reader. Formulas will be avoided in this section as much as possible.

(a) Reflections by a harbor mole. Given is the coastal shape y_0 as function of time for a coast, being bordered on one side by a changing delta and which remains at rest at the other side (Fig. 3.7). At time $t = 0$ the coast is straight. The direction of wave incidence is parallel to the y-axis (wave crests parallel to the x-axis).

Question: how changes this coast, if at time $t = 0$ a harbor mole is constructed which hampers all transport?

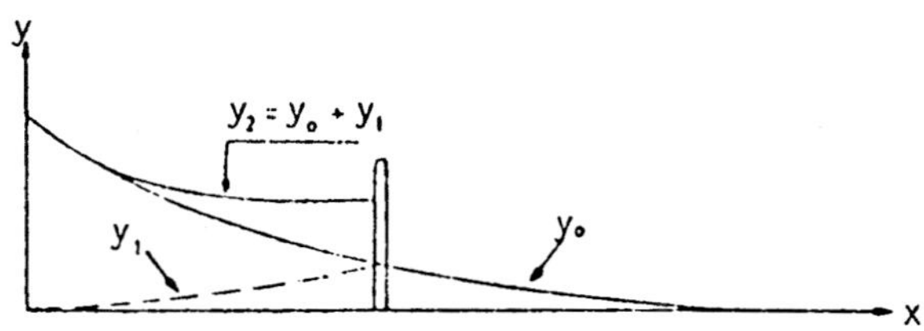

Fig. 3.7 Reflection of the coastal shape by a harbor mole.

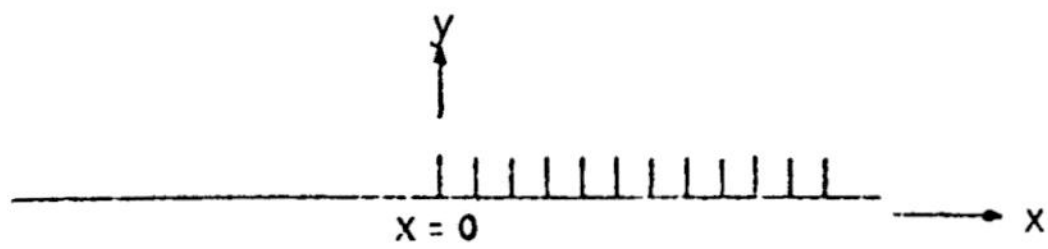

Fig. 3.8 Coast with a groin system.

Solution: at the site of the mole the transport should be zero; this can only be the case if the angle of wave incidence at that site is zero, i.e. if the coast at that site is parallel to the x-axis.

According to Theorem 2 (with substitution $A = -1$) the reflected coastline y_1 is also a solution of the coastal equation; furthermore, as the coastal equation is linear also y_2, equal to $y_0 + y_1$ suffices with respect to this equation (Theorem 1). The solution y_2 obeys as well the condition that the coast as the site of the harbor mole is parallel to the x-axis. Therefore, y_2 is the requested solution.

(b) Partial reflection by a groin system. Instead of constructing a harbor mole, which stops all transport, now a series of groins is constructed at time $t = 0$.

Question: how does the coastline change?

Solution: groins hamper a part of the transport, occurring at a coast without groin.[e] This implies that the longshore transport changes less when the coastal direction changes, i.e. the coastal constant s/h in Eq. (3.6) decreases.

Assume, that $\bar{s}$ along the protected coast is a factor p smaller than along the unprotected coast ($p < 1$)[f] it looks obvious, that along the unprotected coast reflection will occur. Let the reflection coefficient be r_e.

Take the origin of the coordinate system at the site of the first groin, let y_u be the y-coordinate of the unprotected beach and y_p the y-coordinate of the protected beach. Then the coastline of the unprotected beach can be found from the original coastline y_0 by taking the reflection into account:

$$y_u(-x) = y_0(-x) + r_e y_0(x) \tag{3.10}$$

Thus the y-coordinate at the first groin ($x = 0$) becomes $(1 + r_e)\ y_0$. This boundary condition y_p for the protected part implies the same function of time as $y_0(0)$ only $y_\text{p}(0)$ is $(1 + r_e)$ times as large.

---

[e] In Section 4.5. this process will be considered in more detail.

[f] It will appear, that under peculiar circumstances p can be larger than 1 (Section 3.3.4).

According to Theorems 1 and 4a the solution of the protected part will be:

$$y_{\mathrm{p}} = (1 + r_e)\{y_{\mathrm{o}}(x\sqrt{p})\} \tag{3.11}$$

This function suffices, as well with respect to the boundary conditions as with respect to the coastal equation for the protected part.

Thus the magnitude of the reflection coefficient r_e results from the continuity equation at $x = 0$: the longshore transport left and right of the first groin should be equal.

Because the direction of wave incidence is parallel to the y-axis, S_0^* will be equal to zero; thus is found from Eq. (3.1):

$$-s\left[\frac{\partial y_{\mathrm{u}}}{\partial x}\right]_{x=0} = -ps\left[\frac{\partial y_{\mathrm{p}}}{\partial x}\right]_{x=0}$$

Expressing y_{p} and y_{u} in y_0 with the aid of Eqs. (3.10) and (3.11) results into:

$$(1 - r_e)\left[\frac{\partial y_0}{\partial x}\right]_{x=0} = -\frac{p(1 + r_e)}{\sqrt{p}}\left[\frac{\partial y_0}{\partial x}\right]_{x=0} \tag{3.12}$$

from which r_e is found:

$$r_e = \frac{1 - \sqrt{p}}{1 + \sqrt{p}} \tag{3.13}$$

Fig. 3.9 shows the construction of the requested coastline from y_0.

(c) Accretion of a coast near a harbor mole. Consider a coast which is straight at time $t = 0$ with a wave impinging obliquely on the coast (Fig. 3.10). At a time $t = 0$, a harbor mole is constructed which acts so as to stop all transport.

Required: the Coastal shape.

Solution: on the site of the harbor mole the transport is zero at any time. Therefore at this site the coast should be parallel to the direction of the wave crests. At long distances from the harbor mole the coast is parallel to the x-axis. As the direction of the tangents to the curve may not change, this means that the scale of the coastal shape in the y-direction should increase in the same way as in the x-direction, therefore with $\sqrt{t}$. Therefore, the total area increases with a factor of t. This can also be found using continuity per unit of time: the harbor mole captures an amount of sand S_0^*.

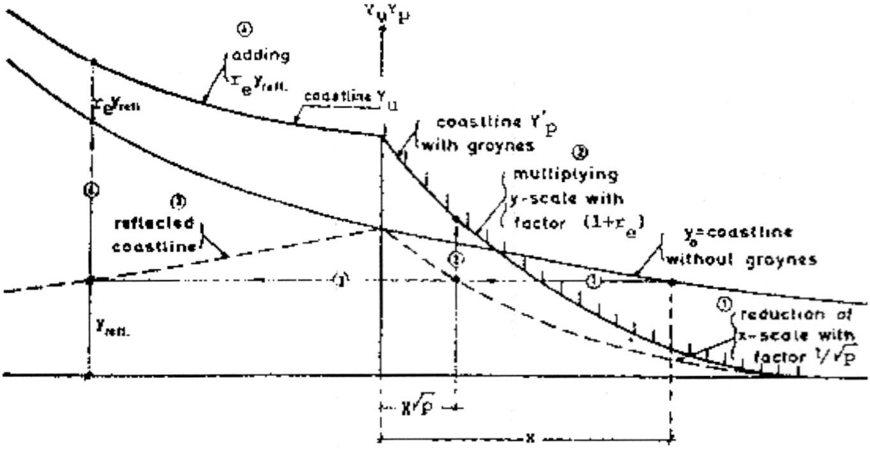

Fig. 3.9 Construction of coastal shape of a partly protected beach from a given shape y_0 which would occur if the coast would not be protected.

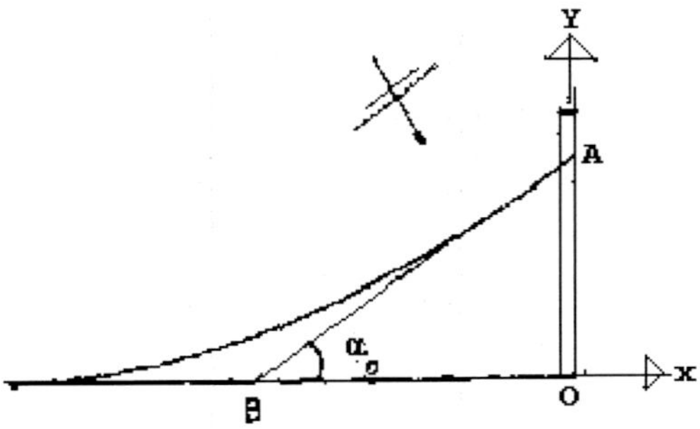

Fig. 3.10 Coastal shape near a harbor mole. The tangents at $x = 0$ and $x = \infty$ are parallel to the wave direction and the x-axis, respectively.

The area, enclosed by the x- and y-axis is approximately 1.5 times as large as area OAB (Fig. 3.10) (or as per the computation in section (a), which follows, $\pi/2$ times as large).

The area is therefore:

$$S_0^* \times \frac{t}{h} = \frac{\pi}{2} \times \frac{1}{2}\frac{OA^2}{tg\alpha_0}$$

from which:

$$OA = \frac{2}{\sqrt{\pi}} \times \sqrt{S_0^* \times tg\alpha_0/h} \times \sqrt{t} \qquad (3.14a)$$

By the use of Eq. (3.2) the following may be found:

$$OA = \frac{2}{\sqrt{\pi}} \times tg\alpha_0 \sqrt{\left(\frac{s}{h} \times t\right)} \tag{3.14b}$$

Alternately:

$$OA = \frac{2}{\sqrt{\pi}} \times S_0^* \sqrt{\frac{t}{sh}} \tag{3.14c}$$

Example: $S_0^* = 200.000\,\mathrm{m}^3/\mathrm{yr}$, $s = 5 \times 10^6\,\mathrm{m}^3/\mathrm{yr}/\mathrm{rad}$, $h = 5\,\mathrm{m}$;

Therefore OA $= 45\sqrt{t}\,\mathrm{m}$

In the Sections (d)–(f) cases are treated which belong to the same "family" of solutions.

(d) Erosion behind a harbor mole (Fig. 3.11a). Figure 3.11a speaks for itself: erosion behind a harbor mole (anti-symmetric with the accretion).

(e) A river delta with constant sand supply S_d^* per unit of time (Fig. 3.11b). The accretion at the site of the river mouth given by:

$$OA = \frac{1}{\sqrt{\pi}} \times S_d^* \sqrt{\frac{t}{sh}} \tag{3.15}$$

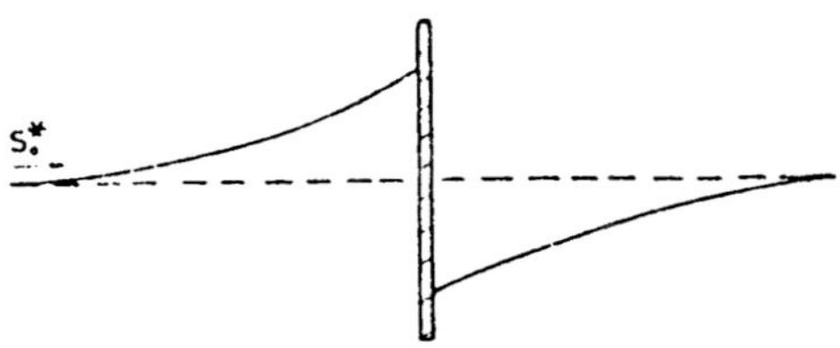

Fig. 3.11a Erosion behind a harbor mole.

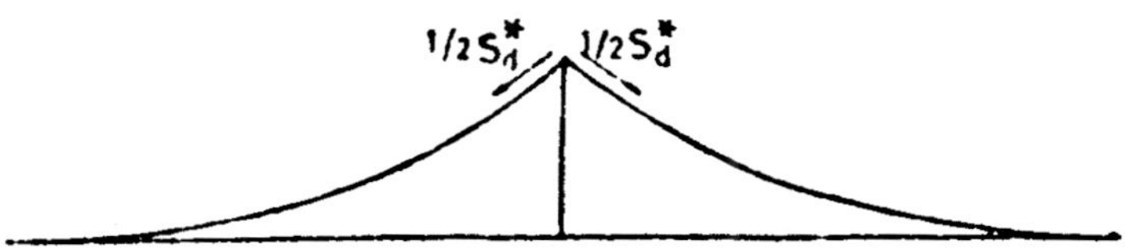

Fig. 3.11b The shape of a river delta.

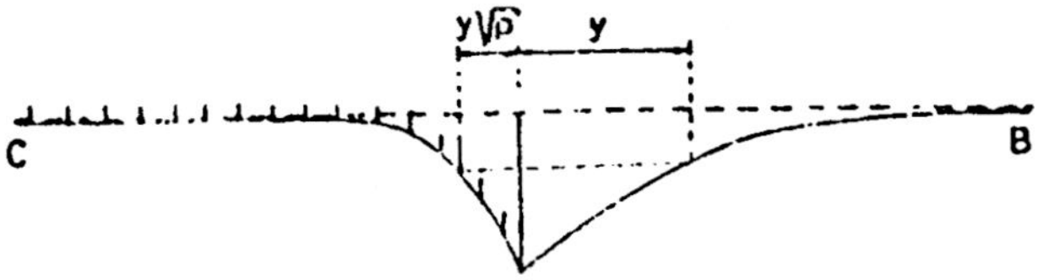

Fig. 3.11c Erosion behind the last groin.

(f) Erosion behind the last groin of a beach groin system (Fig. 3.11c). At the site of the last groin the erosion is $(1-\sqrt{p})$ times the erosion behind a harbor mole (p is defined in (Subsection 3.2.5 ad.b). This is shown in the following: the horizontal scale of the protected part is $\sqrt{p}$ times the scale of the unprotected part (Theorem 4a) Assume that S^*_{unpr} per year erodes from the unprotected part. Therefore $\sqrt{p} \times S^*_{\text{unpr}}$ erodes from the protected part.

Furthermore the transport at point B in Fig. 3.11c (at large distance from the first groin) is according to (3.1) equal to S_0^*. At point C in Fig. 3.11c (also a great distance from the first groin, but on the protected coast) the transport is equal to $p \times S_0^*$. Per unit of time erosion is therefore:

$$(1-p)S_0^* = (1+\sqrt{p})S^*_{\text{unpr}}$$

Thus the eroded pit becomes:

$(1-p)/(1+\sqrt{p}) = (1-\sqrt{p})$ times as deep as in the case of a long harbor mole.

(g) Accretion of a coast, which has reached the head of a groin. Assume that the length of the groin is land the time (after the construction of the groin) for the coastline to reach the head of the groin is t_1. Thus by reference to (3.14b) t_1 may be found as follows.

$$t_1 = \frac{\pi l^2}{4tg^2\alpha_0} \times \frac{h}{s} \tag{3.16}$$

The coast will accrete according to Fig. 3.12a. Note that at a point corresponding to $x=0$ the coast remains at the same site. Thus the coast does not accrete in the y-direction. By application of the Law of scale it is logical to assume a coastal shape whereby the scale increases in the x-direction (in proportion with $\sqrt{t}$), but not in the y-direction.

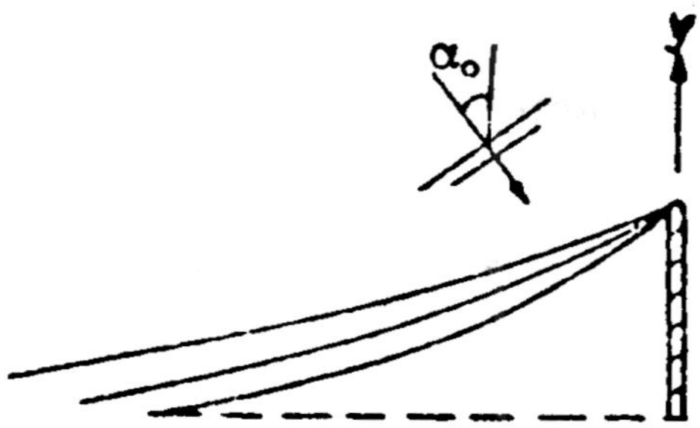

Fig. 3.12a Accretion of a coast, after it reached the head of the mole.

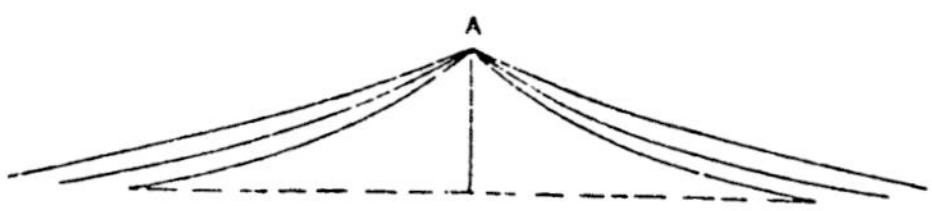

Fig. 3.12b Instantaneous dumping of sand at one place.

The solution y^* of Eq. (3.7) for $n = 0$ could be shown to apply here. However, the coastal shape before $t = t_1$ was a solution of Eq. (3.7) for $n = 1/2$ and both solutions (for $n = 0$ and $n = 1/2$) do not coincide. After a short time past $t = t_1$ the shape $[y^*]_n = 1/2$ will change gradually to the shape $[y^*]_n = 0$.

Once the shape is adapted the gradient ∂_y/∂_x will decrease in proportion to $1/\sqrt{t}$. Therefore, the transport along the groin will increase, following (3.3) and proportional to $\alpha_0\,(1 - \sqrt{t_2/t})$ where t_2 is a proportionality constant with dimension "time" $(t > t_2)$.

Remark. Analogous to Section 3.5.7 it may be demonstrated that the sand-supply needed to keep point A in Fig. 3.12b at its present position, decreases in proportion with $1/\sqrt{t}$.

(h) Instantaneous dumping of sand at one place. At a certain time a quantity of sand S^{**} is dumped on a beach in the shape of a Gauss curve as per Fig. 3.13. The Gauss curve is essentially a solution of Eq. (3.7) for $n = -1/2$.

In the course of time this quantity disperses over a larger part of the coast, i.e. the scale increases with $\sqrt{t}$, the y-scale should thus decrease with $1/\sqrt{t}$ in order to hold the dumped quantity constant. The coastal shape

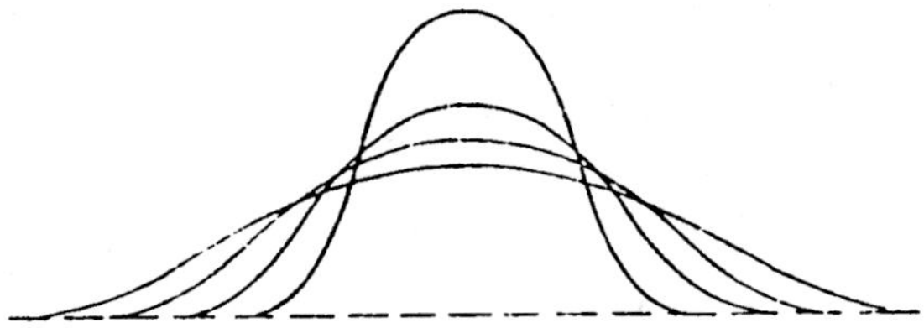

Fig. 3.13 Instantaneous dumping of sand at one place.

according to Section 3.2.6 can be expressed in the following formula:

$$y = \frac{S^{**}}{2\sqrt{\pi sht}} \times e^{\frac{-hx^2}{4st}} \tag{3.17}$$

Example: $S^{**} = 200.000\,\mathrm{m}^3$, $s = 5 \times 10^6\,\mathrm{m}^3/\mathrm{yr}$, $h = 5\,\mathrm{m}$;

$$y = \frac{11.25}{\sqrt{t}} \times e^{\left(\frac{-x}{2000t}\right)^2} \quad (x \text{ and } y \text{ in } m,\ t \text{ in years})$$

Figure 3.14 shows in the same figure the delta according to Section 3.2.5 ad.c and the dumped material as per (3.17). If the quantities S_d^* and S^* dumped in one year are equal, the accretion at the site $x = 0$ is for the delta twice as large as that for the dumped material. the most seaward point of the dump attenuates with $1/\sqrt{t}$.

The points of contraflexure on the Gauss curve are at distance $\pm x_b$ from $y = 0$, where:

$$x_b = \sqrt{2\frac{s}{h}t} \tag{3.18}$$

About 2/3 of the dumped material lies between $x = -x_b$ and $x = +x_b$. In the above mentioned example, x_b equals $1.4\sqrt{t}\,\mathrm{km}$.

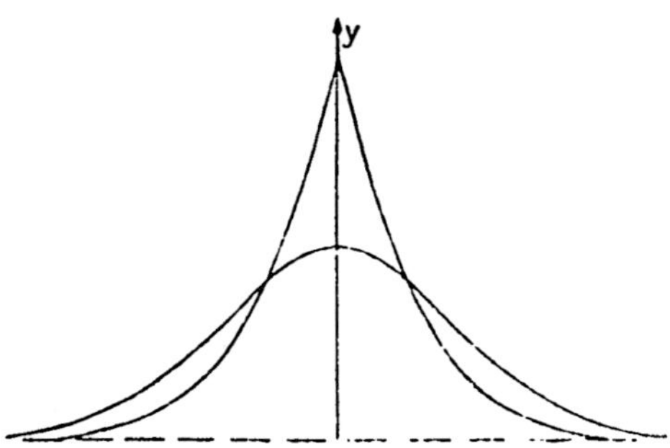

Fig. 3.14 The shape of the coastline after an instaneous dumping and after continous dumping, both after one year (same amount per year).

3.2.6. *Analytical basic solutions*

As shown above, the solution of the coastal equation depends on boundary conditions and initial condition. For practical use, however, the basic "tools" of two kinds of analytic solutions and a numerical method will be mostly sufficient. In this section, these two kinds of analytic solutions will be displayed, where in Section 3.2.7 a numerical solution will be given. This method can be used also in a graphical way.

Analytical solutions can be found as well by use of Laplace transforms.[40] Some of the results, displayed in this chapter, have been derived in this way. For details, refer to Ref. [7].

(a) The error function family. The kind of analytical solutions, which will be mentioned in this section belong to the same "family" because all are integrals or derivatives of each other.

Theorem 5. *If $y(x,t)$ suffices the coastal Eq.* (3.6), *then the following functions suffice this equation as well: the partial derivatives of y to x and t; the integrals of y to x and t, provided that the integration "constants" are indeed constant and are (generally) no functions of t or x, respectively.*

Proof. If Eq. (3.6) is valid, then (3.6) remains valid, when the left-hand side and right-hand side are differentiated (c.q. are integrated under the conditions, mentioned ad 2°). Changing the succession of differentiation (c.q. integration) yields the proof of Theorem 5.

A simple way of getting solutions from (3.6) is solving Eq. (3.7). The simplest solution is found when $n = 0$; in this case the equation reads:

$$\frac{s}{h}\frac{d^2y^*}{dx^{*2}} + \frac{1}{2}x^* \times \frac{dy^*}{dx^*} = 0 \tag{3.19}$$

Assume $dy^*/dx^* = p_1$, then it is found from (3.19):

$$\frac{s}{h}\frac{dp_1}{p_1} = -\frac{1}{2}x^* \times dx^* \to p_1 = Ae^{-\frac{x^{*2}h}{4s}}$$

$$y^* = A\int e^{-x^{*2}h/4s}dx^*$$

Substitution of $y^* = y\,t^{-n}$ and $x^* = x\,t^{-m}$ with $n = 0$ and $m = 1/2$ yields for y:

$$y = A'\int e^{-x^2h/4st}d\left(x \times \sqrt{\frac{h}{4st}}\right) \tag{3.20a}$$

If for A' is chosen: $A' = 2/\sqrt{\pi}$ and if the integration boundaries are chosen as shown below, then (3.20a) becomes the "complementary error function":

$$y = \frac{2}{\sqrt{\pi}} \int_{x\sqrt{h/4st}}^{\infty} e^{-x^2 h/4st} d\left(x \times \sqrt{\frac{h}{4st}}\right) \tag{3.20b}$$

$$y = \operatorname{erfc}\left(x \times \frac{h}{4st}\right) \tag{3.20c}$$

In order to explain the shape of the function at the right-hand side of (3.20c), this function will be differentiated with respect to x; according to Theorem 5, this again yields a solution of the differential equation[g]:

$$y = \frac{A''}{\sqrt{t}} e^{-x^2 h/4st} \tag{3.21}$$

Figure 3.15a displays the solution (3.21). Figure 3.15b gives the curve $y = \exp(-x^2)$. As (3.20b) shows, the hatched area in this figure equals $\sqrt{\pi}/2$ times the y-coordinate of erfc x; last mentioned function is given in Fig. 3.15d. However, in the exponent of (3.21) one finds "x_2/t" instead of purely "x_2" and therefore, when integrating according to (3.20b) the factor "$1/\sqrt{t}$" in (3.21) disappears in (3.20c).

The solution given by (3.20c) and shown in Fig. 3.15c therefore appears to be independent of time at $x = 0$.

Repeated integration of (3.20c) to x gives the solution[h,i]:

$$y = \sqrt{t}\, i^1 \operatorname{erfc} x \times \sqrt{\frac{h}{4st}} \tag{3.22}$$

i^1 erfc x is the first integral of the error function, the hatched area in Fig. 3.15d. According to the theory of the error functions, the total integrated area of erfc x between 0 and ∞ equals $\sqrt{\pi}/2$.

The integration can be continued, giving functions as[j]:

$$y = t^{\frac{1}{2}n'}\, i^{n'} \operatorname{erfc} x \times \sqrt{\frac{h}{4st}} \tag{3.23}$$

[g] Here "y" simply denotes "a solution of the coastal equation". More formally the solutions of (3.20a) and (3.21) should be marked off from each other, by a subscript for instance. However, no confusion is forseen.

[h] The constant factor $2\sqrt{(s/h)}$ is omitted.

[i] i^1erfc x is a usual notation for this kind of functions; any coincidence with the "i" of complex numbers is purely accidental.

[j] The constant factor $(2\sqrt{(s/h)})n'$ is omitted.

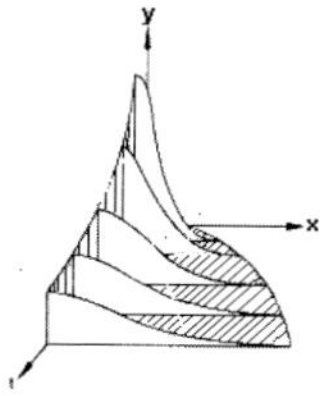

Fig. 3.15a. $y = \frac{1}{\sqrt{\pi t}} e^{-x^2/4\pi t}$

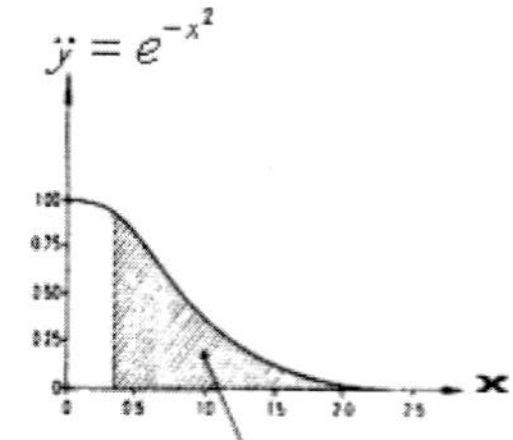

Fig. 3.15b.

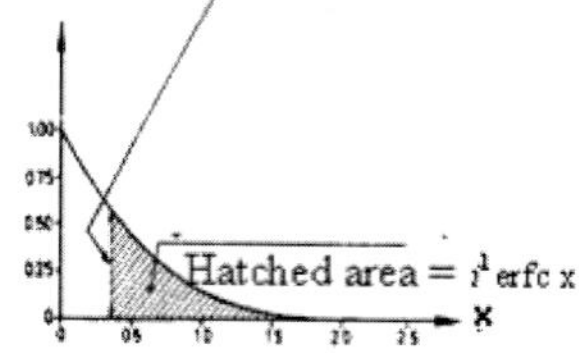

Fig. 3.15c.

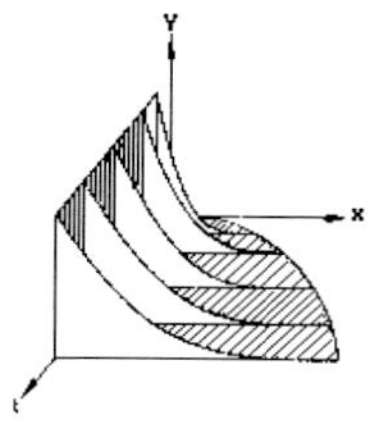

Fig. 3.15d. $y = \text{erfc}\frac{x}{2\sqrt{t}}$

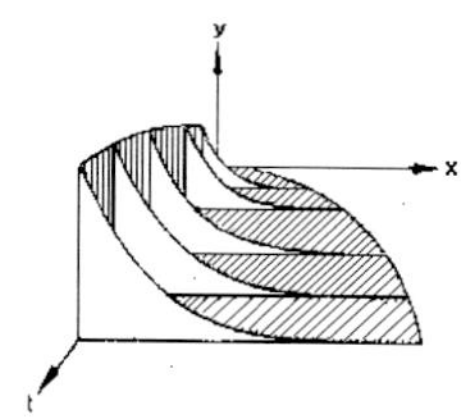

Fig. 3.15e. $y = \sqrt{t}.i^1\text{erfc}\frac{x}{2\sqrt{t}}$

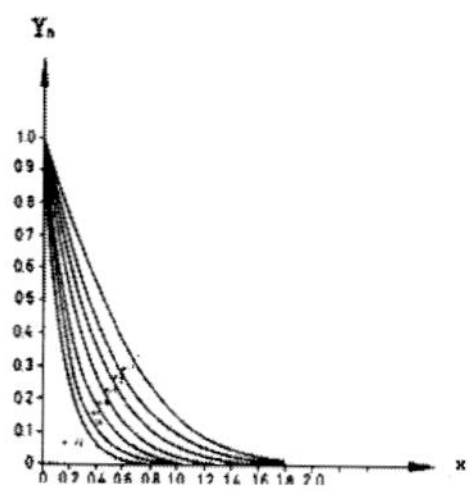

Fig. 3.15f. Repeated integrals of the error function.

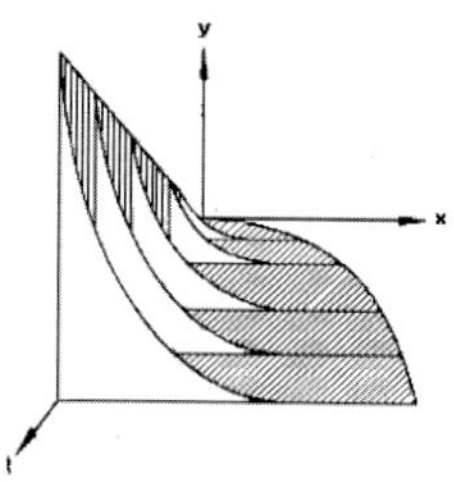

Fig. 3.15g. $y = t.i^2\text{erfc}\frac{x}{2\sqrt{t}}$

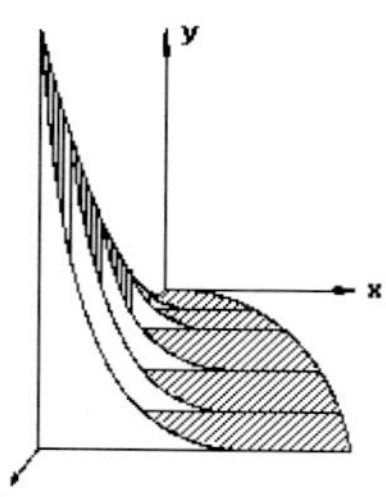

Fig. 3.15h. $y = t\sqrt{t}\times i^3\text{erfc}\frac{x}{2\sqrt{t}}$

Fig. 3.15 Basic solutions of the coastal equation.

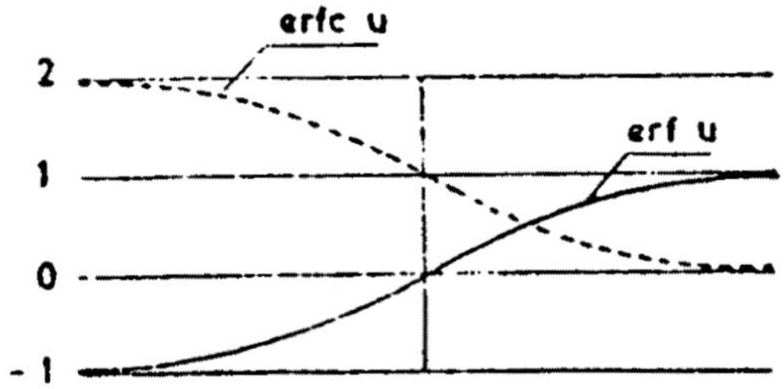

Fig. 3.16 The error function and the complementary error function.

With this function, at $x = 0$ the y-coordinate increases accordingly:

$$[y]_{x=0} = t^{\frac{1}{2}n'}$$

It thus shows, that the solutions of (3.7) for $n = 1/2n'$ are found as:

$$y^* = i^{2n}\,\mathrm{erfc}(x^*\sqrt{h/s}) \tag{3.24}$$

For the mathematical properties of the error function and related functions is referred to Abramowitz and Stegun (1971), Chapter 7, from which the following is derived.

The error function cf. Fig. 3.16 is defined as:

$$\mathrm{erf}\,u = \frac{2}{\sqrt{\pi}}\int_0^u e^{-u^2}\,du \tag{3.25}$$

The error function can be developed in a series in the following way:

$$\mathrm{erf}\,u = \frac{2}{\sqrt{\pi}}\int_0^u e^{-u^2}\,du = \frac{2}{\sqrt{\pi}}\sum_{n=0}^{\infty}\int_0^u \frac{(-u^2)^n}{n!} = \frac{2}{\sqrt{\pi}}\sum_{n=0}^{\infty}\frac{(-1)^n u^{2n+1}}{(2n+1)\,n!} \tag{3.26}$$

The first approximation of erf u for small values of u is therefore $2u/\sqrt{\pi}$.

The function increases when u goes to infinity. It is strongly related to a function often used in error theory: the probability of exceedance of a value x by a normal distributed stochastic variable $\underline{x}$ with mean m and standard deviation σ equals:

$$P_r(\underline{x} > x) = \frac{1}{2}\,\mathrm{erfc}\left(\frac{x-m}{\sigma\sqrt{2}}\right) \tag{3.27}$$

where

$$\text{erfc}\ u = 1 - \text{erf}\ u. \tag{3.28}$$

Integration of erfc from u to ∞ gives i^1 erfc u. Direct integration of the error function, using partial integration gives:

$$i^1\ \text{erfc}\, u = \int_u^\infty \text{erfc}\, u\, du = u\,\text{erfc}\, u\Big|_u^\infty - \frac{2}{\sqrt{\pi}}\int_u^\infty ue^{-u^2} du$$

$$i^1\ \text{erfc}\, u = u\,\text{erfc}\, u + \frac{1}{\sqrt{\pi}} e^{-u^2} \tag{3.29}$$

The integrals of the error function are tabulated by Abramowitz and Stegun,[1] Table 7.4. They are "normalized" to the functions Y_n (Fig. 3.15f), i.e. multiplied by a constant in this way, that $Y_n(0)$ equals 1. The relation between i_n erfc u and $Y_n(u)$ appears to be:

$$Y_n(u) = 2^n\Gamma\left(\frac{n}{2}+1\right) i^n\ \text{erfc}\, u \tag{3.30}$$

where $\Gamma(n/2+1) = (n/2)!$ if n is even $(n \geq 2)$, and $\Gamma(n/2+1) = n/2 \times (n/2-1)(n/2-2)\ldots 3/2\sqrt{\pi}$ if n is odd $(n \geq 3)$.

From Abramowitz and Stegun[1] the following recurrences can be derived:

$$Y_n(u) = Y_{n-2}(u) - 2\frac{u}{n}\frac{\Gamma\left(\frac{n}{2}+1\right)}{\Gamma\left(\frac{n}{2}+\frac{1}{2}\right)} \times Y_{n-1}(u) \tag{3.31}$$

Instead of integrating (3.21), this function can be differentiated as well. Then one finds functions as displayed in Fig. 3.17. However, the velocity of attenuation increases with the number of times, that (3.21) is differentiated: the y-coordinate is proportional to $t^{-n/2}$, when one differentiates n times. □

Example: Let a certain quantity of sand be dumped on the beach in an arbitrary shape. The coastal shape can be thought of to be composed of a Gauss curve plus a number of curves of the kind shown in Fig. 3.17. In course of the time a kind of "survival of the fittest" takes place: the Gauss curve attenuates the slowest and will be visible therefore during the longest time. In whatever shape the sand is dumped, after some time, a Gauss curve will result.

(b) Periodical solutions. Quite another way of solution of the coastal Eq. (3.6) is splitting the equation into variables.

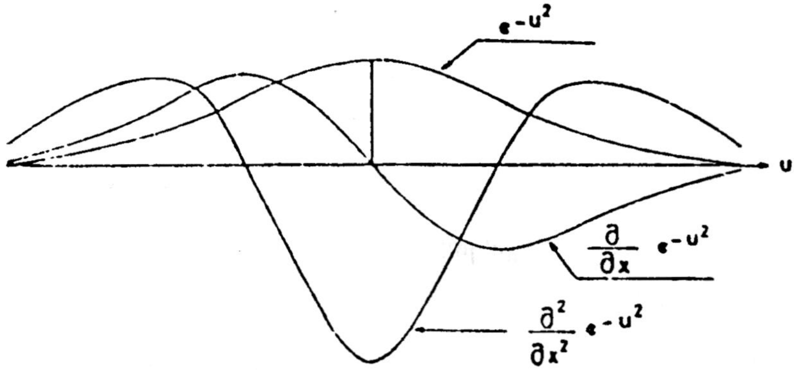

Fig. 3.17 The derivatives of the Gauss curve.

Assume:

$$Y = X \times T \tag{3.32}$$

where X and T are only functions of x and t, respectively. Substitution in (3.6) yields:

$$X\frac{dT}{dt} = \frac{s}{h}T \times \frac{d^2X}{dx^2}$$

$$\frac{h}{s} \times \frac{1}{T} \times \frac{dT}{dt} = \frac{1}{X}\frac{d^2X}{dx^2} \tag{3.33}$$

The left-hand side of (3.33) is a function of t, the right-hand side a function of x; therefore, the equation only holds if the left-hand side and right-hand side are a constant. Mention this constant: $-k_2$. Then one finds for X and T:

$$T = e^{-k^2\frac{s}{h}t}; \quad X = A_1e^{ikx} + B_1e^{-ikx}$$

$$\text{or:} \quad y = (A_1e^{ikx} + B_1e^{-ikx})e^{-k^2\frac{s}{h}t} \tag{3.34}$$

The integration constants A, B, and k can be complex.

Consider first the case, that k is real; an appropiate choice for A_1 and B_1 gives:

$$y = e^{-t/T_0}(A\cos ky + B\sin ky) \tag{3.35}$$

Let k be: $k = 2\pi/L$, then for T_0 it is found:

$$T_0 = \frac{L^2}{4\pi^2} \times \frac{h}{s} \tag{3.36}$$

This function is displayed by a standing, attenuating wave (Fig. 3.18).

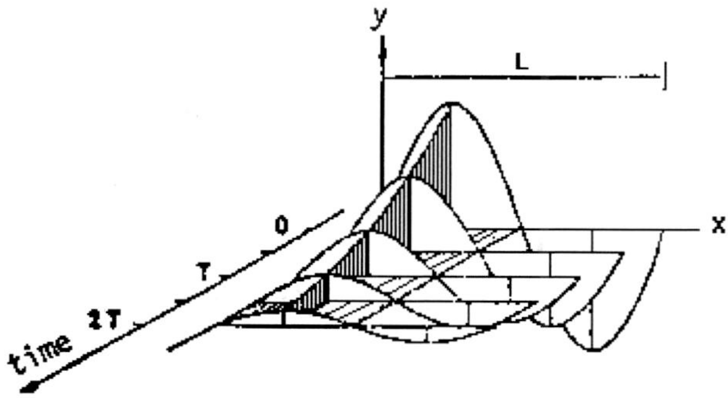

Fig. 3.18 Attenuation of the wave shape of a sinusoidally shaped coastline.

Example: $L = 200\,\text{m}$; $h = 5\,\text{m}$, $s = 5.106\,\text{m}^3/\text{yr}$

The wave attenuates each 10^{-3} year = 8 hours a factor e. If $\lambda = 2\,\text{km}$ this lasts for 10^{-1} year $\cong$ 1 month.

N.B.: If s is negative (large angle of wave incidence; transport increases when the angle of wave incidence decreases) then T_0 will be negative, i.e. y increases for increasing t, which means, that the coast is unstable!

Consider now the case, that k is complex. Let k^2 be $-2ik_1^2$, in which k_1 is real (i.e. k^2 is barely imaginary), then k will be:

$$k = \pm(1 - i)k_1 \tag{3.37}$$

Then (3.34) reads:

$$y = e^{2ik_1 \frac{2st}{h}} \left\{A_1 e^{k_1 x(l+i)} + B_1 e^{-k_1 x(l+i)}\right\}$$

Retain the real part:

$$y = A_1 e^{k_1 x} \cos(k_1 x + \omega_1 t) + B e^{-k_1 x} \cos(-k_1 x + \omega_1 t) \tag{3.38}$$

in which $\omega_1 = \frac{2\pi}{T_1} = 2k_1^2 \times \frac{s}{h}$

or, after substitution of $L_1 = 2\pi/k_1$:

$$L_1 = \sqrt{4\pi \frac{s}{h} T_1} \tag{3.39}$$

The first term represents a sandwave, travelling and attenuating in negative x-direction, the second one a wave, travelling and attenuating in positive x-direction. Figure 3.19 shows the behavior of the second term in the course of the time. After ca 1/2 L_1 the wave is attenuated. The propagation

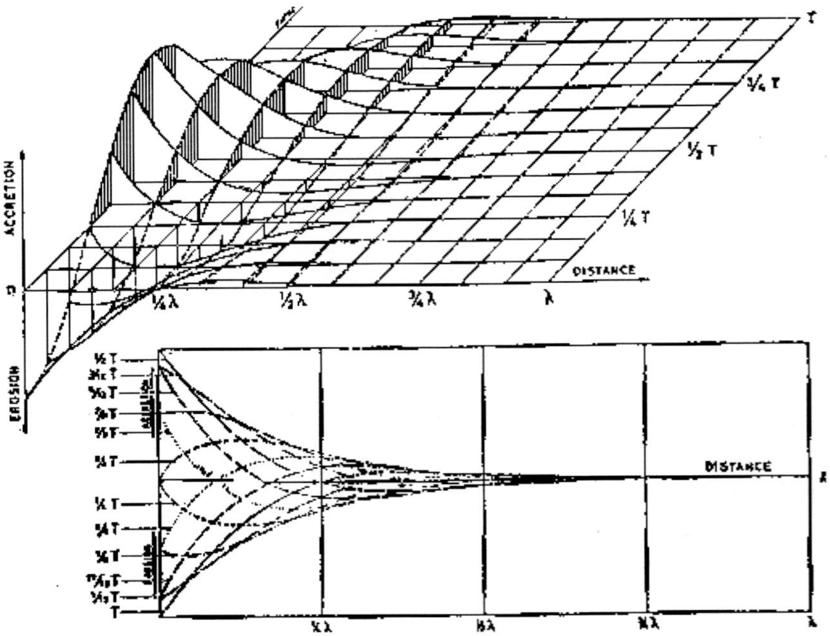

Fig. 3.19 Shape of sandwave as function of time $\left(y = e^{-2\pi\frac{x}{L_1}}\cos 2\pi\left(\frac{x}{L_1} - \frac{t}{T}\right)\right)$.

velocity L_1/T_1 is proportional to $1/\sqrt{T_1}$ according to (3.39). The wave propagates therefore slower, when the period is larger.

Example:

$T_1 = 1$ year, $s = 5.106\,\text{m}^3/\text{yr}$, $h = 5\,\text{m}$

Then: $L_1 = 3.5\,\text{km}$.

3.2.7. *Graphical solutions*

The coastal Eq. (3.6) can be solved very simply numerically or even graphically. In these notes, only a simple graphical method will be explained. This method is mostly credited to Schmidt[38]; the first application is by BINDER.[26,k,l]

[k]Referred is also to Langejan[42] and de Josselin de Jong[40] dealing about ground water flow; much of the exposure given in Section 3.2.7 originates from this paper.

[l]The method is treated here because it gives very much feeling for the feature of diffusion. Needless to say that in the number-chruncher's era there are more accurate ways. However, nature will not see the difference.

It is a forward differential method, where from three points of the coastline at a certain time the situation of one point at a successive time is calculated.

The coastal Eq. (3.6) is written as a differential equation:

$$\begin{aligned} \frac{b_{t+\Delta t} - b_t}{\Delta t} &= \frac{s}{h} \frac{\frac{c_t - b_t}{\Delta x} - \frac{b_t - a_t}{\Delta x}}{\Delta x} \\ \frac{b_{t+\Delta t} - b_t}{\Delta t} &= \frac{s}{h} \frac{a_t - 2b_t + c_t}{(\Delta x)^2} \\ b_{t+\Delta t} &= \frac{s}{h} \frac{\Delta t}{(\Delta x)^2}(a_t + c_t) + \left(1 - 2\frac{s}{h}\frac{\Delta t}{(\Delta x)^2}\right) b_t \end{aligned} \tag{3.40}$$

Here Δx and Δt can be chosen rather arbitrarily; however, if Δt is chosen too large with respect to Δx, then the numerical process becomes unstable. The largest Δt, for which stability is guaranteed equals:

$$\frac{s}{h} \times \frac{\Delta t}{(\Delta x)^2} = \frac{1}{2} \rightarrow \Delta t = \frac{(\Delta x^2)}{2\, s/h} \tag{3.41}$$

In this case, the solution becomes very simply, as is shown by substituting (3.41) into (3.40):

$$b_{t+\Delta t} = \frac{1}{2}(a_t + c_t) \tag{3.42}$$

Every point $B_{t+\Delta t}$ of the coastline at time $t+\Delta t$ now can be found from two neighboring points A_t and C_t lying at the coastline at time t, at a distance Δx from the considered point. For this goal the line $A_t C_t$ is drawn; where this line intersects the x-coordinate of B_t is the point wanted point $B_{t+\Delta t}$.

In the case that a coastline is given as initial condition, and furthermore the y-coordinate of the coast in the course of the time at both boundaries, the method does not lead to difficulties. However, more explanation is wanted when the transport, i.e. the coastal direction is given at a boundary.

Suppose, for instance, that the coast extends in positive x-direction from $x = 0$ and that the coastal direction is given in $x = 0$. Assuming the point "B" from Fig. 3.20 lying on the y-axis (Fig. 3.21), then the validity exists:

$$\left[\frac{\partial y}{\partial x}\right]_{x=0} = \frac{c_t - a_t}{2\Delta x} \tag{3.43}$$

In this case, however, A is an unknown, non-existing point on the wrong side of the boundary, for the same reason $b_{t+\Delta t}$ from (3.42) cannot be

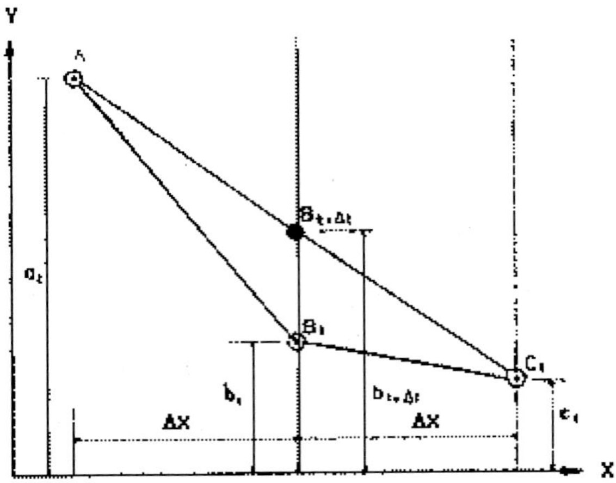

Fig. 3.20 The graphical method of Schmidt.

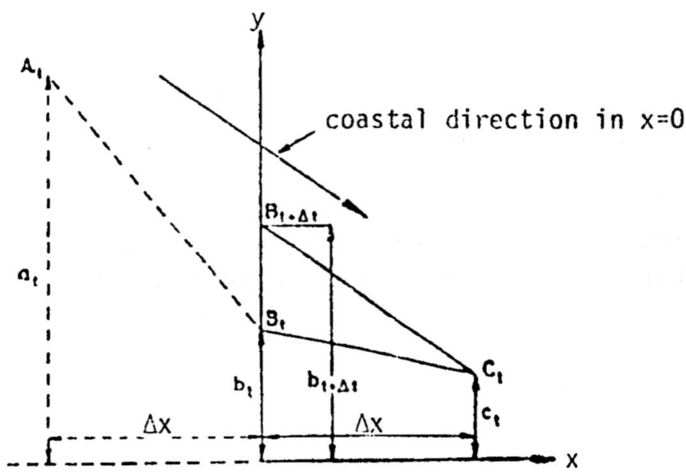

Fig. 3.21 Graphical method of Schmidt for a given transport at one boundary.

determined. However, from Eqs. (3.42) and (3.43) the unknown can be eliminated from which a relation can be derived between the unknown $b_{t+\Delta t}$ and the known values C_t and $[\partial y/\partial x]_x = 0$

$$b_{t+\Delta t} = c_t + \Delta x \left[\frac{\partial y}{\partial x}\right]_{x=0} \quad \text{for } x = 0 \tag{3.44}$$

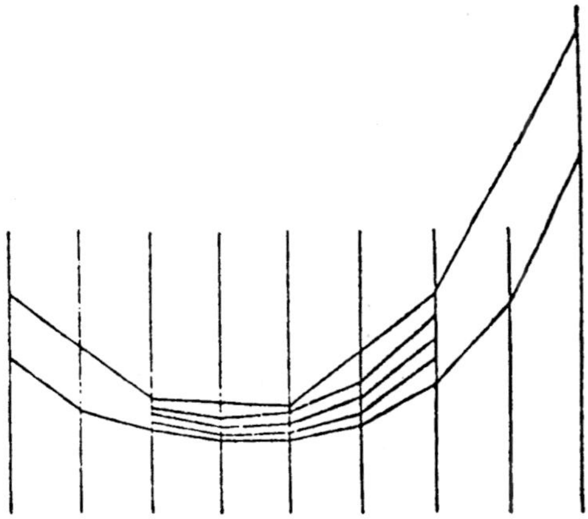

Fig. 3.22 Local refinement of the x-scale.

Therefore (Fig. 3.21), a line drawn through C_t parallel to the tangent to the coast in B_t does not intersect the y-axis in B_t (as one would expect) but $B_{t+\Delta t}$.

The method can be sophisticated in some ways:

- Different values of the coastal constant s/h along the coast.
- Doubling the x-scale, after some time; according to (3.41) this gives a time-scale being 4 times as large (Fig. 3.22).
- Local refining the x-scale. Figure 3.23 gives an example. Only over the refined areas one calculates $y_{t+\Delta t}$, $y_{t+2\Delta t}$ and $y_{t+3\Delta t}$; for the rest only $y_{t+4\Delta t}$.

3.3. 2-Line Theory

3.3.1. *Introduction*

The accretion of a coast all the way up to the end of a groin is highly unlikely to occur. In particular, in the vicinity of the groin the assumption that all the contour lines remain parallel during the accretive process can lead to considerable deviations from reality. This is depicted in Fig. 3.24.

In an investigation of the effect of a structure, which interrupts a part of the littoral drift the capabilities of theory are limited. It is impossible to even qualitively predict anything about factor p from Section 3.2.5b, where p indicates how the transport diminishes when groins are constructed.

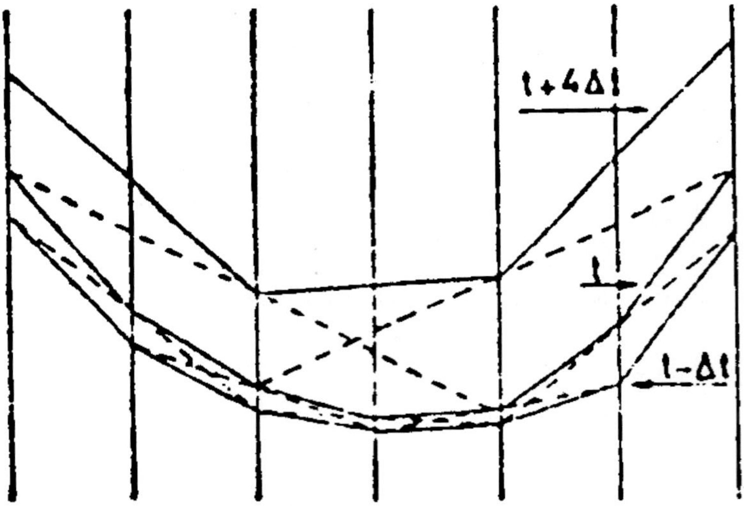

Fig. 3.23 Doubling the x-scale gives a time-scale, being 4 times as large.

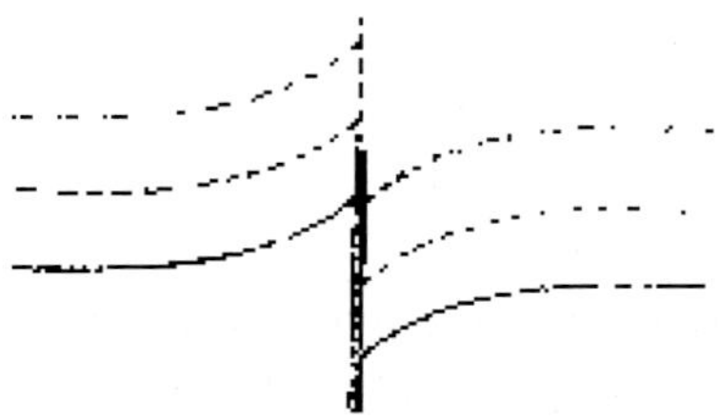

Fig. 3.24a Coastline with parallel depth contours.

In this chapter, we will treat a theory, in which beach and inshore are considered separate and can move separate to each other. However, there is a dependence on each other as offshore transport can occur from beach to inshore and inversely.

3.3.2. *The mechanism of a groin*

The theory of transport near a groin is based on basic concepts developed by Van Bendegom.[22] The concepts are as follows.

Shortly after the construction of a groin on a coast in equilibrium, the groin will stop the sand transport near the beach, but not along the inshore zone. The beach therefore accretes; however the inshore zone does not accrete with result: the profile on the windward side of the groin becoming

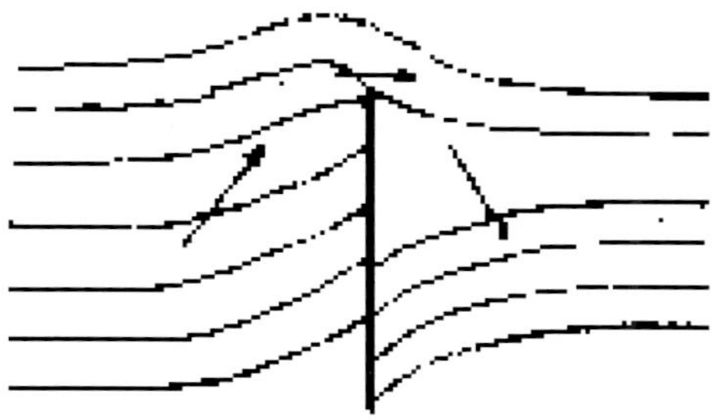

Fig. 3.24b A better approximation. The arrows denote the direction of the sand transport. From Ref. [11].

steeper than the equilibrium profile. On the leeward side of the groin erosion takes place on the beach, causing on this side a profile flatter than the equilibrium profile. Therefore on the windward side sand rolls downward, while on the leeward side it is transported in an upward direction. This is caused by asymmetry of the incident waves, where, high swash velocities alternate with lower backwash velocities (Fig. 3.24b).

Another important mechanism which occurs at the groin is the so-called rip-current. A groin does not only inhibit sandtransport, but also the longshore current. This current will be bent in a seaward direction, and hence a considerable amount of sand will be discharged from beach to inshore.

Both mechanisms will be considered separately. With reference to the rip-current it will only be investigated with regard to the consequences of a certain amount of sand S^*_{rip} being transported from beach to inshore. For the magnitude of S^*_{rip} reference should be made to Chapter 3.5. However for both considered mechanisms, a mathematical model is required which treats beach and inshore separately.

3.3.3. *Definitions and assumptions*

(a) A beach groin is defined as a vertical obstacle which stops sand transport along a beach, but not along the inshore zone (Fig. 3.25).

(b) The coastal profile is divided into three parts, a beach; an inshore zone and a horizontal part in between. By definition the boundary between beach and inshore lies at the construction depth h_1 of the beach groin. The inshore zone lies between h_1 and $h = h_1 + h_2$.

(c) The beach and inshore zone are furthermore schematized as a stepwise profile, according to Fig. 3.26. The areas PQSR and RTUV in Fig. 3.26a are the same as the corresponding areas in Fig. 3.26b. Figure 3.25 merely gives a three-dimensional view.

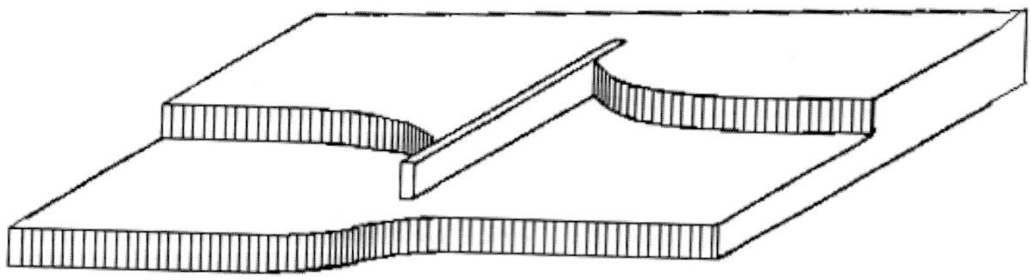

Fig. 3.25 Schematized coastal topography near a groin according to the two line theory. From Ref. [7].

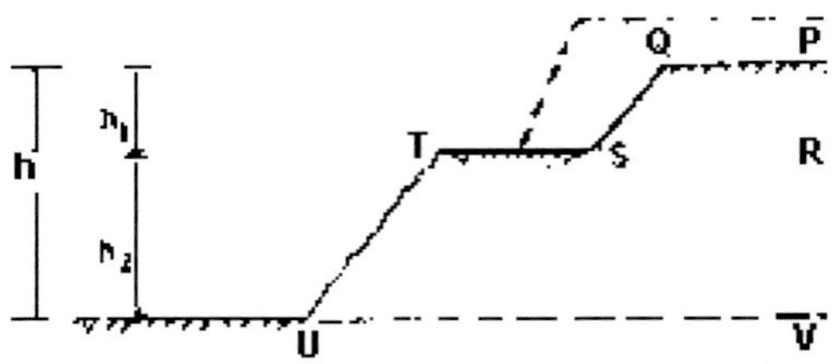

Fig. 3.26a Schematized profile.

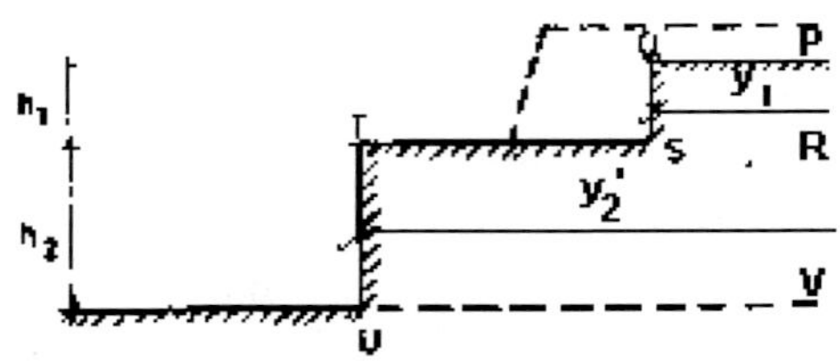

Fig. 3.26b Two-line theory.

(d) In a plan view one observes two lines at a distance y_1 and y_2' from the x-axis. These will be called, respectively, the "line of the beach" and the "line of inshore."

(e) The "equilibrium distance" W is the distance $y_2' - y_1$ between beach and inshore, which exists when the profile is in equilibrium.

(f) The following dynamic equations are assumed. If the distance $y_2' - y_1$ is equal to the equilibrium distance W no interference is assumed. If the distance is smaller than W, then the profile is too steep and an offshore transport will result. A transport in the direction of the coast (onshore) occurs in the inverse case. This relationship is linearized, the offshore transport per unit length is considered to be (Fig. 3.27):

$$S_y = s_y(y_1 - \{y_2' - W\}) \tag{3.45}$$

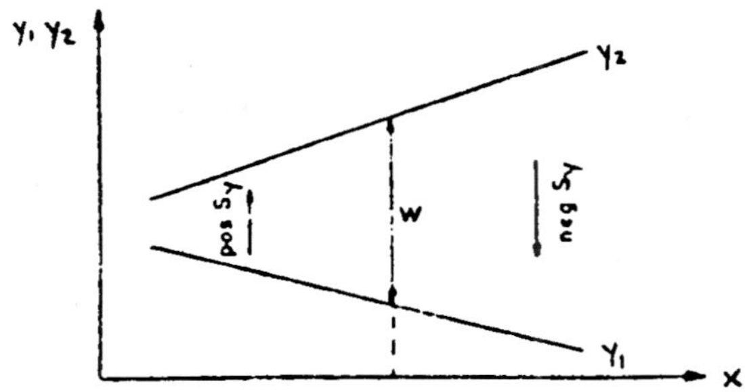

Fig. 3.27 Schematized offshore/onshore transport.

in which s_y is a proportionality constant. The dimension of s_y is (L/t), opposite to s_1, in which the dimension is (L/t^3).
Use of a simple notation gives the following:

$$y_2 = y_2' - W \tag{3.46}$$

$$S_y = s_y(y_1 - y_2) \tag{3.47}$$

(g) With reference to the longshore transport, the assumption of Pelnard-Considère, according to Section 3.2, will be applied for both the beach and inshore zones.

$$\begin{aligned} S_1^* &= S_{01}^* - s_1 \frac{\partial y}{\partial x} \\ S_2^* &= S_{02}^* - s_2 \frac{\partial y_2}{\partial x} \end{aligned} \tag{3.48}$$

Here S_{01}^* and S_{02}^* are the transport when $\partial y_1/\partial x$ and $\partial y_2/\partial x$ are zero. s_1 and s_2 are proportionality constants, analogous to s in Eq. (3.1).

3.3.4. *The stationary case*

(a) General solution. In this section, the stationary case will be tested. With reference to the continuity equation, the following is yielded for, respectively, the beach and inshore:

$$\begin{aligned} \frac{dS_1^*}{dx} + S_y &= 0 \\ \frac{dS_2^*}{dx} - S_y &= 0 \end{aligned} \tag{3.49}$$

After substitution of the transport Eqs. (3.47) and (3.53) one finds the coastal equations:

$$-s_1\frac{d^2y_1}{dx^2} + s_y(y_1 - y_2) = 0$$
$$-s_2\frac{d^2y_2}{dx^2} + s_y(y_2 - y_1) = 0 \qquad (3.50a,b)$$

The solution of these simultaneous differential equations should proceed by substituting y_2 from (3.50a) in Eq. (3.50b) and hence solving this last mentioned equation.

One then finds:

$$y_1 = Ae^{x/L_o} + Be^{-x/L_0} + Cx + E$$
$$y_2 = -\frac{s_1}{s_2}\left(Ae^{x/L_o} + Be^{-x/L_0}\right) + Cx + E \qquad (3.51)$$

In which:

$$L_0 = \sqrt{\frac{1}{s_y\left(\frac{1}{s_1} + \frac{1}{s_2}\right)}} \qquad (3.52)$$

Equation (3.51) can be solved for a number of boundary conditions. For elaboration refer to Refs. [3, 5]. Here only the boundary conditions will be mentioned and the results will be shown.

(b) Rip currents. Rip currents are situated at a distance $2L$ apart, by which an amount of sand S^*_{rip} is transported from beach to inshore (Fig. 3.32).

Boundary conditions:

$$\left[\frac{\partial y_1}{\partial x}\right]_{x=L} = -\frac{S^*_{\text{rip}}}{2s_1}$$
$$\left[\frac{\partial y_1}{\partial x}\right]_{x=-L} = \frac{S^*_{\text{rip}}}{2s_1}$$
$$\left[\frac{\partial y_2}{\partial x}\right]_{x=L} = \frac{S^*_{\text{rip}}}{2s_2}$$
$$\left[\frac{\partial y_2}{\partial x}\right]_{x=-L} = -\frac{S^*_{\text{rip}}}{2s_2}$$

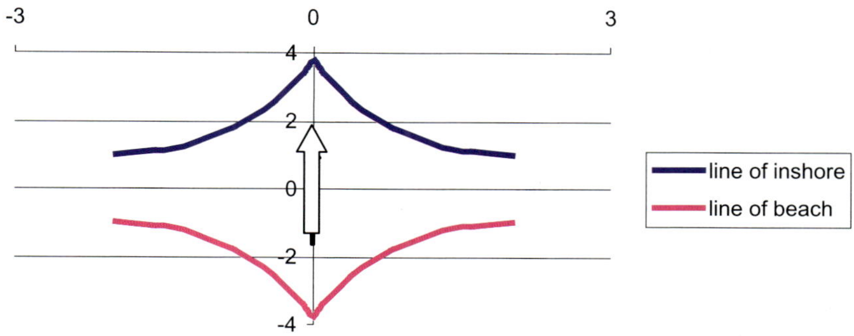

Fig. 3.28 Beach and inshore at a coast with rip-currents ($L = 2L_0$).

Results:

See Fig. 3.28 for $L = 2L_0$ and Fig. 3.37 for $L = \infty$.

In Fig. 3.28, on the horizontal scale is depicted $(x \pm L)/L_0$; the value L/L_0 has been taken equal to 2. The vertical scale is $y_{1,2}$, depicted in a dimensionless way (i.e. $y_{1,2}$ is divided by $S^*_{\text{rip}}/(2s_1L_0 \sinh -2)$).

$$\begin{aligned} y_1 &= \frac{-S^*_{\text{rip}}}{2s_1L_0} \frac{\cosh(x/L_0)}{\sinh(L/L_0)} \\ y_2 &= \frac{S^*_{\text{rip}}}{2s_2L_0} \frac{\cosh(x/L_0)}{\sinh(L/L_0)} \end{aligned} \tag{3.53}$$

The mean distance between the line of the beach and line of inshore will be

$$W + S^*_{\text{rip}}/2s_yL.$$

Rip currents are thus shown to cause a flattening of the profile. The mean distance between beach and inshore can also be derived directly from the offshore transport equation.

In the stationary case as much sand should be transported shorewards as is transported seawards by the rip current.

$$S^*_{\text{rip}} = -\int_{-L}^{+L} S_y dx$$

$$S^*_{\text{rip}} = -s_y \int_{-L}^{+L} (y_1 - y_2) dy$$

Define O_1 and O_2 as the areas in Fig. 3.36, enclosed between original line of the beach (resp. inshore) and the final ones. In other words, $O_1 = 2Ly_1$

mean and $O_2 = 2Ly_2$ mean, in which y_1 mean and y_2 mean indicate the average value of y_1 and y_2 between $x = -L$ and $x = +L$.

Then the following relations can be stated:

$$S^*_{\text{rip}} = -s_y(O_1 + O_2)$$

$$S^*_{\text{rip}} = -2\,L\,s_y(-y_{1\,\text{mean}} + y_{2\,\text{mean}}) \tag{3.54}$$

$$y_{1\,\text{mean}} - y_{2\,\text{mean}} = \frac{S^*_{\text{rip}}}{2\,L\,s_y}$$

(c) Groins[3,5] Groins are situated at a mutual distance of $2L$ apart. Boundary conditions:

$$\left[\frac{\partial y_1}{\partial x}\right]_{x=L} = \frac{S^*_{01}}{s_1}; \quad \left[\frac{\partial y_1}{\partial x}\right]_{x=-L} = \frac{S^*_{01}}{s_1}$$

$$[y_2]_{x=\pm L} = 0; \quad \text{antisymmetry}: \ y_2(x) = -y_2(-x)$$

Results Figs. 3.29 and 3.30:

$$y_1 = \frac{S^*_{01}}{s_1} \times \frac{x + \frac{s_2}{s_1} L \frac{\sinh(x/L_0)}{\sinh(L/L_0)}}{1 + \frac{s_2}{s_1}\frac{L/L_0}{\tanh(L/L_0)}}$$

$$y_2 = \frac{S^*_{01}}{s_1} \times \frac{x - L \frac{\sinh(x/L_0)}{\sinh(L/L_0)}}{1 + \frac{s_2}{s_1}\frac{L/L_0}{\tanh(L/L_0)}} \tag{3.55}$$

For *small* distances between the groins ($L < L_0$), Fig. 3.30, the beach turns in the direction of the wave crests while the inshore remains practically straight. From (3.2) this shows that S^*_{01}/s_1 equals the angle of wave

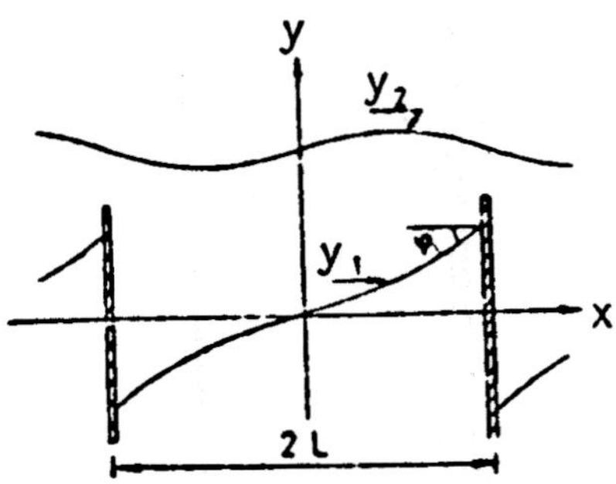

Fig. 3.29 Coast with groins (no rip-currents).

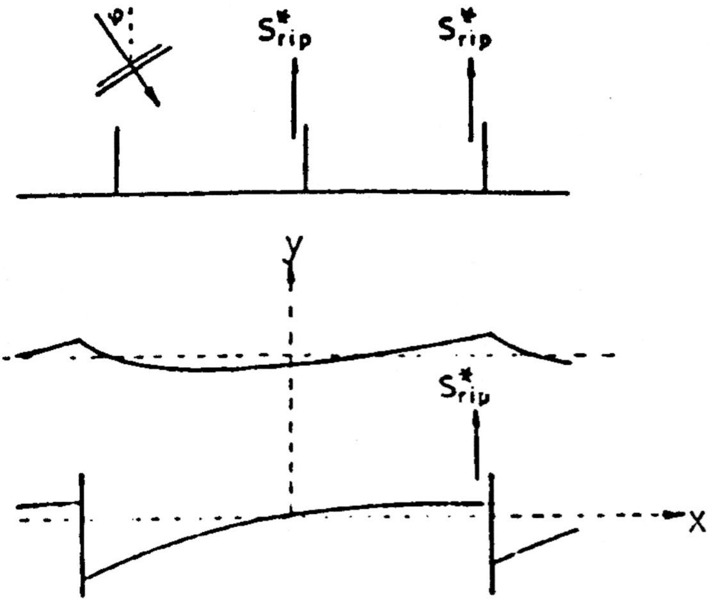

Fig. 3.32 Beach and inshore at a coast with groins.

The coastal shape, as well as the amount of sand that passes through, depend to a great extent on the magnitude of the rip current.

As in this section only stationary cases are considered, the accumulated longshore transport is everywhere equal.

When no groins are constructed this equals:

$$S_0^* = S_{01}^* + S_{02}^*$$

After construction it is changed into:

$$S^* = S_0^* - \left(S_{01}^* - \frac{1}{2}S_{\text{rip}}^*\right)\frac{1+\frac{s_1}{s_2}}{\frac{L}{L_0}\text{cotanh}\frac{L}{L_0}+\frac{s_1}{s_2}} \tag{3.57a}$$

Therefore as long as S_{rip}^* is smaller than 2 S_{01}^*, the transport will decrease after construction of the groins.

Equation (3.57a) can also be written as:

$$S^* = S_{02}^* + S_{01}^*\frac{\frac{L}{L_0}\text{cotanh}\frac{L}{L_0}-1}{\frac{L}{L_0}\text{cotanh}\frac{L}{L_0}+\frac{s_1}{s_2}} + \frac{1}{2}S_{\text{rip}}^*\frac{1+\frac{s_1}{s_2}}{\frac{L}{L_0}\text{cotanh}\frac{L}{L_0}+\frac{s_1}{s_2}} \tag{3.57b}$$

For small values of L/L_0, the value of cotanh (L/L_0) will be practically equal to 1; then the transport shows to be:

$$S^* = S^*_{02} + \frac{1}{2}S^*_{\text{rip}}$$

For large values of L the transport will approach S^*_0.

(d) Transition of two-line theory to one-line theory. In Section 3.2.5, we made the statement that: "Groins hamper a part of the transport which occurs at a coast without groins. This implies that the longshore transport changes less when the coastal direction changes, i.e. the coastal constant s/h in (3.6) changes. It is assumed that s along the protected coast is a factor p smaller than it is along the unprotected coast".

Afterwards it was proved that p influenced the coastal shape considerably. If S^*_{rip} would be taken as zero, then p could be calculated as follows.

Compare Fig. 3.33a with Fig. 3.33b and Fig. 3.33c with Fig. 3.33d.

The transport in Fig. 3.37a is $S^*_{01} + S^*_{02}$, the transport in Fig. 3.33b is given by (3.63) (with $S^*_{\text{rip}} = 0$). In Fig. 3.33c the transport along the beach is $S^*_{01} - s_1\partial y'/\partial x$ to compare with a transport S^*_{01} in Fig. 3.33a. Similarly, the transport along the inshore is $S^*_{02} - s_2\partial y'/\partial x$ instead of S^*_{02} in Fig. 3.33b. Therefore, the transport in Fig. 3.33d can be found from (3.57b) (with $S_{\text{rip}} = 0$), by substituting "$S^*_{01} - s_1\partial y'/\partial x$" for S^*_{01} and "$S^*_{02} - s_2\partial y'/\partial x$" for S^*_{02}. One turns then virtually the x-axis over $\arctan(\partial y'/\partial x)$.

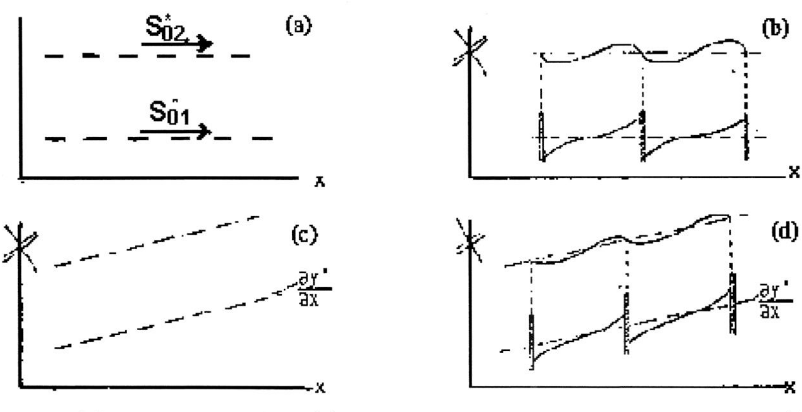

Fig. 3.33 (d) can be derived from (b) by turning the x-axis over an angle arctan $\partial y'/\partial x$, taking the transport along beach and inshore in (c) as "stationary transport" (analogous to S^*_{01} and S^*_{02} in (a)).

Thus one finds, that in the case $S^*_{\rm rip} = 0$, one can write the following for (3.57):

$$S^* = (S^*_0)' - s'\frac{\partial y'}{\partial x} \tag{3.58}$$

in which:

$$(S^*_0)' = S^*_{02} + S^*_{01}\frac{\frac{L}{L_0}\text{cotanh}\frac{L}{L_0} - 1}{\frac{L}{L_0}\text{cotanh}\frac{L}{L_0} + \frac{s_1}{s_2}} \tag{3.59}$$

and

$$s' = \frac{s_1 + s_2}{1 + \frac{s_1}{s_2} \times \frac{L_0}{L}\tanh\frac{L}{L_0}} \tag{3.60}$$

$s' = ps$ in which $s = s_1 + s_2$

$$p = \left(1 + \frac{s_1}{s_2} \times \frac{L_0}{L}\tanh\frac{L}{L_0}\right)^{-1} \tag{3.61}$$

as $S^*_{\rm rip} = 0$

This is therefore the desired result for p; however, this has been found with the assumption that $S^*_{\rm rip} = 0$. This however is a considerable restriction. If one does take $S^*_{\rm rip}$ into consideration, the result will depend very much on the relationship assumed between $S^*_{\rm rip}$ and S^*_{01}. The most logical assumption seems to be that $S^*_{\rm rip}$ is proportional to S^*_{01}.

Hence:

$$S^*_{\rm rip} = s_{\rm rip} \times S^*_{01} \tag{3.62}$$

$S_{\rm rip}$ is therefore a dimensionless coefficient, of which the magnitude can be found by calculating the current velocity in the rip current first (Section 3.8.2.) and then calculate the transport, for instance with the Byker formula.[m] Then in the way as given before one finds the following:

$$(S^*_0)' = S^*_{02} + S^*_{01}\frac{\frac{L}{L_0}\text{cotanh}\frac{L}{L_0} - 1 + \frac{1}{2}s_{\rm rip}\left(1 + \frac{s_1}{s_2}\right)}{\frac{L}{L_0}\text{cotanh}\frac{L}{L_0} + \frac{s_1}{s_2}} \tag{3.63}$$

[m] See also Byker's class notes.[25]

and

$$s' = (s_1 + s_2)\frac{1 + \frac{s_{\rm rip}}{2} \times \frac{s_1}{s_2} \times \frac{L_0}{L}\tanh\frac{L}{L_0}}{1 + \frac{s_1}{s_2} \times \frac{L_0}{L}\tanh\frac{L}{L_0}} \tag{3.64}$$

$$p = \frac{1 + \frac{s_{\rm rip}}{2} \times \frac{s_1}{s_2} \times \frac{L_0}{L}\tanh\frac{L}{L_0}}{1 + \frac{s_1}{s_2} \times \frac{L_0}{L}\tanh\frac{L}{L_0}} \tag{3.65}$$

If $s^*_{\rm rip} < 2s^*_{01}$, p will be smaller than 1 (the coastal processes are slowed down), otherwise p is greater than 1. In the last case, the groins have a deleterious rather than constructive effect.

(e) Permeable groins. Assume, that only a part of the sand transport is inhibited by the groins, for instance because these groins are permeable or are low groins. Assume, that an amount $S^*_{\rm perm}$ is allowed to pass through, where $S^*_{\rm perm}$ is proportional to S^*_{01}:

$$S^*_{\rm perm} = s_{\rm perm}S^*_{01}$$

where $s_{\rm perm}$ is a permeability coefficient.

In case treated in Section 3.3.4 ad.c (no rip-currents) $\partial y_1/\partial x$ in $x = \pm L$ will be a factor $1 - s_{\rm perm}$ smaller. One may consider $S^*_{\rm perm}$ as a part of S'^*_{02}; this means that the formulae keep their validity if read as follows:

$$\begin{aligned} &S^*_{02} + S^*_{\rm perm}, \quad \text{for } S^*_{02} \\ &S^*_{01} - S^*_{\rm perm}, \quad \text{for } S^*_{01} \\ &s_{\rm rip}/(1 - s_{\rm perm}), \quad \text{for } s_{\rm rip} \end{aligned}$$

3.3.5. *Non-stationary processes*

(a) The basic equations. The equations of motion for the unstationary case are:

$$\begin{aligned} s_1\frac{\partial^2 y_1}{\partial x^2} - s_y(y_1 - y_2) &= h_1\frac{\partial y_1}{\partial t} \\ s_2\frac{\partial^2 y_2}{\partial x^2} - s_y(y_2 - y_1) &= h_2\frac{\partial y_2}{\partial t} \end{aligned} \tag{3.66}$$

Adding both equations gives:

$$\frac{\partial^2(s_1y_1 + s_2y_2)}{\partial x^2} = \frac{\partial(h_1y_1 + h_2y_2)}{\partial t} \tag{3.67}$$

which can be writen as:

$$\frac{s}{h}\frac{\partial^2 y}{\partial x^2} + \frac{h_1 h_2}{h^2}\left(\frac{s_1}{h_1} - \frac{s_2}{h_2}\right)\frac{\partial^2(y_1 - y_2)}{\partial x^2} = \frac{\partial y}{\partial t} \tag{3.68}$$

in which:

$$\frac{s}{h} = \frac{s_1 + s_2}{h_1 + h_2} \tag{3.69}$$

$$y = \frac{1}{h}(h_1 y_1 + h_2 y_2) \tag{3.70a}$$

$$y_- = y_1 - y_2 \tag{3.70b}$$

$$h = h_1 + h_2 \tag{3.71}$$

y represents the "average coastline", i.e. the weight average between the line of the beach and the line of the inshore. From (3.68) it shows, that the Eq. (3.6) of Pelnard-Considère is valid as well in the two line theory, if $s_1/h_1 = s_2/h_2$. The analytic elaboration of the Eq. (3.68) has only been carried out for the case, that $s_1/h_1 = s_2/h_2$. In that case one can write (3.68) as:

$$\frac{s}{h}\frac{\partial^2 y}{\partial x^2} = \frac{\partial y}{\partial t} \tag{3.72}$$

and

$$\frac{s}{h}\frac{\partial^2 y_-}{\partial x^2} - \frac{s_y h}{h_1 h_2} y_- = \frac{\partial y_-}{\partial t} \tag{3.73}$$

The two-line problem has now been transferred–with the aid of Eq. (3.72) — to a one-line problem, where (3.73) gives an equation for the cross-shore transport.

Equation (3.73) is of the same kind as (3.72); this shows to be the case by the substitution:

$$y_e = y_- \times e^{-t/T_0}; \quad i.e.:\ y_- = y_e \times e^{-t/T_0} \tag{3.74}$$

with

$$T_0 = \frac{h_1 h_2}{s_y h} \tag{3.75}$$

In this case the equation for y_e gets the same shape as (3.72).

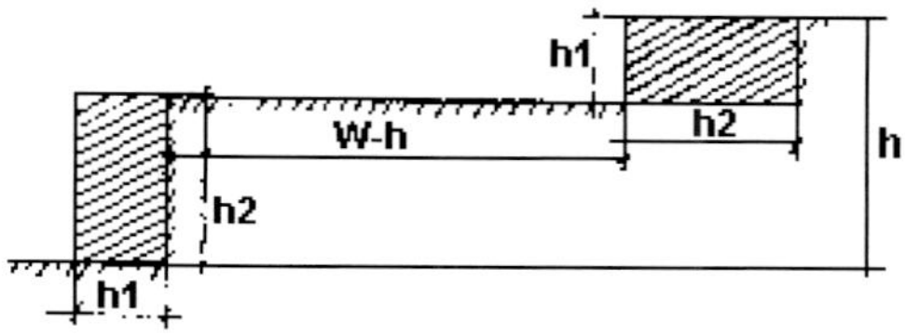

Fig. 3.34 Meaning of T_0.

T_0 is the time, necessary for cross-shore transport S_y for transferring an amount of sand $h_1 h_2$ from the beach to the inshore, when the distance between beach and inshore is kept continuously W-h (Fig. 3.34). Equations (3.72) and (3.73) can only be solved simply in a small number of cases, as the boundary conditions give difficulties.

Finally one then finds y_1 and y_2 from y and y_-:

$$y_1 = y + \frac{h_2}{h} y_- \qquad y_2 = y - \frac{h_1}{h} y_- \tag{3.76}$$

(b) One single rip-current. A relatively simple problem is the morphodynamics of the development of a rip-current on a beach and inshore, which are originally straight. Let the development start at $t = 0$, at the site $x = 0$.

The mean coastline y keeps at its position: $y = 0$. However y_- changes; with the aid of Laplace-transform[40,60] can be found:

$$y_- = \frac{-S^*_{\text{rip}}}{4 s_y L_0}\left[e^{x/L_0}\text{erfc}\left(\frac{x/L_0}{2\sqrt{t/T_0}} + \sqrt{\frac{t}{T_0}}\right) + e^{-x/L_0}\text{erfc}\left(\frac{x/L_0}{2\sqrt{t/T_0}} - \sqrt{\frac{t}{T_0}}\right)\right] \quad (x > 0) \tag{3.77}$$

Fig. 3.35 depicts erfc. function and terms of (3.77). From (3.76):

$$y_1 = h_2 y_- / h \text{ and } y_2 = -h_1 y_- / h$$

For large t, the solution tends to the stationary solution (Fig. 3.36):

$$y_1 = -\frac{S^*_{\text{rip}}}{2 s_1} L_0\, e^{\mp x/L_0} \qquad y_2 = \frac{S^*_{\text{rip}}}{2 s_2} L_0\, e^{\mp x/L_0} \tag{3.78}$$

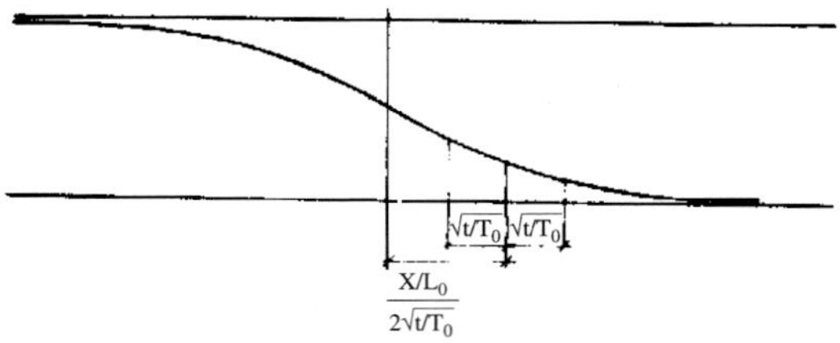

Fig. 3.35 The terms (3.77) of the error function.

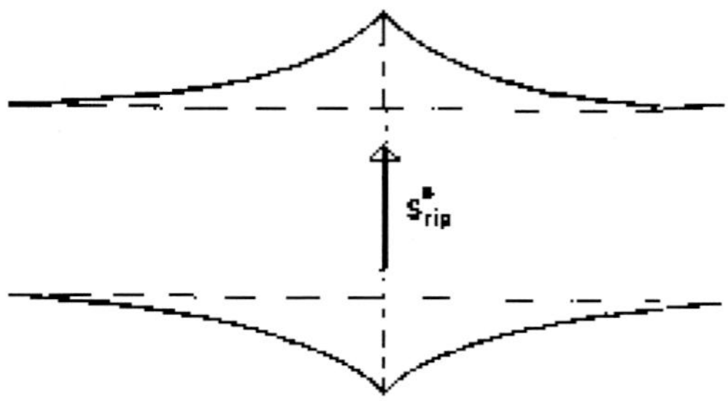

Fig. 3.36 One rip-current (stationar).

The minus sign holds on the right-hand side of the rip-current, the plus sign on the left-hand side.

Solutions like (3.77) can be used in all cases, where as well on the beach as on the inshore the transport in the point $x = 0$ is known and is constant in the time.

The coastline y (in the above-mentioned case) is kept at its place. By superposition of the "delta-solution" according to Pelnard-Considère,[48] one may find solutions for arbitrarily (but stationary) $\partial y_1/\partial x$ and $\partial y_2/\partial x$ at $x = 0$.

(c) One single groin. Only for the case $s_1 = s_2$ and $h_1 = h_2$ it is possible to find an analytical solution for the sedimentation near a groin (in the case of oblique waves); the stationary final situation has already been given (Eq. (3.56)).

The boundary conditions are[n]:

$$\frac{\partial y_1}{\partial x} = \frac{S_{01}^*}{s_1} \quad \text{at } x = 0, \;\; y_2 = 0 \tag{3.79}$$

[n]Symmetry!

The problem has been solved using Laplace-transforms. Bakker,[5,7] gives the solution. The result is shown in Fig. 3.37. The solution for the coastline y consists of a series of terms of the kind $tn - 1/2Y2n - 1$ where the Y-functions are the complementary errorfunctions (Chapter 3.2.6.1).

The solution for y_- (i.e for the slope of the coast) consist of a series of terms $e - t/T_0 \times t - n - t \times Y2n - 1$ and of the "rip-current"-solution (3.77):

$$\begin{aligned}\frac{y_{1,2}}{L_0 \tan \alpha_0} &= \frac{1}{2\sqrt{\pi t/T_0}}\left[\sum_{n=1}^{\infty}(-1)^n \frac{(t/T_0)^n}{\left(h-\frac{1}{2}\right)n!} Y_{2n-1}\left(\frac{x/L_0}{2\sqrt{t/T_0}}\right)\right] \\ &\pm \frac{e^{-t/T_0}}{2\sqrt{\pi t/T_0}}\left[\sum_{n=1}^{\infty}\frac{(t/T_0)^n}{n!} Y_{2n-1}\left(\frac{x/L_0}{2\sqrt{t/T_0}}\right)\right] \\ &\mp \frac{e^{-x/L_0}}{2}\operatorname{erfc}\left(\frac{x/L_0}{2\sqrt{t/T_0}} - \sqrt{\frac{t}{T_0}}\right) \\ &\pm \frac{e^{x/L_0}}{2}\operatorname{erfc}\left(\frac{x/L_0}{2\sqrt{t/T_0}} + \sqrt{\frac{t}{T_0}}\right) \end{aligned} \tag{3.80}$$

For large t the solution approaches the stationary solution (3.56), given in Chapter 3.3.4 (Fig. 3.31).

The erosion, directly behind the groin can be computed for all cases in which $s_1/h_1 = s_2/h_2$ (the reason is, that although the Laplace-transformed $\overline{y}$ and $\overline{y}_-$ can be found for all cases, the inversion to y and y_- has up to now only been performed after the substitution of either $h_1 = h_2$ either $x = 0$).

The total erosion is found from the solution of the stationary problem (3.56). Substitution of $x = 0$ in Eq. (3.56) gives:

$$[y_1]_{x=0} = -L_0 \tan \alpha_0 \frac{s_1 + s_2}{s_2} \tag{3.81}$$

From the non-stationary solution can be derived, after which time $(1 - \delta)$ times the final erosion is observed $(\delta \ll 1)$.

Call this time t_δ; then for small δ the approximation is found:

$$t_\delta = \frac{1}{\pi \delta^2} \times \frac{s_1^2}{s_2 s^2} \times \frac{h}{s_y} \tag{3.82}$$

It thus shows, that the time necessary to reach 0.99 times the final erosion is 100 times as long as the time to find 0.9 times the final value. The initial erosion is equal to the one of the one-line solution.

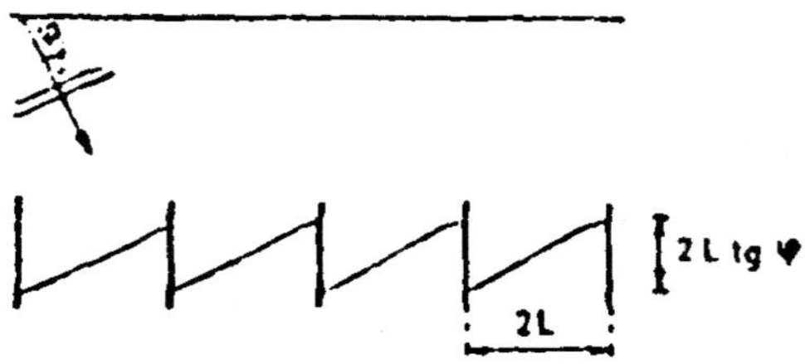

Fig. 3.30 Beach and inshore when the groins are near to each other.

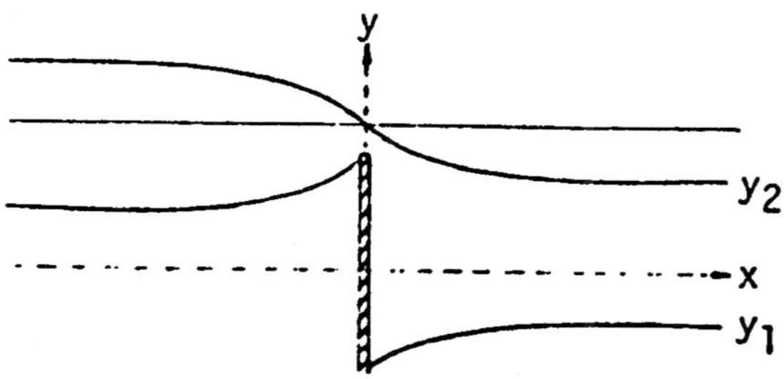

Fig. 3.31 Beach and inshore at a coast with one groin.

incidence on the beach. The difference between the siting of the beach on the left- and right-hand side of the groin amounts to:

$$[y_1]_{x=L} - [y_1]_{x=-L} \approx 2L \tan\alpha$$

In which $\tan\alpha = S_{01}^*/s_1(L < L_0)$.

For *large* distances between the groins (Fig. 3.31) this distance tends to:

$$[y_1]_{x=L} - [y_1]_{x=-L} \approx 2\frac{s_1 + s_2}{s_2} \times L_0 \tan\alpha$$

If only one groin is constructed then the following coastal shape results: (eroding part):

$$\begin{aligned} y_1 &= -\frac{S_{01}^*}{s_1} \times L_0\left(\frac{s_1}{s_2} + e^{-x/L_0}\right) \\ y_2 &= -\frac{S_{01}^*}{s_2} \times L_0\left(1 - e^{-x/L_0}\right) \end{aligned} \tag{3.56}$$

In Fig. 3.32 the coastal shape is sketched for the case that on the lee-side of every groin a rip current is situated. The coastal shape can be calculated by linear superposition of (3.53) and (3.55).

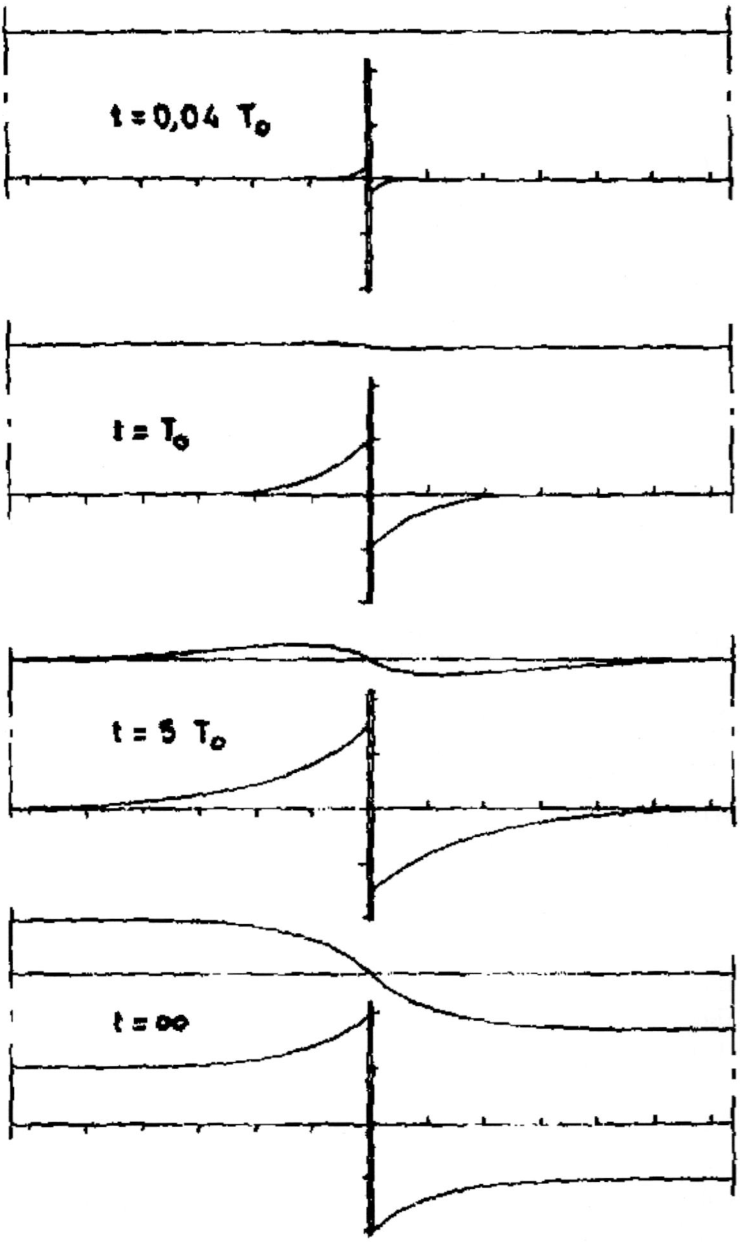

Fig. 3.37 Influence of offshore transport on a coast with one groin.

On the eroding side initially the coastline will develop itself in the way as given by (3.22) (one of the Pelnard-Considère solutions):

$$y_1 = -\tan\alpha_0\sqrt{t}\ \mathrm{erfc}(x \times \sqrt{h/4st}) \tag{3.83}$$

with $\tan\alpha_0 = S_{01}/s_1$

Formally, in (3.83) "h/s" should be replaces by "h_1/s_1", but because of the condition $s/h = s_1/h_1 = s_2/h_2$ this makes no difference.

A special case of periodic solutions is the simulation of the influence of variable wave conditions on the boundary.

When the wave direction changes periodically for instance according;

$$\varphi = \hat{\varphi}\sin\omega_\varphi t \tag{3.84}$$

This generates a sandwave near a groyne (Fig. 3.38).

$$y = \frac{\hat{\varphi}\lambda_\kappa}{2\pi\sqrt{2}} \times e^{-k_\varphi x}\cos(\omega_\varphi t - k_\varphi x) \tag{3.85}$$

in which:

$$\lambda_\kappa = \frac{2\pi}{k_\varphi} = 2\sqrt{\pi\frac{s}{h}T_\varphi} \tag{3.86}$$

T_φ being the period of the fluctuations of φ and λ_φ being the length of the sandwave.

3.4. More-Line Theories

3.4.1. *Equations for a three-line model*

A logical extension moving from one-line to two-line modeling is more line modeling. The end of this Chapter (3.4.3) will give some general considerations. Three-line modeling will be treated first. In the three-line model (used here) the coastal zone is schematized in zones (Fig. 3.39), which will be called: "beach", "inshore" and "deeper part".

In each zone, it is assumed that all contour lines remain parallel and that the profile in this zone only moves horizontally. It is sufficient to compute only an average line in each zone, because it characterizes this zone. There is assumed to be a horizontal separation plane between the zones. The sediment transport between the lines is assumed to be proportional to the

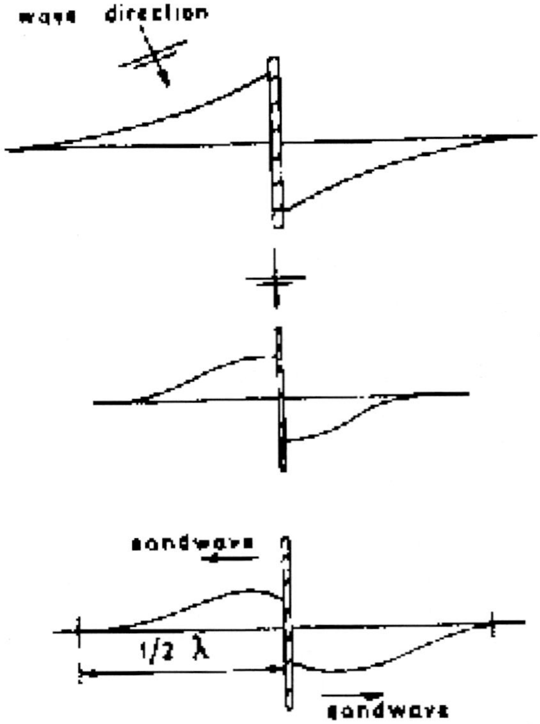

Fig. 3.38 Generation of sandwaves by variable wave direction.

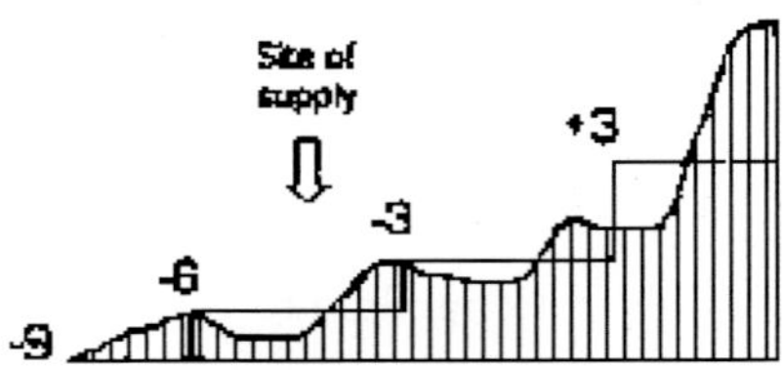

Fig. 3.39 Schematization 3-line.

difference between the equilibrium distance and the actual distance between the characterizing lines.

In the present application, an initial equilibrium position (before any supply takes place) is assumed, where the three lines are straight and

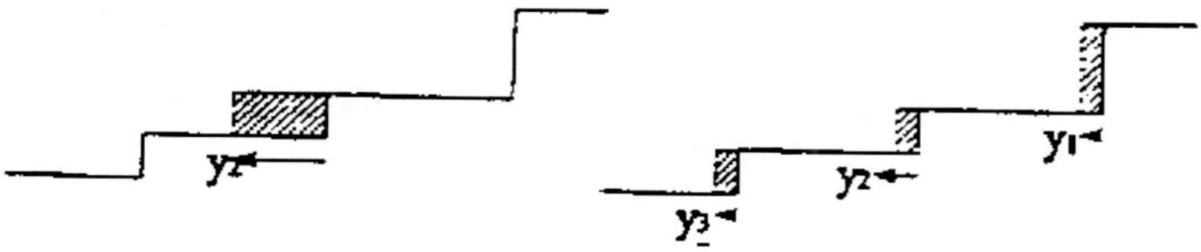

Fig. 3.40 Definition y_1, y_2, y_3.

parallel (Fig. 3.40) Distances $y_{1,2,3}$, are the deviations of the lines of the beach, inshore and deeper part, compared with this initial position. The y-value is computed by dividing the surplus volume in a zone (with respect to the initial position) by its height. In case $y_1 = y_2 = y_3$, the profile again is an equilibrium slope. If longshore transport could be neglected (see Section 3.4.3) the coast would tend to this equilibrium, 'if for instance originally' sand on the inshore would have been supplied (Fig. 3.40).

The process, sketched above results in cross-shore diffusion of sand. Adding longshore diffusion to the cross-shore diffusion, according to the approach of Pelnard-Considère[48] and Bakker,[7] the following dynamic equations can be derived:

$$\frac{dy_1}{dt} = \frac{s_1}{h_1} \times \frac{\partial^2 y_1}{\partial x_1^2} - \frac{s_{y1}}{h_1} \times (y_1 - y_2)$$

$$\frac{dy_2}{dt} = \frac{s_2}{h_2} \times \frac{\partial^2 y_2}{\partial x_1^2} - \frac{s_{y1}}{h_2} \times (y_1 - y_2) - \frac{s_{y2}}{h_2} \times (y_2 - y_3) \qquad (3.88)$$

$$\frac{dy_3}{dt} = \frac{s_3}{h_3} \times \frac{\partial^2 y_3}{\partial x_1^2} - \frac{s_{y2}}{h_3} \times (y_2 - y_3)$$

On the left-hand side is the accretion of line of beach, inshore and deeper part, respectively. On the right-hand side the gradient of the longshore transport is determined by the constants s_1, s_2 and s_3 respectively; the cross-shore transport ("beach to inshore" and "inshore to deeper part") by s_{y1} and s_{y2} respectively. $h_{1,2,3}$ denotes the depth allotted to the three zones. Figure 3.41 shows the diffusion of an (originally rectangular) supply shape in longshore and cross-shore direction in the course of time see Subsection 5.3.6 for an application.

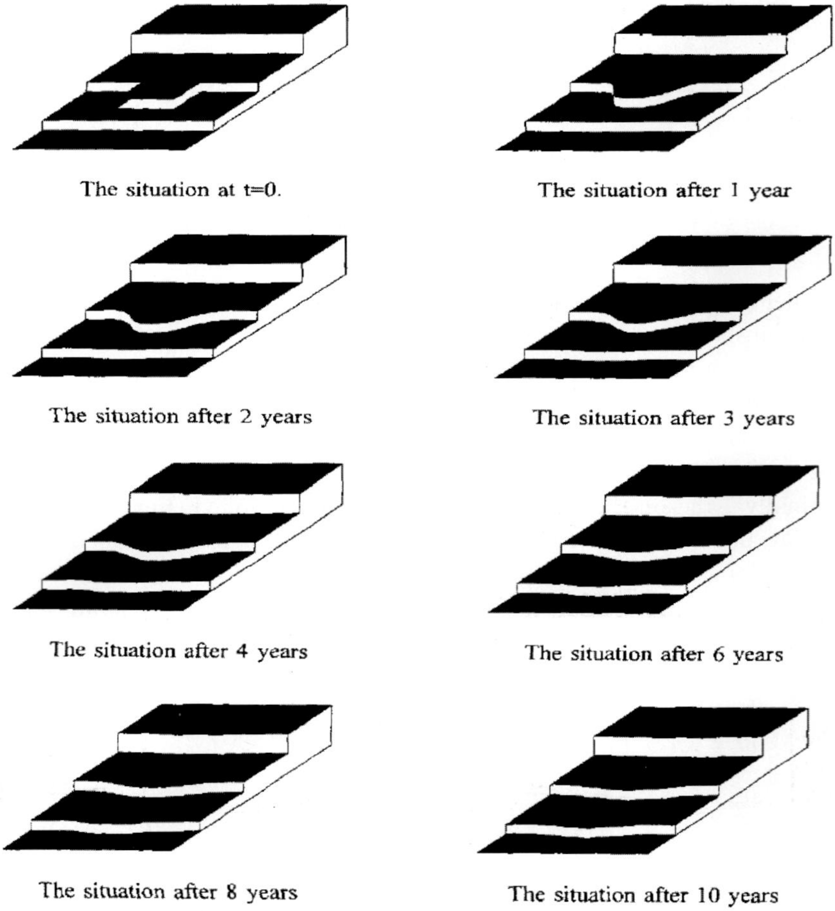

Fig. 3.41 Diffusion of supply (from Ref. [23] in Chapter 5) shows a plan view of such coast. Cases shown here are calculated numerically. Refer furthermore to Subsection 5.6.2.

3.4.2. *Smit's solution*[o] *for the three-line model (2-D problem)*

The dynamic equations for the cross-shore transport S_{yi} between the layers (with $i = 1, 2$) read:

$$S_{yi} = s_{yi}(y_i - y_{i+1}) \tag{3.89}$$

[o] Cf. Refs. [53, 58].

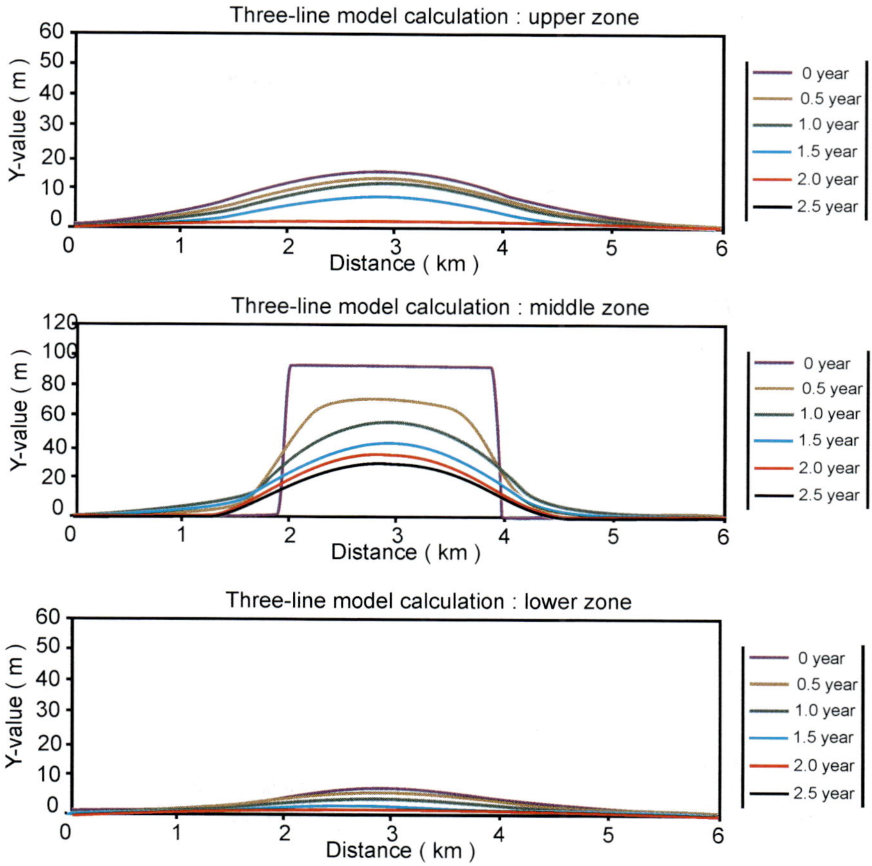

Fig. 3.42 Development with time of a shoreface nourishment in middle layer.

Continuity $\Sigma(h_i, y_i)$ with $i = 1, \ldots, 3$ shows, that y_2 is implicitly determined by y_1 and y_3.

Continuity also yields:

$$S_{y1} = -h_1 \frac{dy_1}{dt}, \quad S_{y2} = +h_3 \frac{dy_3}{dt} \tag{3.90a,b}$$

Combination of (3.89) and (3.90a,b) leads to the vectorial equation:

$$\frac{d\vec{y}}{dt} = \overline{\overline{A}} \times \vec{y} \tag{3.91}$$

where $\vec{y}$ is $y_{1,3}$ and the matrix $\overline{\overline{A}}$ is $(a_{11}, \ldots, a_{22})$, where:

$$a_{11} = -\frac{s_{y1} \times (h_1 + h_2)}{h_1 h_2}, \quad a_{12} = -\frac{s_{y1} \times h_3}{h_1 h_2}$$

$$a_{21} = -\frac{s_{y2} \times h_1}{h_2 h_3}, \quad a_{22} = -\frac{s_{y2} \times (h_2 + h_3)}{h_2 h_3} \tag{3.92}$$

Displaying the solution gives a reason to define five timescales. Two of those are the timescales T_{01} and T_{02} of the two-line systems: "Beach and inshore" and "inshore + deeper part"[5] equal to $-1/a_{11}$ and $-1/a_{22}$, respectively. The third one, T^* is related to the other diagonal of the matrix $\overline{\overline{A}}$:

$$T_*^2 = \frac{1}{a_{12} \times a_{21}} = \frac{h_2^2}{s_{y1} \times s_{y2}} \tag{3.93}$$

The last ones are $1/\lambda_{1,2}$, as occur in the solution of (3.91):

$$\vec{y} = \vec{K} e^{\lambda_1 t} + \vec{M} e^{\lambda_2 t} \tag{3.94}$$

where $\vec{K} = (K_1, K_2)$ and $\vec{M} = (M_1, M_2)$, in which:

$$K_1 = \frac{a_{21}}{\lambda_2 - \lambda_1} y_{3b} + \frac{\lambda_2 + \frac{1}{T_{01}}}{\lambda_2 - \lambda_1} y_{1b}, \quad M_1 = +\frac{a_{21}}{\lambda_2 - \lambda_1} y_{3b} - \frac{\lambda_1 + \frac{1}{T_{01}}}{\lambda_2 - \lambda_1} y_{1b} \tag{3.95a,b}$$

$$K_2 = \frac{1}{a_{21}} \left(\lambda_1 + \frac{1}{T_{01}} \right) K_1, \quad M_2 = \frac{1}{a_{12}} \left(\lambda_2 + \frac{1}{T_{01}} \right) M_1 \tag{3.95c,d}$$

where y_{1b} and y_{3b} are the initial values of y_1 and y_3.

3.4.3. *General considerations on more-line theories*

(a) Introduction to more-line theories. In the 1970s, more-line theories got considerable attention. In the Netherlands, Van Overeem[47] developed such a numerical line model for an arbitrary number of lines for the computation of coastal changes caused by waves and currents. Similarly, in America, Perlin and Dean[49] developed an n-line model, that allows an arbitrary number of contour lines to represent the beach profile. The popularity of those kind of models has remained limited. To the opinion of the present author, one should keep in mind the schematic character of line models. The Coastal Engineering Manual states:

> Schematic three-dimensional models have not yet reached the stage of wide application, due to their complexity, requirement for considerable computational resources and need for expertise in operational applications.

Agreeing with that, the (personal) resistance is also a matter of principle: if more attention is devoted to the detail, it is worthwhile to consider to swap to the side of numerical models, in which serious attention is given to detailed hydraulics as well.

(b) Coastal Genesis and more-line modeling. In the scope of the present book (and anticipating to Chapter 4 on cross-shore modeling) attention is paid to a two-dimensional yz-model, (a "play-model") developed in the scope of the Coastal Genesis project.[58] Here the question was posed to simulate the behavior of the Dutch coast in the last 6000 years.

First the (physical) problem will be stated; next the proposed mathematical approach will be explained. Here only a two-dimensional simulation of some aspects of sea level rise was made. In this context, it is worthwhile to pay the attention as well to the much more sophisticated models, developed by (mainly) Delft Hydraulics (Stive; Eysink) and which have been published in the scope of the Coastal Genesis Project.[54,58]

Figure 3.43 shows the sea level rise in the course of time since 10000 Before Present (BP) according to Jelgersma.[38] Figure 3.44, from Jelgersma *et al.*,[39] shows the submerging of the North Sea in the last milennia. Figure 3.45 shows the development of the Dutch coast according to Jelgersma *et al.*[39] Apart from a regression during the period of fast sea level rise a transgression took place between ca. 5500 BF and 3000 Br.[39,51,55,62] Afterwards, erosion started again.

Is it possible to make a conceptual model of this process of sea level rise? A good model should be able to reproduce as well the coastal transgression from 5500 to 3000 BP as the erosion afterwards.

In the present preliminary stage, it has only been attempted to make a model which gives qualitative agreement, without paying too much attention to quantitative accuracy before explaining this model, some literature concerning this subject will be reviewed.

A very simple model for the response of a coast on sea level rise is provided by Bruun[31] and earlier papers (Fig. 3.46). He starts from the assumption that the sea level rise is so slow, that the upper part of the coast — up to a certain so-called "closure depth" — will always be able to

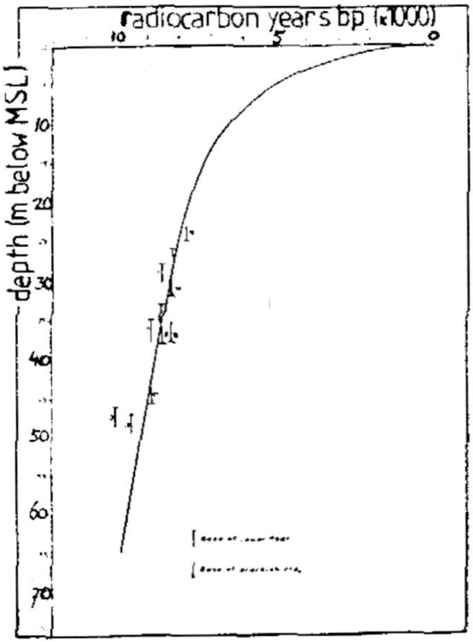

Fig. 3.43 Sea level rise during the last milennia after Jelgersma.[38] T, base of lower peat; ⊥, base of brackish clay.

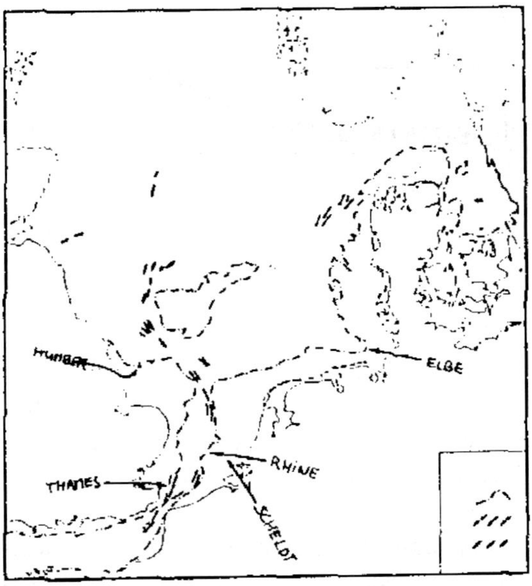

Fig. 3.44 The submerging of the North Sea in the last milennia after Jelgersma *et al.*[39] -.-.-, Coastline 8000 BP sealevel 30 m below present level; ////, old sediment ridges.

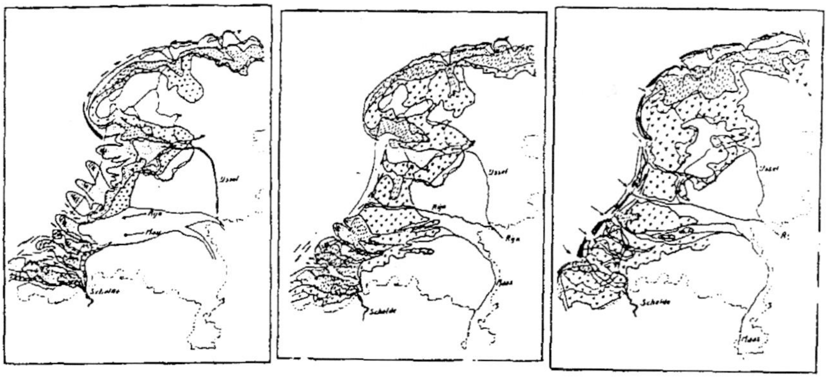

Fig. 3.45 Development of the Dutch coast according to Jelgersma *et al.*[39] Time (years before present): (a) 5500 (b) 4600 (c) 2000 sealevel (m below present level): 7 3.5 1.

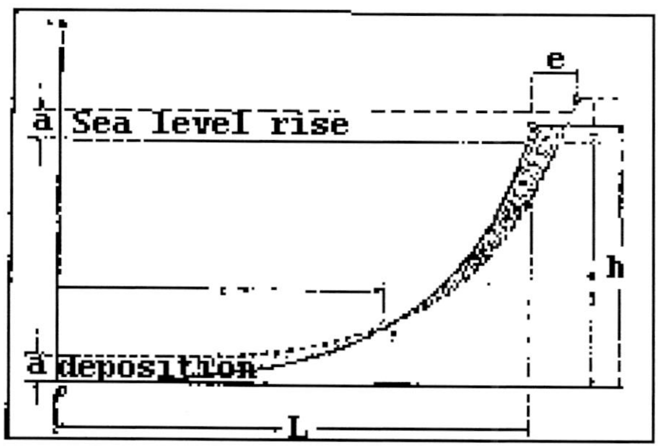

Fig. 3.46 The Bruun model, translation of coastal profile due to sea level rise. After Bruun.[31]

adjust itself and will keep a cross-section according to an equilibrium profile. From simple geometry and continuity it show, that the coastal erosion will be equal to the cotan of the mean slope (L/h in Fig. 3.46) times the mean sea level rise a. Here h denotes the so-called "closure depth", i.e. the depth up to which eroded material settles itself.

According to the authors it will always cost time to attain the equilibrium profile. They agree with Wright et al.[59] that short-term processes take place with a small closure depth where a larger part of the profile will be involved in long-term processes; the time necessary to reshape the profile to an equilibrium profile will be longer. Figure 3.47, after Wright *et al.* elucidates this idea.

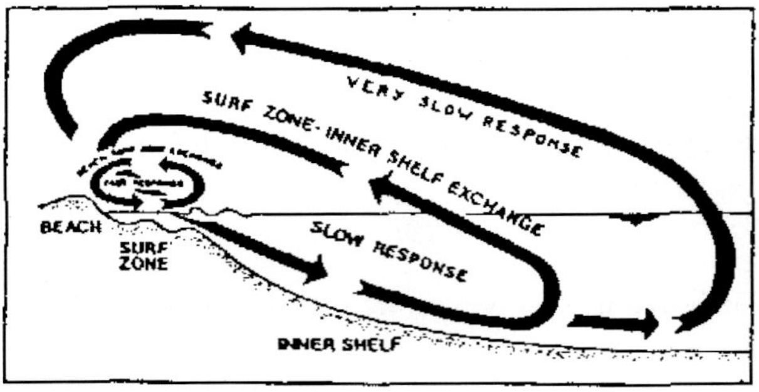

Fig. 3.47 Interaction of different parts of the profile, after Wright *et al.*[59]

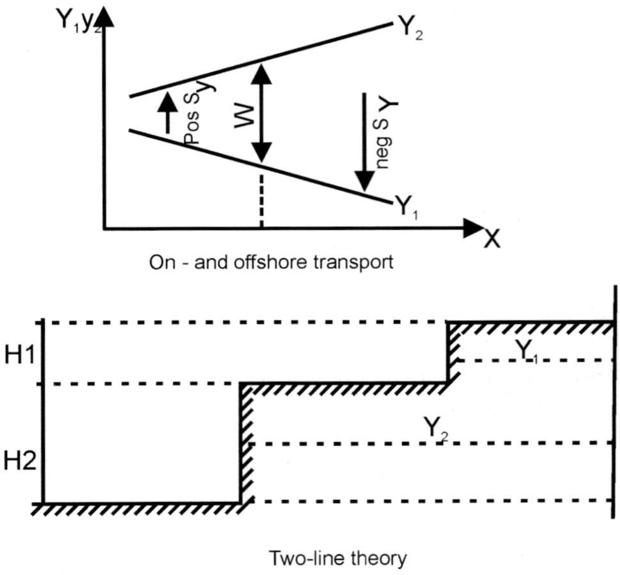

Fig. 3.48 Schematization in beach and inshore time scales according to Bakker[3] and Swart.[56]

The time-dependent adapting of a coastal profile to its equilibrium profile is simulated mathematically by Bakker[5] and Swart.[56]

These authors schematized the coastal profile essentially to two areas, a beach and an inshore, each represented by a line in plan view (Fig. 3.48). Beach and inshore are separated by a horizontal flat between. The region seaward of the inshore is assumed to be horizontal: here no cross-shore transport takes place. According to their schematization and according to experiments of Swart, if longshore transport plays no role, the line of beach and inshore will tend to reach an equilibrium distance W in a negative-exponential way.

The model starts from the assumption, that the cross-shore transport S_y depends on the overal coastal slope:

$$S_y = s_y(y_1 - (y_2 - W)) \tag{3.96}$$

where S_y is positive in seaward direction, y_1 and y_2 denote the position of the line of beach and inshore (Fig. 3.48, i.e. the mean of the positions of the equidistant contour lines of the beach and the inshore, respectively) and W denotes the equilibrium distance.

Basic physical thought behind the model is, that a slope determines the cross-shore transport; that two counteracting mechanisms determine this slope, where one mechanism increases the offshore transport when the slope increases (gravity mostly) and the other determines the onshore transport and increases, when the slope becomes flatter (for instance the asymmetry of the orbital motion). Making a Taylor series of the transport in the vicinity of the equilibrium slope gives a dynamical equation of the kind as given by (3.96). Implicitly it is assumed that the cross-shore transport is not affected by local changes in slope of beach and inshore separately, as long as the mean position of the beach, c.q. the inshore remains the same. Some motivation for this assumption can be found from Fig. 2.19: like a huge rake the breaker bars correct differences in large-scale slope every 3–4 years and the local slope is a very short-term feature, effected by the breaker bars and has no essential importance.

This philosophy also more or less determines the lower edge of the inshore, which can be considered as a kind of "closure depth" as well. The theory only holds when the profile of beach and inshore respectively can adapt itself much faster to changes than at the time scale of the coastal process to be schematized by the two-line theory.

Looking to Fig. 2.17 it might be expected that cross-shore transport below the level, where the breaker bars decay (say 5 m below MSL) will be much smaller than above this level. Therefore, this provides a logical closure depth. The characteristic time scale T_0 for exponential decay according to the two-line theory equals

$$T_0 = \frac{h_1 h_2}{s_y h} \tag{3.97}$$

where h_1 and h_2 are the thickness of the layer of beach and inshore, respectively, and h equals $h_1 + h_2$.

In the case of sea level rise, cross-shore transport will take place between the inshore and the layer below, however at a larger time scale (slower) than between beach and inshore. Furthermore, it might be expected that the slope between the inshore and this lower layer to a great extent determines the cross-shore transport between those layers. In other words: the local slope will be more of importance than the overall slope and therefore a two-line approach will not be sufficient. The ideas of Wright *et al.*[59] (variable closure depth) and of Bakker and Swart (negative-exponential tending to an equilibrium profile) can be combined in the following way. To simulate sea level rise, a 6-line program has been made, starting from six layers with seaward distance with respect to a reference point of y_1, y_2, y_6, each layer having a height of 3 m; the cross-shore transport s_{y1}, $s_{y2}, \ldots, s_{y5}$ between the first and the second, the second and the third, etc. layer is related to $y_1 - y_2$, $y_2 - y_3, \ldots$ in the following way:

$$S_{yi} = s_{yi}((y_i - L_i) - (y_{i+1} - L_{i+1})) \tag{3.98}$$

where $i = 1, \ldots, 5$ and Li denotes the position of y_i in an equilibrium position. The characteristic time T_i is defined according to Bakker and Delver[16]:

$$T_i = \frac{(h_i h_{i+1})}{s_{yi}} \tag{3.99}$$

Figure 3.49 shows the schematization and the assumed time scales T_i. The values of T_4 and T_5 are pure guesses: for some justification of the values of T_1 to T_3 is referred to Bakker and Delver.[16]

It is assumed, that the present profile is an equilibrium profile (Fig. 3.50a) and that 6000 BP the rising of the sea level was so fast, that no time was available to bring the profile in an equilibrium shape; the profile of 6000 years ago is assumed to be the present profile, in which the water level is 6 m lowered (Fig. 3.50c); the water level at 6000 BP was 6 m lower than the present level (Fig. 3.50b,c). This "sudden" exposure of the present coastal area to the waves is caused also for a large part by the degradation and erosion of former coastal barriers in the North Sea, which became submerged. With respect to this concept is referred to Niederoda *et al.* (1985) for the behavior of coastal barriers in general and to van Straaten,[55] for the behavior of the Holland coast. According to the model, the coast would have been accreted if no sea.

Figure 3.51a (in which the third dimension denotes the time) shows how, according to the model, the coast would have been accreted, if the sea

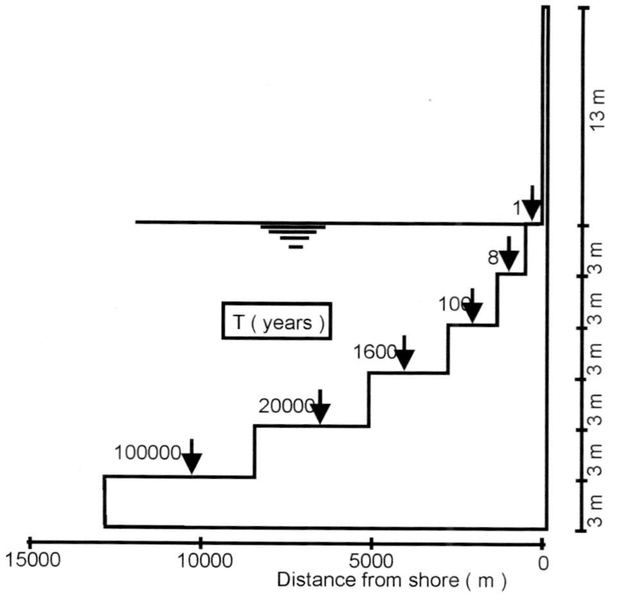

Fig. 3.49 Schematization n-line mode and time scales.

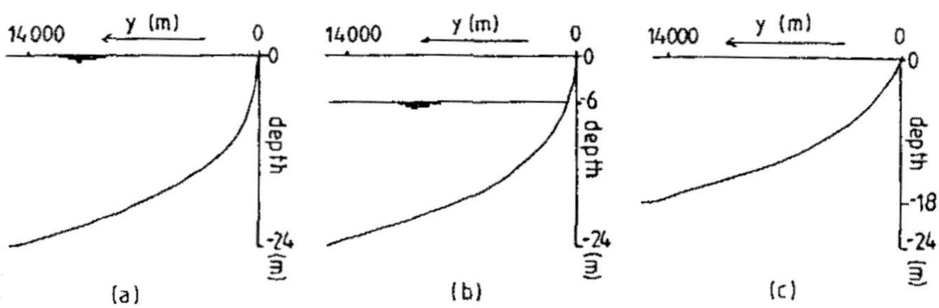

Fig. 3.50 Initial profile of the coast. (a) Present profile. (b) Situation immediatelly after submergence. (c) Assumed profile of 6000 BP.

level had not risen since 6000 BP; in Fig. 3.51b this sea level rise has been taken into account. In Fig. 3.51a the coast accretes, because the profile of Fig. 3.50c is much flatter than the equilibrium profile; in Fig. 3.51b erosion starts again because the upper part of the profile has formed nearly an equilibrium profile and erodes according to Bruun concept (see above). For details concerning the way of computation is referred to Smit.[53]

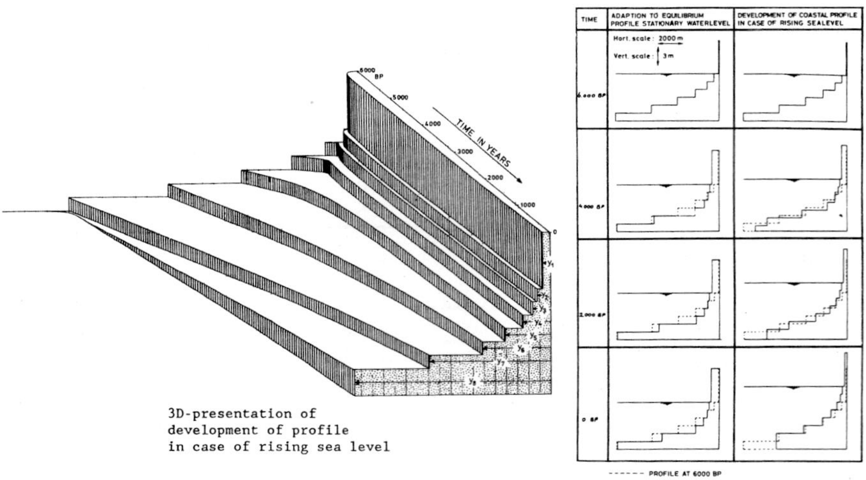

(b) (a)

Fig. 3.51(a, b) The 6-line model of Smit.[53]

Thus it shows, that the model gives qualitative agreement with the features, found in nature.

(c) Conclusions on cross-shore modeling with more-line theories. If coastal behavior is considered at increasing time scales, beginning at a time scale of decades and going to milennia, the effect of sea level rise and cross-shore transport becomes more and more pronounced. Line models may provide (a rough) insight at every time scale.

For the coast of Holland, coastal constants derived on one hand from data concerning the last 15 years (with the aid of Kalman filtering[p]) and on the other hand from data with regard to the last 100–1000 years are not contradictory. For larger time scales, a schematization of the coast in one or two layers (one- or two-line model) will not be sufficient. Small amounts of cross-shore transport from deeper water to the inshore, as well as the local slope of these deeper regions become important at this time scale. In this subsection, for a case like this the use of, for instance, a six-line model is proposed. This simulates as well a closure depth dependent on the time-scale of the feature corresponding to the measurements of Wright *et al.*[59] as a negative-exponential trend to an equilibrium profile according to ideas and experiments of Bakker[5] and Swart.[56] The method is more sophisticated

[p]Cf. Section 7.8.

than the Bruun rule.[31] Accretion of the Holland coast between 4000 years and the Roman times and erosion later on can be simulated this way.

In Fig. 3.51b the left-hand figures show with time steps of 2000 years the calculated adaptation of the coastal profile in absence of sea level rise. The dotted line gives a comparison with the initial profile (6000 BP).

On the left-hand side of Fig. 3.51b it is shown, that the original coastal profile (a result of the submergence of the Doggersbank and the Northsea) is too flat, compared to the equilibrium profile, to be shaped by the prevailing waves. So, if sea level rise would not be taken into account, gradual accretion of the dunes could have been expected. However, because of sea level rise, adjustment to the equilibrium profile at a higher level leads to erosion of the dunes and sedimentation on the inshore. The whole process creeps up to a higher level. Combination of those two processes (Fig. 3.51a and right-hand side of Fig. 3.51b) results in the facts, i.e. that since the time of the crucification of Christ the initially accreting Dutch coastal dunes started to erode again.

3.5. Symbols Concerning Chapter 3

Symbol		Introduced in
$(a_{11}, \ldots, a_{22})$	Elements of matrix $\overline{\overline{A}}$	(3.92)
A,B	Integration constants	(3.35)
A,B	Constants	Theorem 1, page 116
$\overline{\overline{A}}$	Matrix $(a_{11}, \ldots, a_{22})$	(3.91)
A, B, C, E	Integration constants	(3.51)
A'	Integration constant	(3.20a)
A''	Integration constants	(3.21)
$a_t, b_t, c_t,$	Integration constants	(3.40)
e	Basic number neperian Logarithmic system	(3.17)
erf	Error function	(3.25)
erfc	Complementary error function	(3.20)
i^1erfc, i^2erfc,	First (second, third, nth) integrals	(3.22)
i^3erfc, i^nerfc	of the error function	(3.23)

Symbol		Introduced in
h	Depth of movable area of coast	Fig. 3.1
h_1, h_2	Depth of beach, c.q. inshore	Fig. 3.26b
h_3	Depth of third layer	(3.90)
i	$\sqrt{-1}$ Also: counter (Subsection 3.4.3)	(3.34).
k	$2\pi/L$	(3.34)
k_1	real part of k	(3.37)
k_φ	$2\pi/\lambda_k$	(3.85)
$\vec{K}$	Vector in (3.94) (cf. (3.95))	(3.94)
l	Length of groin	(3.16)
L	Wave length of a sinusoidal shaped beach Also: half the distance between two groins	Fig. 3.18, Fig. 3.28
L_0, L_1	Characteristic lengths, depending on (c.q. determining) coastal constants	(3.52), (3.9)
L_i	y-Coordinate of yn in equilibrium position	(3.98)
m	Exponent; also: mean value	Theorem 3, page 117 (3.27)
n	Exponent; also: index (counter)	Theorem 3, page 117
n'	Counter ($=2n$)	(3.23)
O_1, O_2	Areas of accretion/erosion cf. Fig. 3.28 and 3.36	(3.54)
p	Reduction factor of s, Caused by groins	(3.11), (3.61)
Pr(..)	Probability of ...	(3.27)
r_e	Reflection coefficient	(3.10)
s	Coastal constant	(3.1)
s'	Coastal constant (protected area)	(3.58)
s_1, s_2	Coastal constants (beach, inshore)	(3.48)
s_y	Coastal constant (cross-shore)	(3.45)
$s_{\rm rip}$	$S^*_{\rm rip}/S^*_{01}$	(3.62)
$s_{\rm perm}$	Permeability coefficient	3.3.4e, page 149
$s_{y1,2}$	Coastal constant (cross-shore; 3 lines)	(3.88)
s_{yi}	(idem; more than 3 lines)	(3.89)
S	Sand transport (m^3/m;yr)	

Symbol		Introduced in
S_y	Cross-shore transport	(3.45)
S^*	Sand transport (total; in a coastal zone; m^3/yr)	(3.1)
S_0^*	Sand transport (when coast is parallel to x-direction)	(3.1)
$(S_0^*)'$	Overall sand transport S_0^* along a protected beach	(3.59)
S_{01}^*	Sand transport along the beach (when coast is parallel to x-direction)	(3.48)
S_{02}^*	Sand transport along the inshore (when coast is parallel to x-direction)	(3.48)
S_1^*	Sand transport along the beach	(3.48)
S_2^*	Sand transport along the inshore	(3.48)
S_d^*	Sand, supplied to a river delta	3.2.5e, page 123
S_{unpr}^*	Sand, eroding from an unprotected stretch next to a protected stretch	3.2.5f, page 124
S_{perm}^*	Transport along a permeable beach	3.3.4e, page 149
S_{rip}^*	Sand transport by rip current	(3.53)
S_{yi}	Cross-shore transport between layer i and $i+1$	(3.89)
S^{**}	Amount of sand (m^3)	(3.17)
t	Time	(3.4)
Δt	Increment of time	(3.4)
t'	$t/T1$	(3.8)
t_1	Time, in which the coast reaches the head of a groin	(3.16)
t_δ	Time, when accretion reaches (1-delta) Times its final value	(3.82)
T	Function of t	(3.33)
T_0	(a) One-line theory: time, representative for decay of a disturbance along the coast	(3.36)
	(b) Two-line theory: time, representative for restauration of equilibrium profile (two-line theory)	(3.75)

Symbol		Introduced in
T_1	Period of sand wave	(3.9)
T_i	$hnhn + 1/syn$	(3.99)
u	Argument of erf	(3.25)
W	Equilibrium distance	(3.45)
x	Axis along mean coastal direction	(3.1)
x^*	xt^{-m}, independant variable of y^*	(3.7)
x'	x/L_1	(3.9)
Δx	Elementary unit of x	
x_b	x-Coord. point of contraflexure Gausscurve	(3.18)
X	Function of x	(3.33)
y	Coordinate in seaward direction	(3.1)
$y_0(\mathrm{x})$	y-Coordinate in seaward direction	
$y_{1,2}$	y-Coordinates of "depth contours" in two-line theory ($y_2 = y_2' - W$)	(3.45)
$y_{1,2,3}$	y-Coordinates of "depth contours" in three-line theory	(3.88)
y_2'	y-Coordinate of inshore	
y^*	yt^{-n}	(3.20)
$\vec{y}$	(y_1, y_3)	(3.91)
y_e	$y_-.e^{-t/T_0}$	(3.74)
y_-	$y_1 - y_2$	(3.70b)
$y_{(p)}(u)$	y-Coordinate of (un)protected beach	(3.11)
y_n	y-Coordinate of nth layer	(3.98)
$y_u(x)$	y-Coordinate of unprotected area of a protected beach	(3.10)
$y_{1,2\mathrm{mean}}$	Length-averaged distance of y_1, respectively y_2	(3.54)
$Y_n(u)$	Standarized nth integral of error function	(3.30)
α_0	Angle of wave incidence	(3.2)
δ	Arbitrary number, small compared to 1	(3.82)
λ_k	Wave length of a sinusoidal disturbance caused by variable wave conditions	(3.86)
$\lambda_{1,2}$	(Inverse) time scales	(3.94)
π	4*atan(1)	
σ	Standard deviation	(3.27)

Symbol		Introduced in
φ	(Momentaneous) angle of wave incidence	(3.84)
$\hat{\varphi}$	Amplitude of periodical wave change	(3.84)
ω_1	$2\pi/T_1$	(3.38)
ω_φ	Wave number (periodical changing wave direction)	(3.84)

References

1. Abramowitz, M. and I. A. Stegun, *Handbook of Mathematical Functions*, In: Eds. M. Abramowitz and I. A. Stegun. Dover Publ Inc., New York, 1970.
2. Bakker, W. T. and T. Edelman, *The coastline of riverdeltas*, Proc. of the 9th Conf. on Coastal Engineering, Lisbon, 1964.
3. Bakker, W. T., One aspect of the dynamic of a coast, partly protected by a row of groynes, Rijkswaterstaat, Dept. for Coastal Res., Study Report W.W.K. 67-5.
4. Bakker, W. T., *A mathematical theory about sand waves and its application on the Dutch Wadden isle of Vlieland.* Shore and Beach, May 1968.
5. Bakker, W. T., *The dynamics of a coast with a groyne system*, Proc. of the 11th Conf. on Coastal Engineering, London, 1968.
6. Bakker, W. T., *De invloed van rip-currents op een kustvorm (in Dutch: the influence of rip-currents on the coastal shape)*, Rijkswaterstaat, Dept. for Coastal Res., Memo 68-3.
7. Bakker, W. T., The influence of offshore transport on the dynamics of a coast with a harbour mole, Rijkswaterstaat, Dept. for Coastal Res., Study Report W.W.K. 69-2.
8. Bakker, W. T., Berekening van het langstransport door golven met de methode van evenwijdige dieptelijnen (in Dutch: Calculation of the longshore transport, using the method of: "parallel depth contours"), Rijkswaterstaat, Dept. for Coastal Res., Study Report W.W.K. 69-7.
9. Bakker, W. T., The influence of diffraction near a harbour mole on the coastal shape, Rijkswaterstaat, Dept. for Coastal Res., Study Report W.W.K. 70-2.
10. Bakker, W. T., Over de invloed van golven en getij op het zandtransport in de brandingszone (in Dutch: About the effect of waves and tide on the sand transport in the surf zone), Rijkswaterstaat, Dept. for Coastal Res., Study Report W.W.K. 70-9.
11. Bakker, W. T., E. H. J. Klein Breteler and A. Roos, *The dynamics of a coast with a groyne system*, Rijkswaterstaat, Dept. for Coastal Res., Study

Report W.W.K. 70-13, Also: Proc. of the 12th Conf. on Coastal Engineering, Washington, 1970.

12. Bakker, W. T., Littoral drift in the surf zone, Rijkswaterstaat, Dept. for Coastal Res., Study Report W.W.K. 70-16.
13. Bakker, W. T., H. G. H. ten Hoopen and G. H. Grieve, Berekening van het zandtransport volgens de methode Svasek bij een strand en een vooroever die een hoek met elkaar maken (in Dutch: Calculation of the sand transport according to the method of Svasek for a beach and an inshore, making an angle with each other), Rijkswaterstaat, Dept. for Coastal Res., Study Report W.W.K. 71-18.
14. Bakker, W. T., The influence of longshore variation of the wave height on the littoral current, Rijkswaterstaat, Dept. for Coastal Res., Study Report W.W.K. 71-19.
15. Bakker, W. T., *The influence of longshore variation of the wave height on the littoral drift*, 13th Conf. on Coastal Engineering, Vancouver, 1972.
16. Bakker, W. T. and G. Delver, *Kustveranderingen door een lage zanddam voor de kust van Delfland (in Dutch: Coastal changes caused by a shallow sanddam in front of the Delfland coast)*, Delft University of Technology; Coastal Engineering Group, 1986.
17. Bakker, W. T., C. v. d. Kerk and J. H. de Vroeg, *Determination of coastal constants in mathematical line models*, 2nd European workshop on coastal zones as related to physical processes and coastal structures, Loutraki, Greece, 1988.
18. Bakker, W. T. and J. H. de Vroeg, *Coastal modeling and coastal measurements in the Netherlands*, In: Eds. A. de Graauw and L. Hamm, Proc. Seminar "Prototype Measurements to validate numerical Models of coastal Processes", May 1988.
19. Bakker, W. T., J. L. Xu and O. Koshizawa, *Niigata offshore breakwaters: coastal shape*, 5th Int. Conf. on Coastal and Port Engineering in Developing Countries, Cape Town, 1999.
20. Bateman, H., *Tables or Integral Transforms — Volume I*, Cal. Inst. of Technology, Bateman Manuscript Project.
21. Battjes, J. A., Radiation stress in short crested waves. *J. Marine Res.* **30**(1), 1972.
22. Bendegom, L. van, *Beschouwingen over de grondslagen van de kustver dediging. (in Dutch: considerations concerning the basics of coastal defense)*, Rijkswaterstaat, Directorate Benedenrivieren — Study Department "North", 1949.
23. Bijker, E. W. and J. N. Svasek, *Two methods for determination of morphological changes near coastal structures*, XXIInd Int. Nav. Congress, Paris, Sect. II, Subj. 4, pp. 180–202, 1969.
24. Bijker, E. W., Longshore transport computations. *J. Waterways Harbor Coastal Eng. Div., ASCE*, 97(WW4), 687–701, 1971.
25. Bijker, E. W., *Lecture Notes for Topics in Coastal Engineering*, Delft University of Technology, 1972.

26. Binder, L., *Uber äussere Wärmeleistung und Erwarmung electrischen Maschine (in German: about external heat transfer and heating of electrical maschines)*, Dissertatie Techn. Hochschule, München, 1910.
27. Bochove, G. van and H. G. H. ten Hoopen, *Internal report (M.Sc. Thesis).*. Delft University of Technology; Dept. of Civil Engng; Sect. Coastal Engng, 1971.
28. Bowen, A. J., B. L. Inman and V. P. Simmons, Rip-currents. *J. Geophys. Res.* **74**(23), 5469–5490, 1968.
29. Bowen, A. J., The generation of longshore currents on a plane beach. *J. Marine Res.* **72**(2), 1969.
30. Bowen, A. J., The generation of longshore currents on a plane beach. *J. Marine Res.* **72**(2), 1969.
31. Bruun, P., *Review of conditions for uses of the Bruun rule of erosion. Coastal Eng.* **7**, 77–89, 1983.
32. Churchill, R. I., *Operational Mathematics.* McGraw-Hill, 1958.
33. Edelman, T., Vormverandering aan zandige kusten — deel I: Algemene inleiding en afleiding van de kustvergelijking (in Dutch: Adaption of shape of sandy coasts: part 1: General introduction and derivation of the coastal equation), Rijkswaterstaat, Dept. for Coastal Research; Study report 64-1.
34. Edelman, T., De basisoplossing voor een lineaire partiële djfferentiaalvergelijking (in Dutch: the basic solution of a linear partial differential equation), Rijkswaterstaat, Dept. for Coastal Research; Study report WWK 71-17.
35. Grijm, W., *Theoretical forms of shorelines*, Proc. of the 7th Conf. on Coastal Engng, The Hague, 1960.
36. Grijm, W., *Theoretical forms of shorelines*, Proc. of the 9th Conf. on Coastal Engng, Lisbon, 1964.
37. D. L. Inman, P. O. Komar and A. J. Bowen, *Longshore transport of sand*, Proc. of the 11the Conf. on Coastal Engineering, London, 1968.
38. Jelgersma, S., *Sea level changes in the North Sea basin*, The quaterny history of the North Sea, pp. 233–248, 1979.
39. Jelgersma, S., D. Beets and R. Schuttenhelm, *Kustmorfologie in ontwikkeling (A geological view on the coast, in Dutch)* Pt/Civ. Techn. nr. 11, pp. 3–8, 1985.
40. de Josselin de Jong, G., *Consolidatie in drie dimensies, II. (in Dutch: Consolidation in three dimensions)*, Laboratory for Soil Mechanics Mededelingen (Communications), part VIII no. 2, October 1963.
41. Klein Breteler, E. H. J., Getijberekening voor de kustzone tussen Hoek van Holland en IJmuiden met behulp van het getijmodel van Booij (in Dutch: Tidal computation for the coastal zone between Hook of Holland and IJmuiden using the tidal model of Booij) Rijkswaterstaat, Dept. for Coastal Research, Study report 70-10.
42. Langejan, *Oplossing van consolidatieproblemen met behulp van de grafische methode van Schmidt. (in Dutch: Solution of consolidation problems with the graphical method of Schmidt)*, Laboratory for Soil Mechanics Mededelingen (Communications), part VII, no. 2, November 1962.

43. Larras, J., *Plage et cotes de sable (in french: Sandy beaches and coasts)*, Collections du Laboratoire National d'Hydraulique.
44. McCowan, J., On the highest wave of permanent type, *Phil. Mag.* **55**, **38**(233), 351–358, Oct. 1894.
45. Opdam, H. J., Een mathematisch model van de kustzone, betreffende de invloed van golven, getij, wind en corioliskracht op de stroming. langs de kust (in Dutch: a mathematical model of the coastal zone, concerning the effect of waves, tides, wind and the coriolis force on the current along the coast) Rijkswaterstaat, Dept. for Coastal Res., Study Report W.W.K. 70-11.
46. Opdam, H. J., Experimentele fluorescëine metingen betreffende de invloed van golven en getij op de stroming langs de kust (in Dutch: Experimental fluerescine dye measurements concerning the effectof waves and tide on the current along the coast) Rijkswaterstaat, Dept. for Coastal Res., Study Report W.W.K. 71-5.
47. Overeem, J. V., Numeriek model voor de berekening van kustlijnveranderingen t.g.v. golven en stroom. (Numerical model for the computation of coastal changes caused by waves and currents). Delft University of Technology; Dept. of Civil Engng; Sect. Coastal Engng, 1978.
48. Pelnard-Considère, R., *Essai de théorie à l'evolution des formes de rivages en plages de sables et de galets (in French::Preliminary theory for the evolution of shapes of sandy and pebble beaches)*, Quatrième Journées de l'Hydraulique, Paris 13-15 Juin 1954 Les Energies de la Mer, Question 3.
49. Perlin, M. and R. G. Dean, *A Numerical Model to Simulate Sediment Transport in the Vicinity of Coastal Structures*, Newark, Coastal and Offshore Engineering and Research, 1983.
50. Rijn, L. C. van and A. J. de Leeuw, Zandtransportmodel Westerschelde: stochastisch onderzoek zandtransportformule (in Dutch: Sandtransport model Western Scheldt: stochastical investigation on a xand transport formula), Delft Hydraulic Laboratory, Rep. S 582., Oct. 1985.
51. Roep, T. B., Progradation, erosion and changing coastal gradient in the coastal barrier deposits of the Western Netherlands. *Geologie en Mijnbouw* **63**(3), 225–336, 1984.
52. Schmidt, E., *Uber die Anwendung der Differenzenrechnung auf technische Anheiz- und Abkülungsprobleme (in German: On the applicationof differential calcuklation to technical heat- and cooling problems)* Beitragen zur technischen Mechanik und technischen Physik, (August Töpp Festschrift) Berlin, pp. 179–189, 1924.
53. Smit, E. S. P., *De rol van het sediment dwarstransport in de ontwikkeling van de Hollandse kust (in Dutch: the role of the sediment crossshore transport in the development of the Holland coast)*, M.Sc. Thesis; Delft University of Technology; Dept. for Coastal Research, 1988.
54. Stive, M. J. F. and W. D. Eysink, Voorspelling ontwikkeling kustlijn 1990–2090; part of: Dynamisch model van het Nederlandse kustsysteem (In Dutch: Prediction development coastline 1990-2090; part of: Dynamic model of the Dutch coastal system), Rijkswaterstaat & WL Delft Hydraulics, Coastal Management report, underlying investgation report 3.1.

55. Straaten, L. J. M. U. van, Coastal barrier deposits in South and North Holland. *Mededelingen Geologische Stichting*, **17**, 41–75, 1985.
56. Swart, D. H., *Offshore sediment transport and equilibrium beach profiles*, Ph.D Thesis; Delft Univ. of Technology, 1974.
57. Voogt, J. B. and A. J. Romein, *M.Sc. Thesis,* Dept. Coastal Engngn., Delft University of Technology, 1971.
58. Vroeg, J. H. de, E. S. P. Smit and W. T. Bakker, *Coastal genesis*, Proc. 21st Int. Conf on Coastal Engng, Malaga, 1988.
59. Wright, L. D., S. K. May, A. D. Short and M. O. Green, Beach and surf zone equilibria and response times. *Coastal Engineering*, 1984.
60. Wylie, C. R. Jr., *Advanced Engineering Mathematics.* McGraw-Hill Book Company, 1960.
61. Zaat, J. A., *Numerieke Analyse (in Dutch: Numerical Analysis).* Delft, 1969.
62. Zagwijn, W. H., The paleo-geographic evolution of the Netherlands during the Quaterny. *Geologie en Mijnbouw*, **53**, 369–385, 1974.

Chapter 4

Determination of Coastal Constants

4.1. General Considerations

4.1.1. *Introduction to Chapter* 4

Chapter 4 (and Chapter 5) form the closing link between Chapters 2 and 3. Remember that in Chapter 2 the Dutch coastal history was related. Chapter 3 gave mathematical theory on coastlines (or lines of beach and inshore), moving as they were governed by formulae and coastal constants like "s" or "Q".[a]

However, apart from the mathematical formulation, the results of mathematical line modeling depend upon the magnitude of those "coastal constants", which serve as calibration constants in those models. The magnitude of these constants as found in nature (a topic linking Chapters 2 and 3) is the substantial topic of this chapter.

The subject "Kalman filtering" is more universal than depicted in this table: it covers "plan shape" as well.

Refer to the next chapter concerning experimental verification; evidently, in that chapter the *practice* of finding the coastal constants also will be focused.

Two ways of subdivision of the present chapter can be distinguished (Fig. 4.1):

A: concerning the kind of data on which is focused on, i.e. "historical data concerning topography" versus "data on wave climate":

B: concerning the characteristic of the shore on which it is focused on, i.e. "on the plan shape of the beach" versus "on the profile".

[a]Still, a remark from 1968 of Dr. Kemp of London University echoes in the ears of the author: "This is a ***very*** nice way of solving problems by just calling them Q!"

	B. Shore characteristics	
	a. Plan shape	b. Cross-shore
A. Data		
1. Historical topography (Section 4.2)	Curve fitting	Kalman filtering
2. Wave climate	Deterministic calculation of coastal constants (Section 4.3)	Calculation of coastal profiles and cross-shore coast.const. (Section 4.4)

Fig. 4.1 Topics of Chapter 4.

Ad A: one way of finding coastal constants is:

A1: to match calculated coastal behavior with historical data and find the coastal constants in this way; another one is:

A2: to calculate those constants theoretically from wave theory, wave data and sand transport theory and thus derive formulae concerning those coastal constants, expressed in wave climate.

Ad B: a logical example of this kind of investigation is: find the cross-shore constant s_y (occurring in, for instance, Eq. (3.50a,b)) from the behavior of coastal profiles.

There is a distinct difference in philosophy between "method A1" and "method A2".

Method A1 is mainly a statistical method. *A priori* it is assumed, that the coastal theory as developed in Chapter 2 is correct. In this philosophy, truth is only hidden by statistical noise, which has to be minimized.

Method A2 returns back to the basics. Sand transport formulae like the CERC-formula[22] concerning longshore transport and formulae like the one of Stive,[36] concerning cross-shore transport are assumed to be "true". It is investigated (Section 4.3), whether those theories imply mathematically the assumptions of Pelnard Considère,[33] being at the foundation of Chapter 3.

This method used also gives feeling, whether "all" sensible sand transport theories can be summarized by the Pelnard Considère theory. Remind the footnote on page 1: "all" sensible theories could lead to different formulae for the coastal constants, but still those formulae could all agree concerning the validity of linear dependance of transport, as related to (the tangens of) the angle the coastline makes with the x-axis!

This would lead to certainty on shapes, only uncertainty on length- or time scales. See for instance Subsection 3.2.4 on laws of scale.

However, more deviations between reality and applied coastal schematization can be expected. For instance, mind the assumption of a stationary angle of wave incidence, being questionable in prototype.[b]

Thus, Section 4.3, which focuses on "method A2" (cf. Fig. 4.1), described above, starts from a more realistic schematization than the one of Pelnard Considère.

The following build-up of the present chapter results from the considerations above.

In the second part of this general introduction (Section 4.1), it is considered, *which* of the variables, occurring in Chapter 3 (line theories) should be specificly quantified (Subsection 4.1.2).

Constants, which appear in these models deal with the relation between the curvature of the line(s) and the accretion; furthermore, in the two-line theory, an additional constant s_y determines the relation between the steepness of the profile (i.e. the distance between the lines) and the cross-shore transport.

First, in Section 4.2 (cf. "subdivision A1", Fig. 4.1), it is considered, how those coastal constants can be derived from historical data. Two methods will be applied:

A1&Ba: "CAD-like method": matching results of calculated coastlines to nature by trial-and-error changing of coastal constants, where deviation between calculation and nature is minimized.

A1&Bb: Kalman filtering.

With respect to the aspect "B" ("longshore" or "cross-shore") it may be added, that the Kalman filtering method (A1Bb, as will be treated in Subsection 4.2.2) allows determining of the longshore- and cross-shore components simultaneously.

Next, in Section 4.3, attention is given to plan shapes and is focused on the constants S_0 and s, occurring in Chapter 3 and their sub-divisions S_{01} and S_{02}, s_1 and s_2 over beach and inshore; those constants are expressed in the wave climate.

Subsequently, Section 4.4 deals with cross-shore transport.

For formulation of the constants, determined by longshore transport (Section 4.3), in the one-line approach the CERC-formula has been implemented. With respect to the two-line theory the Svasek variation of the CERC-formula was used.

[b]Refer to Eq. (3.84) to get an impression of the effect on the coastal shape of a variation of this wave direction in the course of time.

With respect to the constant determining the cross-shore transport (i.e. concerning the time scale of restoration to the equilibrium profile of a non-equilibrium profile) an effort is made to link with the Stive theory.[36] Dimensionless equilibrium profiles of the offshore zone result in this way, however only of the part seaward of the outer breaker bar.

As in the first row of Fig. 4.1, the separation line between the "Ba-" and "Bb-" (planshape versus profile) calculation shows some vagueness: two-line theories will be catagorized as "plan-shape" theories, although the existance of cross-shore elements forms the most important reason for their existance. The present considerations with respect to the one- and two-line theories are confined to the calculation of the coastal constants in longshore direction (S_0 and s, and their distribtion over beach and inshore) from simple wave and sand transport theory. However, effect on the beach shape of complicated refraction on the inshore is not implemented in the line theories. This effect on the sand transport along the beach, by a change of the direction on the *inshore* will be considered in Subsection 4.3.2 (although not in-depth).

In the holy temples of diffusitivity, from which sound and scent reflects in Chapter 3, breaker bars form a disturbing, if not annoying element. Their behavior forms the best kept bar secret (quoting Knaapen[31]). Fortunately, as has been displayed in Chapter 2, the breaker bars behave like modest alien guests. Bar-period averaged, diffusion theory keeps its validity.

However, this implies, that on the time scale, according to which a non-equilibrium profile restores itself to an equilibrium profile, only global information is obtained.

The approach "A2" ("wave climate", Fig. 4.1) does not lead always to better results than A1, as much depends on the accuracy of the used sand transport formula and the schematization. Anyway it gives a relation between more sophisticated theories and the simple line theories.

In this chapter, some knowledge of the reader with the use of the coastal models is presumed. With respect to this is referred, for instance to Chapter 3 and to Pelnard Considère,[33] Bakker,[5,6] Swart,[38] Casteleyn,[21] Pilarczyk and v.Overeem,[34] Bruning,[19] Andringa.[1]

Chapter 5 will treat the subject in extenso. Results are displayed in Table 4.1.

4.1.2. *Considerations on coastal equations and coastal constants*

In this subsection of the general introduction (Section 4.1), it is considered, *which* of the variables, occurring in Chapter 3 (line theories) should be specificly quantified.

Table 4.1 Comparison of values of s/h, found from hindcasting of field data.

		Sand layer between (m w.r.t. MSL)		h	s/h (m^2/yr)	Found from
1	Sylt	−8	+4	12	55.100	Analysis supply 1978 (cf. Subsection 5.3.3)
2	Ameland	−5	+8	13	110.000	Wave climate calculations (cf. Subsection 5.3.6)
2'	Terschelling	−6	+3	9	510.000	
3	Vliehors	−5	0	5	400.000	Analysis sand wave analysis coastal behavior 1860–1960
4	Texel	−5	+10	15	136.000	Analysis coastal behavior 1860–1960
5	Holland coast	−5	+20	25	71.300	Wave climate calculations analysis coastal behavior 500–1980 (cf. Subsection 5.3.5)
6	Rijnland coast	−5	+15	20	50.000	Kalman filtering
7	Scheveningen	−7	+18	25	120.000	Analysis coastal behavior 1900–1930 + wave climate computations
8	Cadzand	−8	+4	12	40.500	Analysis coastal nourishment 1978 (cf. Subsection 5.3.4)

Shaping a coastal model is as well feeling as mathematics: apart from substitution of coastal constants proper boundary conditions should also be applied.

Both aspects (constants and boundary conditions) will come into focus below.

The coastal equations for the two-line theory are (cf. Eqs. (3.45) and (3.66)):

$$h_1 \frac{\delta y_1}{\delta t} = s_1 \frac{\delta^2 y_1}{\delta x^2} - s_y(y_1 - (y_2' - W)),$$
$$h_2 \frac{\delta y_2}{\delta t} = s_2 \frac{\delta^2 y_2}{\delta x^2} + s_y(y_1 - (y_2' - W)). \qquad (4.1a,b)$$

Generally, the indices "1" and "2" refer to beach and inshore, respectively. The x-axis is in the main coastal direction and y_1, y_2' denote the y-coordinate of beach and inshore; W is the equilibrium distance between beach and inshore, t is the time; h_1 and h_2 are the thickness of the sand layers, representing beach and inshore. In the following, $y_2' - W$ will be denoted by y_2.

For the definition of beach and inshore and for the derivation of Eq. (4.1a,b) is referred to Section 3.3 and to Bakker,[5] or to Pilarczyk and v.Overeem, 1986.[34]

Refer to Chapter 5 concerning practical examples of solutions.

The first part of these equations indicate, that accretions at beach and inshore are proportional to their curvature, with proportionality constant $s_{1,2}$. The second part of these equations covers the effects of cross-shore transport from beach to inshore, contained in the last term on the righthand side of Eq. (4.1a,b). This transport S_y is linear with the distance between beach and inshore (proportionality constant s_y) and constant zero when this distance is the equilibrium distance W.

Equations (4.1a,b) result (Chapter 3) from continuity and the dynamic equations for the transport S_1 along the beach:

$$S_1 = S_{01} - s_1 \frac{\partial y_1}{\partial x} \tag{4.2}$$

and a similar equation for the transport S_2 along the inshore; in Eq. (4.2) S_{01} is the transport, when the beach is parallel to the x-axis.

The one-line theory (see for instance Section 3.2 or the initiating paper of Pelnard Considère,[33] and further Bakker and Edelman[4]) only deals with a coastline y, governed by Eq. (4.1) without indices and without the last term; as proportionality constant one just finds s.

The one-line theory starts from a similar equation as Eq. (4.2) for the transport S along the coast, which is found by omitting all indices "1" in Eq. (4.2).

Consider the two-line theory; most of the following can be translated easily to the one-line theory as well, which is left to the reader. The depths h_1 and h (i.e. $h_1 + h_2$) should be estimated from prototype data; on the effect of this choice shall be referred to in the Sylt case (next chapter; Subsection 5.3.3).

Constants to be found are: the proportionality constants $s_{1,2}$, s_y and the equilibrium distance W. Generally (with proper boundary conditions) knowledge concerning S_{01} and S_{02} is not necessary for solving Eqs. (4.1a,b); inversely, one may not expect, in general, to get information about those constants from the coastal behavior.

Furthermore, for solving the equations, the boundary conditions should be known.

For hindcasting, one can start from known behavior at the boundaries (case "Vlieland", to be treated in Chapter 5; cf. Bakker[6]); generally however, after hindcasting and calibration one also wants to forecast.

The accuracy, with which the future boundary conditions can be determined can be decisive for the accuracy of the prediction.[24] One should estimate those conditions with an eye on the physics of the process.[19,34] Possible boundary conditions to apply are:

- no changes at infinity ($y = 0$ for $x = \inf$);
- extrapolation of existing trends at a certain site $x = x_1$;
- all sand transport hampered at $x = x_1$.

Mathematically, the first two cases do not give difficulties, but in the last case (using the one-line theory as example) one should know s as well as the transport S_0 when the coast is in x-direction, in order to find S_0/s, being (according to Eq. (4.2)) the coastal direction at $x = x_1$.

While often the coastal constants are uniform along the coast, in the case of harbor moles the change of wave height and direction around the harbor results in variation of the coastal constants along the coast.

The line-theories in coastal dynamics are based upon a philosophy of wave-induced transport in the surf zone. However, tidal-induced transport can be taken into account to a certain extent by introducing source- or sink-terms "p" in Eq. (4.1). Also p is a coastal constant, it is non-uniform in longshore direction. Introduction of those source- and sink-terms is a subject of the paper of Dijkman *et al.*[24] Refer to the next chapter: Subsection 5.3.5: "the Holland coast".

4.2. Calculation of Coastal Constants from Hindcasting of Field Data

4.2.1. *Direct comparison coastline computations with field data*

Investigating sand waves on the Dutch Wadden island (Vlieland;[6] cf. Subsection 5.2.1) Bakker derives the coastal constant s/h of the one-line theory from nature by correlating the coastal curvature with the accretion and erosion of the coast.

In a more sophisticated way, de Vroeg[40] and de Vroeg *et al.*[41] calculate the coastal constants $(s/h)_{1,2}$ and S_y from the coastal data of beach and inshore by application of Kalman filtering. In Subsection 4.2.2, more details are given. Kalman filtering combines the advantage of deterministic and statistical approaches. It is an algorithm which combines the measurements and physical knowledge of the system in order to achieve an optimal estimate of the behavior of the system. The method has been applied for the area between IJmuiden and Scheveningen.

Bakker *et al.*[11] Bruning,[19] and Andringa,[1] from which results are described in Chapter 5, use a kind of CAD by displaying as well "natural coastal behavior in the shape of the lines of beach and inshore" as "coastal model results" on the same screen and calibrating the results by changing the coastal constants, using the method of the least squares. Hindcasts concerning the coastal behavior could be obtained in this way, accurate up to 10 m.

Table 4.1 gives values of the coastal constant s/h, found from various investigations. Also some results, found from wave climate calculations (Section 4.3) are mentioned.

Values of s_1 of the order of 1–2 million m^3/yr seem usual, with Sylt and Cadzand at the low side and Scheveningen on the high side. This last value might be overestimated somewhat.

Values found for s_y — taking the separation plane between beach and inshore at the level at the vicinity of the low-water line — vary between 1.5 m/yr (Sylt; Subsection 5.3.3) to 3 m/yr (Cadzand; Subsection 5.3.4) for sand supplies (laid down by Pilarczyk *et al.*[34]). For long period features smaller values appear to be most probable (order of 0.5 m/yr, Fig. 4.2). Some non-linearity of s_y is not unlikely. Janssen[27] finds for Scheveningen, for a separation plane at about NAP-3 m, values of the order 0.1–0.2 m/yr.

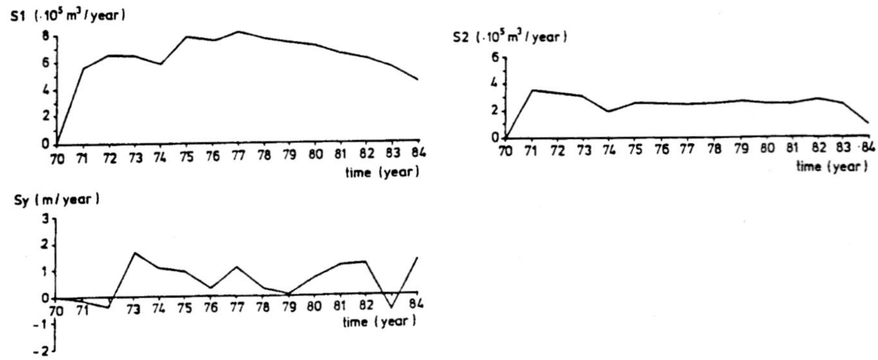

Fig. 4.2 Calculated estimates for the coastal constants (s_1, s_2, s_y) as function of time, using Kalman filtering. Data Rijnland 1970–1980. From de Vroeg.[40]

4.2.2. *Kalman filtering*

4.2.2.1. *Kalman filtering in general*

The Kalman filter combines the advantages of the deterministic and the statistical approach, and has proven to be very suitable for recursively updating parameter estimates of time-varying model coefficients, based upon the latest data information. It is an algorithm which uses measurements, a physical model of the system and knowledge of the inaccuracy (noise) of both the measurements and the model, in order to achieve an optimal estimate of system behavior.[29] An application of the method has been explained by de Vroeg *et al.*,[41] concentrating on the application of the one-line theory. In anticipation, here the application for the two-line theory of linear Kalman filtering will be highlighted.

The method is based on the concept of on the one hand the existence of a "system", not necessary stationary, but anyway systematically changing and on the other hand periodical measurements concerning some characteristics of the system, blurred by "noise". The noise has two reasons:

- reality is more complicated than the assumed system;
- measurements show errors.

It is assumed, that prognostics of the characteristics to be measured can be found from a linear transformation of the system. This linear relationship implies on the other hand, that "backtransformation" of information on the characteristics involves information on the system.

The filtering procedure consists of cycles of linear transformations of respectively:

- the system (being transformed from the state at one time step to the state at the next time step);
- the anticipated result of the measurements at this "next timestep" (found from the state of the system);
- comparison between measurements at this "next timestep" and anticipated result; this gives a gain of information on the system, which is fed back by making a correction on the state of the system at the considered time. The "weight" assigned on the information derived from the measurements in relation to the weight assigned to the anticipated result depends on the given noise characteristics. It is derived in a mathematical way, starting from the least-square error standards.

All the information, as well concerning the backtransformation of measured characteristics to system as well concerning the relative importance of the measurements compared to "theory" is contained in the "Kalman gain matrix" $K(t)$, which (linearly) transforms the difference between measured and anticipated characteristics to a correction (valid for the present time step) of the system. This gain matrix forms the essence of the Kalman filtering. From it results the gain of information on the system, found from the last measurements.

With the last-mentioned transformation the cycle is closed, apart from the fact, that also the gain matrix is corrected after each time step, depending on the ratio between the noise in the measurements and the noise in the latest estimate (the larger the noise in the measurements, the less information the measurements give about the system).

Numerically, the use of Kalman filtering involves, that all measurements are elaborated in a sequential way, which means, that the computer may "forget" all previous measurements and all previous states of the system when elaborating "actual" measurements. This means, that only a confined amount of storage is required. All the relevant information about these former measurements is contained in the present state of the gain matrix, which is successively improved.

An application of Kalman filtering will be treated in this book, ad b. below an application for estimation of the coastal constants in the two-line theory will be exposed.

For the general equations concerning the Kalman filtering the reader is referred to de Vroeg *et al.*[41]

4.2.2.2. *Application on the two-line model*

Applying Kalman filtering for the two-line theory, as "system" has been defined: the system of constants s_1, s_2, s_y having principally a "noise" var(s_1), var(s_2), var(s_y). No covariance among s_1, s_2, s_y is assumed.

This definition of the system involves a simplified application of the Kalman filtering theory, for in the present case the system in principle is stationary, where, generally, the system is allowed to change in the course of the time. Still, estimates of the system may vary in course of time, because new information is delivered by the various measurements of the characteristics of the system.

These "characteristics" are defined as the y-coordinates of the lines of beach and inshore in the various measuring points.

Thus, defining a measuring equation as:

$$\underline{z}(t) = H(t-1) \times \underline{x}(t) + \underline{v}(t), \tag{4.3}$$

where $\underline{z}(t)$ is the measured state, $\underline{x}(t)$ is the system state and $\underline{v}(t)$ denotes the inaccuracy of the measurements,[c] the matrix H equals[d]:

$$\left|\begin{matrix} A_1(y1_1 - 2y1_2 + y1_3) & 0 \\ 0 & A_1(y2_1 - 2y2_2 + y2_3) \\ A_1(y1_{n-2} - 2y1_{n-1} + y1_n) & 0 \\ 0 & A_2(y2_{n-2} - 2y2_{n-1} + y2_n) \end{matrix}\right.$$

$$\left.\begin{matrix} B_1(-y1_2 + y2_2 - W) & y1_2 \\ B_1(-y2_2 + y1_2 + W) & y2_2 \\ B_1(-y1_{n-1} + y2_{n-1} - W) & y1_{n-1} \\ B_1(-y2_{n-1} + y1_{n-1} + W) & y2_{n-2} \end{matrix}\right|,$$

where the column vectors $\underline{z}$ and $\underline{x}$ are, respectively (written for typographical reason as row vectors):

$$\underline{z} = (y1_2, y2_2, \ldots, y1_{n-1}, y2_{n-1}) \tag{4.4a}$$

and

$$\underline{x} = (s_1, s_2, s_y, 1). \tag{4.4b}$$

In the matrix H, the constants $A_{1,2}$ and $B_{1,2}$ are equal to $\Delta t/(\Delta x^2 h_{1,2})$ and $\Delta t/h_{1,2}$ respectively.

In the system definition of $\underline{x}$, the last unit "1" is added for numerical reasons.

The coefficient W, defining the equilibrium distance between beach and inshore, is not filtered at all. If s_y as well as W would have been filtered, a non-linear filtering technique should have been applied. The equilibrium distance is estimated from measurements.

The sensibility for the applied value of W will be evaluated when considering the results of the Kalman filtering (Subsection 4.2.2.3).

For the way of calculating the gain matrix is referred to de Vroeg *et al.*[41]

---

[c]i.e. "New measured state = calculated (new) state of the system plus noise (cf. Subsection 4.2.2.1).

[d]Discretized y-coordinates of the system are indicated by (elements of) arrays $y1_{(n)}$ and $y2_{(n)}$ respectively (the accent of y_2 has been omitted for topographical reasons).

4.2.2.3. *Results*

Kalman filtering has been applied[40] for determining the coastal constants of the Rijnland coast. Later on the method has been applied on the total uninterrupted coast of Holland; this last investigation has been reported on by Dijkman *et al.*[24]

Figure 4.2[40] shows some results for the two-line theory, applied on the Rijnland coast (range number RSP km 58.00–98.00); for results of application for the same area of Kalman filtering starting from the one-line theory is referred in Refs. 40 and 41.

The "beach zone" included the dunes as well; it was schematized as the area between NAP (about MSL) +15 m and the low-water line (NAP −0.72 m); the inshore zone between NAP −0.72 m and NAP −5 m.

Figure 4.2 shows the values of the system (s_1, s_2, s_y) in course of time; the original estimate (1970) of (s_1, s_2, s_y) was $(0, 0, 0)$. Measuring noise has been estimated as 125 m^2 for the beach zone and 700 m^2 for the inshore zone. The result did not change considerably with assumptions of other values for this noise. The value of s_y showed sensibility for the magnitude, assumed for W (taken 280 m in Fig. 4.2); the effect of this variation on s_1 and s_2 was much less.

N.B.: mind, that values for the coastal constants depend not only on the site, but also on the schematization of the coast with respect to the level of the separation plane between beach and inshore. Evidently, as well the fact of including the dunes (if present) in the "beach area" plays an important role. Theoretically, this matter will be considered in Subsection 4.3.3, to which the present remark may refer already.

4.3. Computation of Coastal Constants in Longshore Direction from the Wave Climate

4.3.1. *Introduction to Section* 4.3

Naturally, the coastal constants should be related to the wave climate. Consider first the constants, determining the longshore transport.

However, trying to calculate these constants one has to cope with one of the nasty errors of nature: this shows often only a far and maybe accidental resemblance to mathematical feasible schematizations.

In the one- and two-line theories there exists a distinct level, above which all contour lines are unstationary; being all parallel to each other according to the one-line theory or parallel to either the "line of the beach" or "line of the inshore" respectively according to the two-line theory. Below this level, there should be, theoretically, a (stable) horizontal plane (Section 3.2 or Bakker,[5] Pelnard-Considère[33]). Along the Dutch beaches, such a stable horizontal plane does not occur. In the mathematical model, one may mitigate the condition of such a plane, by assuming a stable sloping area below the dynamic area. Thus one finds two sloping areas (the lower one stable and the upper one dynamical) for the one-line theory (Fig. 4.3) and three (beach, inshore and a stable one, Fig. 4.4) for the two-line theory. In Subsection 4.3.3 it will be shown, that the choice of the level h between mobile and immobile coastal zone (or of $h_{\text{tot}}(= h_1 + h_2)$ in the case of the two-line theory) is crucial for the result of the calculated "s".

There is physics in this choice: mind the assumption of the two-line theory, that the groin reaches to a depth h_1.

As will be suggested by Fig. 4.5, a large angle of wave incidence and a "deep" level of the foot of the harbor mole[e] or groin (($c_1/c_0 \approx 1$), Fig. 4.5) might trigger instabilities in beach shapes.[f]

In the following (Subsection 4.3.2), first for the one-line theory the magnitude of the coastal constants S_0 and s as depending on deep water wave conditions is calculated, starting from the assumption of stationary and uniform wave conditions. Next, the effect of the choice of the level h (the lower boundary of the movable zone) on the constant s will be investigated.

Furthermore, for the two-line theory, some remarks will be made on the effect of the direction of the inshore on the longshore transport along the beach.

After this (Subsection 4.3.3), the way of calculation of the coastal constants in the case of variable wave conditions (i.e. in the case of a "wave climate") is elucidated. Next (Subsection 4.3.4), a program is described for calculation of the coastal constants, where the wave conditions near the beach do not need to be uniform along the beach, but may vary, for instance because of the vicinity of a harbor mole.

Some "final" remarks concerning Section 4.3 are made in Subsection 4.3.5; Subsection 4.3.6 may be considered as an Appendix to Section 4.3, giving the underlying computations, which are referred to in Subsection 4.3.2.

---

[e]Disregarding features as variable wave directions, diffraction, shelter, effects of tidal currents, etc., cf. Subsection 5.3.5 for the "real" thing.

[f]Negative coastal constants; cf. Chapter 3, Eqs. (3.35) and (3.36).

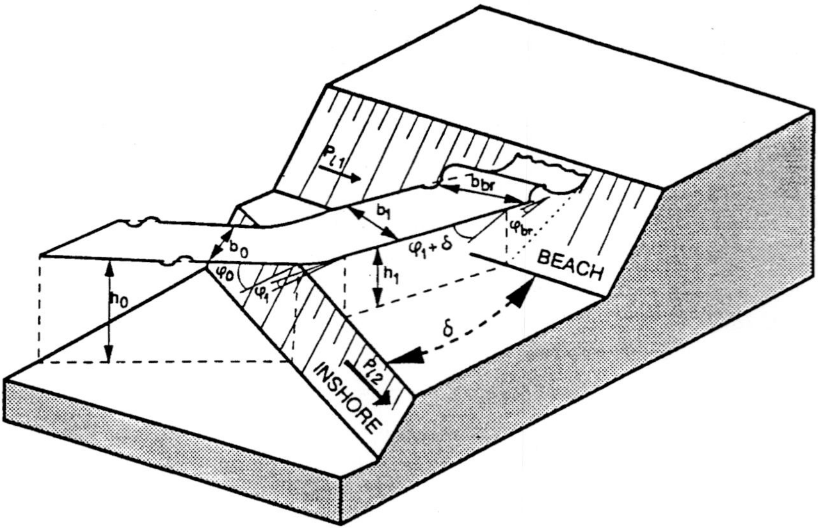

Fig. 4.3 Three-dimensional view of beach profile with a wave ray approaching the beach. P_{l1} = longshore component of the wave energy along the beach. P_{l2} = longshore component of the wave energy along the inshore.

With respect to the constant s_y concerning cross-shore transport, refer to Section 4.4.

4.3.2. *Waves stationary and uniform along the coast*

Ten Hoopen and Bakker[26] schematize[g] the coastal topography as shown in Fig. 4.3, consisting of a beach and an inshore, having parallel depth contours, with a horizontal plane (at depth h_1) in between. Beach and inshore make an angle δ with each other.

Starting from uniform and stationary wave conditions alongshore in deep water, this schematization allows for calculation of longshore transport along beach and inshore respectively. Use is made of a concept, originating from Svasek and Bijker.[20,37] This leads to longshore transport, proportional to "the longshore component of the loss of energy flux between two depth

[g]They treat a more general case than considered in the following, i.e. the "beach" is bordering a shoal, which eventually may be submerged.

contours",[h] which for beach (Pl_1) and inshore (Pl_2) can be specified by the following formulation[26] (taking the cosine of the breaker angle equal to 1):

$$\begin{aligned} Pl_1 &= Kh_1^3 \frac{\sin(\varphi_1+\delta)}{c_1}, \\ Pl_2 &= K(h_{\mathrm{br}}^3 - h_1^3)\frac{\sin(\varphi_0)}{c_0}, \end{aligned} \qquad \text{when } h_{\mathrm{br}} > h_1, \tag{4.5a}$$

$$\begin{aligned} Pl_1 &= Kh_{\mathrm{br}}^3 \frac{\sin(\varphi_1+\delta)}{c_1}, \\ Pl_2 &= 0, \end{aligned} \qquad \text{when } h_{\mathrm{br}} < h_1. \tag{4.5b}$$

Here K is a proportionality constant, expressed in wave characteristics and specified in Eq. (4.18b) in Subsection 4.3.6;[i] c denotes the wave celerity and φ the angle of wave incidence; the indices "0" and "1" indicate the site, respectively, at deep water and at the upper boundary of the inshore. Also the value of the breaker depth h_{br} in Eq. (4.5) will be expressed in wave conditions in Subsection 4.3.6 (cf. Eq. (4.18a)).

Mind, that Eq. (4.5b) was quoted already in Section 2.4; it is Eq. (2.2) in another shape.

Equations (4.5a,b) form the basis of most computations, carried out by Van Bochove,[18] Casteleyn,[21] Bakker,[12] and Jas[28] on the coastal constants $S_0, S_{01}, S_{02}, s, s_1, s_2$. When Eq. (4.5b) is valid (small waves), one finds back the stable lower zone and the dynamical upper zone of the one-line theory. Computations concerning S_0 and s with respect to the one-line theory[j] are given into some detail in Subsection 4.3.6; calculations concerning the other mentioned parameters are carried out in a similar way, cf. Bakker[12]; those calculations start from a topography given in Fig. 4.4 (three slopes) instead of the one given in Fig. 4.3 (two slopes).[k]

Results (Fig. 4.5) give dimensionless expressions for the mentioned coastal constants, where $A_1 H_0^2 c_0$ is taken as reference; A_1 being a proportionality constant (Eq. (4.23)) relating sediment transport to longshore

---

[h] As Battjes points out, a flux is a scalar. Thus it cannot have a "longshore component". Here the meaning of the used term will be specified as a vector in longshore direction with length $\Delta \sin\varphi \cos\varphi$.

[i] Subsection 4.3.6 can be seen as an Appendix with computations, added to the present Section 4.3.

[j] More accurate than mentioned before: putting $(\cos\varphi_{\mathrm{br}}^2)^{1/5}$ to 1 instead of $\cos\varphi_{\mathrm{br}}$.

[k] Thus accounting for turning of depth contours of the inshore zone by dynamical shore processes, which is not involved in Eq. (4.5a).

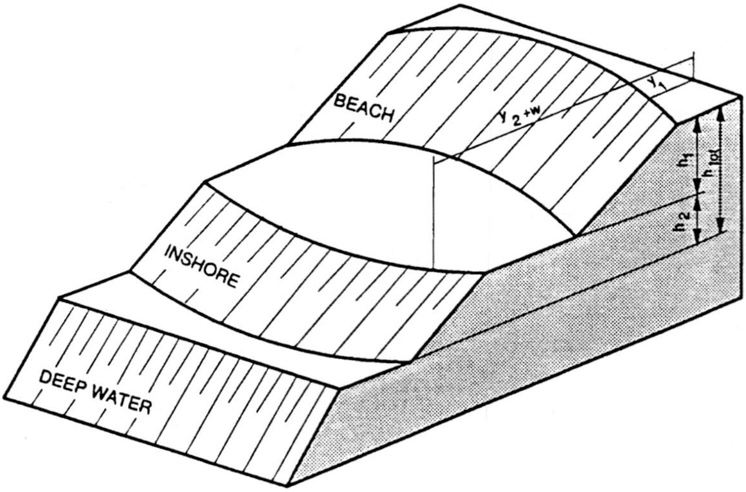

Fig. 4.4 Three-dimensional view of the bottom schematization, used for the calculation of the coastal constants according to the Svasek-formula.

component of energy flux and H is the wave height. Independent variables are wave direction, wave steepness and (for s) the location of the level h of the toe of the beach area.

Figure 4.5 (one-line theory) shows S_0 (dotted line) and s (drawn lines) as function of the deep water angle of wave incidence; curves are valid for various values of c_1/c_0.

The indicated values are valid for such wave steepness S_0', that the ratio c_{br}/c_0 is equal to 0.5 in the case that the waves are normally incident to the inshore as well as to the beach. For other wave steepness S_0' (only) the vertical scale of Fig. 4.5b has to be adapted linearly.

As will be figured out in Subsection 4.3.6: this adaptation is ruled by Eqs. (4.22) and (4.23), giving (for $\varphi = 0$) the relation between wave steepness S_0' and c_{br}/c_0.

For large values of as well angle of wave incidence as well c_1/c_0, negative s-values are found. Bakker and Edelman[4] point out, that coastlines can be unstable under those conditions. This may be one reason for the origin of beach cusps.

For a given deep water wave steepness, the constant s increases when c_1 decreases, relative to c_0 (Fig. 4.5), i.e. when the depth h up to which the contour lines are parallel to the shoreline decreases (Fig. 4.5).

Figure 4.5 depicts a physical explanation. In Fig. 4.5, the contour lines up to very deep water are parallel to the shoreline. Before breaking, wave crests have the opportunity to refract, becoming more or less parallel to the plan-shape of the shoreline. Therefore breaker angles will be small and so will be the transport gradients. The opposite is the case when the wave crests do not have the opportunity to adapt itself to the beach planform (Fig. 4.5).

For small values of c_1/c_0 only positive values of s are found (Fig. 4.5), because also at large angles of deep water wave incidence, the angle of incidence at the shoal of depth h_1 does not become large.

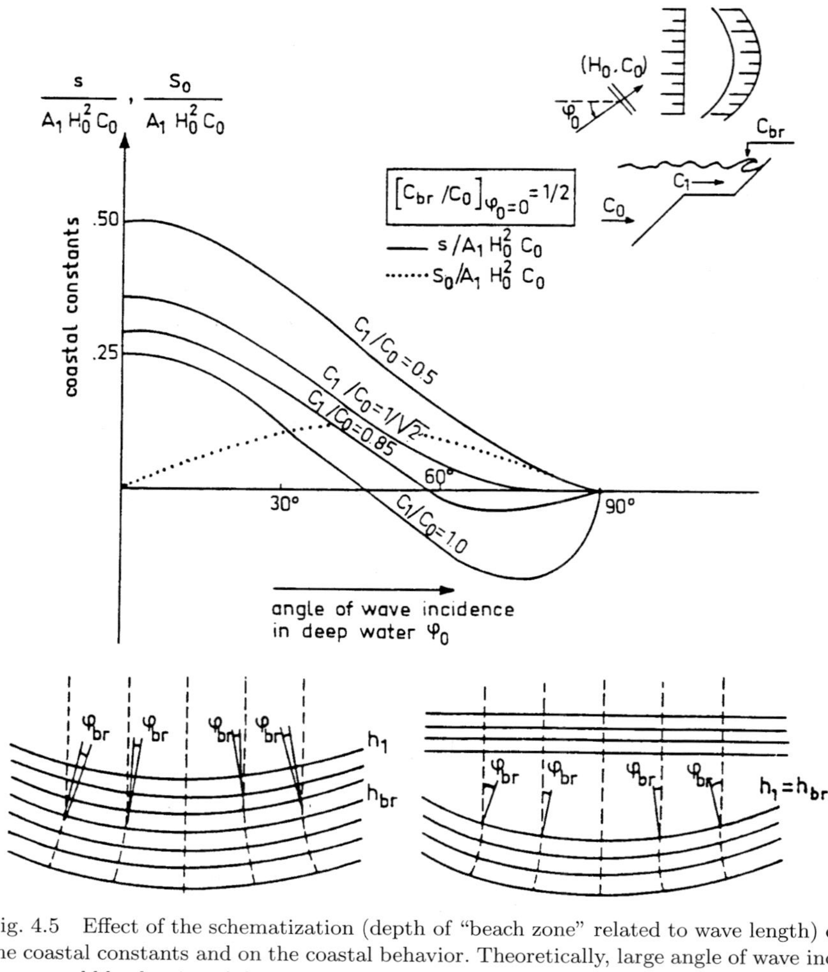

Fig. 4.5 Effect of the schematization (depth of "beach zone" related to wave length) on the coastal constants and on the coastal behavior. Theoretically, large angle of wave incidence could lead to instability of the coast (negative coastal constant s). Mind Chapter 5 concerning other aspects.

Depicted is: $s/A_1^2H_0^2c_0$ and $S_0/A_1H_0^2c_0$ as function of the angle of wave incidence and the ratio c_1/c_0. The curves are valid for a ratio $(c_{\rm br}/c_0)_{\varphi=0} = \frac{1}{2}$; the vertical scale is proportional to this last mentioned ratio.

4.3.3. *Effect of wave climate (uniform along the coast)*

Contrarily to Subsection 4.3.2 (stationary waves), the present Subsection 4.3.3 will take the effect of the wave climate into account. The assumpion (Subsection 4.3.2) that, far from the coast, this climate is uniform will still keep its validity. Locally, conditions can vary, because of, for instance, coastal curvature.

Thus, the influence of variable wave conditions on coastal constants will be considered.

Bakker *et al.*[8] and Bruning[19] calculate the constants s_1 and s_2 for the longshore transport from the wave climate, starting from Eq. (4.5a,b), taking the probability of each wave class into account (Subsection 4.3.4).

By giving the angle δ between beach and inshore, respectively, a positive value and a negative value of the same magnitude, one finds (by subtracting the values for the transport along the beach in both cases and dividing by 2δ) the rate of change of the transport s_1 along the beach, depending on the beach orientation.

The derivation used for the PELNARD-CONSIDERE-formula (3.6) keeps its validity when the littoral drift S, the stationary transport S_0 and the constant s are averages over a year instead of instantaneous values. However, it will not be directly clear, which value has to be taken for $h_{\rm tot}$. In order to estimate $h_{\rm tot}$ it is useful to compute first the distribution of the littoral drift in a range, perpendicular to the coast. An example of such a distribution gives Fig. 2.25. The way of calculation will be described below.

First, use is made of the Svasek assumption (cf. Subsection 4.3.2): the littoral drift *between two depth contours* is proportional to the longshore component of the loss of energy *between those contour lines*. Then Svasek's assumption can be written as:

$$\Delta S = A_1' \Delta(H^2 c) \sin\varphi \cos\varphi$$

Assuming[1] in the breaker zone proportionality between local depth h and wave height H, i.e. $H = \gamma h$, and proportionality between wave celerity C

[1]Taking the ratio group velocity/phase velocity $n_{\rm br}$ in the breaker zone equal to 1.

and $\sqrt{(gh)}$, i.e. $C = F_r\ \sqrt{(gh)}$, in which F_r is a Froude number (assumed constant) the difference in H^2C between two adjacent depth contours equals:

$$\Delta(H^2C) = [d(\gamma^2 F_r g^{1/2} h^{5/2})/dh] = (5/2)\gamma^2 F_r g^{1/2} h^{3/2} \Delta h$$

Now first the stationary transport S_0 will be computed, according to the Svasek theory. In this case, all contour lines are parallel and Snel's law is valid:

$$\sin\varphi/\sin\varphi_{\text{br}} = c/c_{\text{br}} = \sqrt{(h/h_{\text{br}})}$$

Thus ΔS can be expressed in φ_{br}:

$$\Delta S = \frac{5}{2} A_1' \gamma^2 F_r g^{1/2} h^2 h_{\text{br}}^{-1/2} \Delta h \sin\varphi_{\text{br}} \cos\varphi_{\text{br}}$$

where $\cos\varphi$ has been assumed equal to $\cos\varphi_{\text{br}}$.

The total transport S_0 will be the integration over the breaker zone, i.e.:

$$S_0 = \int_0^{h_{\text{br}}} (\Delta S/\Delta h) dh = \frac{5}{6} A_1' \gamma^2 F_r g^{1/2} h_{\text{br}}^{5/2} \sin\varphi_{\text{br}} \cos\varphi_{\text{br}}$$

Relating the *total* littoral drift to the *total* energy flux (CERC formula; cf. Eq. (2.4)) the equation above shows, that those are related with a proportionality constant : $(5/6)\ A_1'$; per definition, this coefficient is denoted with "A_1":

$$A_1 = \frac{5}{6} A_1' \Rightarrow A_1' = \frac{6}{5} A_1$$

As the former derivation demonstrates, A_1' differs from A_1, which is caused by the physical fact, that refraction proceeds in the surf zone: the angle of incidence decreases, which affects (in the Svasek assumption) the littoral drift.

In case of variable wave conditions (taking $\cos\varphi_{\text{br}} \cong 1$ and writing $\sin\varphi_0/C_0$ for $F_r^{-1} g^{-1/2} h_{\text{br}}^{-1/2} \sin\varphi_{\text{br}}$), for the yearly littoral drift between two depth contours $h - \Delta h/2$ and $h + \Delta h/2$ is computed:

$$\Delta S = 3A_1\gamma^2 F_r^2 g h^2 \Delta h \sum_{\substack{\text{all wave classes} \\ \text{for which } h < h_{\text{br}}}} (\sin\varphi_0/c_0)\Pr(H, T, \varphi) \tag{4.6}$$

in which $\Pr(H,\ T,\ \varphi)$ denotes the probability of occurrence of a wave class for which H, T and φ are between certain margins (for instance

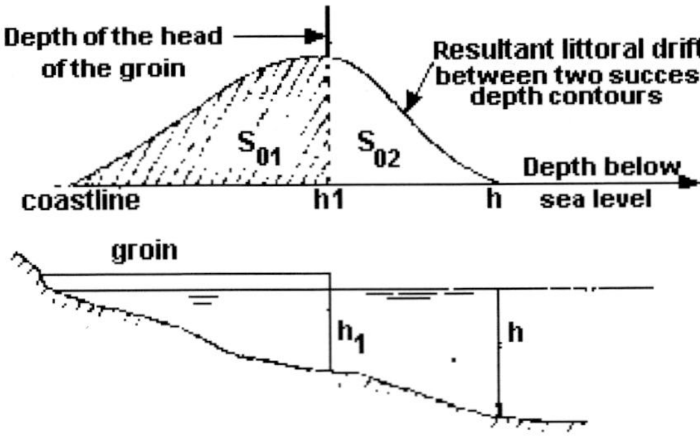

Fig. 4.6 S_{01} is longshore transport hampered by the groin; S_{02} is undisturbed longshore transport.

$1/2\,\mathrm{m} < H < 1\,\mathrm{m}, 5\,\mathrm{s} < T < 6\,\mathrm{s}, 30° < \varphi < 60°$). More details about the computation are given by Bakker.[7]

As an example, Fig. 4.6 depicts ΔS for a unit $\Delta h = 0.5\,\mathrm{m}$ as function of the waterdepth h; depicted are respectively the partial sums over all (wave class-)directions giving a transport in Northward, respectively Southward direction and (subtracted) its resultant: the distribution of the transport over the depth. Mind, this is only the wave-driven transport.[m]

From the distribution of the transport S_{01} and S_{02} are found (Fig. 4.6):

$$S_{01} = \Sigma\, \Delta S \tag{4.7a}$$

for $0 < h < h_1$

$$S_{02} = \Sigma\, \Delta S \tag{4.7b}$$

for $h > h_1$

After that, s_1 and s_2 can be found by repeating the computation according to (4.6), but with a "wrong'" coastal direction, which has been turned over an angle δ say 15°.

[m]Mind, this is only *wave-driven* transport. An impression, how this relates to the *tide-*driven transport is given in Fig. 2.25. Remember, that in Fig. 2.25 the transport is expresssed per m of beach width and in Fig. 2.36 per half meter of depth. Where the last-mentioned is (with a slope of ca 1:100) ca. 50 times as much. Refer to Fig. 2.29b for the location of Scheveningen.

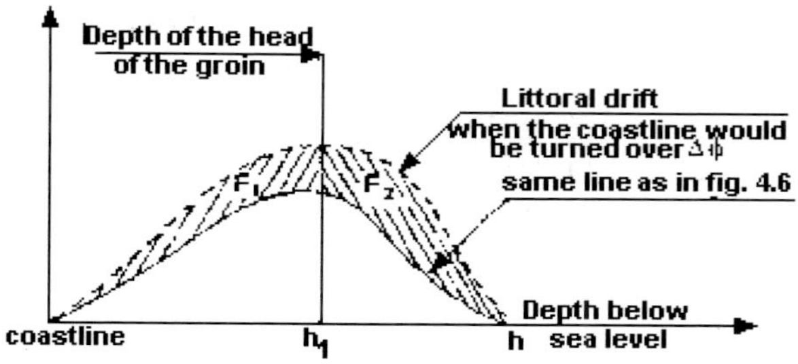

Fig. 4.7 Definition of s_1 and s_2 as transport differential.

This gives the interrupted line in Fig. 4.7, instead of the solid line, which represents the transport distribution for the original coastal direction.

Now s_1 equals F_1/δ (F_1 is the hatched area in Fig. 4.7 at the left-hand side) and $s_2 = F_2/\delta$.

From the transport distribution, also a reasonable guess about h_{tot} can be made.

4.3.4. *The KC/KL programs*

A further extension, permitting variability of wave conditions alongshore (Subsection 6.3.2), is given by the KC/KL-programs.

Constants can be calculated with the "KC"-program of Rijkswaterstaat and Delft University of Technology ("KC" stands for "Kustconstanten", Dutch for "Coastal Constants"). Coastlines afterwards are calculated with the "KL"-program (-"kustlijn" - "coastline"). This gives the possibility to calculate the coastline near harbour moles.

The effect of eddies, occurring during the tidal cycle can be incorporated in introducing a sources-and sinksterm "p" in Eq. (4.1a,b). An application for the coast of Holland is treated in Subsection 5.3.5.

The KC/KL-program has been developed in cooperation between Rijkswaterstaat and Delft University of Technology. It is based upon investigations of Bakker,[5] Bakker *et al.*,[8] ten Hoopen and Bakker[26] and Bochove.[16–18] Programming and documentation has been performed initially by A.Roos and H.W.M.Bots where the final version of the program has been made by A.Casteleyn, who also reported on this subject.[21]

The programs KC and KL are built up in a modular way.

The program KC starts from a bottom topography, which can be described in the following way.

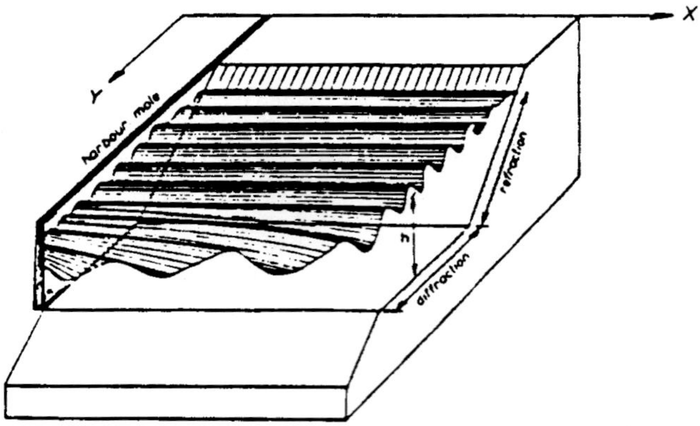

Fig. 4.8 Diffraction around a harbor mole.

Assumed are parallel depth contours from deep water to a certain level (arbitrarily to choose), where the head of a harbourmole is situated (the existance of this harbourmole is optional); at this depth one finds a horizontal area, where diffraction takes place; from there landward, again one finds an upward slope with contour lines parallel to the contour lines in deep water; this up to a level h'_1 (with respect to mean sea level) where the waves are about to break. Here again one finds a flat area; from here upward, contour lines make an angle δ with the contour lines in deeper water (δ to be chosen arbitrarily, Fig. 4.3).

For an arbitrary number of equidistant points along the x-axis (parallel to deep water) the wave height is calculated, taking as well as diffraction on the shoal at the level the contour lines in refraction into account, of the harbour mole (Fig. 4.8).

As each computation at such is rather simple, the program allows for calculating in one run for:

- various values of δ;
- various places along the x-axis;
- various angles of deep-water wave direction;
- various water level elevations;
- various wave periods;
- three different fractions of the time, during which the tide level is 1 m lower, c.q. higher than the mean level, c.q. is at the mean level;
- various values of the deep water wave height.

Each of the various topics, mentioned above means a loop in the program, where the loop of the first-mentioned topic embraces the loop of the later-mentioned topic.

The wave climate is characterized by considering various wave classes, each characterized by a certain deep-water wave height H, period T and direction φ and a certain probability of occurrence $P_r(H, T, \varphi)$.

Calculated is the expectation of the longshore component of the wave energy flux Pl_1 along the beach:

$$E(Pl_1) = \sum Pl_1 \times P_r(H, T, \varphi) \tag{4.8}$$

where Pl_1 is given by Eqs. (4.5a) or (4.5b), depending on the breaker depth. Here h_1 is the water depth, i.e. h_1' plus the waterlevel elevation.

In the North-Sea, deep-water wave direction and wave height are determined more or less by wind-direction and windforce. These in turn determine the set-up by storm surge of the North Sea level along the Dutch coast (overviewing for a moment the effect of tide). The relevant relations have been derived from data from Dorresteyn[23] from the Royal Dutch Meteorological Institute.[23]

During the "water-level elevation"-loop the various possibilities that the water reaches the given level are considered, i.e. as well as (for instance) the possibility of small waves during flood tide as high waves (high North sea elevation) during ebb tide.

Assuming a linear relation between the littoral drift S and Pl_1 and calling $S(\delta_1)$ the value of S, when δ gets a certain value δ_1 one can find the value of the coastal constant S at a certain place as:

$$s = \frac{S(\delta_1) - S(-\delta_1)}{2\delta_1} \tag{4.9}$$

The KL- program is a program, which calculates explicitely the y-coordinate of the coastline given the coastal constants (for instance found from the KC-program). Use is made of Eq. (4.1a,b). There are various options, starting from various boundary conditions. One option is also to compare the calculations with given data concerning the behavior of the coastline.

4.3.5. *"Final" remarks*[n]

The present chapter gives the order of magnitude of coastal constants to be expected in nature and some ways of determination. However, before

[n]Subsection 4.3.6 may be considered as an appendix.

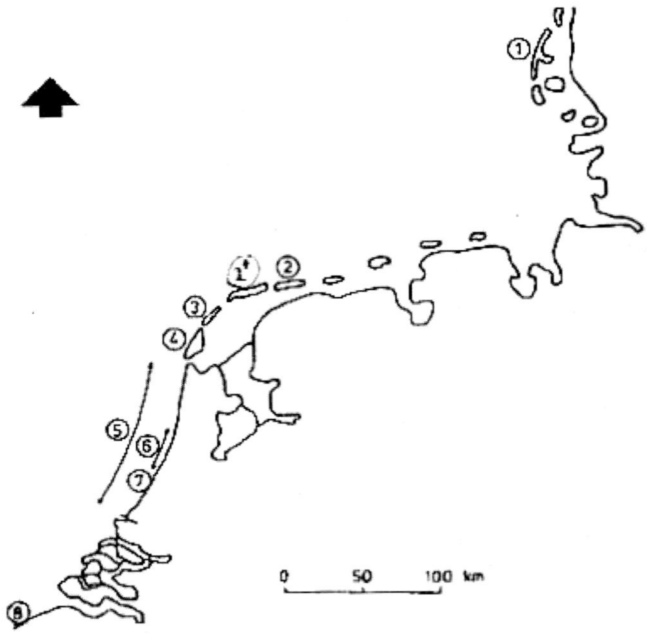

Fig. 4.9 Site investigations and coastal constants found.

(Subsection 4.3.6) diving into the details of the computations, referred at in this Section 4.3, some concluding remarks should be made:

In Table 4.1 page 183, a summary is given concerning the magnitude of the coastal constant s/h as resulted from various Dutch investigations°

Some warning should be added.

Equation (4.1) start from the assumption that beach and inshore together form a self-contained system and that no interaction between the offshore zone and the inshore takes place. If this condition is not fulfilled in nature, also sophisticated mathematical methods (as for instance Kalman filtering) can give spurious results. Also calculation of coastal constants from the wave climate gives results, depending on the schematization. For instance, Subsection 4.3.2 shows, that the choice of the level in the one-line theory, up to where the bed is assumed to be mobile does not influence not only h but also s in the constant s/h.

Coastal engineering remains an art, apart from being a science.

°A similar table has been presented earlier by Pilarczyck and Overeem *et al.*[34]

4.3.6. *Effects of schematization: the effect of the determination of the depth h1 and h2 on the coastal constants*

In this subsection, the underlying computations will be given, leading to the dimensionless transport relations, as displayed in Fig. 4.5[p] This figure depicts the coastal constants s and S_0, i.e.: made dimensionless with the unit $A_1 H_0^2 c_0$, as function of the angle of wave incidence and of the ratio c_1/c_0, the propagation velocity at the level of the separation plane at the lowest level of the beach related to the deep water propagation velocity.

In this subsection, first dimensionless ratio's will be given, relating S_0 and s of the one-line theory to the deep-water wave conditions. Schematization is according Fig. 4.3, where h_1 is assumed to be somewhat larger than h_{br}.

Later on, some results will be given concerning computations, in which part of the waves break on the inshore, starting from a schematization, given in Fig. 4.4. Under those circumstances it is assumed, that the waves start to break on the inshore, where Svasek 's assumptions concerning the longshore transport are used.

The following calculation of the transport S along the beach starts from the schematization given in Fig. 4.3 and from the assumptions, mentioned by van Bochove.[16] These include for instance the use of the CERC-formula: the transport S is proportional to the longshore component of the energy flux Pl_1:

$$S = A_1 \times (8Pl_1/\rho_{\mathrm{w}}) \tag{4.10a}$$

with:

$$8Pl_1/\rho_{\mathrm{w}} = n_{\mathrm{br}} H_{\mathrm{br}}^2 c_{\mathrm{br}} \sin\varphi_{\mathrm{br}} \cos\varphi_{\mathrm{br}} \tag{4.10b}$$

n being the ratio group velocity/phase velocity, H the wave height, c the phase velocity and φ the angle of wave incidence; the index "br" denotes breaking and $\rho_{\mathrm{w(s)}}$ the specific density of water (sand).

Including in S also the volume of pores, when the material settles, one easily finds the constant A_2 in the equation $I_1 = A_2 \times Pl_1$, where I_1 is the immersed weight of transported material equal to:

$$A_2 = \frac{8(\rho_{\mathrm{s}} - \rho_{\mathrm{w}})a_1}{\rho_{\mathrm{w}}} A_1 \tag{4.11}$$

---

[p] page. 4–11.

$a_1$being the ratio of content of solid material (1-porosity ratio). Wave conditions are assumed uniform and stationary.

Furthermore, it is assumed, that the breaker depth is smaller than h_1. Then for the "longshore component of the wave energy flux" Pl_1 along the beach one finds (making use of conservation of energy flux between wave rays at a mutual distance b; cf. Fig. 4.3):

$$Pl_1 = \frac{b_0}{b_{\text{br}}} n_0 E_0 c_0 \sin\varphi_{\text{br}} \cos\varphi_{\text{br}} \tag{4.12a}$$

where $n_0E_0c_0$ is the wave energy flux in deep water; E denoting the wave energy per unit of area, c the phase-velocity and nc the group-velocity. The index "0" refers to deep water and "br" to breaker conditions.

$$E = \rho_{\text{w}} g H^2/8 \tag{4.12b}$$

In (4.12b), g denotes the acceleration of gravity.

Using Snell's law, $\sin\varphi_{\text{br}}$ can be expressed as:

$$\sin\varphi_{\text{br}} = \frac{c_{\text{br}}}{c_1}\sin(\varphi_1 + \delta) \tag{4.13}$$

where the subscript "1" refers to the values at the level of the "toe" of the beach (Fig. 4.3), and where φ_1 is the angle of wave incidence on the upper side of the inshore.

Applying Snell's law once again, one may replace c_1 in (4.13) by the deep water value c_0, finding:

$$\sin\varphi_{\text{br}} = \frac{c_{\text{br}}}{c_0}\frac{\sin\varphi_0}{\sin\varphi_1}\sin(\varphi_1 + \delta) \tag{4.14}$$

Make use of van Bochove[18]:

$$\frac{b_0}{b_{\text{br}}} = \frac{\cos\varphi_0}{\cos\varphi_1}\frac{\cos(\varphi_1 + \delta)}{\cos\varphi_{\text{br}}} \tag{4.15}$$

Substitution of (4.14) and (4.15) into (4.12) gives for Pl_1:

$$Pl_1 = n_0 E_0 c_{\text{br}} \frac{\sin 2(\varphi_1 + \delta)}{\sin 2\varphi_1} \sin\varphi_0 \cos\varphi_0 \tag{4.16}$$

Assume a ratio γ between the wave height H_{br} and the water depth h_{br} at the time of breaking and take the celerity of wave propagation of a nearly breaking wave equal to:

$$c_{\text{br}} = F_r\sqrt{g h_{\text{br}}} \tag{4.17}$$

where F_r is a kind of Froude number, of the order 1. From the conservancy of wave energy between wave rays one finds (using (4.17)):

$$h_{\rm br} = \left[\frac{n_0 H_0^2 c_0 b_0}{\gamma^2 F_r \sqrt{g}\, b_{\rm br}}\right]^{\frac{2}{5}} \tag{4.18a}$$

Here H_0 is the deep water wave height (in irregular waves: the wave height representative for the local wave energy,[q] i.e. $H_0^2 = H_{\rm sign}^2/2$).

Firstly, using (4.10b), (4.13) and (4.17) it is easy to reconstruct (4.5b); taking $\cos\varphi_{\rm br}$ by approximation equal to 1. This yields an expression for K, occurring in (4.5a,b). One then finds for K:

$$K = n_{\rm br}\, \gamma^2 F_r^2 g \tag{4.18b}$$

However, a more accurate expression for the transport S (expressed in deep water wave conditions) is possible as well, where on the other hand also the magnitude of s is investigated.

With the aid of the developed formulae it is possible to present the following kind of relations:

$$S_0/A_1 H_0^2 c_0 = f(H_0/L_0, \varphi_0) \tag{4.19}$$

$$s/A_1 H_0^2 c_0 = f(H_0/L_0, c_1/c_0, \varphi_0) \tag{4.20}$$

where H_0/L_0 denotes the initial wave steepness and c_1 the wave celerity at the transition level between beach and inshore. However, a clear-cut relationship with the initial wave steepness requires assumptions concerning the magnitude of F_r and γ and thus loss of generality. Therefore, the independent variable H_0/L_0 will be replaced by the ratio between the deep water wave celerity c_0 and the wave celerity $[c_{\rm br}]_{\varphi=0}$ when the wave breaks, under the condition, that all contour lines are parallel and that the waves are normally incident.

With respect to this, the following relation can be found from Eqs. (4.17) and (4.18a), taking $\delta = 0$:

$$\frac{c_{\rm br}}{c_0} = \left[\left(\frac{F_r\sqrt{g}}{c_0}\right)^4 \frac{n_0 H_0^2}{\gamma^2} \frac{\cos\varphi_0}{\cos\varphi_{\rm br}}\right]^{\frac{1}{5}} \tag{4.21}$$

[q] It is logical to take also γ in such a way, that the wave $H_{\rm br}$ is representative for the local wave energy. In [17], the formulas have been adopted in this way that they refer to $H_{\rm sign}$ (i.e., there also τ relates to the significant wave height in the breaker zone). Generally, only the relationship "H_0/γ" is important. When H_0 is replaced by a wave height with a certain probability of occurrence, γ should be adapted correspondingly.

When $\varphi_0 = 0$ (and thus $\varphi_{br} = 0$) one thus finds the following relation between c_{br}/c_0 and the wave steepness[r]:

$$\left[\frac{c_{br}}{c_0}\right] = \left[\frac{4\pi^2 F_r^4 n_0}{\gamma^2}\right]^{\frac{1}{5}} \left(\frac{H_0}{L_0}\right)^{\frac{2}{5}} \tag{4.22}$$

For calculation of S_0 take the x-axis parallel to the depth contours of the inshore. Neglecting the difference between $(\cos\varphi_{br})^{1/5}$ and 1, from Eqs. (4.10a,b), (4.15), (4.21) and the law of conservation of wave energy flux between wave rays follows directly for the transport S_0 when all contour lines are parallel ($\delta = 0$):

$$\frac{S_0}{A_1 n_0 H_0^2 c_0} = \left\{\frac{c_{br}}{c_0}\right\}_{\varphi=0} \sin\varphi_0(\cos\varphi_0)^{\frac{6}{5}} \tag{4.23}$$

The value of s can be found by differentiating (4.15) to δ and putting $\delta = 0$. With the aid of (4.10), (4.11) and (4.15) one finds:

$$s = A_1 n_0 F_r H_0^2 \sqrt{g h_{br}} \cos 2\varphi_1 \frac{\sin 2\varphi_0}{\sin 2\varphi_1} \tag{4.24}$$

which can be expressed in the wanted variables:

$$\frac{s}{A_1 c_0 H_0^2} = n_0 \frac{(c_{br}/c_0)_{\varphi=0}}{c_1/c_\circ} \left\{1 - 2\left(\frac{c_1}{c_0}\right)^2 \sin^2\varphi_0\right\} \frac{(\cos\varphi_0)^{6/5}}{\cos\varphi_1} \tag{4.25}$$

Figure 4.5 shows $s/A_1 c_0 H_0^2$ as function of φ_0(n_0 has been taken equal to 1/2).

The various curves are valid for values of c_1/c_0 equal to 1, 0.85, $1/\sqrt{2}$ and 0.5 respectively. The value of $[c_{br}/c_0]$ has been chosen equal to 0.5; it shows from (4.25), that another value of $[c_{br}/c_0]_{\varphi=0}$ only influences the vertical scale of Fig. 4.5.

Similar calculations[12] have been carried out concerning the two-line theory.

In that case, there has been started from a bottom topography, depicted in Fig. 4.4 and from the longshore transport equations, given by (4.5a) (i.e. $h_{br} > h_1$).

---

[r]Replacing c_0^2/g by $L_0/(2\pi)$.

The result is:

$$
\begin{aligned}
S_{01} &= A_1 \gamma^2 F_r g^{\frac{1}{2}} h_1^{2\frac{1}{2}} \sin\varphi_1' \cos\varphi_1' \\
S_{02} &= A_1 \gamma^2 F_r g^{\frac{1}{2}} (h_{\mathrm{br}}^3 - h_1^3) h_{\mathrm{tot}}^{\frac{1}{2}} \sin\varphi_2' \cos\varphi_2' \\
s_1 &= A_1 \gamma^2 F_r g^{\frac{1}{2}} h_1^{2\frac{1}{2}} \cos 2\varphi_1' \\
s_2 &= A_1 \gamma^2 F_r g^{\frac{1}{2}} (h_{\mathrm{br}}^3 - h_1^3) h_{\mathrm{tot}}^{\frac{1}{2}} \cos 2\varphi_2'
\end{aligned}
\qquad (4.26\text{a–d})
$$

where φ_1' and φ_2' are the angles of wave incidence (when all contour lines are parallel) at depth h_1 and h_{tot},; last-mentioned depth being $h_1 + h_2$. However, Eqs. (4.26a–d) have a confined accuracy, as there is an effect of the refraction on the inshore on the transport along the beach, which is neglected in those equations. The rate of change of the transport along the beach as function of the direction of the inshore, divided by s_1 is depicted in Fig. 4.10, being a function of h_1/h_{tot} (on the horizontal axis in Fig. 4.8) and of the angle of wave incidence φ_2' at the depth h_{tot} (angles written near the various curves in Fig. 4.8).

Figure 4.8 is derived from the following equation.

Call: δ_2 the angle of the inshore with respect to the depth contours in deeper water;

δ_1 the angle of the beach with respect to those contours;

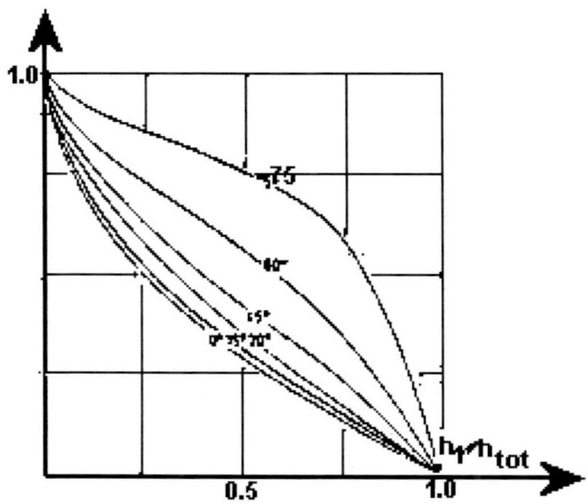

Fig. 4.10 Influence direction inshore/ influence beach versus h_1/h_{tot}. Indicated angles are angles of wave incidence at the level h_{tot}.

Then for the transport S_1 along the beach is found:

$$S_1 - S_{01} = A_1\gamma^2 F_r g^{\frac{1}{2}} h_1^{2\frac{1}{2}} \cos 2\varphi_1' \left[\left\{1 - \frac{\cos\varphi_2'}{\cos\varphi_1'}\sqrt{\frac{h_1}{h_{\text{tot}}}}\right\}\delta_2 - \delta_1\right] \quad (4.27)$$

The term between accolades, as function of h_1/h_{tot}, is depicted in Fig. 4.10.

When the direction of the beach changes, the transport along the beach changes. However, when the direction of the inshore changes, the transport along the beach changes as well, because of different refraction on the inshore.

Indicated in Fig. 4.10 is the proportion between the rate of change of the transport along the beach, caused by the change of direction of the inshore, compared to the rate of change of the transport along the beach, caused by a change of direction of the beach.

Variables are the level of the separation plane between beach and inshore (i.e. h_1/h_{tot}) and the angle of wave incidence at the level h_{tot} (written by each line separately).

Neglecting the term between accolades, from (4.27) one directly can derive (4.26c).

For assumptions made during the derivation the reader is referred to.[12] Important is the assumption of uniform conditions along the coast, i.e. no effects of curvature of the inshore and of gradients in wave set-up.[9]

Effects of curvature of the inshore on refraction are investigated by Jas.[28]

Another effect of the curvature of beach and inshore, being conciously overlooked, is the following feature. The used equation of sand motion is developed for a straight beach. As water entrains the sand, one has to take into account the continuity of water when applying straight-beach formula for curved beaches. Apart from cross-shore transport because of deviations of the equilibrium profile, cross-shore transport because of rip-currents, initiated by discharge of surplus of water to the inshore may occur. The order of magnitude of this feature is estimated by Overeem.[32]

4.4. Focussing on cross-shore interaction

4.4.1. *Introduction to section 4.4*

Similar to the approach, given in Section 4.3, for computation of coastal constants (concerning the *longshore* direction), attempts have been made to express the constant s_y (concerning the *cross-shore* direction) in the wave conditions.

Remind from Section 3.3.5 (Eqs. (3.74) and (3.75)), that s_y is related quite simple to the restauration time T_0 of an equilibrium profile; inversely, sophisticated theories concerning coastal profile dynamics could lead to physical expressions for T_0, and thus for s_y.

Firstly, there may be referred to Swart's Ph.D.-thesis, 1974.[38] Prototype data (Chapter 4.2) yield smaller values of s_y than resulting from his method.

A second line of thought,[30] exposed in outline below (Subsection 4.4.2), can be summarized in the following way.

Using the Baillard-transport[3] formula and following a simplified Stive concept,[36] the dimensionless equilibrium profile outside the surfzone can be expressed in wave steepness S_0' and in the dimensionless fall velocity of the sediment (Fig 4.11a,b). With respect to the length scales, the deep water wave length is the reference. With respect to the fall velocity w, the reference is H_0/T, where H_0 is the rms-deep water wave height and T the wave period.

However, as the interaction between beach and inshore is a process, dominating mainly in the surf zone, up to now the developed method has practical limitations: it yields astronomical time scales, only of importance when dealing with problems as predicting the effect of sea level rise on coastal behavior. Maybe the method can be extended to the surf zone in the future.

4.4.2. *The V.d.Kerk method[30] into more detail*

Starting from the concept of an equilibrium profile, it is logical to assume, that a deviation from this equilibrium profile leads to a restoration to this profile at a certain time scale. This is the line of thought, leading to the two-line theory, where the distance between the line of the beach and the line of the inshore determines the cross-shore transport, being larger when the deviation from the equilibrium distance is larger. For uniform long-shore conditions this yields a negative-exponential decay of the deviation of the equilibrium profile. Independent of any assumptions on the mechanism which causes the equilibrium, the method essentially is right, if the restoring mechanism indeed may be linearized over the range of deviations to be expected; beach and inshore itself show some restoring capacity of the equilibrium profile over that part of the equilibrium profile (i.e. over the beach and inshore itself) with a time scale small compared to the time scale,

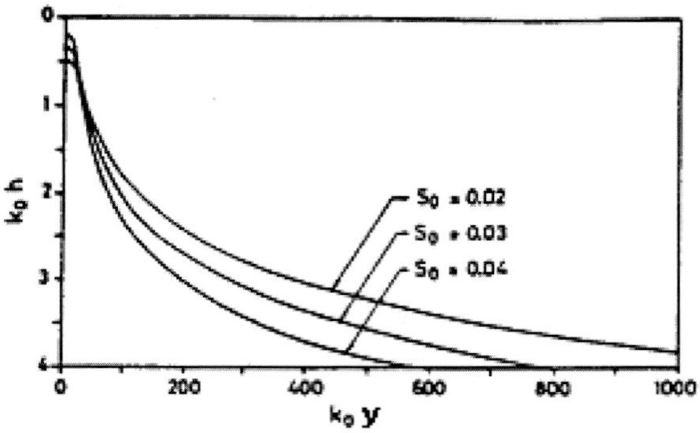

(a)

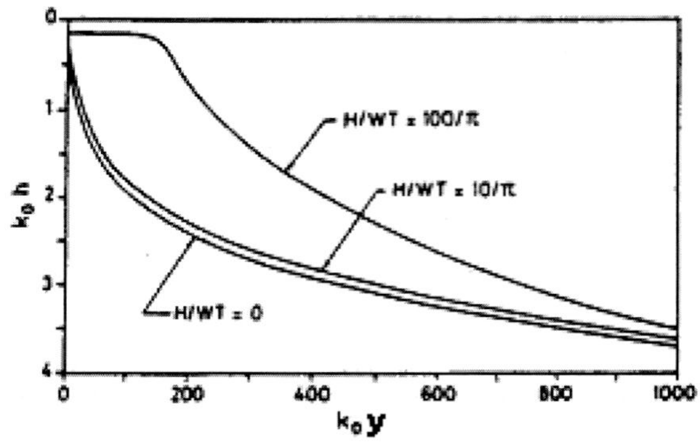

(b)

Fig. 4.11 (a) Influence of wave steepness S_0' on the equilibrium profile ($H_{\rm rms}/wT = 30/\pi$) from.[30] (b) Influence of the dimensionless fall velocity $H_{\rm rms}/wT$ on the equilibrium profile. $S_0' = 0.02$, from.[30]

determining the restoration of the overall profile. In other words, the theory assumes that the volume of sand, contained in the beach area determines the cross-shore transport from beach to inshore (assuming fixed position of the inshore), not depending on the fact whether the sand is stored on the upper side or the lower side of the beach. Unsaid, the theory implies more or less the assumption, that beach and inshore itself have an equilibrium profile.

If this is not the case, a more-line theory should be used.[5] In this section, one possible mechanism for restoration of the equilibrium profile is considered more into detail, i.e. the one, on which Stive's Crostran-model[34] is based.

The method is described into more detail by V.d Kerk.[30]

Using the concept of Bailard,[3] the local equilibrium slope β_e can be expressed as:

$$\tan\beta_e = \frac{\langle q_{as}\rangle + \langle q_{un}\rangle}{\langle q_{s1}\rangle / \tan\beta_e} \tag{4.28}$$

in words: the equilibrium slope is found as the sum of the transports q_{as} (caused by the asymmetry of the waves) and q_{un} (caused by the undertow), divided by the rate of change of the gravity transport q_{s1} as function of the slope. The indication "⟨⟩" deals on expectations. Stive,[36] gives expressions for the various kinds of transport (relating those to the Bagnold/Bailard concept), for instance:

$$\frac{\langle q_{s1}\rangle}{\tan\beta_e} = \frac{F_b}{(u3^*)\tan\phi} + F_s\frac{u_{rms}}{w}\varepsilon_s(u5^*) \tag{4.29}$$

where F_b and F_s refer to the effect of bottom and suspended load; (un^*) denotes:

$$(un^*) = \frac{\langle |u|^n\rangle}{u_m^n} \tag{4.30}$$

where u_m denotes the amplitude of the first harmonic of the orbital velocity. The angle ϕ is the angle of repose of the material. The variables F_b and F_s depend on the bottom friction, exerted by the wave motion, Bagnold's[2] efficiency factor $\varepsilon_{b(s)}$ for bottom transport and suspended transport, respectively, the specific density of the sediment and the rms-orbital velocity u_{rms}. Furthermore, F_b depends on ϕ and F_s on the ratio u_{rms}/w, where w is the fall velocity of the material. Mind, that the expression mentioned in (4.28), i.e. the denominator on the right-hand side of (4.28) is not a function of β_e, as $\langle q_{s1}\rangle$ is proportional to $\tan\beta_e$.

At the top of the bar (of which the level will be called h_{bar}) $\langle q_{as}\rangle$ and $\langle q_{un}\rangle$ will counterbalance each other and a horizontal crest of the bar will result.

Because the phase shifts between first- and second-harmonic of the orbital velocity can amount in the breaker zone up to 90°, in which case $\langle q_{as}\rangle$ has a value zero, Stive[36] schematizes, that $\langle q_{as}\rangle$ is proportional to the percentage non-breaking waves. On the other hand, $\langle q_{un}\rangle$ is directly proportional to $\langle u\rangle$, the mean velocity of the undertow. For non-breaking waves, this (onshore) undertow can be approximated by Longuet–Higgins' "conduction solution"; for breaking waves, a first approximation of the (offshore) velocity has been given by Stive and Wind,[35] later improved and extended by de Vriend and Stive[39] and by Battjes and Stive.[14]

Making some short-cuts, V.d. Kerk,[30] succeeded in writing (4.28) in the shape:

$$\frac{dh}{dy} = f\left(k_0 h, \frac{H_0}{wT}, S_0'\right) \tag{4.31}$$

S_0' is the deep-water steepness of the waves and wT/H_0 a dimensionless fall velocity of the sediment; T being the wave period belonging to the largest wave energy density in the spectrum and H_0 the rms-deep water wave height. The wave number k_0 equals:

$$k_0 = \frac{4\pi^2}{gT^2} \tag{4.32}$$

Clearly, y being the offshore direction, the left-hand sides of (4.28) and (4.31) denote the same variable; furthermore the attractiveness of the presentation according to (4.31) is evident, inviting to solve the differential equation, thus yielding the dimensionless equilibrium profile:

$$k_0 h = f\left(k_0 h, \frac{H_0}{wT}, S_0'\right) \tag{4.33}$$

Unfortunately, shortcuts are necessary. Sophisticated present wave height formulae as used in Crostran[34] take into account, that also at horizontal shoals waves decay. Therefore, formula of the kind (4.33)[15] are outdated:

$$\frac{H_{\mathrm{rms}}}{h} = f(kh, S_0') \tag{4.34}$$

where k is the local wave number. In fact, the ratio H_{rms}/h can have various values for the same values of kh and S_0'. However, for a gentle downward slope in seaward direction an approximation like the one, given by Battjes and Janssen[15] is reasonable; it has been used here because an approximation like (4.33) is essential for the derivation of an equation like (4.30).

Then it is possible to write all the equations of the Stive model[36] in a dimensionless way,[30] thus resulting in equilibrium profiles, depending on H_0/wT and S_0' (Fig 4.11a,b).[s]

[s]For the undertow the formulation of[35] has been used as[36] was not yet published at the time of the investigation.

As $\langle q_{\text{as}} \rangle$ and $\langle q_{\text{un}} \rangle$ counterbalance each other when the fraction of breaking waves increases ($\langle q_{\text{un}} \rangle$ will then be seaward and increases), from (4.28) it shows that moving from the seaward side into the breaker zone, the bottom slope will decrease and become zero at a certain level. Landward of this point, the approximation has no sense any more, as (4.34) looses its validity. Thus anyway the existence of (an outer) breakerbar(s) can be explained, in a way resembling to the qualitative description of Dyhr Nielsen and Sorensen.[25]

Also Bailard,[3] calculates equilibrium slopes as function of similar parameters as given here. However, as Bailard does not take seaward undertow by breaking waves into account (as done by Stive) he finds very steep equilibrium slopes for small values of $k_0 h$. For further comparison the reader is referred to.[30]

A global comparison with Dutch coastal slopes in the offshore zone, of which data is presented by Postma and Kroon[10] shows a reasonable correspondence, when waves of small steepness are presumed.

Unfortunately, one cannot find cross-shore interactions between beach and inshore, starting from the above calculated equilibrium profiles, as the equilibrium profiles found are only valid outside the breaker zone, i.e. seaward of the inshore. However, one can investigate the behavior of the offshore zone, following the ideas of the two-line theory. This is in order to get an impression of the time-scales involved in the cross-shore processes in this zone.

Consider the dimensionless profiles of Fig. 4.11 somewhat closer. These will be schematized as shown in Fig. 4.12.

The upper boundary is taken at the site of the outer breaker bar, (depth: $k_0 h = k_0 h_{\text{bar}}$); the lower at the boundary of initiation of motion, which has been tentatively taken as: $k_0 h = 4$.

The origin of the y-axis will be chosen at the mean position of the depth contours, i.e. in the "hinge point" at a level $k_0 h = k_0(h_{\text{bar}} + h_1)$ in such a way, that the content of the water landward of the vertical plane through the origin balances the volume of the (wet) sand seaward of that plane.

Call in the terms of the two-line theory the area between the levels $k_0 h = k_0 h_{\text{bar}}$ and $k_0 h = k_0(h_{\text{bar}} + h_1)$ the "beach" and the area below (down to the level $k_0 h = 4$) the "inshore" (although these terms are here not totally appropriate).

Let h_{tot}, being $4/k_0 - h_{\text{bar}}$ be the height of the total considered area[t] (Fig. 4.12).

[t]Returning to nature and omitting the reference "k_0".

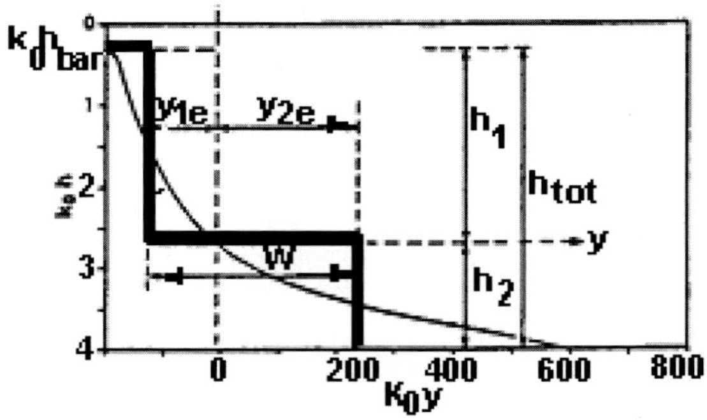

Fig. 4.12 Schematization of the offshore zone in a "beach" and an "inshore". For shortness, *inside* the figure, the factor "k_0" has been omitted. The origin of the y-axis has been taken at the hinge point.

Schematize beach and inshore as a stepwise profile, thus creating vertical planes with coordinates y_{1e} and y_{2e}. The equilibrium distance W in the considered case thus is $|y_{1e}| + y_{e2}$.

From continuity ($|h_1 y_{1e}| = h_2 y_{2e}$, where h_2 is defined as $h_{tot} - h_1$) one finds:

$$|y_{1e}| = \frac{h_2}{h_{tot}} W \tag{4.35}$$

Consider now a profile, having at every height a slope, being a factor α larger than the equilibrium slope β_e, where α is a fraction, somewhat larger than 1. If such slopes would have been integrated to give a profile, one would have found a distance between beach and inshore being W/α.

The distance $|y_1|+y_2$ between beach and inshore thus becomes (in linear approximation) $(\alpha - 1)W$ smaller than the equilibrium distance, thus giving an offshore transport S_y (Eq. (4.1)) of:

$$S_y = s_y(\alpha - 1)W \tag{4.36}$$

Generalizing (4.35), one finds for y_1:

$$y_1 = -\frac{h_2}{h_{tot}} \frac{W}{\alpha} \tag{4.37}$$

The continuity equation reads:

$$-h_1 \times \frac{dy_1}{dt} = S_y \tag{4.38}$$

Differentiating (4.37) with respect to t and substituting the result and (4.36) into (4.38) gives a differential equation in $\alpha - 1$, showing (for small values of $\alpha - 1$, compared to 1) a negative-exponential decay of $\alpha - 1$:

$$(\alpha - 1)_t = (\alpha - 1)_0 \exp(-t/T_0) \tag{4.39}$$

with:

$$T_0 = h_1 h_2 / (s_y h_{\text{tot}}) \tag{4.40}$$

From (4.36) it can be derived that:

$$\frac{dS_y}{d\alpha} = s_y W \tag{4.41}$$

On the other hand, the gradient of S_y with respect to α is given by (4.29), reduced by V.d. Kerk (cf. pp. 4.22–4.31). By dividing this transport gradient by g^2T^3 (T being the wave period), (4.29) can be written in a dimensionless way. Let the dimensionless $\langle q_{s1} >$ be: $\langle q_{s1} >_{\text{dim}}$. Then Eq. (4.29) can be written as a known function f_1 of the following dimensionless variables:

$$\frac{\langle q_{s1} \rangle_{\text{dim}}}{\tan \beta_e} = f_1 \left(S_0', \frac{H_{\text{rms}}}{wT}, k_0 h \right) \tag{4.42}$$

For the comparison with (4.41) the value of $k_0 h$ at the hinge point, Fig. 4.11, (i.e. $h = h_1 + h_{\text{bar}}$) will be relevant. Equalizing the right-hand sides of (4.41) and (4.42) gives (cf. (4.31)):

$$f_1 \left(S_0', \frac{H_{\text{rms}}}{wT} \right) = \frac{s_y W}{g^2 T^3} \tag{4.43}$$

Substituting s_y, as given in (4.43), into (4.40) yields (cf. Fig. 4.13):

$$\frac{T_0}{T} = \frac{\left(\frac{W h_1 h_2}{h_{\text{tot}}} \right)_{\text{dim}}}{16\pi^4 \times f_1(S_0', H_{\text{rms}}/wT)} \tag{4.44}$$

where "[ ]$_{\text{dim}}$" indicates dimensionless variables (all the variables between the brackets, having the dimension "length", should be multiplied by k_0).

Thus the procedure for calculating the dimensionless time of decay of a disturbance is the following. Start from one of the dimensionless profiles, given in Fig. 4.11, valid for a certain value of S_0' and H_{rms}/wT. Calculate the area of "sand" in this profile and divide this by $4 - k_0 h_{\text{bar}}$ where

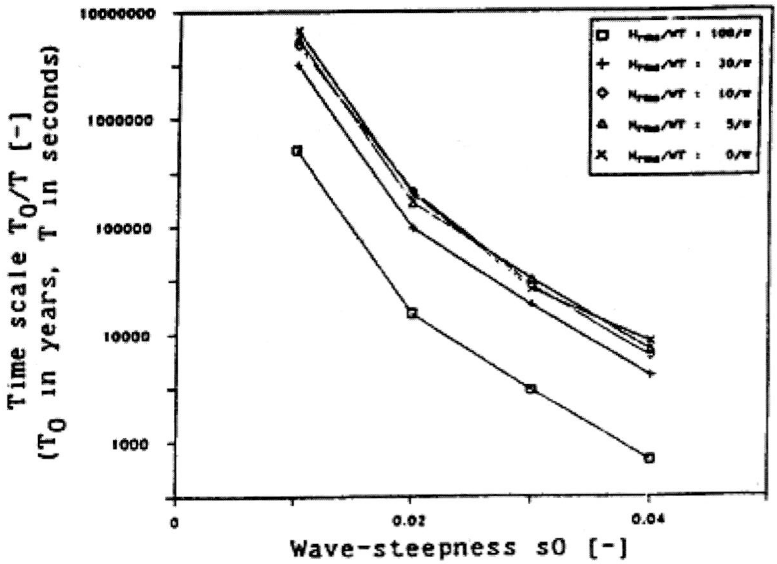

Fig. 4.13 Dimensionless time scales T_0/T of restoration to the equilibrium profile, from.[30]

$k_0h_{\rm bar}$ denotes the dimensionless height of the outer breaker bar. Let the division result in a distance k_0y'. Draw a vertical line at a distance k_0y' from the right-hand side of the figure; on the left-hand side of the line is the "beach", on the right-hand side the "inshore". Determine the height k_0h_1 (with respect to $k_0h_{\rm bar}$) of the intersection point of the vertical line with the profile.

Determine the area of sand above the level k_0h_1 and divide this by k_0h_1. This gives $k_0y'_{1\rm e}$ (y-coordinate of the line of the beach with respect to the left-hand side of the figure). Determine the coordinate $k_0y'_{2\rm e}$ of the inshore in a similar way; the distance between the line of the beach and the line of the inshore gives k_0W. In this way, the numerator of Eq. (4.44) is determined; for the calculation of the denominator the reader is referred to.[30]

Very large time scales result from (4.44).

The following concluding remarks should be added.

During the explanation, given above, the effect of the change of the undertow (and thus of the offshore transport) by a change of the coastal slope has not been considered. Not only the rate of change of the gravity transport (4.29) determines the equilibrium slope. By V.d. Kerk,[30] this effect has been included in the computations by an iteration procedure.

Only the sand balance between beach and inshore has been considered during the derivation of (4.44). It has not (yet) been investigated, whether it is allowed to assume a common time scale over the whole profile.

It would be interesting to combine the results of the present study with the investigations on more-line theories, given by De Vroeg *et al.*[41]; however, it is still more relevant to investigate the interaction in the surf zone with the breaker bars.

4.5. Symbols Concerning Chapter 4

Symbol		Introduced in
a_1	Ratio of content of solid material	Eq. (4.11)
$A_{1,2}$	coefficients $\frac{\Delta t}{\Delta x^2 h_{1,2}}$; also: coefficients in CERC eqn	Eq. (4.3) (4.10;11)
A'_1	Coefficient in Svasek's equation	p. 4–12; 196
b_0, b_{br}	Distance between wave rays; Fig. 4.3	Eq. (4.12a)
$B_{1,2}$	$\frac{\Delta t}{h_{1,2}}$ coefficients	Eq. (4.3)
c_0, c_1, c_{br}	Propagation velocity (deep water; depth h_1; breaker line)	Eq. (4.5)
e	Basic number neperian Logarithmic system	Eq. (4.39)
$E_{0,br}$	Wave energy per unit of area (deep water; breaker line)	Eq. (4.12a,b)
$E(\ldots)$	Expectation of ...	Eq. (4.8)
F_b, F_s	Coefficients in Stive's equation	Eq. (4.29)
F_r	Froude number	Eq. (4.17)
g	Acceleration of gravity	Eq. (4.6)
h	Depth	Table 4.1
h_1, h_2	Depth of beach, c.q. inshore	Fig. 4.3
h'_1	Depth of the level, up to where the contour Lines follow the line of the beach	Fig. 4.8
h_{bar}	Top level of of breaker bar	p. 4–23
h_{br}	Depth of breaking waves	Eq. (4.5a)
h_{tot}	$h_1 + h_2$	Fig. 4.4

Symbol		Introduced in
$\triangle h$	Depth interval	p. 197
H	Wave height	p. 196
H_0	Deep water wave height	p. 495
$H(t-1)$	Matrix, linking past $(t-1)$ and present (t) state of the system	Eq. (4.3)
H_{br}	Breaker height	Eq. (4.10b)
H_{rms}	Root-mean square wave height	Eq. (4.42)
H_{sign}	Significant wave height	p. 196
$\triangle\ (H^2C)$	Loss of energy flux	p. 196
k_0	Wave number	Eq. (4.32)
K	Proportionality constant	Eq. (4.5a)
$K(t)$	Kalman gain matrix	p. 188
L_0	Deep water wave length	Eq. (4.19)
n	Ratio group-phase velocity	Eq. (4.10b)
n_0	n in deep water $(= 1/2)$	Eq. (4.12a)
	Symbols concerning ch.4 (con'd)	
n_{br}	n at breaker depth	Eq. (4.10b)
p	Source- or sink term in (4.1)	p. 185
P_1,P_{11}, P_{12}	Longshore component of the energy flux	Eq. (4.5a,b)
$P_r(H,T,\varphi)$	Probability of waves with characteristics (H,T,φ)	Eq. (4.6)
q_{as}	Sand transport caused by asymmetry of the waves	Eq. (4.28)
q_{sl}	Sand transport caused by slope of the coastal profile	Eq. (4.28)
q_{un}	Sand transport caused by the undertow	Eq. (4.28)
s	Coastal constant	p. 179
$s_{1,2}$	Coastal constant at beach, c.q. inshore	Eq. (4.1a,b)
s_y	Coastal constant (cross-shore)	Eq. (4.1a,b)
S	(Integrated) sand transport along a zone	Eq. (4.10a) p. 181
S_0	Sand transport, when the coast is parallel to the x-axis	p. 181

Symbol		Introduced in
S_{01}, S_{02}	Sand transport along beach (inshore), when the coast is parallel to the x-axis	Eq. (4.2)
$S_{1,2}$	Transport along beach (inshore)	Eq. (4.2)
S_y	Cross-shore transport per unit of beach width	p. 184
S'_0	Wave steepness	Fig. 4.11
$\triangle S$	Longshore transport between 2 depth contours	Eq. (4.7a,b)
$S(\delta)$	Longshore transport, when the beach makes an angle δ with the inshore	Eq. (4.9); Fig. 4.3
t	time	Eq. (4-1a,b)
Δt	Small time element	p. 189
T	Wave period	Eq. (4.6)
T_0	Restauration time	Eq. (3.74)
u_{rms}	rms-Orbital velocity	Eq. (4.29)
$u3^*$; $u5^*$	Orbital velocities, defined in (4.30) (third, c.q. fifth moments)	Eq. (4.30)
un^*	nth Moment of the velocity	Eq. (4.30)
W	Equilibrium distance beach vs inshore	Eq. (4.3); Fig. 3.48
w	Fall velocity of the grains	Eq. (4.29) p. 199
x	Longshore direction	Eq. (4.1a,b)
$\underline{x}(t)$	Vector $(s_1, s_2, s_y, 1)$	Eq. (4.4b)
Δx	Grid length in x-direction	p. 189
$\underline{x}(t)$	System state (cf. (4.6b) at time t	Eq. (4.3)
y	Coordinate, perpendicular to the coast	Eq. (4.1a,b)
$y_{1,2}$	y-Coordinate of beach, c.q. inshore	Eq. (4.1a,b)
	Coordinate inshore corrected by equilibrium distance W	
y'	Representative y-coordinate of the mean position of the depth contours in a coastal profile; below the bar level and above the level $k_0h = 4$ (cf. p. 4.24 and Fig. 4.12).	Eq. (4.1a,b)
y'_2	y-Coordinate of inshore	Eq. (4.1a,b)

Symbol		Introduced in
y_{1e}, y_{2e}	y-Coordinate of the beach, c.q. inshore (equilibrium position)	Eq. (4.36)
$y1_{1,\ldots,n}$	y-Coordinate of the beach at n successive positions along the coast	Eq. (4.4)
$y2_{1,\ldots,n}$	y-Coordinate of the inshore at n successive positions along the coast	Eq. (4.4)
$\underline{z}(t)$	Vector (y_1, y_2) at $n-2$ sites at time t	Eq. (4.3)
α	A fraction, somewhat larger than 1	Eq. (4.36)
β	Local beach slope	Eq. (4.28)
β_e	Equilibrium slope	Eq. (4.28)
γ	H/h	p. 196
δ, δ_1	(Arbitrary) angle, over which the beach, c.q. the inshore is turned	Eq. (4.9) Eq. (4.27)
$\varepsilon'_{b,s}$	Efficiency factor for bottom, c.q. suspended transport	Eq. (4.29)
$\underline{\nu}(t)$	Vector, indicating the inaccuracy of the measurements at time t	Eq. (4.3)
$\rho_{s,w}$	Density of water (sand)	Eqs. (4.10) and (4.11)
φ	Angle of wave incidence	Fig. 4.3
$\varphi_{0,1}$	Angle of wave incidence (deep water, c.q. level h_1	Eq. (4.5)
φ_{br}	Breaker angle	Fig. 4.5
$\varphi'_{1,2}$	Angle of wave in incidence (when all contours are parallel) at depth h_1 and h_{tot} respectively	Eq. (4.26a,d)
$\rho_{w,s}$	Density of water, sand	Eq. (4.10a)

References

1. Andringa, A. J., *Sand Nourishment at Zeebrugge*, Delft University of Technology, Coastal Engineering Group, M.Sc. Thesis.
2. Bagnold, R. A., The flow of cohesionless grains in fluids. *Phil. Trans. Roy. Soc., Ser. A*, **966**, 249, 1956.
3. Baillard, J. A., An energetics total load sediment transport model for a sloping beach. *J. Geophys. Res.* **86**(C11), 10938–10934, 1981.
4. Bakker, W. T. and T. Edelman, The coastline of river delta's, Tokyo, *Proc. 9th Conf. On Coastal Eng.* pp. 199–218, 1963.

5. Bakker, W. T., *The dynamics of a coast with a groyne system*, London, 11th Conf. on Coastal Eng., Ch. 31, 1968.
6. Bakker, W. T., *Sand Waves on the Vliehors*. Shore and Beach, 1968.
7. Bakker, W. T., Computation of the wave induced littoral drift according to the Svasek-method, adapted to parallel depth contours. Rijkswaterstaat, Dept. of Coastal Res., Rep. WWK 71-2, 1971.
8. Bakker, W. T, H. G. H. ten Hoopen and G. R. M. Grieve, Computation of sandtransport according to the method of Svasek for a beach and inshore, making an angle with each other, Rijkswaterstaat, Dept. of Coastal Res., Rep. WWK 71-18, 1971.
9. Bakker, W. T., The influence of longshore variation of the wave height on the littoral current, Rijkswaterstaat, Dept. of Coastal Res., Rep. WWK 71-19, 1971.
10. Bakker, W. T. and E. S. P. Smit, Possible truth about coastal behavior, Tidal Waters Division, Rep. GWAO 87.004, 1987.
11. Bakker, W. T., C. P. T. van der Kerk and J. H. de Vroeg, *Determination of coastal constants in mathematical line models*, Loutraki, Second European Workshop on Coastal Zones, 1988.
12. Bakker, W. T., Coastal constant computations, Tidal Waters Division, Rep. GWAO 89.309, 1989.
13. Battjes, J. A., *Computation of set-up, longshore currents, run-up and overtopping due to wind-generated waves*, Delft University of Technology, Ph.D. Thesis, 1974.
14. Battjes, J. A. and M. J. F. Stive, Calibration and verification of a dissipation model for random breaking waves. *J. Geophys. Res.* **90**(C5) 569–578, 1985.
15. Battjes, J. A. and M. P. F. M. Janssen, *Energy loss and set-up due to breaking of random waves*, Proc. 16th Int. Conf. on Coastal Eng., pp. 569–578, 1987.
16. Van Bochove, G., *The GB4 refraction program*, Rijkswaterstaat, Dept. of Coastal Res., Memo 71-19, 1971.
17. Van Bochove, G., *The GB2 diffraction program*, Rijkswaterstaat, Dept. of Coastal Res., Memo 72-18, 1972.
18. Van Bochove, G., *The GB6 refraction program*, Rijkswaterstaat, Dept. of Coastal Res., Memo 73-6, 1973.
19. Bruning, E. T. F. M., *The accuracy of two-line models, tested on the German Wadden island Sylt*, Delft University of Technology, Coastal Engineering Group, M. Sc. Thesis, 1988.
20. Bijker, E. W. and J. N. Svasek, *Two methods for determination of morphological changes induced by coastal structures*, Paris, XXIInd Int. Nav. Congress, section II, item 4, 1969.
21. Castelyn, J. A., Numerical computations of the changes of coastal shapes under the influence of wave attack, Delft University of Technology, Coastal Engineering Group, Rep. 75-1 (part I, II and III), M.Sc. Thesis.
22. CERC *Shore Protection Manual*, Washington D.C., US Army Coastal Eng. Res. Center, US Govern. Printing Office, 1987.
23. Dorrestein, R., *Wind and wave data of the Netherlands Lightvessels since 1949*, KNMI, Med. en Verhand. ("*communications*") no. 90, 1965.

24. Dijkman, M. J., W. T. Bakker and J. H. de Vroeg, *Prediction of coastline evolution for the Holland coast*, Delft, 22nd Int. Conf. on Coastal Eng., 1990, Session "The Dutch Coast"; paper No. 7, (preprints of this Session were edited as well by Rijkswaterstaat and Delft Hydraulics).
25. Dyhr-Nielsen, M. and T. Sorensen, Some sand transport phenomena on coasts with bars, *Proc. 12th Int. Conf. on Coastal Eng.*, pp. 855–866, 1970.
26. Ten Hoopen, H. G. H. and W. T. Bakker, Erosion problems of the Dutch island of Goeree, Copenhagen, *Proc. 14th Conf. On Coastal Eng.*, pp. 1213–1231, 1974.
27. Janssen, J., *Investigation of the coastal development and the sand transport at Scheveningen*, Delft University of Technology, Coastal Engineering Group, M.Sc. Thesis, 1974.
28. Jas, D., *Detailed research concerning a new harbor entrance near Lagos*, Delft University of Technology, Coastal Engineering Group, M.Sc. Thesis, 1976.
29. Kalman, R. E., A new approach to linear filtering and prediction problems. *J. Basic Eng.*, pp. 35–46.
30. Van der Kerk, C. P. T., *Theoretical calculations concerning the coastal constant* s_y, *starting from the Crostran* conception, Delft University of Technology, Coastal Engineering Group, M.Sc. Thesis, 1987.
31. Knaapen, M. A. F., *Predicting large waves in erodible sand beds*, Twente University of Technology, Ph.D. Thesis, 2001.
32. Van Overeem, J., *Numerical model for the computation of coastal changes, caused by waves and tides*, Delft University of Technology, Coastal Engineering Group, M.Sc. Thesis.
33. Pelnard-Considère, R., *Essai de théorie à l'evolution des formes de rivages en plages de sables et de galets*. Paris, Quatrième Journées de l'Hydraulique, Les Energies de la Mer, Question 3, 13–15.
34. Pilarczyk, K. W., S. van Overeem, *et al.*, *Manual on Artificial Beach Nourishment*, Rijkswaterstaat, CUR and Delft Hydraulics, 1986.
35. Stive, M. J. F. and H. G. Wind, Cross shore mean flow in the surf zone. *Coastal Eng.* **4**, 325–340.
36. Stive, M. J. F., *A model for cross-shore sediment transport*, Taipei, Proc. 20th Int. Conf. On Coastal Eng., 1986.
37. Svasek, J. N., The influence of breaking waves on the stability of sandy coasts, Deltadienst, Hydraulics Dept., Rep. W 68.083, 1968.
38. Swart, D. M., Offshore sediment transport and equilibrium profiles, WL|Delft Hydraulics, Rep. 131, Ph.D. Thesis, 1974.
39. De Vriend, H. J. and M. J. F. Stive, Calibration and verification of a dissipation model for random breaking waves. *J. Geophys. Res.* **90**(C5), 9159–9167, 1985.
40. De Vroeg, J. H., *The use of Kalman filtering for investigation of coastal behavior*, Delft University of Technology, Coastal Engineering Group, M.Sc. Thesis, 1987.
41. De Vroeg, J. H., E. S. P. Smit and W. T. Bakker, *Coastal genesis*, Malaga, Proc. 21st Int. Conf. On Coastal Eng., 1988.

Chapter 5

Experimental Verification

5.1. Introduction

This chapter, dealing on experimental verification of the theory, displayed in Chapters 3 and 4, will be subdivided as:

(a) Description of experimental verification by (two) hydraulic laboratories, in Delft (Delft Hydraulics) and in the Laboria National de Engenharia Civil in Lisbon will be mentioned (Section 5.2[a]).
(b) Description of in-situ experiments, c.q. artificial nourishments, at a number of sites along the coasts, as depicted in Fig. 5.28 (Section 5.3).

Apart from the research, reported in this chapter, (other) results of the intensive and appreciated collaboration in the "Bad Weather Club" (joined research of the Dutch Rijkswaterstaat, Delft Hydraulics with the coastal authorities and Hydraulic Laboratories in England, Denmark (Danish Hydraulic Institute) and Germany during the years 1970–1990 should be mentioned, laid down in many papers. Some quotations will be given in the following sections.

For instance, the intensive and appreciated cooperation with the Hydraulic Research Station Wallingford, which used coal models for groin tests should be gratefully mentioned. Refer to a summarizing paper of Tomlinson.[47]

Remark: in this chapter, test results like "T [number] refer to tests in WL|Delft Hydraulics.

[a]The text follows nearly literally Hulsbergen *et al.*[28]

5.2. Experiments in Laboratories

5.2.1. *Experiments in WL|Delft Hydraulics*

5.2.1.1. *Model facility and test conditions*

Groin theory as displayed in Chapter 3 has been experimentally tested in Delft Hydraulics[28]; cf. as well the M918 reports of WL|Delft Hydraulics.[52,53] Text below follows the text concerning M918 rather explicitly.

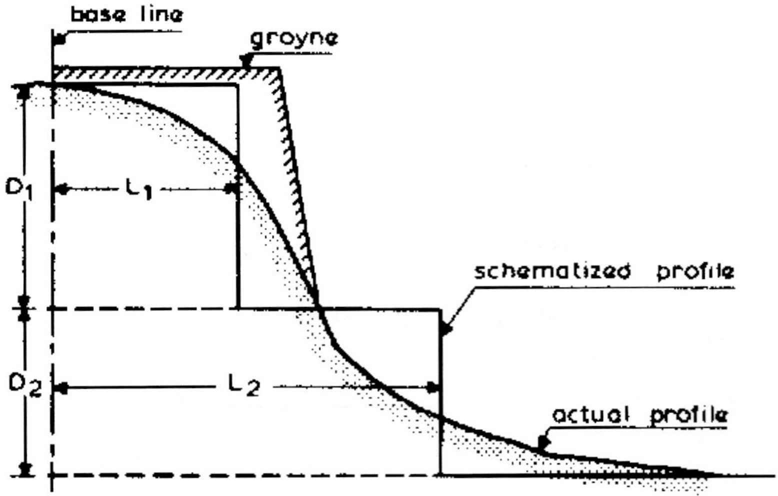

Fig. 5.1 Schematization of coastal profile.

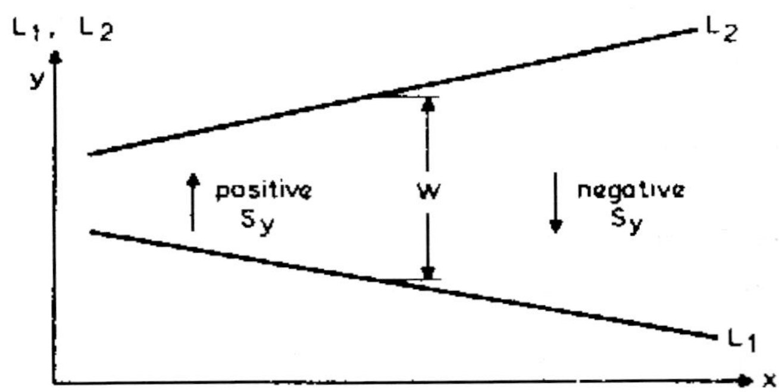

Fig. 5.2 Top view of schematized coast lines.

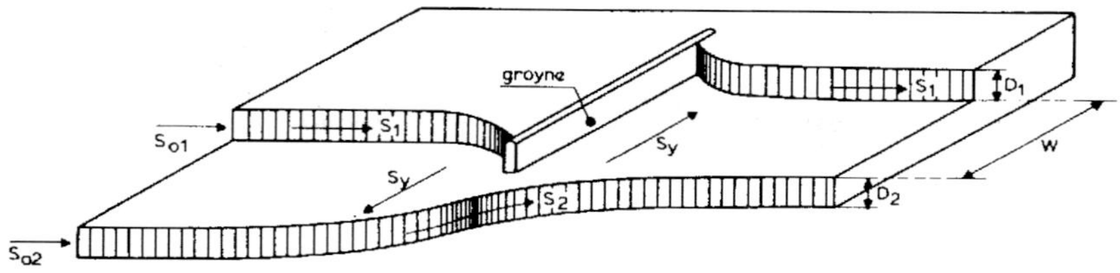

Fig. 5.3 Visualization of theoretical coast.

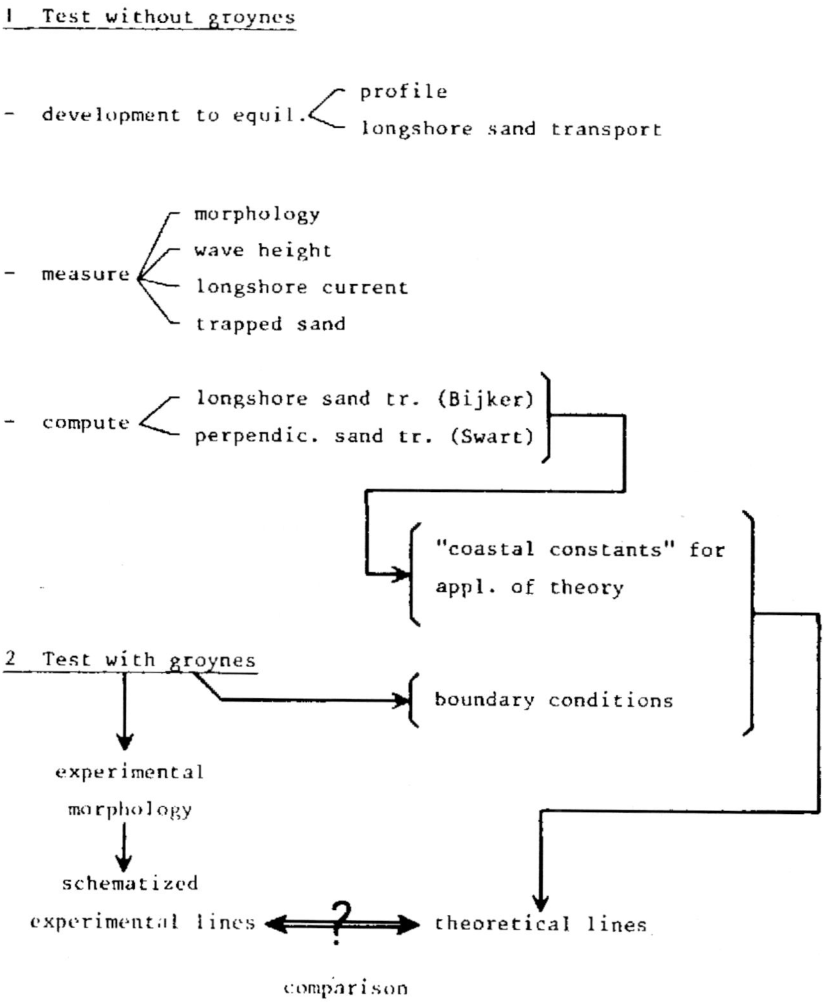

Fig. 5.4 Procedure of verification.

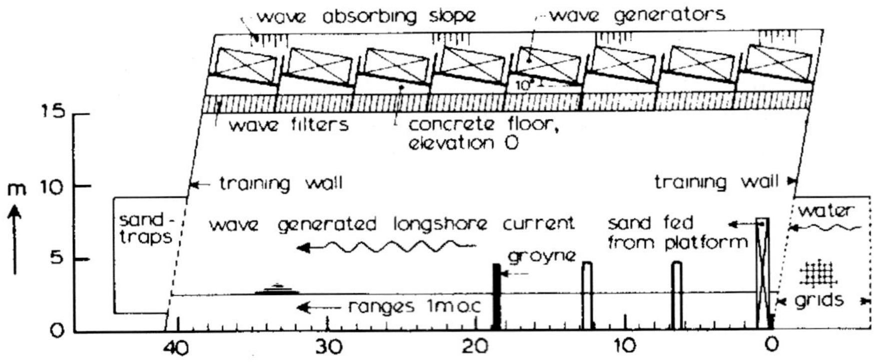

Fig. 5.5 Lay-out of model facility.

First in this subsection, the model facility and test procedure will be described.[b] After that, the result of tests will be reported, i.e.:

(a) with a single groin
(b) with a row of three groins

Subsequently, results will be displayed.

The lay-out of the model basin is given in Fig. 5.5. A large amount of various conditions were installed, details of which are extensively discussed by Delft Hydraulics.[52] As bed material dune sand was used with $\rho = 2650\,\mathrm{kg\,m^{-3}}$ and $D_m = 0.220\,\mathrm{mm}$. Either 0, 1 or 3 impermeable groins were present, with their crests well above the water level, with rough 1:1 slopes and aligned perpendicularly to the coast. The spacing was 6 m and their length so that they obstructed the longshore current partly or completely. The regular waves had a period of 1.55 s or 1.15 s, and a height between 0.07 m and 0.14 m.

Depending on the wave steepness and the local beach slope, either spilling or plunging breakers occurred, under an angle of approx 5° with the shore line. Water and sand were fed upstream and caught downstream in order to represent an inifinitely long, straight beach.

In short, the procedure was to try and establish a dynamic equilibrium along a straight beach without groins, and to assess meanwhile the development of the wave heights, the current velocity, the morphology and the trapped sand distribution. Then, after adding a groin system, the new

[b]Figures 5.1–5.26, (also derived from Hulsbergen *et al.*[28]) will be shown at the end of this section.

hydraulic and bathymetric development were recorded and subsequently compared with the theory. Here only a limited, though typical, selection of the numerous test results is presented, while in Subsection 5.2.3 a discussion is given, making use of some selected measuring data which are deemed useful for a good interpretation of the results.

5.2.1.2. *Tests with a single groin*

Representative as a "good" test is T22, where the main conditions were: wave height 0.075 m, wave period 1.55 s, water depth 0.38 m in front of the wave generators. The average rate of sand nourishment was 71 $dm^3\,h^{-1}$, while 55 $dm^3\,h^{-1}$ was caught in the traps. The rate of water supply was constant at 30 $dm^3\,h^{-1}$ in order to feed the longshore current. The groin extended seaward to 4.2 m, whereas the waves broke at approx. 3.5 m. Figure 5.6a and b depicts the beach, looking upstream, after 0.5 h and 50 h. They show a quite regular accretion upstream, and erosion downstream of the groin. This is also represented by Fig. 5.6c, a difference chart displaying the extent and the locations of erosion and accretion during the 50 h of the test. In an attempt to monitor the actual longshore sand transport by local measurements, the wave height and the longshore current distribution were frequently measured in sections perpendicular to the shore.

Fig. 5.6a Test T22 after 0.30 h.

Fig. 5.6b Test T22 after 50 h.

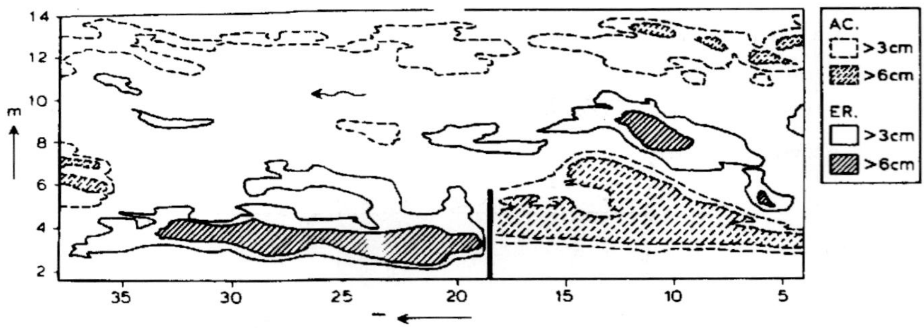

Fig. 5.6c Test T22, difference chart 0–50 h.

These data, together with the local depth and the bed material characteristics, are enough to compute the longshore sand transport according to Bijker. Examples are given in Fig. 5.7a and b for a Section 8 m upstream of the groin and at the groin itself, respectively. The obvious decrease of the longshore sand transport capacity is in good accordance with the observed accretion of the beach. Figure 5.7c shows both rates of longshore sand transport in the course of time. Typical accreting and eroding coastal profiles are given in Fig. 5.7d and e, respectively, the latter together with the distribution of the longshore sand transport as caught in the sand traps.

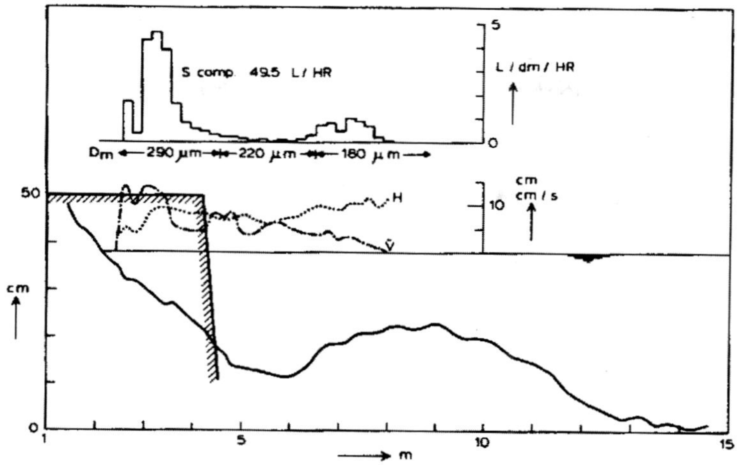

Fig. 5.7a Test T22, 8 m upstream of groin after 3 h.

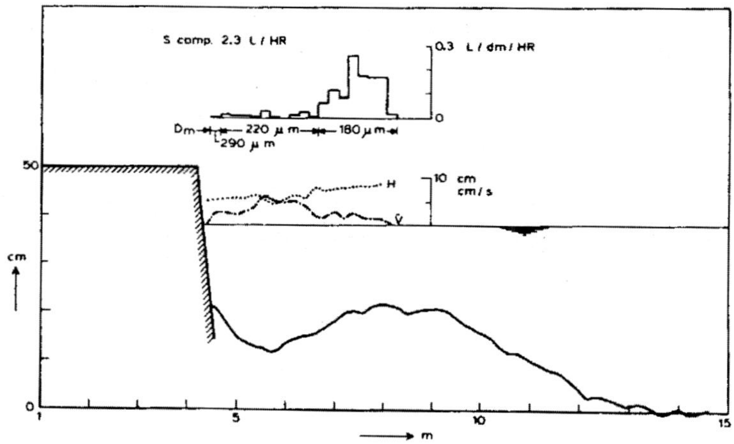

Fig. 5.7b Test T22, at groin after 4 h.

From Bijker's and Swart's theories, together with the appropriate data from the experiment, the following quantities were derived for application of the theory:

Upper level of beach	: 0.51 m	
Elevation of interface	: 0.24 m	(*with respect to the level*
Lower level of inshore	: 0.18 m	*of the wave generator floor*).

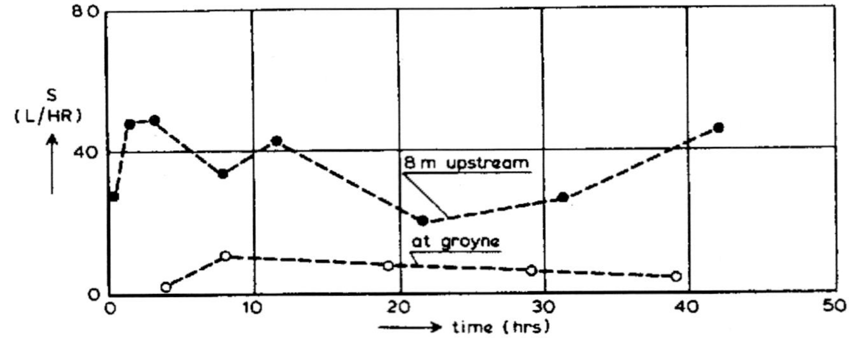

Fig. 5.7c Test T 22; longshore transport according to Bijker.

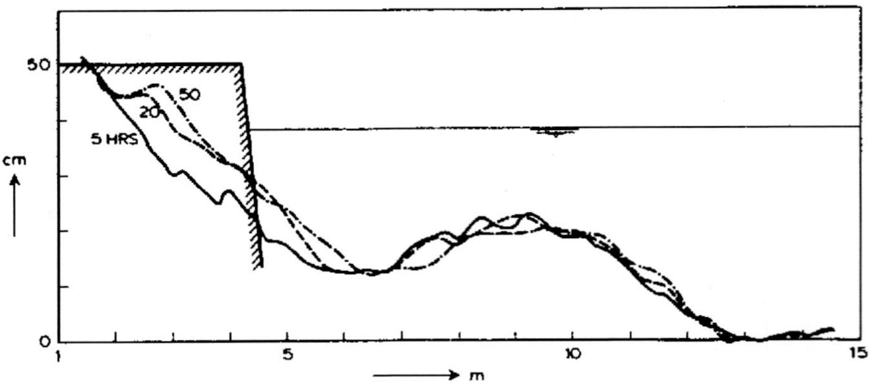

Fig. 5.7d Test T22, 3.5 m upstream of groin.

$$S_{01} = 87\,\mathrm{dm}^3\,\mathrm{h}^{-1};\; s_1 = 994\,\mathrm{dm}^3\,\mathrm{h}^{-1};\; S_{02} = 29\,\mathrm{dm}^3\,\mathrm{h}^{-1};$$

$$s_2 = 332\,\mathrm{dm}^3\,\mathrm{h}^{-1};\; W = 7.62\,\mathrm{m};\; s_y = 0.0338\,\mathrm{dm}\,\mathrm{h}^{-1}$$

The resulting schematized beach- and inshore-lines for 0.10 and 50 h are given in Fig. 5.7f, together with the comparable theoretical lines. The fit is quite good, especially for the beach-lines. It is thought that this favorable result was reached under the influence of an extraordinarily regular breaker type (plunging), which was accompanied by a very regular longshore current. More over the groin interrupted the longshore current almost entirely, thus obstructing the longshore transport of sand to a great extent.

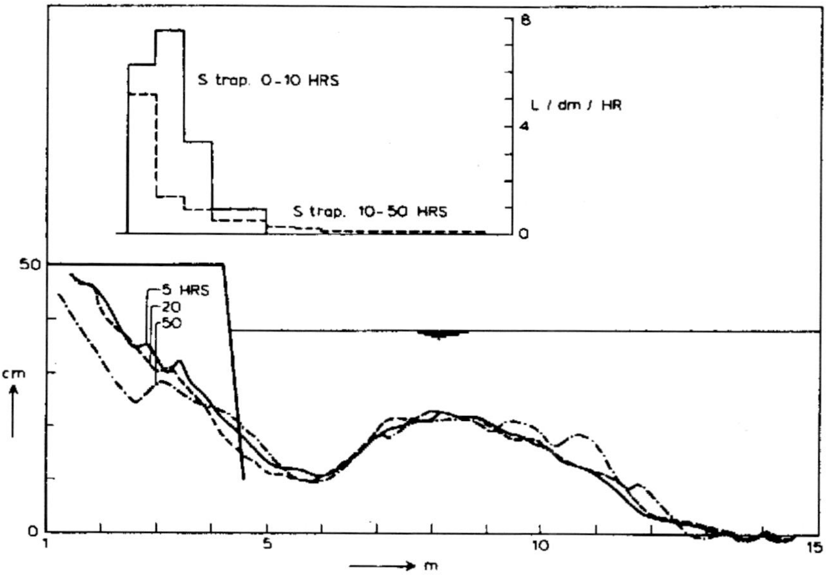

Fig. 5.7e Test T22, 3.5 m downstream of groin.

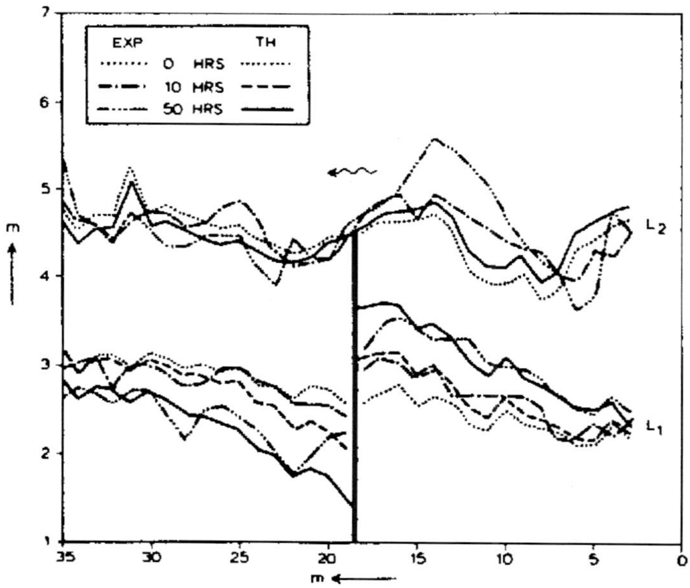

Fig. 5.7f Test T22, comparison of beach- and inshore-lines.

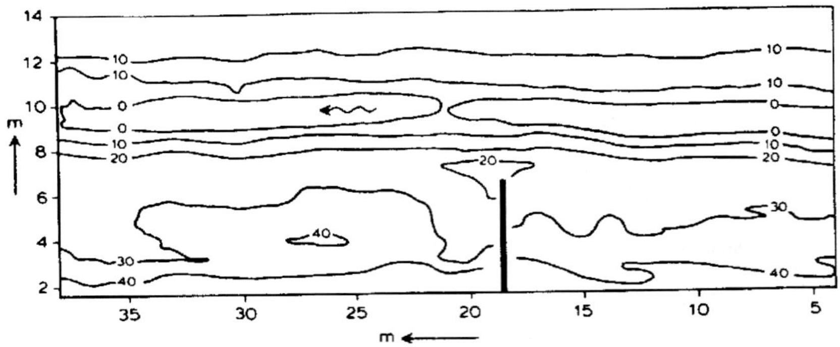

Fig. 5.8a Test T18, 60 h, contours in cm above model datum.

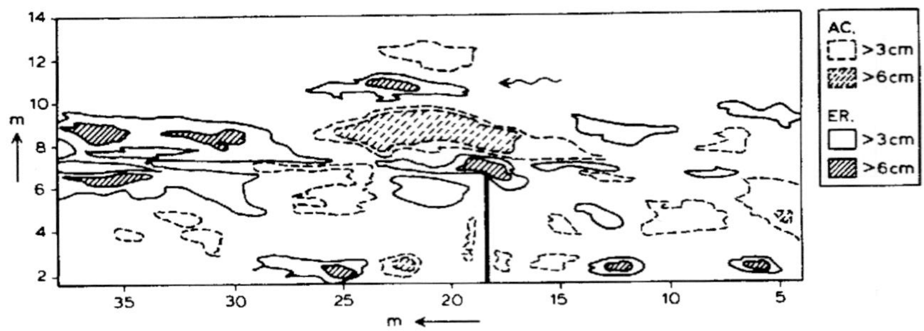

Fig. 5.8b Test T18, difference chart 0–20 h.

Another test with a single groin a T18, Figs. 5.8 and 5.9, contrasted sharply with the above results and the main conditions were wave height, 0.10 m, wave period, 1.55 s; water depth, 0.38 m in front of the wavegenerators; average rate of sand nourishment, 88 $dm^3\,h^{-1}$, while 106 $dm^3\,h^{-1}$

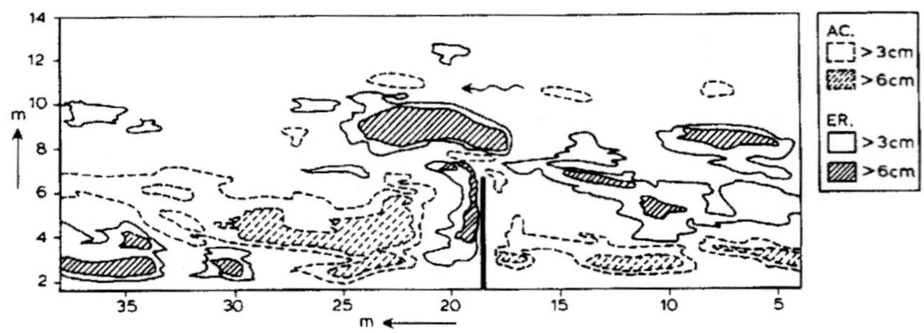

Fig. 5.8c Test T18, difference chart 20–60 h.

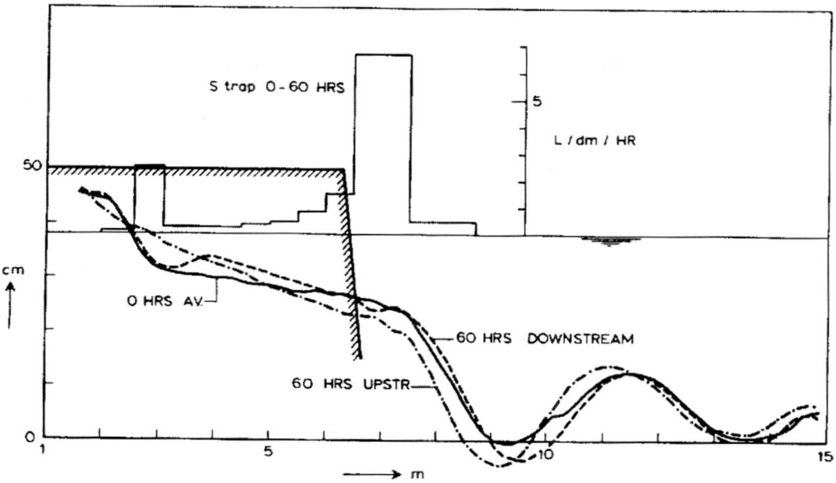

Fig. 5.8d Test T18, average upstream and downstream profiles.

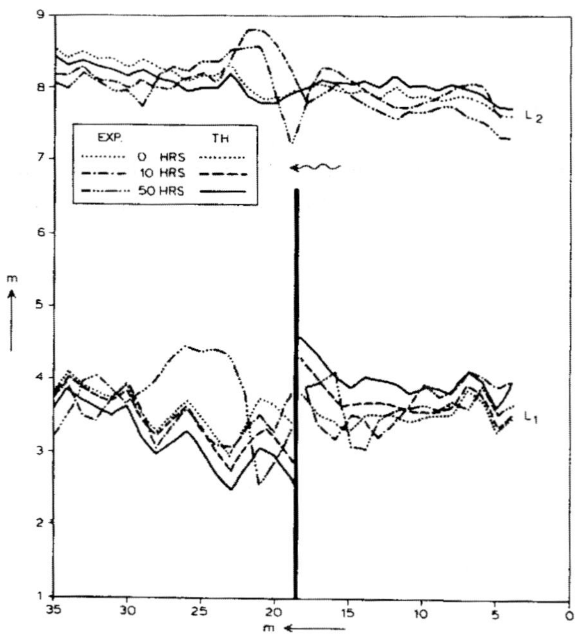

Fig. 5.9 Test T18, comparison of beach- and inshore-lines.

was caught in the traps. The water discharge was constant at 55 dm^3 s^{-1} In Fig. 5.8a the contours after 60 h are shown, where the attention is drawn to the clear accretion, but this time *down* stream of the groin. Locally even a new beach line appeared seaward of the original one. The development can be followed in Fig. 5.8b and c, where different charts of 0–20 h and 20–60 h are showing how the bulk of the longshore sand transport moved around the groin in two stages. In the intermediate stage, the maximum accretion amounted to 20 cm at some distance downstream and seaward of the head of the groin. After 20 h the bulk of this sand deposit was removed and shifted in shoreward direction, and somewhat more downstream. Figure 5.8d shows that neither the upstream nor the downstream overall profile was subject to large changes. It also points towards a possible reason for the remarkable developments encountered here: the trapped sand distribution, averaged over 60 h, indicates that the groin could hardly be expected to stop the longshore transport. However, the average trapped sand distribution in the preceeding test T17 (without groin), was used to design this groin length such that it would stop approximately 50% of the existing longshore sand transport.

After the above explanation it is not surprising that the experimental results are not well covered by the theoretical beach- and inshore-lines (Fig. 5.9), which are based on the following data:

Upper level of beach	: 0.46 m	
Elevation of interface	: 0.235 m	*(with respect to the level*
Lower level of inshore	: 0.06 m	*of the wave generator floor).*

$$S_{01} = 73.5\,\text{dm}^3\text{h}^{-1};\ s_1 = 840\,\text{dm}^3\text{h}^{-1};\ S_{02} = 65\,\text{dm}^3\text{h}^{-1};$$

$$s_2 = 742\,\text{dm}^3\,\text{h}^{-1};\ W = 10.08\,\text{m};\ s_y = 0.029\,\text{dm}\,\text{h}^{-1}$$

The impression exists from various observations (see also Subsection 5.2.3) that secondary wave formation, caused by the bar at a distance of 11 m (Fig. 5.8d), may have played an important part in re-establishing the original coastal profile after 20 h. It is felt that this type of secondary waves has an important influence on coastal profile dynamics.

5.2.1.3. *Tests with a row of three groins*

In Fig. 5.10a the contours after 50 h of test T23 with three groins are presented. The conditions were: wave height, 0.075 m; wave period, 1.55 s;

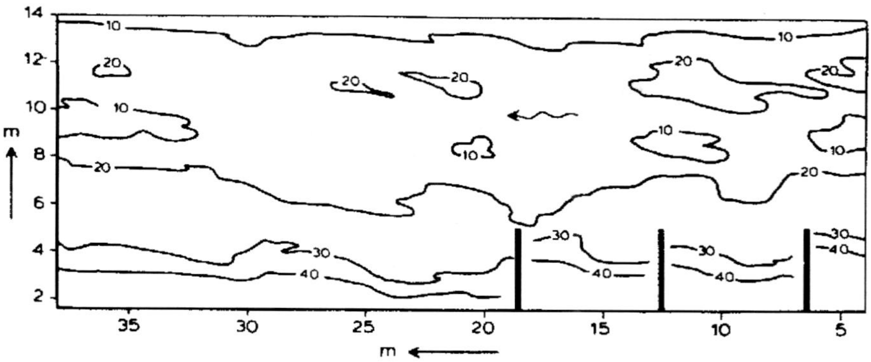

Fig. 5.10a Test T23, 50 h, contours in cm above model datum.

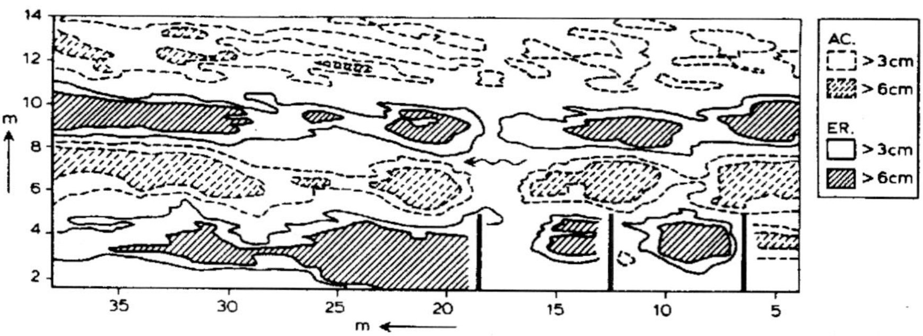

Fig. 5.10b Test T23, difference chart 0–50 h.

water depth, 0.38 m in front of the wave generators. The average rate of sand nourishment was 44 dm^3h^{-1} while 27 dm^3 h^{-1} was caught in the traps. The average rate of water supply was 25 dm^3 h^{-1}. The groins extended to 3.7 m. Until 20 h a plunging breaker occurred at approx. 3.7 m, whereas later on the spilling type dominated, with variations in breaker location. In Fig. 5.10b a clear rhythmic behavior is revealed by the difference chart, both in longshore direction (as caused by the groins) and in perpendicular direction (as caused by an overall coastal profile development). The various quantities with respect to the theoretical computations are:

Upper level of beach	: 0.51 m	
Elevation of interface	: 0.255 m	(*with respect to the level*
Lower level of inshore	: 0.160 m	*of the wave generator floor*).

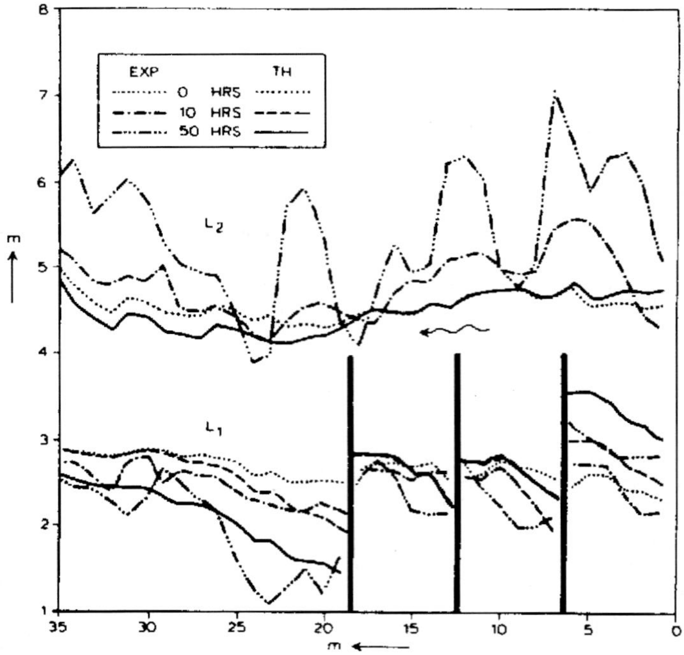

Fig. 5.10c Test T23, comparison of beach- and inshore-lines.

$$S_{01} = 87\,\mathrm{dm^3\,h^{-1}};\ s_1 = 995\,\mathrm{dm^3h^{-1}};\ S_{02} = 29\,\mathrm{dm^3h^{-1}};$$
$$s_2 = 331\,\mathrm{dm^3h^{-1}};\ W = 8.10\,\mathrm{m};\ s_y = 0.032\,\mathrm{dm\,h^{-1}}.$$

Figure 5.10c, displaying the experimental and theoretical lines representing the beach and the inshore, reveals a reasonable equality as far as the beach is concerned. However, large deviations occur in the inshore, where the typical rhythmic varations are not found back in the theory.

Test T34 (Fig. 5.11) had unusual conditions in so far that the water level fluctuated in a 1-h cycle 0.025 m plus and minus the average of 0.38 m above the wave generator floor. This was done in order to shift also the type and the location of the breaking waves, in an attempt to avoid rip-current formation at fixed locations. The other conditions were: wave height, 0.115 m; wave period, 1.15 s; average rate of caught sand, 74 dm^3 h^{-1}. The rate of sand nourishment was constant at 70 dm^3 h^{-1}, and the rate of flow was constant at 50 dm^3 h^{-1}. The groins extended to 3.27 m, whereas the breaker location varied widely during the test.

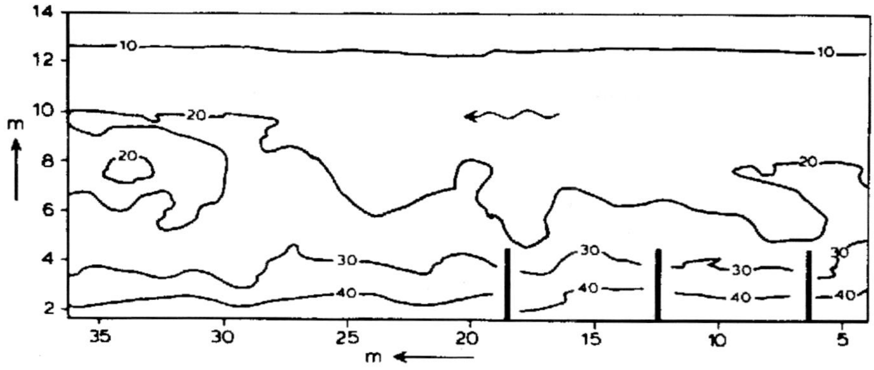

Fig. 5.11a Test T34, 45 h, contours in cm above model datum.

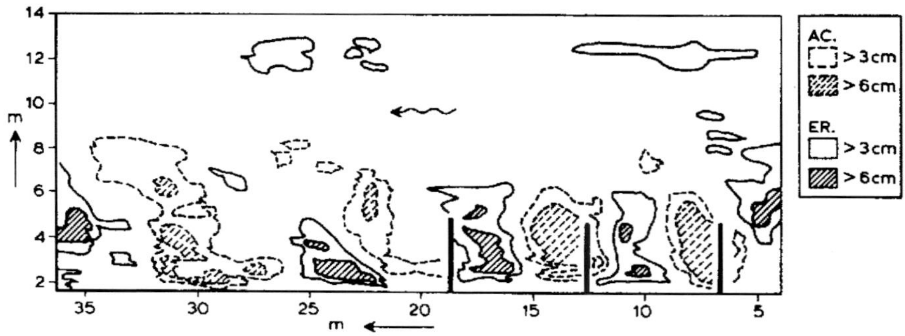

Fig. 5.11b Test T33/34, groin effect 30 h.

The quantities for the theoretical analysis were:

Upper level of beach : 0.49 m
Elevation of interface : 0.275 m (*with respect to the level*
Lower level of inshore : −0.01 m *of the wave generator floor*).

$$S_{01} = 32.4\,\text{dm}^3\,\text{h}^{-1};\ s_1 = 370\,\text{dm}^3\,\text{h}^{-1};\ S_{02} = 35.6\,\text{dm}^3\,\text{h}^{-1};$$

$$s_2 = 407\,\text{dm}^3\,\text{h}^{-1};\ W = 8.11\,\text{m};\ s_y = 0.108\,\text{dm}\,\text{h}^{-1}$$

The result was not much better than that of test T23, as can be seen on Fig. 5.11a–c. Figure 5.11b is a special chart, displaying the difference between test T34 and test T33 (without groins but further with exactly the same conditions as T34) after 30 h. The pattern of alternating erosion and accretion between the groins is opposite to what should be expected

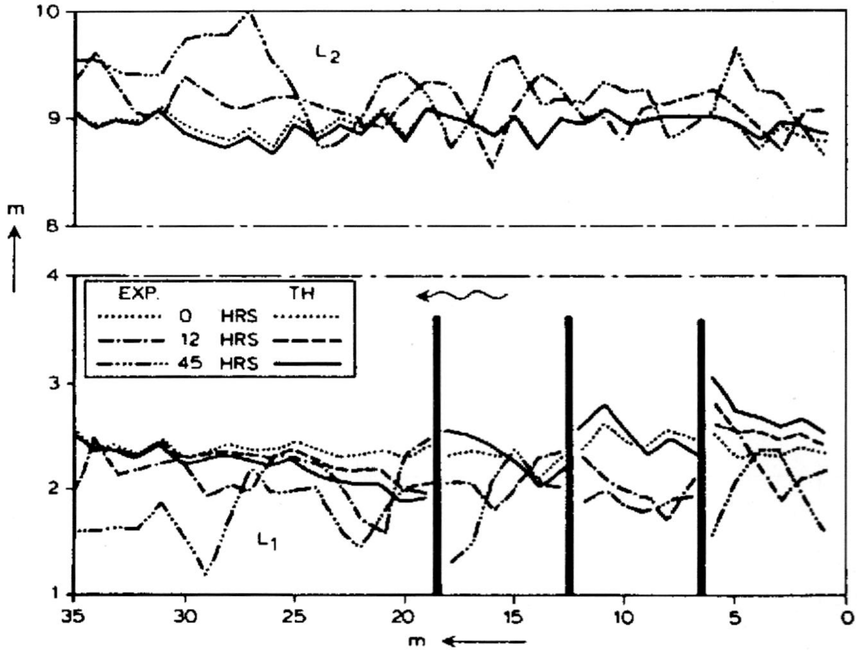

Fig. 5.11c T34, comparison of beach- and inshore-lines.

from the theory. This is most probably caused by an intricate and unstable pattern of currents, stagnation areas, rip currents and regions of lower and higher waves.

5.2.2. *Experiments in Laboratori Nacional de Engenharia Civil*

5.2.2.1. *Model facility and test conditions*

For details, the reader is referred to.[14] Figure 5.12 shows the lay-out of the model. A convex beach profile of pumice-stone ($\rho = 1670\,\text{kg}\,\text{m}^{-3}$, $D_\text{m} = 1.25\,\text{mm}$) was molded between two rockfill groins E1 and E3. These groins had 2:1 absorbent slopes and acted as training walls for the beach. A smaller groin E2 in between with steep 1:5 slopes completely obstructed the littoral drift. The waves were 0.02 m high, had a period of 1 s and were generated under an angle of 20° in a water depth of 0.40 m.

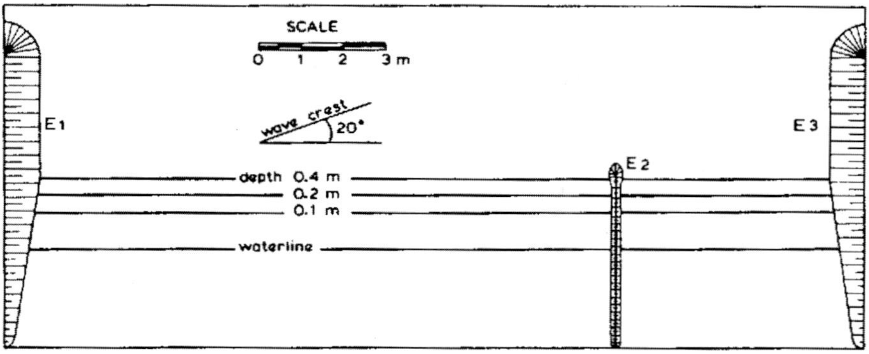

Fig. 5.12 LNEC test, lay-out of model basin.

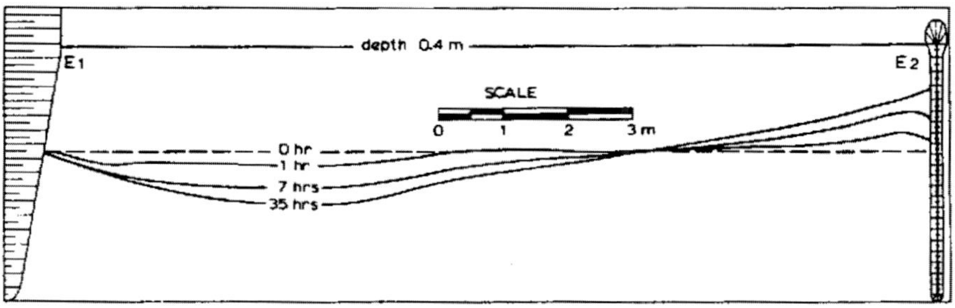

Fig. 5.13 LNEC test, experimental beach lines.

5.2.2.2. *Test results and theoretical beach lines*

Various beach line positions were recorded during the 35 h test, as presented in Fig. 5.13. These are compared to theoretical lines, which were determined as follows.[17] The active profile was schematized to a single beach line, because perpendicular transport across the 0.063 m depth contour was assumed negligible. This depth is twice the breaker depth as calculated for straight, parallel depth contours. Diffraction calculations were performed in the deeper, horizontal part of the basin, where after refraction was calculated up to the 0.063 m contour line, again assuming straight and parallel depth contour lines, remaining so during the whole test. The resulting values for the wave height H_1 and the angle of wave attack ϕ_1 at the 0.063 m depth contour are given in Fig. 5.14.

The final position of the beach is now assumed to be defined by ϕ_1 and H_1 in a first approximation. By ϕ_1 alone, the final beach line is determined

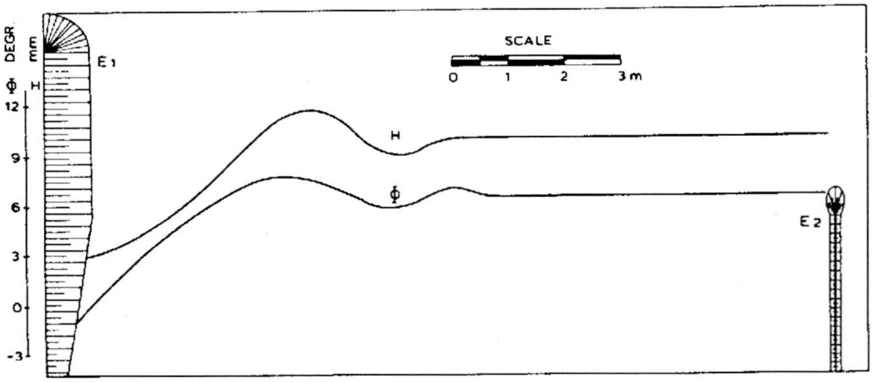

Fig. 5.14 LNEC test, wave height and angle of wave incidence.

by being parallel to the local wave crest as defined by ϕ_1 because only then the net transport along any part of the beach is equal to zero. By H_1 alone on the other hand, differences in wave set-up will develop, causing a varying longshore current velocity V_1 along the beach, where V_1 is a function of various parameters including the beach line orientation.[8] By substituting V_1 equal to zero, the final beach alignment is found. After adding the partial solutions for H_1 and ϕ_1 the final coastline is found as Y_0 in Fig. 5.15. By comparing this line to the experimental 35 h line in Fig. 5.13, some differences can be found:

- In the experiment a certain amount of material does disappear into deeper water. Because Barcelo reports a rip-current near groin E2, the theoretical line is adapted by taking a sand transporting rip-current into account near E2 with an adequate transport capacity. This will at

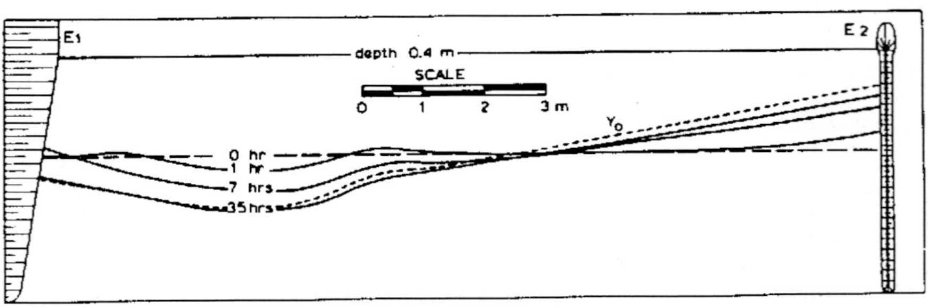

Fig. 5.15 LNEC test, theoretical beach-lines.

the same time increase the theoretical beach line angle near groin E2, bringing it closer to the measured position.

- Differences near groin E1 are likely to be caused by a very low transport capacity in the model, whereas in the calculation no critical velocity for beginning of movement is used.

For the calculation of the coastline after 1, 4, 7, 15 and 35 h the dynamic equation is used (with $s_y = 0$), together with Eq. (3.48). Relevant quantitative parameters are: $S_{\text{ripcurrent}} = 3.4\,\text{dm}^3\,\text{h}^{-1}$, $S_{\text{ol}} = 27.8\,\text{dm}^3\,\text{h}^{-1}$ acc. to Barcelo, $s_1 = 150\,\text{dm}^3\,\text{h}^{-1}$.

The resulting theoretical lines are given in Fig. 5.15; they fit quite well with the experimental lines of Fig. 5.13.

5.2.3. *Discussion of the results*

The above examples are typical in so far that they indicate the rather large variation in the ability of the theory to fit with the experimental results. One should not be too surprised, in view of the wide deviations which were often present in the actual experimental conditions, as compared to the ideal conditions which were (tacitly) assumed in the theory. The main causes of deviation are discussed below.

- Variations in wave height as measured in cross-sections very close to each other, were frequently noted and were judged to have a strong negative effect on the homogeneous conditions sought. Figure 5.16 shows a rather dramatic example, in which the wave height variations are at least partly due to the varying bed profiles. Of course there exists a strong mutual interaction between wave behavior and coastal profile development.
- An example of the progressively deteriorating homogeneity in longshore direction of the coastal profiles is given in Fig. 5.17a. It shows the envelopes of all cross-sections after 10, 30 and 50 h in test T19 (without groin).
- An effect, and simultaneously a cause, of the above-mentioned non-homogeneity is present in the chaotic current pattern in the same test, as presented in Fig. 5.17b. This test, too, began with a nice homogeneous and steady longshore current.
- Apart from being non-homogeneous, the current pattern was unsteady, as is shown in Fig. 5.18. It gives values of the longshore current in test T25, measured in two points 1 m apart perpendicular to the coast. The measurements were performed by timing floats over a distance of 1 m.

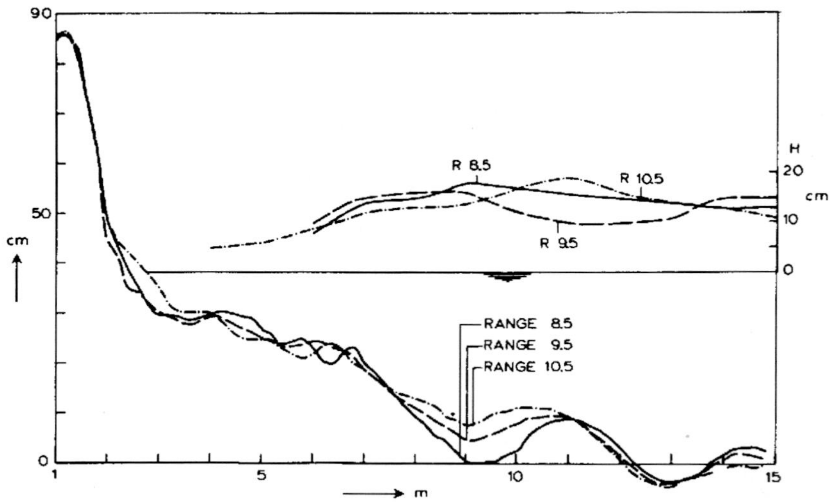

Fig. 5.16 Test T2, variety in beach profiles and wave heights.

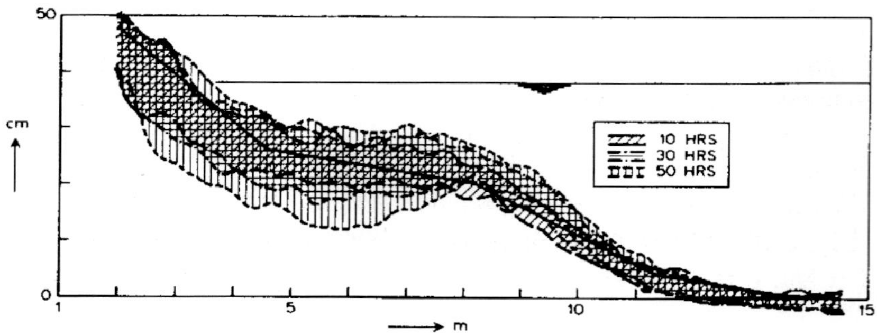

Fig. 5.17a Test T19, envelopes of beach profiles.

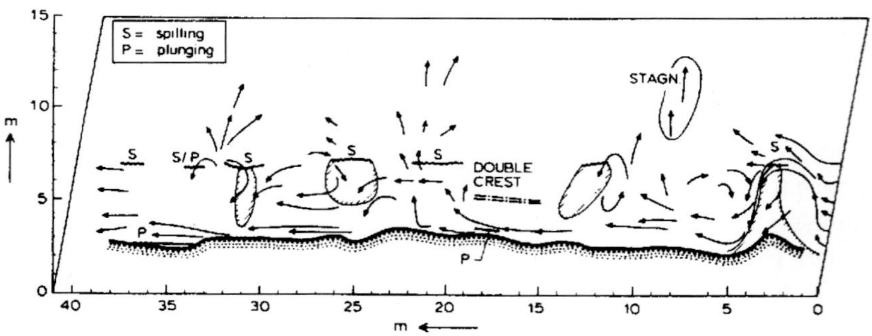

Fig. 5.17b Test T19, current pattern 77 h.

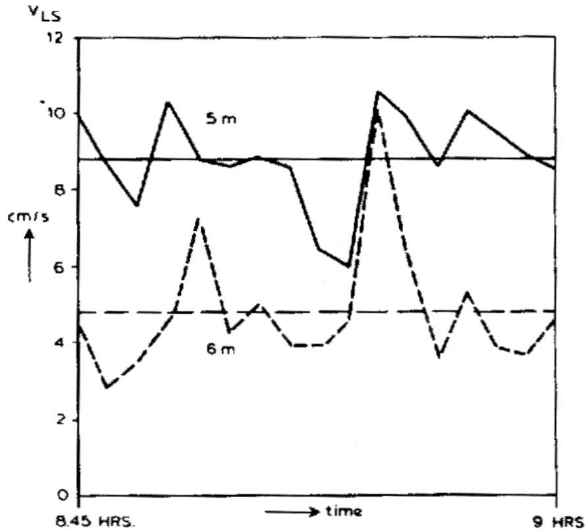

Fig. 5.18 Test T25, unsteadiness in longshore current velocity.

- Secondary waves were present under certain conditions. Apart from those, generated by the sinusoidally moving wave board,[27] also secondary waves were generated while the regular waves passed over a large bar without breaking. In this respect it is interesting to note that Byrne,[15] reports the same wave behavior (long regular swell conditions) in the prototype (Fig. 5.19). Figure 5.20a presents a coastal profile of test T19, with a bar and with the wave height over it; the wave breaks only at a distance

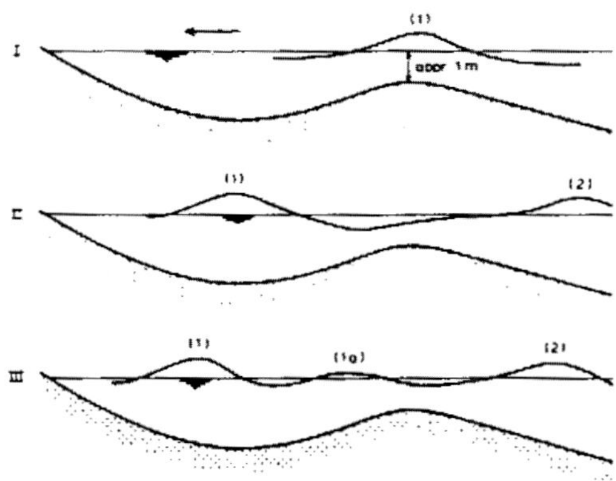

Fig. 5.19 Prototype observation of induced multiple gravity waves (after Byrne).

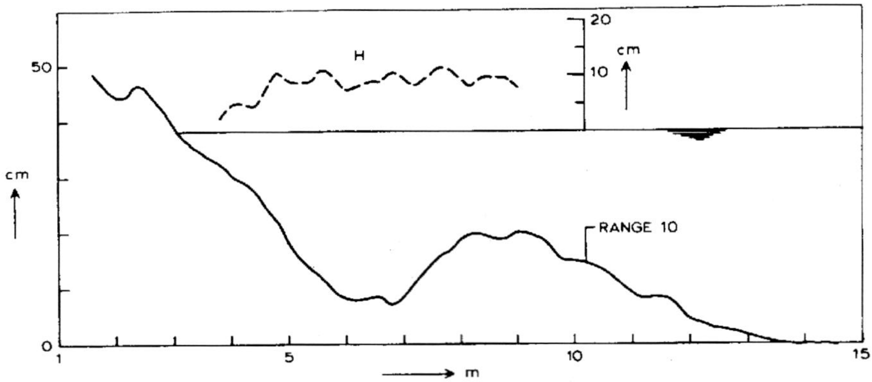

Fig. 5.20a Test T19, 105 h, bar profile and wave height.

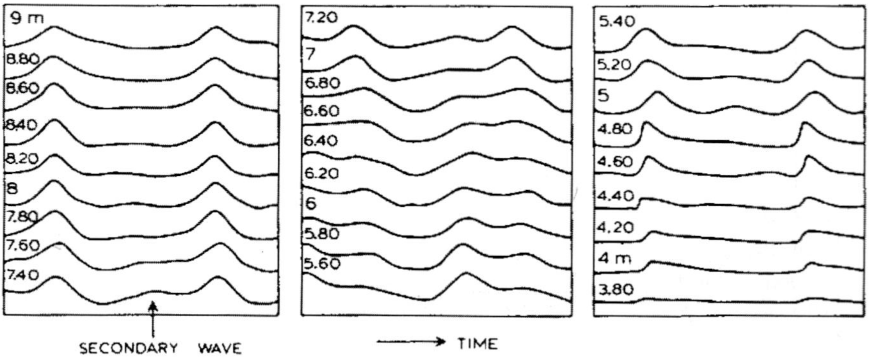

Fig. 5.20b Test T19, wave forms over bar profile from Fig. 20a.

of approx. 5 m. The respective wave forms at successive sites are shown in Fig. 5.20b. These clearly show the presence of secondary waves (with period $T/2$) which are overtaken gradually by the faster main waves. Their presence has undoubtedly a large influence on the perpendicular sand transport,[27] and further on the type of breaking and the kinematic results thereof. It is felt that the presence of those secondary waves can has a dominating effect on the formation and the stability of coastal profiles.

- The pair of Fig. 5.21a and 5.21b illustrates the strong relationship between secondary waves, bar formation, type of breaker, and subsequently the rate and distribution of the longshore transport. The

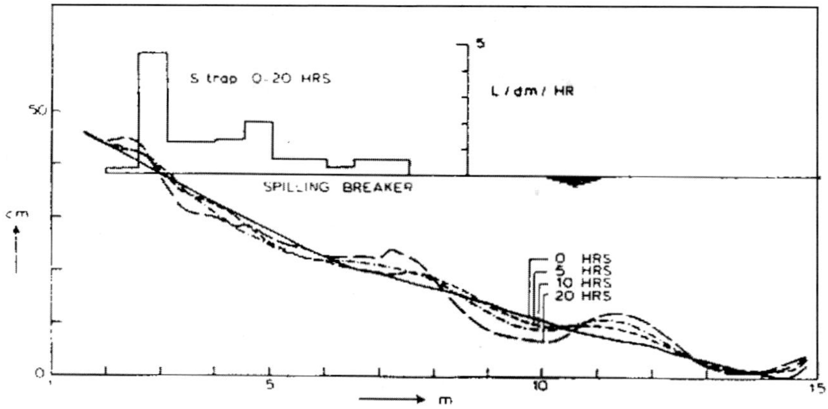

Fig. 5.21a Test T17, spilling breaker conditions 0–20 h.

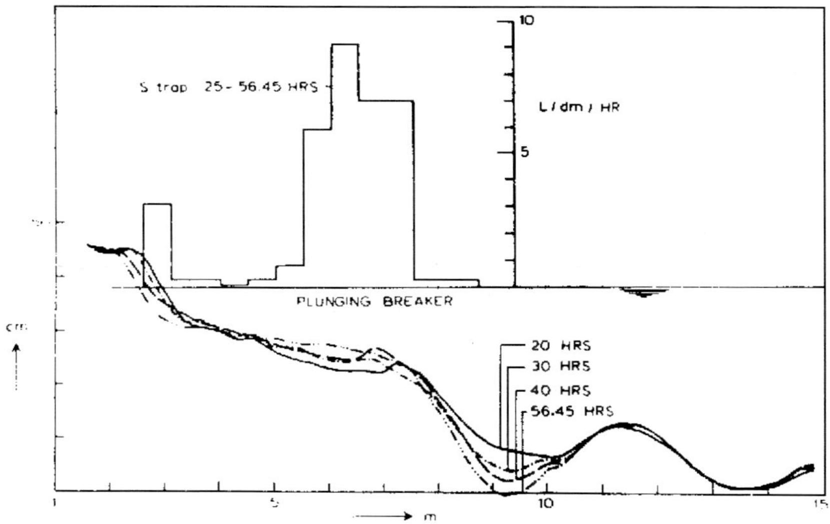

Fig. 5.21b Test T17, plunging breaker conditions 20–56.45 h.

conversion of the breaker type from spilling to plunging was in very good agreement with Galvin's results as far as the influence of the seaward slope of the breaker bar is concerned.[22] After this conversion the rate of longshore transport was approximately three times as large as before, and much more concentrated.

- The most obvious visible result of the increasingly non-homogeneous conditions in many tests was the formation of beach cusps. Both small

Fig. 5.22 Test T19, small scale beach cusps.

Fig. 5.23 Test T27, large scale beach cusps.

(Fig. 5.22) and large (Fig. 5.23) examples have been recorded. It goes without saying that this is most unwelcome in tests which are meant to verify the effect of groins on a sandy beach. Fortunately, the presence of a groin appeared to induce a more stable condition, in which the cusp formation was partly suppressed.

Conclusions concerning the model experiments

The conclusions which follow from these tests and their comparison with the theory can be summarized as follows.

- The theory is good in cases with a stable, neat, and well defined longshore current system, and if the groin(s) intercept a substantial part of the long shore sand transport.
- In other, more complex systems, the theory in its present form is not adequate on a forecast basis. However, if the current pattern is well enough defined and specified, the theoretical result may be improved by taking into consideration the effects of stream refraction, diffraction, and rip-currents.
- The coastal current system should be studied in more detail in order to improve the theory of Bakker.[c]

5.3. Results of Groin Application in Nature

5.3.1. *Introduction*

In this Section 5.3 the validity of the line theories will be verified with measurements along sandy coasts[d] (cf. Fig. 5.21). Consider (Fig. 5.24) Vlieland, the Vliehors[e] (one-line theory: Subsection 5.3.2); the German

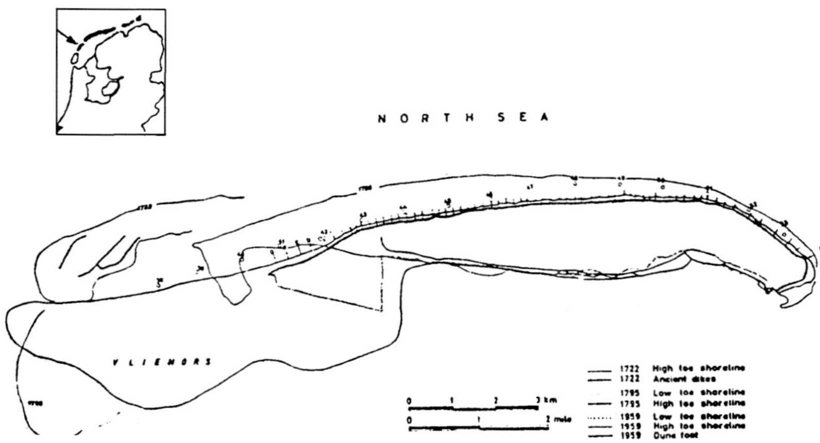

Fig. 5.24 Vlieland. Former and present situation.

[c]This recommendation has been acknowledged (cf. Subsection 5.4 below).

[d]Most of the present cases were documented earlier in the Manual on Artificial Beach Nourishment[16] and its appendix "Background Information".[16] Research leading to the specific part of that publication presented here and its editing was carried out by "Bakker and Friends".

[e]Cf. Fig 2.2 for an areal view of the site and Figs. 5.24 and 5.25 for a map.

Wadden island Sylt (Subsection 5.3.3; one- and two-line theories); Cadzand (Subsection 5.3.4, two-line theory); the total Holland coast (Subsection. 5.3.5).

The last case is specificly difficult as the various harbor moles each give their own boundary conditions. There should be matching between the assumptions of the left harbor mole for one harbor and the right one of the adjacent one, which reduces the number of the degrees of freedom. The investigation of *one* harbor mole only gives more possibilities for assumptions on conditions, which should be given the benifit of the doubt.

Finally, in Subsection 5.3.6 cases treated in the scope of the international project "NOURTEC" on coastal nourishment (Terschelling, Norderney) will be discussed.

5.3.2. *The Vliehors*

In this subsection, the previous theory will be applied to the Dutch Wadden Island of Vlieland. For more details is referred to Bakker.[2] In this section a practical application of the previous theory will be mentioned, on the North side, the Netherlands are bordered by a chain of "Wadden Islands". Between the islands and the continent is a region of tidal flats connected with the North Sea by tidal channels with a depth up to 60 ft. (Fig. 5.24). This section deals with the second island from the West, called Vlieland. In the seventeenth century a dune ridge on the North side all along the island protected the two villages, West-Vlieland and East-Vlieland, against North Sea. However, since then, a very strong erosion occurred, especially on the West-side (about 1 mile in 250 years!); the village West-Vlieland had to be abandoned and the dune ridge all along the Western part of the island disappeared (Fig. 5.24). Only a sandbank, called Vliehors, a little bit southward of the original island reminds of the past. Nowadays, only seals and soldiers play their games there.

On the Eastern part, Rijkswaterstaat (the Governmental Board for Ways, Waterways and Harbors) constructed 53 groins during the years 1854–1885, six groins from 1916 till 1923 (near the Eastend) and finally the five groins C, D, E, F, G on the extreme West (1957–1959).

One hundred years ago, Rijkswaterstaat made a fixed range system all along the Dutch coast and measured yearly the distance from the measuring poles to the High Water line and the Low Water line. The distance between two neighboring poles is 1 km.

This section will give an analysis of the results of these measurements on the sandbank called Vliehors, the Western part of the island.

5.3.2.1. *Measurements_and visual analysis*

Figure 5.25 gives the range system and the measurements in the respective ranges 37–42, as a function of time. The upper line always represents the line of MLW,[f] the lower line the line of MHW and therefore the grey fields represent the tidal area. The tidal difference is about 1.50 m (5 ft). A rising of the lines means accretion and descending of the lines means erosion.

Clearly, one observes the general erosion of this part of the coast, about 5 m/yr (interrupted lines in Fig. 5.25). But superponed, one also notices a kind of "sandwave" with a period of about 60 years.

Twice, in 1870 and 1930 a shoal of the outer delta of the inlet between Vlieland and the neighboring island, Texel, grew together with the Vliehors. Afterwards, this shoal seemed to move along the coast with a velocity of about 1/6, resp. 1/3 km/yr.

5.3.2.2. *Relating transport to accretion, c.q. erosion*

The coastline will be defined as the shoreline at mean water level, for which the mean value between the y-coordinate of the line of MLW and MHW will be taken.

In every year the littoral drift Q_L through range 37 1/2 is, according to (5.1):

$$S_L^* = S_0^* - s\frac{y_{38} - y_{37}}{1000} \tag{5.1}$$

in which $y_{37(38)}$ denotes the y-coordinate of the coastline in range 37(38) in metres, at time t. The factor "1000" in the denominator is the distance in metres between range 37 and range 38.

Likewise, one finds for the littoral drift through range 41 1/2:

$$S_R^* = S_0^* - s\frac{y_{42} - y_{41}}{1000}$$

Therefore, the supply of sand in one year should be:

$$S_L^* - S_R^* = \frac{s}{1000}(y_{37} - y_{38} - y_{41} + y_{42})$$

[f] Mean low water.

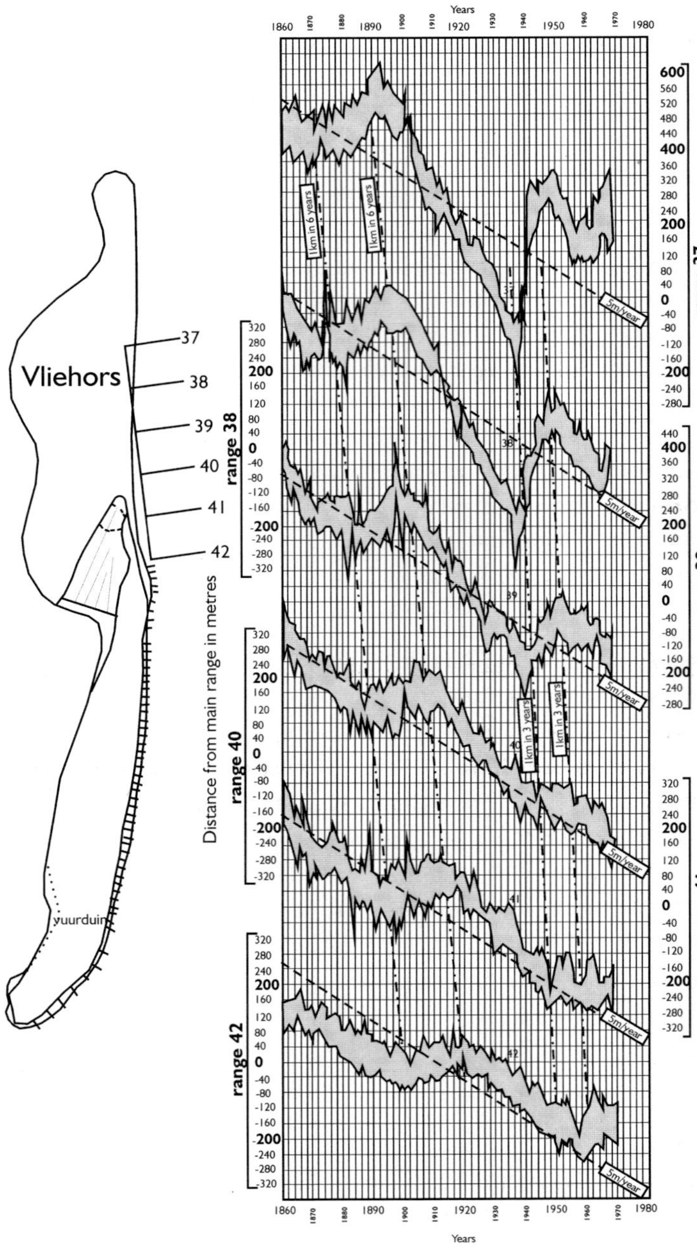

Fig. 5.25 Accretion and erosion of the coast of Vlieland. Shown is the situation of the line of HW and LW in variouis ranges as a function of time.

If one assumes a cross-section of the coast, according Fig. 5.1, then the mean accretion of the area should be:

$$\frac{S^*_{\mathrm{L}} - S^*_{\mathrm{R}}}{4000\,\mathrm{h}} = \frac{1}{4}\frac{s}{h} \times 10^{-6}(y_{37} - y_{38} - y_{41} + y_{42}) \tag{5.2}$$

in which h is the depth up to where accretion occurs, and "4000" is the length of the area.

In fact, the mean accretion of the area is:

$$\text{mean accretion} = \frac{1}{4}(y_{38}+y_{39}+y_{40}+y_{41})_{t+1} - (y_{38}+y_{39}+y_{40}+y_{41})_t \tag{5.3}$$

Equalising (5.2) and (5.3) gives:

$$\begin{aligned}&\frac{s}{h} \times 10^{-6}(y_{37} - y_{38} - y_{41} + y_{42})_{t+1/2}\\&\quad = (y_{38} + y_{39} + y_{40} + y_{41})_{t+1} - (y_{38} + y_{39} + y_{40} + y_{41})t\end{aligned} \tag{5.4}$$

This relation must be valid at every time t.

By correlating the right-hand side with the last factor on the left-hand side of (5.4), the author found for the proportionality constant of MLW and MHW will be taken at s/h the Vliehors:

$$s/h = 0.4 \times 10^6\,\mathrm{m}^3/\mathrm{m\ depth/yr/rad}$$

In Fig. 5.26 the left-hand side (black line) and the right-hand side (grey line of equation (5.4) are plotted separately as a function of time. The correspondence between the black line and the greyline is rather good. However, the lines are shifted vertically with respect to each other. Even if S^*_{L} equals; S^*_{R} (straight coastline), still an erosion occurs of 14.5 m/yr. We will call this "the offshore transport", because one of the causes for it can be transportation of material to deeper regions, where it is beyond the reach of the waves and not available for accreting of the coast.

Other reasons may be:

1. transportation of material over the shoal in case of storm surges (this is the most probable reason, given the set-up at the seaside during storm conditions);
2. transportation of sand by wind;

influence of currents. These currents are not very large (about 1 knot). In the case of the Vliehors, the sand wave solution gives for the wavelength L, and the propagation velocity respectively ($T = 60$ years, $s/h = 0.4\,\mathrm{km}^2/\mathrm{yr}$):

$$L_1 = \sqrt{4\pi \times 0.4 \times 60} \approx 17.3\,\mathrm{km} \qquad c = 17.3/60 = 0.29\,\mathrm{km/yr}$$

In reality, the velocity was 1/6–1/3 km/yr over a distance of 4 km, the rate of attenuation is: $e^{-kx} = e^{-\frac{2\pi\times 4}{17.3}} = e^{-1.44} = 24\%$.

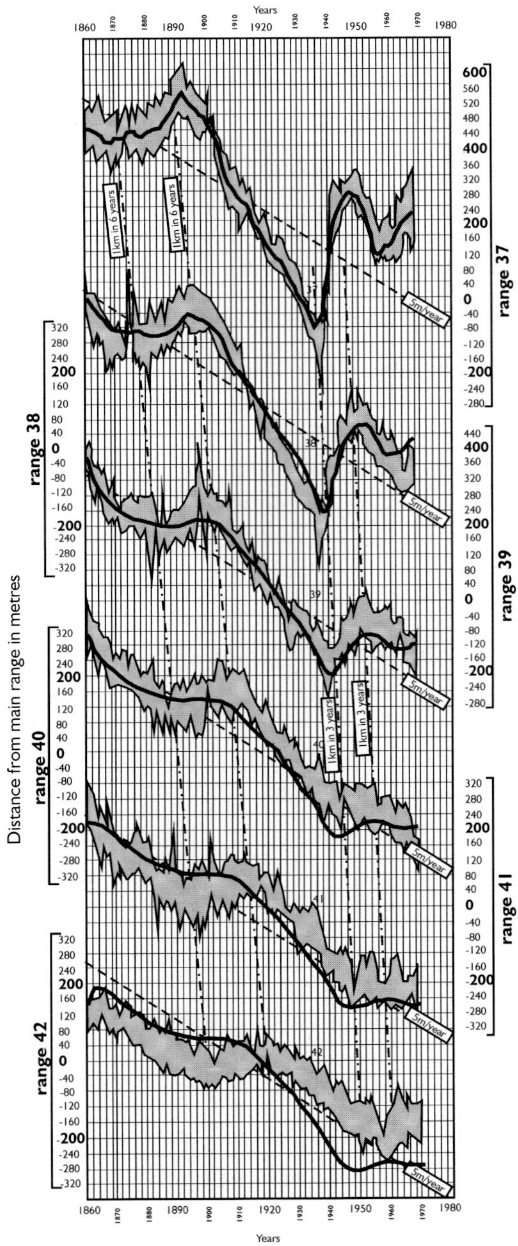

Fig. 5.26 Results of the computation: simulation of the cioastal behavior of the coast of Vlieland.

In reality, the wave attenuated from 300 m in range 37 (1936) to 76 meters in range 41 (1947). In fact, 25% remains.

Essentially, the wave form corresponds with the sandwaves as observed at the Vliehors, but there are some differences:

1. Evidently an offshore transport exists.
2. The sandwave does not propagate until infinity, but finds the groins of the protected part on its way.
3. The erosion and accretion are not exactly harmonically.

The first and last points will be considered more into detail.

Offshore transport. Without offshore transport, the coastal equation (3.6) was found. With offshore transport, this equation becomes:

$$\frac{s}{h}\frac{\partial^2 y}{\partial x^2} - p = \frac{\partial y}{\partial t} \tag{5.5}$$

in which p is the erosion in m/yr, caused by the offshore transport, if the coastline is straight A solution of (5.5) is:

$$y = -At + \left(\frac{p-A}{2} \times \frac{h}{s}\right) x^2 + Bx + C \tag{5.6}$$

in which A, B and C are auxiliary constants. If $A = p$, then the coastline is straight and erodes with a velocity of p m/y. If $A = 0$, then the coastline is a stable parabola. If $A < p$, then the coastline is a slightly eroding parabola. With the proper choice of constants and boundary conditions, for which is referred to[1] the method of Schmidt results in the predicted erosion and accretion in the course of time as given in Fig. 5.26.

5.3.3. *Sylt*

On the central area of Island of Sylt artificial beach nourishment projects have been under-taken in 1972 and 1978, in which the fills were placed as a stockpile groin. In the following the performance of such a sand groin is analyzed.

Only the 1978-nourishment is thoroughly analyzed, since the available data for this nourishment is more extensive than the one of the 1972-nourishment.

General part: scheme of this subsection

Goal of the present investigation is to evaluate the applicability of various schematizations, by finding a value for the overall standard deviation between nature and theory, using the various options, shown in Fig. 5.30. Results (to make a shortcut) are given in Table 5.5.

For the study on morphological processes the coastal engineer has different mathematical models at his disposal. The question arises concerning the advantages and disadvantages of those various models. This subject is treated in the present section. A comparative evaluation of mathematical theories, i.e. the one- and two-line theories will be given, using the accuracy as measure and starting from the data of the 1978 beach nourishment at Sylt.[g] A number of options of schematization (five totally) will pass the review (page 265–280).

For the evaluation, it is necessary to have a clear picture in mind concerning the requirements of such a theory. One has to distinguish:

- the accuracy of the theory; and
- the relevance of the theory.

For example, a theory may be perfectly accurate with respect to its mathematical predictive value, but when the assumed initial shape does not match the actual initial shape of the coast, the theory is not relevant. Therefore, it is necessary to define the standards with respect to *relevance* first, before considering the accuracy. The following standards are proposed:

(a) The mathematical formulation of the initial shape should be able to cover the initial shape of the coastline, as usually found in practical cases, as close as possible.
(b) For the prediction of future developments no information should be required other than that available at the time of the prediction.
(c) The known behavior of the coast before nourishment (for instance gradual erosion, sandwaves) should be incorporated in the mathematical formulation.

With respect to the *accuracy* (and also more or less in connection with the relevance) the following standards seem to apply:

(d) The theory should reproduce the overall development of beach and inshore as close as possible.
(e) The close simulation of the behavior of the beach is of more importance than the close reproduction of the behavior of the inshore.

[g]For situation, cf. Fig. 5.27. For details concerning topography and its development after the supplies, wave climate, etc. refer,[38] Annex II, pp. 2–18.

Fig. 5.27 Location of the project of Sylt, as well of other sites, being in the focus in this Chapter 5.

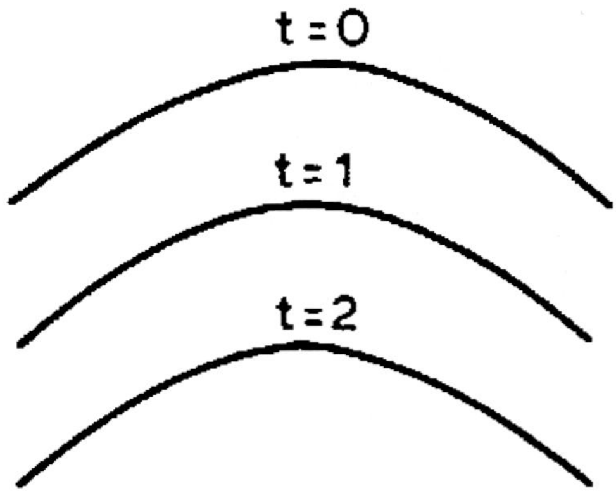

Fig. 5.28 Coastline before nourishment: a gradual retreating parabola.

Considering these standards in relation to the one-line and two-line theories, it has to be kept in mind, that within the scope of one theory various possibilities exist, namely:

- with respect to numerical or analytical application;
- with respect to the schematization.

With respect to numerical or analytical application, the first standard (a) would lead to a preference for numerical methods, which allow the exact covering of the initial conditions. However, these methods require generally a known behavior of the boundary conditions in the course of time, which is in contradiction with standard (b).

Therefore, for all applied schematizations, the initial condition has been schematized in such a way, that a general analytical concept of the anticipated approximate behavior of the coast was assumed; this includes assumptions concerning the boundary conditions for the future situations, which have to be calculated. The boundary conditions, following from this concept might be rather intricate. For instance, in the considered case of Sylt this involves a uniform erosion at the boundaries only up to the moment, that the widening and flattening Gauss-curve (see next specification) influences those boundaries perceivably.

It has been assumed, that the coast of Sylt before nourishment consisted of a retreating parabola (i.e. parallel depth contours of parabolic shape). The curvature of the parabola has been assumed in such a way, that the predicted erosion corresponds with the erosion occurring in nature before the beach nourishment, thus covering standard (c).

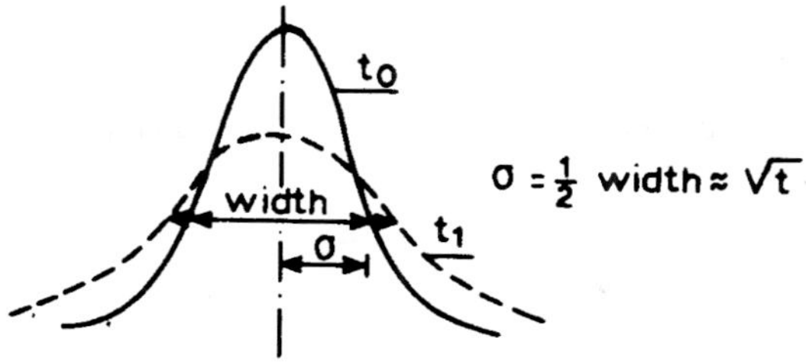

Fig. 5.29 Expanding and decaying Gauss-curve. Cf. Fig. 3.13.

Furthermore it has been assumed, that the initial shape of the nourishment consisted of one or two Gauss-curves for the one- or two-line theory, respectively (see Fig. 5.29).

After the foregoing rather general considerations, it is useful to concentrate somewhat more on Sylt, by considering the standards with respect to the accuracy and its consequences for the schematization.

As the theory should reproduce the overall development of beach and inshore as close as possible (standard d; page 254)), a logical schematization appears to be: a one-line approximation where the "coast" covers the total profile between MSL + 4 m and MSL − 8 m. This will be the A-option (see below) to be considered in the following.

From these assumptions and the assumed coastline theory the boundary conditions as a function of time are determined.

Especially the accuracy of the assumption "initial curve =Gauss curve" gives only information on the usefulness (relevance) of the theory and tells nothing on its predictive value. Therefore, the accuracy of this initial condition, called S_{di} ("standard deviation-initial") will be kept separate from the calculated accuracy of the predicted curves at later times, symbolized by S_{d}. The values of S_{di} and S_{d} should be considered in relation to each other. If S_{d} is smaller than S_{di} then in the course of time the coastline tends to fulfil more and more the assumptions of the theory. Theoretically for the one-line theory it can be shown, that a coast being straight and at rest at infinity, will resemble more and more a Gauss-curve, independent on the initial condition.

A. Option:("1_total):

As the theory should reproduce the overall development of beach and inshore as closely as possible (standard "d)" on page 254), a logical schematization appears to be: a one-line approximation where the "coast" covers

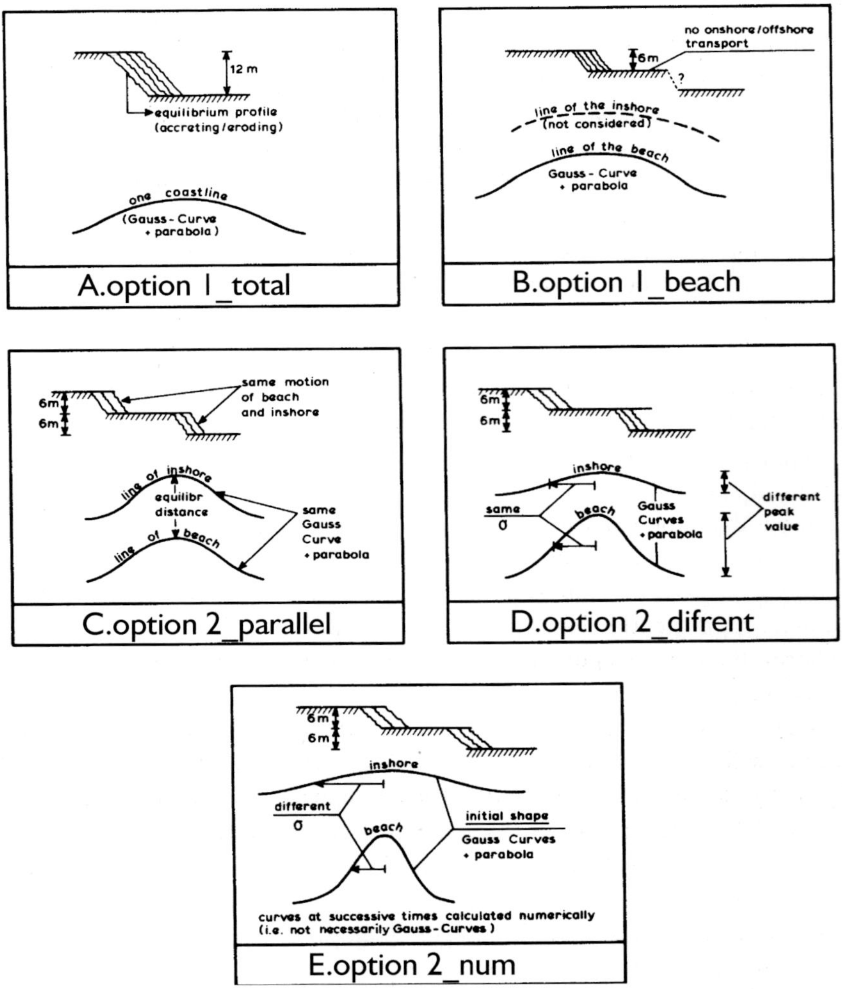

Fig. 5.30 The five options of schematization, which are considered.

the total profile between MSL + 4 m and MSL − 8 m. This will be the A-option (Fig. 5.30) to be considered in the following.

As to be described later on, for this case the initial accuracy $S_{\rm di}$ and the overall accuracy $S_{\rm d}$ of the predicted coastlines after one month and one year will be calculated.

It is not clear, whether the above schematization covers also standard (e), i.e. a close simulation of the behavior of the *beach*, to be defined as the area between MSL + 4 m and MSL − 2 m. One approach to fulfil this condition is to "forget" about the inshore (the area between MSL − 2 m and MSL − 8 m) and forget standard (d), page 254: apply the one-line theory only to the beach area. This will be the *B-option: "1-beach"*. (page 269 and Fig. 5.30). When the same values should be found for S_{di} and S_{d} in the case of the first option and the second one, it is clear that the second one should be preferred, since a good reproduction of the beach behavior is more important for a coastal manager than a good reproduction of the rest of the profile. However, it is not likely that the same values are found, since the offshore losses, to be expected in the second option, are not taken into account.

After this, options C–E, the two-line theory (Section 3.3) will be applied and tested. Again, the Gauss-curve, Fig. 5.29 will play a dominant role.

Referring to Chapter 3: this gives the opportunity to take the variation in coastal slope into account. The lines of beach and inshore are ruled by the same dynamic equations as in the one-line theory, however, sand loss from beach to inshore (offshore transport) is simulated when the profile is steeper than the equilibrium profile, and onshore transport when the profile is more gentle than the equilibrium profile. The relation between onshore/offshore transport and the mean slope of the coastal profile is linearized (Eq. page 183).

The assumption of the one-line theory, that all depth contours are parallel will be maintained during the *C-option 2_parallel* (page 272 and Fig. 5.30). It is a transition between the one-line options, treated before and the "real" two-line options D and E. Option C sticks to the assumption, that the "line of the beach" (the mean of the contour lines between MSL + 4 m and MSL − 2 m) and "the line of the inshore" (the mean of the contour lines between MSL − 2 m and MSL − 8 m) have the same shape as the coastline found from the first option. These two lines can be found by moving this coastline in landward or seaward direction.

Finally, *option D: 2_difrent* and *E: 2_num*: give the "real" two-line theory: the results differ, because of the use of, respectively, an analytical and a numerical approach.

Option *E: 2_num*: does not imply the restriction, that σ (Fig. 5.30) for beach and inshore are the same. Therefore, the initial inaccuracy S_{di} will be less for option *E: 2_num* than for option *D: 2_difrent*, which means that the relevance of option *E: 2_num* will be greater. The initial shape of the nourishment for option *E: 2_num* is schematized by two Gauss-curves, one for the beach and one for the inshore.

The hypothesis is that solutions can be derived from the basic analytic theoretical solutions given in Subsection 3.2.6 (cf: Fig. 3.15a) specificly.

Results: The *results* of the present investigations are summarized in Table 5.5 and are discussedx below. *Thus the following paragraph gives in this way an overall picture of this whole Subsection 5.3.3.*

Graphically, comparison between theoretical and measured lines of beach and inshore is given in Fig. 5.36 (A_option:1_total); Fig. 5.38 (only the line of the beach (B_option: 1_beach)); Fig. 5.39 (theoretical lines of beach and inshore; parallel (C_option:2_paralel); Fig. 5.40: analytical 2-line theory (D_option: 2_diffrent; Fig. 5.41) and numerical 2-line theory (E-option; Fig. 5.41).

This is substantiated in the (numerical) summary, concerning accuracies, given in Table 5.5, page 284.

From a comparison of the standard deviation between the measured and computed coastlines, both for the initial shape and for the shape after one month and one year, it is concluded that for the case of Sylt the *two-line* theory gives the best results. The analytical solution (Option D) gives the most accurate prediction of the evolution of the beach, while the overall picture is best predicted by the numerical solution (Option E). A standard deviation of 10 m with respect to the seaword coordinate line of beach and of 20 m with respect to the line of the inshore, resulting in an overall accuracy of 15 m, could be obtained.

A fast prediction of the evolution of the beach, which is not much worse than the two-line options, can be given by Option B, where the one-line theory has been applied to the beach only. This shows that in the case of the 1978-nourishment on the Isle of Sylt, the dynamics of the beach are determined more by longshore transport than by cross-shore transport.

The use of Option C with parallel lines for beach and inshore should not be recommended for this case, as it shows a rather large inaccuracy. In fact, the latter option starts from the assumption, that any deviation from the equilibrium profile is immediately corrected by cross-shore transport (up to a depth of 12 m!), which implies that this type of transport should have a large effect on the evolution of the beach and inshore. This assumption is not substantiated by nature.

The major conclusion of the computational efforts is that the considered coastlines agree reasonably well with a Gauss-curve and that the actual deformation speed of the sand groin, represented by the increasing width at the points of contraflexure, corresponds with the theoretical widening, which is proportional to the square root of the time.

The hindcast computations for the 1972-nourishment[h] resulted in a coastal constant $s = 0.7 \times 10^6\,\mathrm{m}^3/\mathrm{yr}/\mathrm{rad}$ and $s = 0.9 \times 10^6\,\mathrm{m}^3/\mathrm{yr}/\mathrm{rad}$ for the nourishment of 1978/1979 presently in focus. The predicted value of s, based upon wave data is between 0.4×10^6 and $0.8 \times 10^6\,\mathrm{m}^3/\mathrm{yr}/\mathrm{rad}$. The conclusion is that the difference between hindcasted and predicted values is rather small. Further it was observed that both fills behaved more or less similarly, apart from heavy losses during the first month after placement of the 1978-fill.

The analysis of the deformation speed $\sigma \sim \sqrt{t}$ (on which the results given above are based) is independent of the actual volume of the fill. The behavior of the fills differs if the analysis is based upon the position of the peak of the Gauss-curve. In that case the actual volume plays a role and the resulting values of s are less consistent. For the 1978-fill, s is about half of the value for the 1972-fill. Therefore and because of the above correspondence between hindcasted and predicted values of s, it is recommended to determine the coastal constant from the observed growth of the width of the beach fill. Intermediate losses due to onshore–offshore transport are only of minor significance to the value of the coastal constant.

The results of the A- and C-option are somewhat difficult to compare. When S_d (or S_di) (cf. page 268) in the first and third options have the same value, this implies, that the line of the beach and the line of the inshore in the third option have the same accuracy as the coastline in the first option, i.e. "the profile fits at two places" instead of "in the middle"; under those circumstances the third option naturally should be preferred.

One might compare the accuracy, with which the line of the beach can be predicted according to the second and third options; this gives a validation of the schematizations concerned.

Two options will be considered. In both options, the mean coastline, i.e. the mean between the line of the beach and the line of the inshore will be the same as in the one-line theory (first option). In the *fourth option* the analytical two-line theory is used. This implies, that due to the limitations of the analytical approach, the values σ (half distance between the points of contra-flexure) for beach and inshore should be equal. However, contrary to option 3, the peak values of the Gauss-curve, representing beach or inshore, are not necessarily the same.

[h] 1972: Not considered here.

Apart from the initial situation and the boundary conditions, the solution of a mathematical theory depends on calibration constants. The following analysis is confined to hindcasting. This means that the optimal values of the calibration constants are derived from the same data, as for which the theory is applied. Consequently, the accuracy will be flattered compared to forecasting.

It would be worthwhile to use the constants, obtained in this way, for nourishments at the same site in order to find the accuracy for forecasting. Care is needed, since the calibration only covers a period of one year. It is possible that heavy storms might alter the interaction between beach and inshore considerably.

The used calibration constants are treated when considering the various options. The values are optimized by minimizing the value of S_{d}, i.e. the overall accuracy is as large as possible in this way. For the options involving two lines, the inaccuracy is calculated both for beach and inshore separately, and for beach and inshore together. As mentioned before, the values of S_{di} and S_{d} found for one-line theories are difficult to compare with those, found for two-line theories. Comparable are the values of the inaccuracy of the beach for options 2–5, and the values of S_{di} and S_{d} for the one-line options (1 and 2) or two-line options (4 and 5) separately. The final results are summarized in Table 5.5.

Anticipating the analysis of the options A–E on the following pages and Table 5.5, it can be concluded that the accuracy of the results of a mathematical method depends on:

(a) The accuracy of the physical concept.
(b) The accuracy of the physical constants, used in the theory.
(c) The accuracy of the assumed initial condition and the boundary conditions.
(d) The accuracy of the numerical computation.
(e) The accuracy of the measured data.

In the analysis of the 1978-beach nourishment, the accuracy concerning items (d) and (e) is neglected. The accuracy concerning item (c) depends on the assumed schematization.

Figure 5.31 shows the one-line and two-line schematizations of the coastal profile for Sylt. The coordinate y of the coastline has been taken with respect to a straight reference line, following the general longshore direction.

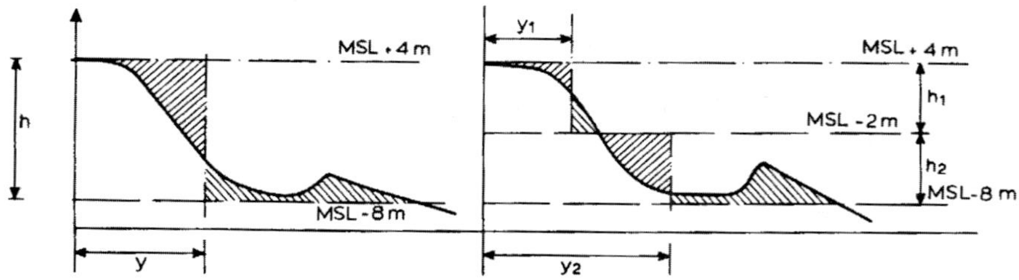

Fig. 5.31 One- and two-line schematization for Sylt.

Figure 5.33 shows the coastline according to measurements at four successive time steps, the first and second time steps just before and after the nourishment. The third and fourth time steps are a month, respectively a year after the nourishment. Figure 5.37 shows the line of the beach and the line of the inshore at the same moments as mentioned in Fig. 5.33. The calculated curves according to the various options, compared to the measurements, are displayed in Fig. 5.36 and Figs. 5.38–5.41 (see Table 5.5) page 284.

On the following pages 5.31–5.47 the five different options are treated in detail. In all those schematizations, it has been assumed that:

- the coast before the nourishment had the shape of a retreating parabola;
- the nourishment is subject to an initial loss L_l of 25%; and
- the gradient sof the longshore transport with respect to the coastal direction equals 660,000 m^3/yr/rad.

The curvature of the parabola and the magnitude of I_L and s has been determined by curve-fitting (cf. "Option A and Fig. 5.35).

The assumption (option 1), that the mean between all depth-contours between MSL + 4 m and MSL − 8 m just after the nourishment can be schematized according to a Gauss-curve, superposed on the initial parabola, is rather accurate (inaccuracy S_{di} = 15.6 m, Table 5.5, column 1).

With the optimal calibrated value of the before-mentioned gradients, it is found that the coastline at two successive times, a month and a year after the nourishment, can be predicted with an inaccuracy S_d = 11.3 m (column 1). It thus shows, that in the course of time the assumption "the coastline is a Gauss-curve", tends to improve, as it should do theoretically.

The assumption (option 2, column 2), that the initial line of the beach is a Gauss-curve, where the line of the beach is defined as the mean of the contour lines between MSL + 4 m and MSL − 2 m, is worse than the assumption, that the coastline is a Gauss-curve (column 1). $S_{\rm di} = 21.2$ m instead of 15.6 m. However, the prediction of the line of the beach (assuming no loss from beach to inshore) after a month or a year is not significantly worse than the prediction of the coastline, $S_{\rm d} = 11.8$ m instead of $S = 11.3$ m.

Considering line of beach and inshore together, the assumption (option 3, column 3) that during the nourishment the coast keeps its equilibrium profile is definitely not according to reality. The included assumption, that the line of beach and inshore have the same shape as the coastline, has an initial inaccuracy $S_{\rm di} = 39.5$ m, and an overall inaccuracy $S_{\rm d} = 23.7$ m after a month/year.

Taking the coastal profile into account, the analytical two-line theory (option 4, column 4) gives a much better approximation. Initially, for the line of the beach and the inshore an inaccuracy $S_{\rm di} = 22.5$ m and $S_{\rm di} = 23.2$ m is found and an "overall" inaccuracy $S_{\rm di} = 22.1$ m. After a month/year the overall inaccuracy diminishes to 16.5 m and the inaccuracy of the line of the beach to 9.6 m. Of all options, the analytical two-line theory predicts the beach shape best (standard e).

The numerical two-line solution (option 5, column 5) is a "taylormade" solution for the behavior of the coast. Now both the different alongshore shape of beach and inshore, and the offshore transport is taken into account. Of course, the initial inaccuracy of the beach in column 5 and the initial value of "$y_{\rm b}$" and "$\sigma_{\rm b}$" are equal to those in column 2 (the optimal shape of the Gauss-curve through the beach). Both the initial overall inaccuracy $S_{\rm di}$ and the overall inaccuracy $S_{\rm d}$ after a month/year are the lowest for the numerical two-line option (option 5), when compared with the other two-line options (3 and 4). Standard (d) is fulfilled the best in this way. The accuracy of the prediction of the beach is somewhat less than the one found in option 4 (inaccuracy is 10.2 m instead of 9.6 m). The prediction of the inshore is better.

Summarizing, it is shown, that for the case of Sylt the best options are the analytical and numerical two-line theory (options 4 and 5); the analytical option is a little better with respect to the prediction of beach behavior, the numerical one with respect to the overall picture.

A fast prediction of the behavior of the beach, which is not much worse than the two-line option, can be given by using option 2 (considering the beach according to the one-line theory). This shows, that in the case of Sylt the dynamics of the beach is determined more by longshore transport than by offshore transport.[i]

For this reason the use of option 3 (parallel depth contours) should not be recommended for this case: it shows a rather high inaccuracy.[j]

- In should be noted, that this conclusion may not be valid, when another initial shape than the one of 1978 is used. Relative effect of longshore and onshore/offshore transport can be calculated with the two-line theory.
- In fact, this option starts from the conception, that any deviation of the equilibrium profile is "immediately" corrected by onshore or offshore transport, i.e. large effect of onshore/offshore transport.

A. Option 1_total: One line theory applied for total profile

From the plan views given by Patzold[36] the cross-section was determined between the levels MSL + 4.0 m and MSL − 8.0 m. The level of MSL − 8.0 m has been chosen in order to take all available data into account.

On the landward side, the boundary consists of a vertical reference plane, parallel to the mean coastal direction (Fig. 5.31).

By using a straight reference plane, instead of starting from differences between successive surroundings, the original curvature of the coast can be taken into account.

From each cross-section, the average y-coordinate y of the depth averaged coastal profile can be found by division of the area by $h = 12\,\text{m}$. Figure 5.33 shows a plot of y for the successive sections, thus yielding the "coast-line" at different times.

First, the natural coastal development has to be reproduced, on which the development of the beach nourishment has to be superimposed. The natural development is simulated by a parabola retreating linearly in the course of time (Fig. 5.32).

$$y = \frac{1}{2}p' \times x^2 + p' \times \frac{s}{h} \times t \tag{5.9}$$

where p' is a negative constant, to be found.

[i]It should be noted, that this conclusion may not be valid, when another initial shape than the one of 1978 is used. Relative effect of longshore and onshore/offshore transport can be calculated with the two-line theory.

[j]In fact the option starts from the conception, that any deviation of the equilibrium profile is "immediately" corrected by onshore of offshore transport, i.e. large effect of onshore/offshore transport.

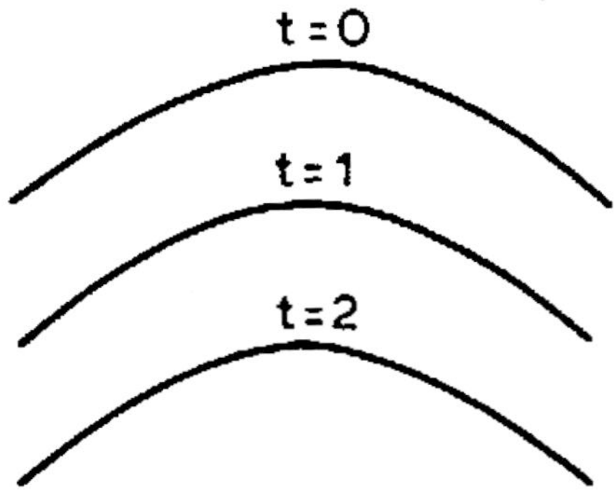

Fig. 5.32 Coastline before nourishment: a gradual retreating parabola.

According to (5.9), y is a solution of the coastal equation (3.6), as one readily finds. This coastal equation is based on continuity. However, the data show volumetric changes in the area considered, which can be caused by initial losses due to settling of material and washing out of fine sediment, offshore losses by heavy storms, etc.

Although the present investigation gives a hindcast instead of a forecast, the purpose is to provide information on the predictive value of the theory. Offshore losses by heavy storms are not predicted, and therefore, it was decided to consider the initial loss as a variable in the curve-fitting procedure.

The optimization was carried out starting from three degrees of freedom:

p': the curvature of the parabola

I_{L}: the initial loss (percentage)

s: the rate of change of the transport S as a function of the coastal direction.

The data has been elaborated in the following way.

(a) A certain value of p' is assumed. The area confined between the coastline before nourishment (April 20, 1978) and the reference line is known, which determines the location of the top of the parabola at $t = -0$ (minus denotes before nourishment). The area contained between the coastline and the reference line is assumed equal to the area contained between parabola and reference line.

Using a curve-fitting procedure, the value of σ_i is determined, being the standard-deviation of the Gauss-curve of the first coastline

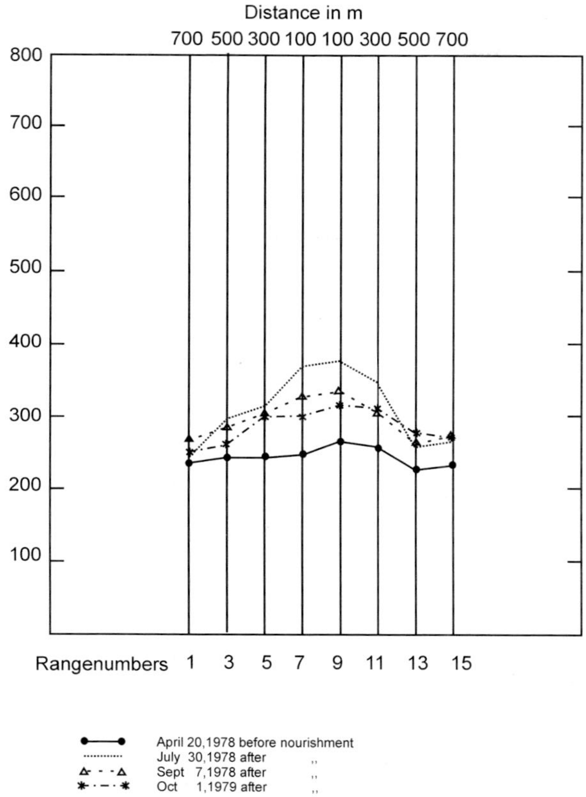

Fig. 5.33 Coastline at successive time steps (one-line theory).

$y(x)$ after the nourishment. Thus this coastline is approximated by (Fig. 5.34):

$$y_i = y_{i0} + \frac{1}{2}p'x^2 + \hat{y}_i e^{\frac{-x^2}{2\sigma_i^2}} \tag{5.10}$$

where

$$y_i = y_{i0} + \frac{1}{2}p'x^2$$

denotes the original parabola at time $t = -0$.

Given the value of σ_i, the value of $\hat{y}_i$ is determined by the area enclosed between the coastline at $t = +0$ (July 30, 1978) and the reference line.

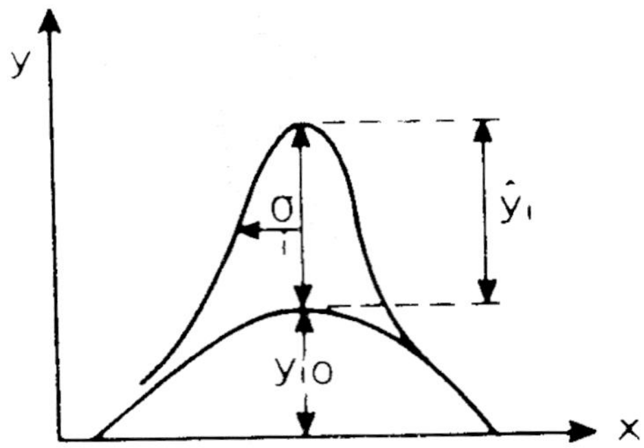

Fig. 5.34 Initial shape of nourishment: Gauss curve superposed on parabola.

The optimization of σ_i is accomplished in the following way. Determined is the value of the initial inaccuracy S_{di}, being:

$$S_{\mathrm{di}} = \sqrt{\frac{\left\{\sum_{n=1}^{m}(y_{\mathrm{meas}}^{(n)} - y_{\mathrm{calc}}^{(n)})^2\right\}}{(m-1)}} \tag{5.11}$$

where m is the number of sections $(m = 8)$

(b) A value of the initial loss I_{L} and s is assumed.
From the relationship $\sigma_i = \sqrt{(2(s/h)T_1)}$ (Eq. (3.18) the value of T_1 results directly from the assumed value of s and the value of σ_i found under b.
The coastline at time t is calculated as:

$$y_{\mathrm{cal}} = y_{i0} + \frac{1}{2}p'x^2 + p'\frac{s}{h}t + (1 - I_{\mathrm{L}})\hat{y}_i\sqrt{\frac{T_1}{(T_1+t)}}e^{\frac{-x^2}{2\sigma^2\frac{T_1+t}{T_1}}} \tag{5.12}$$

After this, the deviation between measured and calculated profiles can be determined from:

$$S_{\mathrm{d}} = \sqrt{\frac{\sum_{k=1}^{2}\sum_{n=1}^{m}(y_{\mathrm{meas}}^{(k,n)} - y_{\mathrm{calc}}^{(k,n)})^2}{(2m-1)}} \tag{5.13}$$

where k denotes the first, respectively, second measured coastline after nourishment. (i.e. the values of September 7, 1978 and October 1, 1979, the coastline at the "second" and "third" time).

(c) From the foregoing, it appears that S_{d} is a function of three variables:

$$S_{\mathrm{d}} = f(p', s, I_{\mathrm{L}}) \tag{5.14}$$

Therefore, the whole procedure is repeated for other values of p^1, I_{L} and s.

It must be kept in mind, that $S_{\rm di}$ has a different meaning than $S_{\rm d}$. The value of $S_{\rm di}$ indicates, how the initial coastal profile corresponds with the assumption according to the theory. This value does not give any information on the value of the theory as such. The value of $S_{\rm d}$ gives the standard deviation between the theoretical, predicted coastline and the measured coastline, assuming sufficient calibration of the relevant constants. The relation between $S_{\rm di}$ and $S_{\rm d}$ gives insight in the relevance and usefulness of the theory for forecasting.

The results of the above described procedure are summarized in Fig. 5.35. The variable on the horizontal axis is $-p' \times s/h$, which denotes the assumed rate of erosion of the coast in m/yr before the nourishment took place (see Eq. (5.9)). On the vertical axis the value of the inaccuracy, $S_{\rm d}$ according to Eq. (5.13) is plotted.

Curves are given for various values of $I_{\rm L}$.

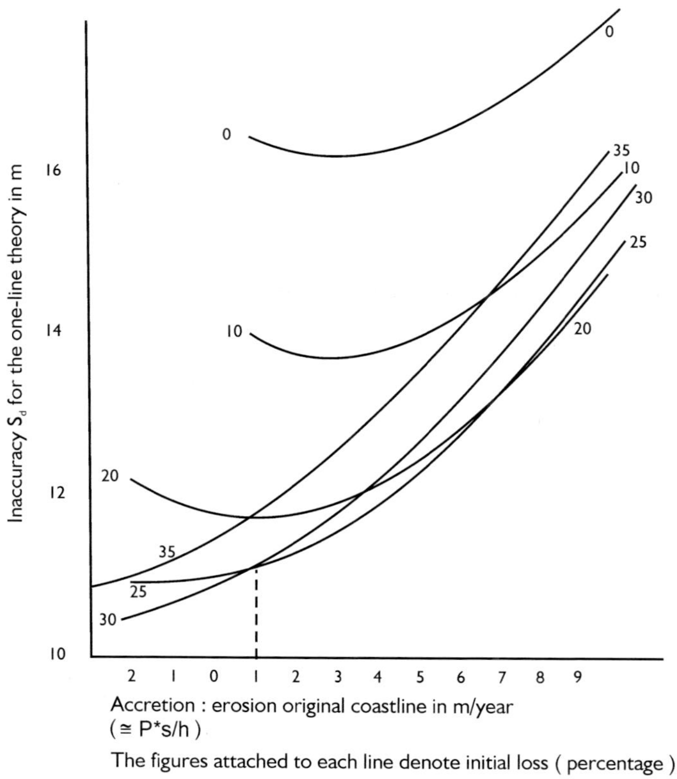

Fig. 5.35 Accuracy of one-line approximation as function of assumed rate of erosion of original coastline and assumed initial loss ($I_{\rm L}$).

Analyzing these results, one has to return to the physics. As Patzold states, the general retreat of the coast is of the order of 1 m/yr.

From Fig. 5.35, the most logical combination appears to be:

$$
\begin{aligned}
I_{\rm L} &= 25\% \\
s/h &= 55{,}000\,{\rm m}^2/{\rm yr}/{\rm rad} \qquad s = 660{,}000\,{\rm m}^3/{\rm yr}/{\rm rad} \\
\sigma_i &= 294\,{\rm m} \\
p^l &= -2\times 10^{-5}\,{\rm m} \\
S_d &= 11.3\,{\rm m} \\
S_{\rm di} &= 15.6\,{\rm m}
\end{aligned}
$$

Figure 5.36 shows the results for the various times of the analysis ($t = -0; t = +0, t = t_1$ and $t = t_2$).

B. Option 1_beach: One line applied for the beach

This section deals with the second option (Fig. 5.31), where the attention is focussed on the beach. It is investigated, to which accuracy the behavior of the beach can be predicted by applying the one-line theory only to the beach area (between the levels MSL + 4 m and MSL − 2 m), and neglecting the off-shore transport between beach and inshore. Therefore the same procedure, as treated by Option 1 has been applied for the line of the beach instead of the coastline, assuming that the curvature of the parabola as found in the former case (see Fig. 5.35) is correct, which means that no further tuning of shape of the parabola took place. Thus when applying Eq. (5.10) as the basic equation for the initial line of the beach, it was assumed that $p' = -2\times 10^{-5}$ m.

The value of y_{i0} differs from the one, used for Option 1. Assuming an equilibrium profile before the nourishment took place, the equilibrium distance W between the line of the beach and the line of the inshore can be found from the lines of beach and inshore at April 20, 1978 (see Fig. 5.37). A mean distance of 336.6 m is found.

The coastline is situated halfway between the line of the beach and the line of the inshore, since the representative "beach-layer" h_1, equals the "inshore layer" h_2, i.e. $h_1 = h_2 = 6$ m. The parabola representing the beach should be shifted a distance $W/2 = 168.3$ m in landward direction with respect to the parabola representing the coastline. Thus, when applying Eq. (5.10) to the line of the beach, the value of $\hat{y}_i$ should be reduced with 168.3 m with respect to the value, used for Option 1.

The values of $\hat{y}_i$ and σ_i (Fig. 5.34) are related in such a way, that the area under the Gauss-curve should be equal to the area, enclosed by the

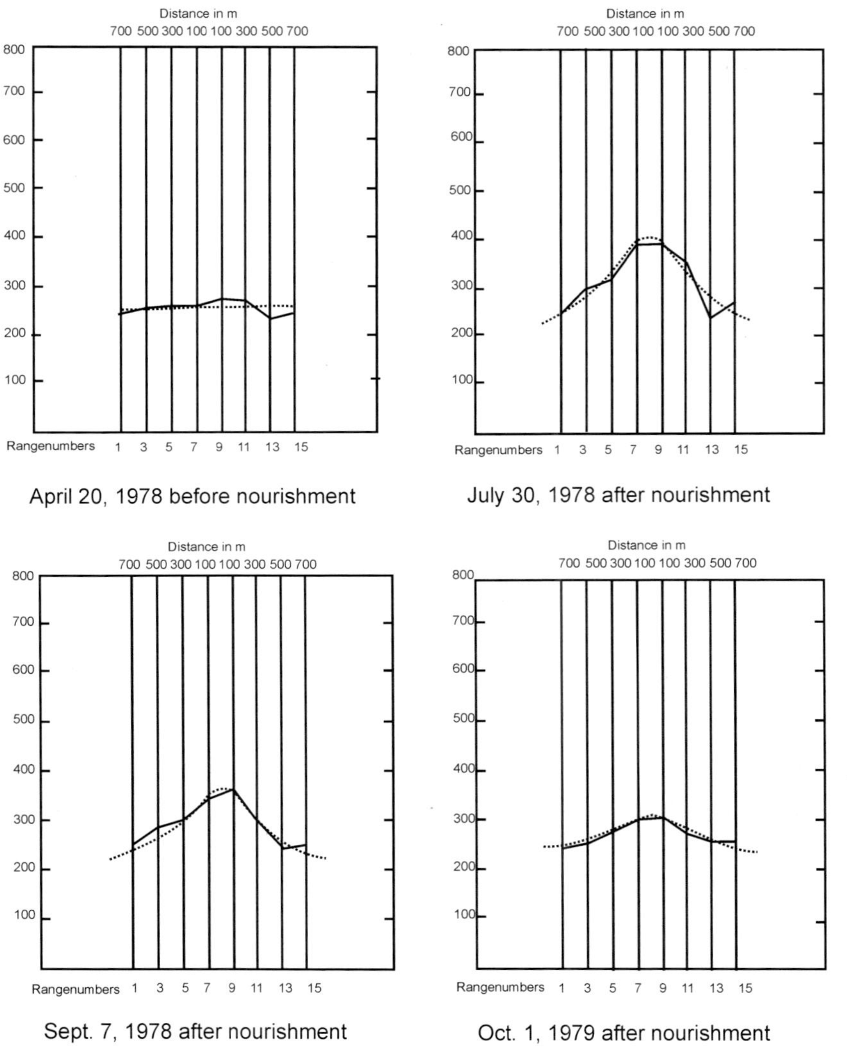

Fig. 5.36 Option A: total comparison between measured coastline and theoretical coastline (optimal hindcast).

parabola and the line of the beach just after nourishment (July 30, 1978, Fig. 5.37). This gives one degree of freedom in determining σ_i by minimizing S_{di} according to Eq. (5.11).

In Fig. 5.38 the calculated and measured beach-lines are given at various time steps.

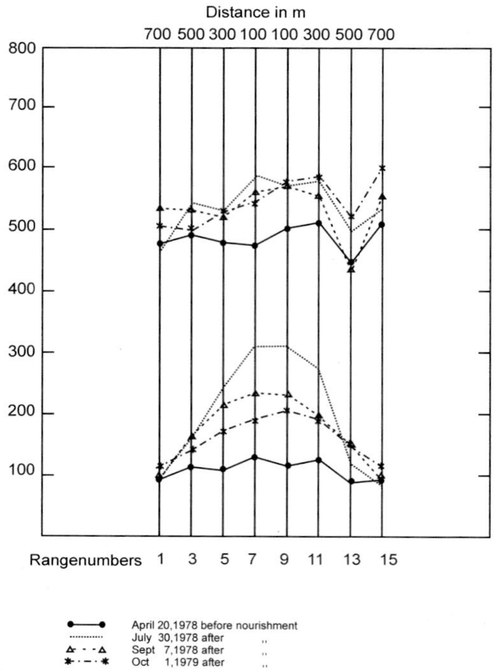

Fig. 5.37 Line of beach and inshore at successive steps.

The calculations result in a value $s_1/h_1 = 55{,}000\,\mathrm{m}^2/\mathrm{yr}/\mathrm{rad}$ and a standard deviation $S_\mathrm{d} = 11.8\,\mathrm{m}$ (cf. Table 5.5). The value of s_1/h_1 is equal to s/h, found in Section 5.2.3. Beach and inshore are equally sensitive to changes in longshore direction.

C. Option 2_ parallel: Parallel lines of beach and inshore

This section deals with the third option (Fig. 5.30). It will be considered, whether better results can be obtained, when the coast is schematized according to two lines instead of one. According to one of the assumptions of the one-line theory, all depth contours should be parallel. Consequently, the line of the beach and the line of the inshore should have the same shape as the coastline and are only shifted with respect to each other.

Analyzing these results, one has to return to the physics. As Patzold,[36] states, the general retreat of the coast is of the order of 1 m/yr.

In Fig. 5.38 the final results of these one-line calculations have been plotted, together with the measured lines of the beach and the inshore

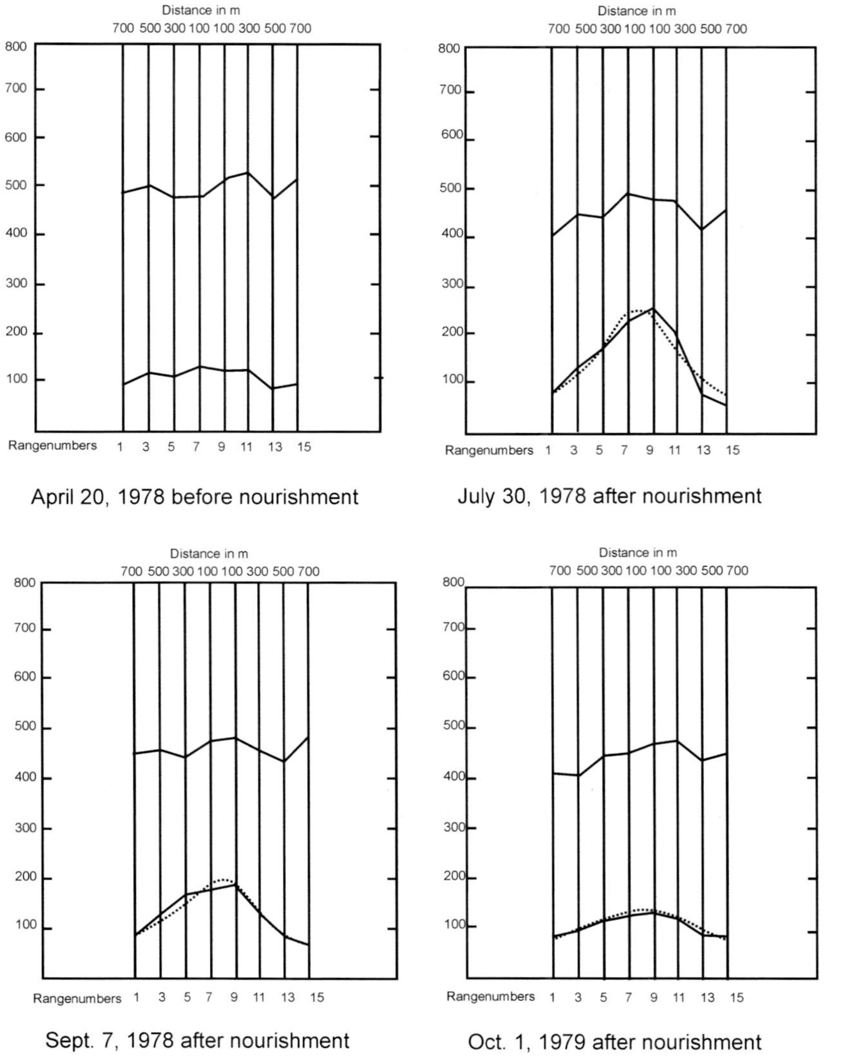

Fig. 5.38 Option B: 1_Beach: one-line theory, applied for the beach only. Comparison of measured and calculated line of the beach, using the one-line theory for the beach only.

at the corresponding time steps. It is assumed, that the distance between the lines of beach and inshore before the nourishment (i.e. at April 20, 1978) is an equilibrium distance. Therefore, the theoretical line of the beach at successive time steps is found by shifting the coastline, obtained in Section 5.2.3, over a distance of 168.3 m in a landward direction. The

Table 5.1 Standard deviation (in m) of the beach and inshore using third option (parallel lines of beach and inshore).

	Beach	Inshore
July 30, 1978	41.95	39.83
Sept. 7, 1978	27.22	34.35
Oct. 1, 1979	11.67	20.52

theoretical line of the inshore is found by shifting the coastline over the same distance in seaward direction.

Table 5.1 shows the standard deviation (calculated according to (5.11)) for each of the successive time steps and for beach and inshore, respectively.

The values on the first row of Table 5.1 (July 30, 1978, just after the nourishment) are combined to give the "overall initial standard deviation" S_{di}. This value S_{di} informs on the relevance of the theory (Section 5.2.3). The values on the second and third rows (Sept 7, 1978 and Oct 1, 1979) are combined to give the "overall standard deviation" S_{di}, which informs on the accuracy of the theory (Section 5.2.3). The initial accuracy S_{di} is calculated for beach and inshore simultaneously:

$$S_{\mathrm{di}} = \sqrt{\frac{\sum_{j=1}^{2} \sum_{n=1}^{m} \left(y_{\mathrm{meas}}^{(n)}(j) - y_{\mathrm{calc}}^{(n)} \right)^2}{(2m-1)}} \tag{5.15}$$

where $j = 1$ (2) indicates the beach (inshore), respectively.

The inaccuracy S_{d} for the predicted lines of beach and inshore is calculated analogous to (5.13) and (5.15):

$$S_{\mathrm{d}} = \sqrt{\frac{\sum_{j=1}^{2} \sum_{k=1}^{2} \sum_{n=1}^{m} \left(y_{\mathrm{meas}}^{(k,n)}(j) - y_{\mathrm{calc}}(j) \right)^2}{(4m-1)}} \tag{5.16}$$

where k and n have the same meaning as in (5.14).

D. Option 2_difrent: analytical two-line theory

Now it will be investigated, whether the result can be improved by taking the coastal slope and offshore transport into account. For this purpose, the two-line theory will be used.

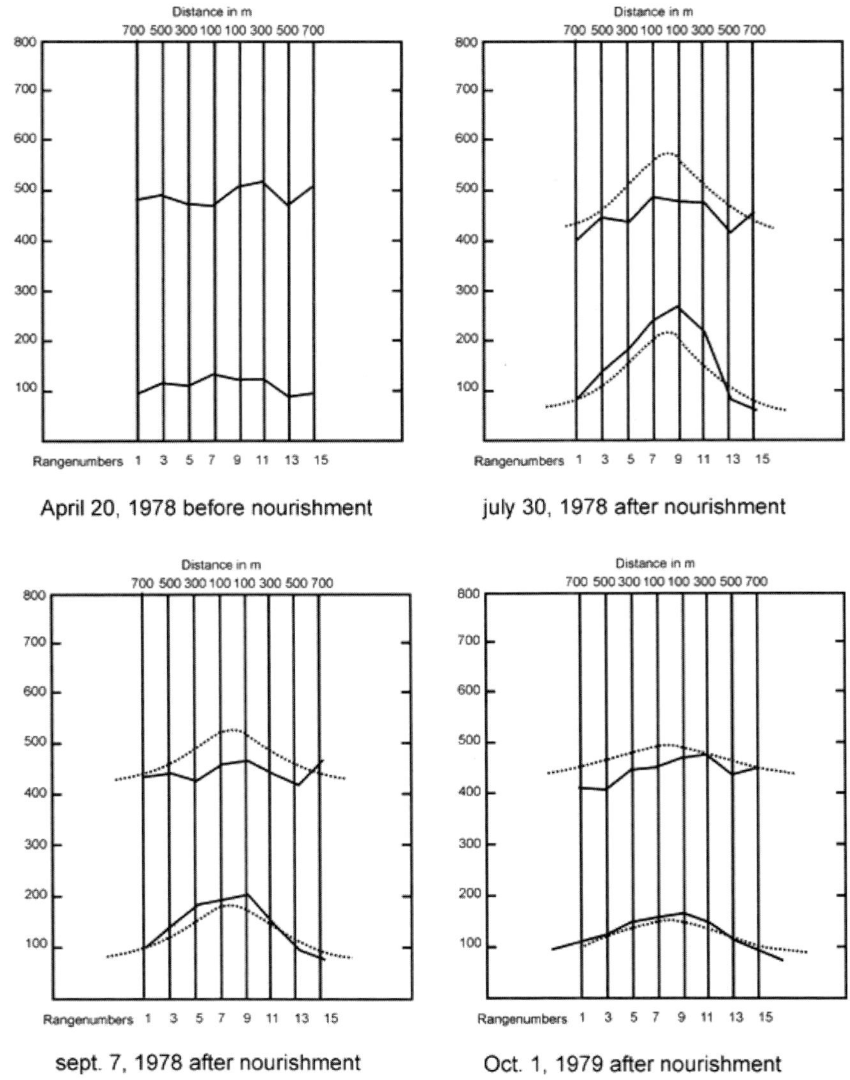

Fig. 5.39 Comparison of measured and calculated line of beach and inshore from the one-line theory assuming parallel depth contours.

Two approaches will be treated successively. According to the first approach, (option 2_difrent) analytical curves will be assumed for beach and inshore. The analytical solutions are, however, only available when $s_1/h_1 = s_2/h_2$ and therefore equal to s/h. As will be shown below, in this context the two-line theory gives no extra degree of freedom. Namely starting from the constants found in the one-line concept and from the mean position of beach and inshore with respect to the reference line, one just finds the accuracy (standard deviation of line of beach and inshore) as a result.

According to the second approach (option E, 2_num) no conditions concerning the ratio between s_1/h_1 and s_2/h_2 are given. In this case, a numerical method is used for calculation of the lines of beach and inshore. The ratio between s_1 and s_2 gives a new degree of freedom. Always, the value of s_2 is taken equal to the value of s, found for the one-line theory.

Concerning the offshore transport, the value of s_y can be obtained in the following way. The equation for the offshore transport S_y reads:

$$S_y = s_y\{y_1 - (y_2^1 - W)\} \tag{5.17}$$

where y_1, y_2^1, coordinates of beach and inshore with respect to the reference line; W, the equilibrium distance; and s_y, a proportionality constant.

If the length of the coastal stretch is considered to be equal l, integration of Eq. (5.17) over l yields:

$$\int_l S_y \times dl = s_y\left\{\int_l y_1 dl - \int_l y_2^1 \times dl + W \times l\right\} \tag{5.18}$$

Assume, that the coast before nourishment is in equilibrium:

$$W \times l = \int_l [y_2^1]_{t=0} \times dl - \int_l [y_1]_{t=-0} \times dl \tag{5.19}$$

Therefore, if O_1 equals the area contained between the line of the beach before nourishment and the same line after nourishment and if O_2 is defined in an analogous way with respect to the inshore, then substitution of Eq. (5.19) into Eq. (5.18) gives for the total transport per unit of time along the whole stretch:

$$\int_l S_y \times dl = s_y(O_1 - O_2) \tag{5.20}$$

where O_1 and O_2 are functions of time.

For small Δt, the left-hand side of the equation may be written as:

$$\int_l S_y \times dl = ([0_1]_t - [0_1]_{t+\Delta t})h_1/\Delta t \tag{5.21}$$

as follows from continuity. In (5.21) h_1 denotes the height of the beach.

Table 5.2 Quantities in m^3 of sandsupply beach and inshore, between Section 1.15.

	30 Aug. '1978	7 Sept. '1978	1 Oct. '1979
Beach inshore	791.802	603.368	431.880
	373.198	281.139	277.920

Table 5.3 Quantities of sand supply at beach and inshore compensated for alongshore losses.

	30 Aug. '1978	7 Sept. '1978	1 Oct. '1979
Beach inshore	798.602	615.751	549.227
	376.403	286.910	353.434
Total	1.175.005	902.661	902.661

From (5.17) it is found that:

$$s_y = \frac{-\Delta O_1 \times h_1}{\Delta t(O_1 - O_2)} \tag{5.22}$$

where $\Delta 0_1 = [0_1]_{t+\Delta t} - [0_1]_t$

From the data, the following quantities of sand supply at the beach and the inshore over the considered stretch of the coast can be calculated (Table 5.2).

Because of the initial loss, the first period between August 30, 1978 and Sept. 7, 1978 is not useful for computation of s_y. On the other hand, the second period between Sept. 7, 1978 and Oct. 1, 1979 is too long for the application of (5.22). A better calculation can be made by first extending the computation from the length l of the coastal stretch to an infinite length, thus compensating for losses alongshore.

From the curve-fitting computations, given ad Option C:2_parallel, the percentage of material, disappearing alongshore can be calculated. The values mentioned in Table 5.2. have been corrected for this effect. The results are given in Table 5.3.

Application of (5.22) yields for s_y, when O_1 and O_2 are taken at Sept. 7, 1978:

$$s_y = -(549{,}227 - 615{,}751) \times 6/(1.026(615{,}751 - 286{,}910)) = 1.14\,\text{m/yr}.$$

If O_1 and O_2 are considered on Oct. 1, 1979 one finds $s_y = 1.92\,\mathrm{m/yr}$.

Therefore an average value of $s_y = 1.53\,\mathrm{m/yr}$ will be assumed.

Now the accuracy of the computation of beach and inshore will be calculated, using the before-mentioned "analytical approach" (option 2_difrent).

For the two-line model the following relations apply:

$$y_1 = y_{1,\mathrm{par}} + y_{1,\mathrm{gauss}}$$

$$y_2 = y_{2,\mathrm{par}} + y_{2,\mathrm{gauss}}$$

where y_1 and y_2 denote the line of the beach and the inshore, respectively, and where in:

$$y_{i,\mathrm{par}} = \frac{1}{2} p^1 x^2 + p^1 \times \frac{s}{h} \times t + y_{i,0} \tag{5.23}$$

$$y_{1,\mathrm{gauss}} = \left\{1 + \frac{h_2}{h_1} \times e^{-\frac{t+T_2}{T_0}}\right\} \frac{S^{**}/h}{\sigma\sqrt{2\pi}} e^{-\frac{x^2}{2\sigma^2}} \tag{5.24}$$

$$y_{2,\mathrm{gauss}} = \left\{1 - e^{-\frac{t+T_2}{T_0}}\right\} \frac{S^{**}/h}{\sigma\sqrt{2\pi}} e^{-\frac{x^2}{2\sigma^2}} \tag{5.25}$$

Here, the following symbols are used: S^{**}, the total amount of sand supply; T_1, constant with dimension "time", referring to a hypothetical moment (before $t = 0$) when the nourishment consisted of an infinite long dam with zero width; T_2, idem, referring to a hypothetical moment, (before $t = 0$), when the inshore was straight; $Y_{i,0}$, y-coordinate of the top of the parabola at time $t = 0$.

$$T_0 = \frac{h_1 h_2}{s_y h} \tag{5.26}$$

$$\sigma = \sqrt{2\frac{s}{h}(t + T_1)} \tag{5.27}$$

Integration of (5.24) and (5.25) over x yields:

$$O_1 = \int_{-\infty}^{+\infty} y_{1,\mathrm{gauss}}\, dx$$

$$O_1 = \left\{1 + \frac{h_2}{h_1} e^{-\frac{t+T_2}{T_0}}\right\} \frac{S^{**}}{h} \tag{5.28}$$

and in the same way:

$$O_2 = \left\{1 - e^{-\frac{t+T_2}{T_0}}\right\} \frac{S^{**}}{h} \tag{5.29}$$

The value T_0 is found from (5.26):

$$T_0 = 6 \times 6/(1.53 \times 12) = 1.96\,\text{year}$$

As $h_1 = h_2$, one finds from subtraction:

$$O_1 - O_2 = 2e^{-\frac{t+T_2}{T_0}} \frac{S^{**}}{h} \tag{5.30}$$

from which:

$$T_2 = T_0 \ln \frac{2S^{**}}{h[O_1 - O_2]_{t=t_2}} - t_2 \tag{5.31}$$

For the derivation of T_2 the "second time" t is used in order to avoid complications with the initial loss (which would have occurred if $t_1 = 0$ had been used).

Thus one finds from Table 5.2, page 277:

$$T_2 = 1.96 \times \ln \frac{2 \times 902661}{12 \times (651.751 - 286.910)/6} - 0.10411 = 1.875\,\text{year}$$

From (5.30) we can also derive:

$$\frac{[O_1 - O_2]_{t=0}}{[O_1 - O_2]_{t=t_2}} = e^{t_2/T_0} \tag{5.32}$$

Therefore, this ratio should be: $\exp(0.10411/1.96) = 1.054$. According to Table 5.3 this ratio equals:

$$\frac{[O_1 - O_2]_{t=0}}{[O_1 - O_2]_{t=t_2}} = \frac{798{,}602 - 376{,}405}{615{,}751 - 286{,}910} = 1{,}284$$

Assuming the sand loss between $t = 0$ and $t = t_2$ evenly distributed between beach and inshore, this ratio reduces to $(902{,}661/1{,}175{,}005) \times 1.824 = 0.986$, where the first two numbers refer to the total quantities of sand at $t = t_2$ and $t = +0$, respectively (Table 5.3).

This shows, that the ratio is of the same order of magnitude, as found from (5.32), and that the assumption of proportional initial sand loss on beach and inshore cannot be directly rejected. Therefore, the same straightforward procedure as used for the one-line theory is applied also in this case:

(a) "tune" the constants with the available degrees of freedom;
(b) consider the best fit for the initial situation (inaccuracy S_{di});
(c) take the initial loss into account;

(d) calculate the line of the beach and inshore at time $t = t_2$ and $t = t_3$ and find the inaccuracy S_d.

The tuning has been treated before. The best fit for the initial situation is found by superposition of Gauss-curves and parabola. The curvature of the parabola is found from the one-line theory (option A).

The y-coordinate of the parabola for beach and inshore are found from the mean y-coordinate of beach and inshore at $t = 0$:

$$y_{1,\mathrm{par}} = y_{\mathrm{par}} - 168.3\,\mathrm{m} \tag{5.33}$$

$$y_{2,\mathrm{par}} = y_{\mathrm{par}} + 168.3\,\mathrm{m}$$

where y_{par} denotes the y-coordinate of parabola for the coastline, which is the mean between the line of the beach and inshore.

From (5.23) to (5.27), the y-values of the lines of beach and inshore at $t = 0$, $t = t_2$ and $t = t_3$ are calculated; no curve-fitting takes place. Figure 5.40 shows the lines of beach and inshore at time $t = -0$, $t = +0$, $t = t_2$ and $t = t_3$.

Using Eqs. (5.15) and (5.16) for the overall initial inaccuracy S_{di} and for the overall inaccuracy S_d, the values $S_{\mathrm{di}} = 22.05\,\mathrm{m}$ and $S_d = 16.51\,\mathrm{m}$ are found. Furthermore, the inaccuracy for beach and inshore separately are calculated using Eqs. (5.11) and (5.13). The results are shown in Table 5.5. For the beach an inaccuracy of 9.62 m is found.

E. Option 2_num: numerical two-line theory

The results may still improve, when in the initial condition the standard deviation of the Gauss-curves, representing beach and inshore respectively, are allowed to deviate from each other. However, under those circumstances an analytical solution is not known and a numerical approach is necessary.

As in the analytical case, the results of the one-line theory will be the starting-point for the two-line approach. The value of the gradient of the longshore transport, s will be derived from the one-line approximation, but the subdivision in s_1 and s_2 will be optimized. This provides an extra degree of freedom.

The theory itself and the way of numerical elaboration is treated in Chapter 3. The time step Δt and the distance between the gridpoints Δx are chosen as:

$$\Delta t = 0.05\,\mathrm{year} \quad \text{and} \quad \Delta x = 200\,\mathrm{m}.$$

The positions of the initial parabola for beach and inshore are taken equal to the ones, used for the analytical two-line approach. Thus, the area A confined between the measured line of the beach in its initial position ($t = +0$) and the beach-parabola is known. The Gauss-curve representing the initial beach position is optimized by varying its standard deviation, keeping the area between this Gauss-curve and the parabola equal to A.

In the same way the Gauss-curve, representing the inshore is determined. Starting from the equation for the superposition of parabola and Gauss-curve, for the initial curves is found:

$$y(1, \text{par}) = 1/2(-2 \times 10^{-5})x^2 - 2 \times 10^{-5} \times (55 \times 10^3)t + 77.1$$

$$y(2, \text{par}) = y(1, \text{par}) + 336.6$$

and for the initial Gauss curves:

$$y(1, \text{gauss}) = 195.0 \exp\{-x^2/(2 * 276^2)\}$$

$$y(2, \text{gauss}) = 60.7 \exp\{-x^2/(2 * 398^2)\}$$

For the initial inaccuracy S_{di} defined according to (5.15) is found:

$$S_{\text{di}} = 20.22\,\text{m}$$

Now the initial loss is taken into account by reducing the y-coordinates of the initial Gauss-curves for beach and inshore with 25%. Next, the lines for beach and inshore are numerically calculated at time $t = t_2$ and $t = t_3$. Here, the offshore transport is determined by the constant s_y, which is taken equal to 1.53 m/yr, as treated at Option 2_difrent. Curve-fitting takes place by optimization of the ratio between s_1 and s_2. A minimum value of the inaccuracy is found when $s_1 = s_2 = 330{,}000\,\text{m}^3/\text{yr}/\text{rad}$.

The inaccuracy S_{d}, defined according to (5.16) appears to be:

$$S_{\text{d}} = 15.10\,\text{m}$$

Furthermore, the inaccuracy of beach and inshore are calculated separately, using Eqs. (5.11) and (5.13) for the initial situation and for the combination of second and third moments, respectively. The results are given in Table 5.4.

The lines of beach and inshore for $t = +0$, t_2 and t_3 are given in Fig. 5.41 (Option E: 2_num).

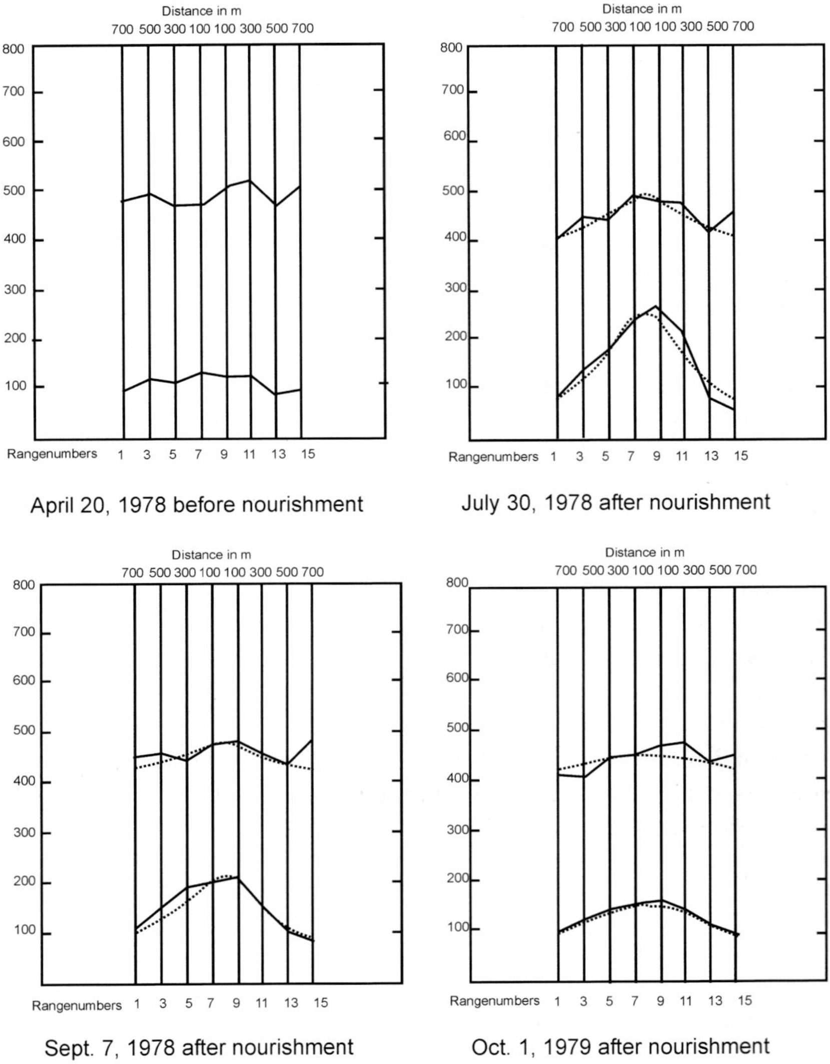

Fig. 5.40 Comparison of measured and calculated line of beach and inshore using the analytical two-line theory (Option D: 2_difrent).

5.3.4. *Cadzand*

5.3.4.1. *Introduction*

In the framework of the extension of the harbour of Zeebrugge, a beach fill of 8.4×10^6 m^3 of sand has been placed between this harbour and the border between Belgium and The Netherlands over a distance of approximately

Table 5.4 Inaccuracy of numerical two-line computation for beach and inshore, respectively.

	Beach	Inshore
Initial situation (S_{di})	21.25	21.06
Second/third moment (S_{d})	10.20	19.16

8 km. This beach nourishment project was carried out between December 1977 and March 1979.

With respect to situation and development, refer to Section 2.6 (p. 2.66–2.75) and to Figs. 2.72–2.75.

The present subsection deals on the preliminary investigations (carried out by the (Dutch) Study Department at Flushing of the Rijkswaterstaat in collaboration with the (Belgian) Dienst der Kust[k] Prior to the execution of the afore-mentioned beach nourishment various studies have been carried out, among which a series of coastline computations with the one-line coastal model. On the basis of wave observations the coastal constant s in the diffusion equation (3.6): $(\partial y/\partial t = (s/h) \times (\partial^2 y/\partial x^2))$ was determined. It was found that the value varied strongly along the coast, but the average value was approximately $1.5 \times 10^6\,\text{m}^3/\text{yr}/\text{rad}$.

In the present analysis, the deformation of the eastern limit of the beach fill has been considered. Assuming that the original coastline in 1977 was straight and that the original schematized coastline of the beach fill is rectangular (see Fig. 5.41), the following analytical solution of the one-line equation for a one-side decay of rectangular corners can be applied (a symmetrical composition, based upon Eq. (3.20c)):

$$y = \frac{y_0}{2}\text{erf}\left(\frac{b-x}{\sqrt{4(s/h)(t+t_0)}}\right) \tag{5.34}$$

where t is time, t_0 is a constant and erf is the error function, y_0 is the width of the fill and b is half of the length of the fill. Furthermore, x is the position with respect to the center of the fill. It is assumed, that the left-hand border does not affect the right-hand border.

[k]The Dienst der Kust has been advised as well by the "Waterbouwkundig Laboratory Borgerhout" as well as "Delft Hydraulics". Cf. for instance the report of Peerbolte.[54] Thus the input of those institutions and the (general) collaboration should be gratefully mentioned, especially concerning the one-line part of the present investigations: the text of pp. 47 and 48 originates from the afore-mentioned report.

The present study has been presented earlier in "Background Information on Artificial Nourishment'.

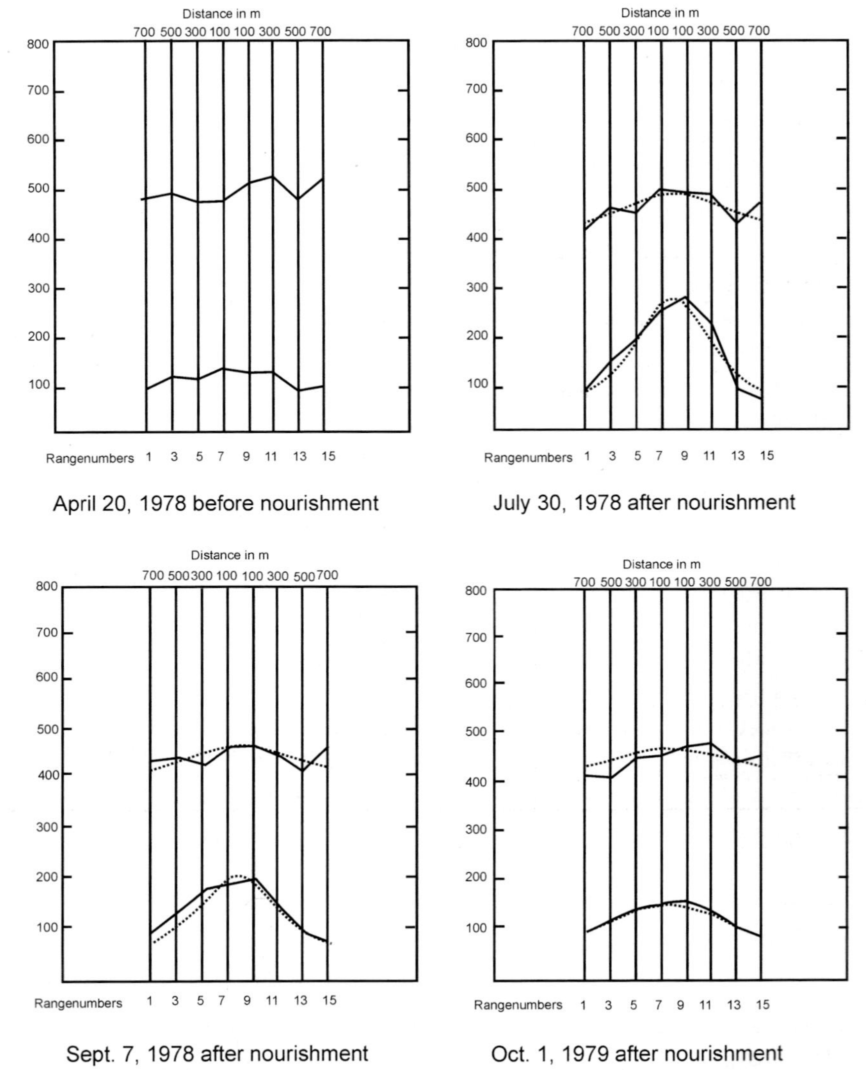

Fig. 5.41 Comparison of measured and calculated line of beach and inshore using the numerical two-line theory (Option E: 2_num).

From the measured coastal profiles it was found that the height of the active profile $h = 12\,\text{m}$. The optimum values of y_0 and b were determined by comparing the computational results with the measured schematized coastlines in the period 1978–1983. With values of:

$$y_0 = 77\,\text{m} \quad \text{and} \quad b = 3970\,\text{m}$$

Table 5.5 Summary of results.

	one-line theory		two-line theory		
number of schematization (option)	1	2	3	4	5
treated in Section Figure	2.5.2 2.12	2.5.3 2.14	2.5.4 2.15	2.5.5 2.16	2.5.6 2.17
way of computation	analytical				numerical
schematization coastal area (MSL: +4 m, −2 m, −8 m)	coastline	beach	beach inshore	as in 3	as in 3
additional assumptions	coastline	line of beach	line of beach and inshore		initial: line of beach and inshore
		has (have) the shape of a Gauss-curve ($\hat{y}$, σ)			
		no sand loss from beach to inshore	beach and inshore have same shape as coastline found in 1 shifted in on/offshore direction	σ for beach and inshore as coastline, found in 1 $\hat{y}_{beach}$* $\hat{y}_{inshore}$	coastal shape follows from numerical computation
initial value(s) of $\hat{y}_{beach}$ and $\hat{y}_{inshore}$ in m (before initial loss)	$\hat{y}$ = 132.87	$\hat{y}_b$ = 192.5	$\hat{y}_b$ = 132.87 $\hat{y}_i$ = 132.87	$\hat{y}_b$ = 170.0 $\hat{y}_i$ = 80.4	$\hat{y}_b$ = 192.5 $\hat{y}_i$ = 68.3
initial value(s) of σ_{beach} and $\sigma_{inshore}$ in m	σ = 294	σ_b = 276	σ_b = 294 σ_i = 294	σ_b = 294 σ_i = 294	σ_b = 276 σ_i = 398
initial inaccuracy (m)	S_{di} = 15.55	beach: 21.25	beach: 41.95 inshore: 39.83 S_{di} = 39.51	beach: 22.45 inshore: 23.20 S_{di} = 22.05	beach: 21.25 inshore: 21.06 S_{di} = 20.44
inaccuracy of calculation (m)	S_d = 11.31	beach: 11.77	beach: 20.23 inshore: 27.33 S_d = 23.65	beach: 9.62 inshore: 21.71 S_d = 16.51	beach: 10.20 inshore: 19.16 S_d = 15.10

found in this way, the optimum value of $\sqrt{(4(s/h)t_0)}$ was determined for the year 1979, taking $t_0 = 1$ year. It was found that $\sqrt{(4(s/h)t_0)} = 402.5\,\mathrm{m}$ and thus $s = 0.5 \times 10^6\,\mathrm{m}^3/\mathrm{yr}/\mathrm{rad}$.

With this value of s, the coastlines for the years 1980–1983 were computed, and the results agreed reasonably well with the measured schematized coastlines,

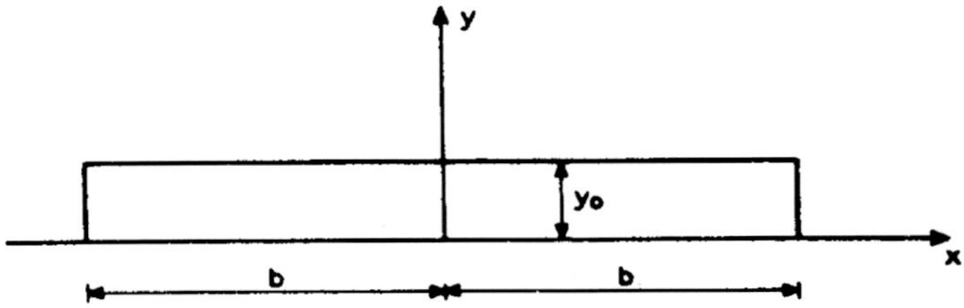

Fig. 5.42 Schematization of beach-fill.

Computation of beach fill deformation according to the one-line theory; data and analysis

For the analysis of the deformation of the eastern limit of the beach fill the area between km. 15.0 and km. 16.8 has been considered (for situation refer to Fig. 2.73). On the basis of beach levellings and sounding charts the coastal profiles for the year 1978, 1979, 1980, 1981, 1982 and 1983 have been schematized according to the one-line theory. For the average height of the active profile a value of 12 m has been assumed.

Figure 5.43 shows the schematized coastlines plotted with respect to the 1977-position of the coast. The area depicted is "the right-hand side area of Fig. 5.42", cf. Fig. 2.75. The development is rather irregular, in particular in the section km 15.4 and eastward thereof (Fig. 2.73), where the morphologic evalution are influenced by the tidal area Zwin. Therefore it was decided to restrict the analysis to the coastal zone between km. 15.6 and km. 16.8 (cf. Fig. 2.73)

In first instance the optimum values of y_0 and b in Eq. (5.34) were determined by comparing the computational results with the schematized coastlines in 1978, 1979, 1980, 1981, 1982 and 1983. For the considered coastal area between the sections km 15.6 and km 16.8, which correspond with the model coordinates 4600 and 3400 (Fig. 5.44), the following values were found:

$$y_0 = 77\,\text{m}$$
$$b = 3970\,\text{m}$$

With these figures the optimum value of $\sqrt{4st/h}$ was determined for the year of 1979, for which it was assumed that $t = 1$ year, since the eastern part of the beach fill was placed in the period July 1977–June 1978. The optimization for this year yielded the following value $\sqrt{4st/h} = 402.5\,\text{m}$.

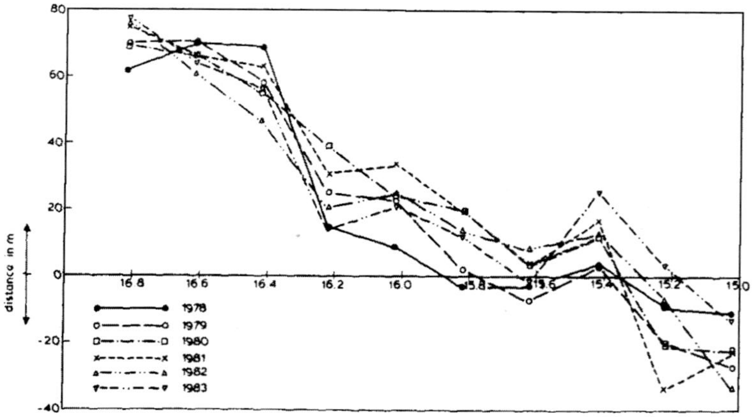

Fig. 5.43 Location of analyzed beach replenishment. Schematized coastlines with respect to coastline 1977. Cf. Fig. 2.73 for location range system.

Since $t = 1$ year it follows that $s/h = 40.500\,\mathrm{m}^2/\mathrm{yr}$.

The average value of h is: $h = 12\,\mathrm{m}$ and then the coastal constant $s \approx 500{,}000\, m^3/\mathrm{yr}/\mathrm{radian}$.

With the above values of $y_0 = 77\,\mathrm{m}$, $b = 3970\,\mathrm{m}$ and $s/h = 40500\,\mathrm{m}^2/\mathrm{yr}$ the coastlines for the subsequent years have been computed and compared with the measured, schematized coastlines.

The results are shown in Fig. 5.44 for the years 1979, 1980, 1981, 1982 and 1983. It should be observed that the measured positions of the coast east of km 15.6 have not been taken into account.

Evaluation The predicted coastlines for the years 1980–1983 agree reasonably well with the measured position of the coast. It should be borne in mind that the analysis has been limited to a relatively short coastal stretch and that only the alongshore deformation of the coast has been considered. In line with other cases it might be concluded that the theory gives a fair estimate.

As far as the coastal constants concern, again large variations are encountered. In a previous study of Delft Hydraulics the coastal constant s was determined on basis of the longshore transports computed with the CERC-formula. It was found that $s \approx 1{,}500{,}000\,\mathrm{m}^3/\mathrm{yr}/\mathrm{rad}$, but according to the computations the value varied strongly along the coast. If this value is introduced in the present computations instead of $s = 500{,}000\,\mathrm{m}^3/\mathrm{yr}/\mathrm{rad}$, then the predicted coastline position y would differ with roughly 20% from the results presented in Fig. 5.44.

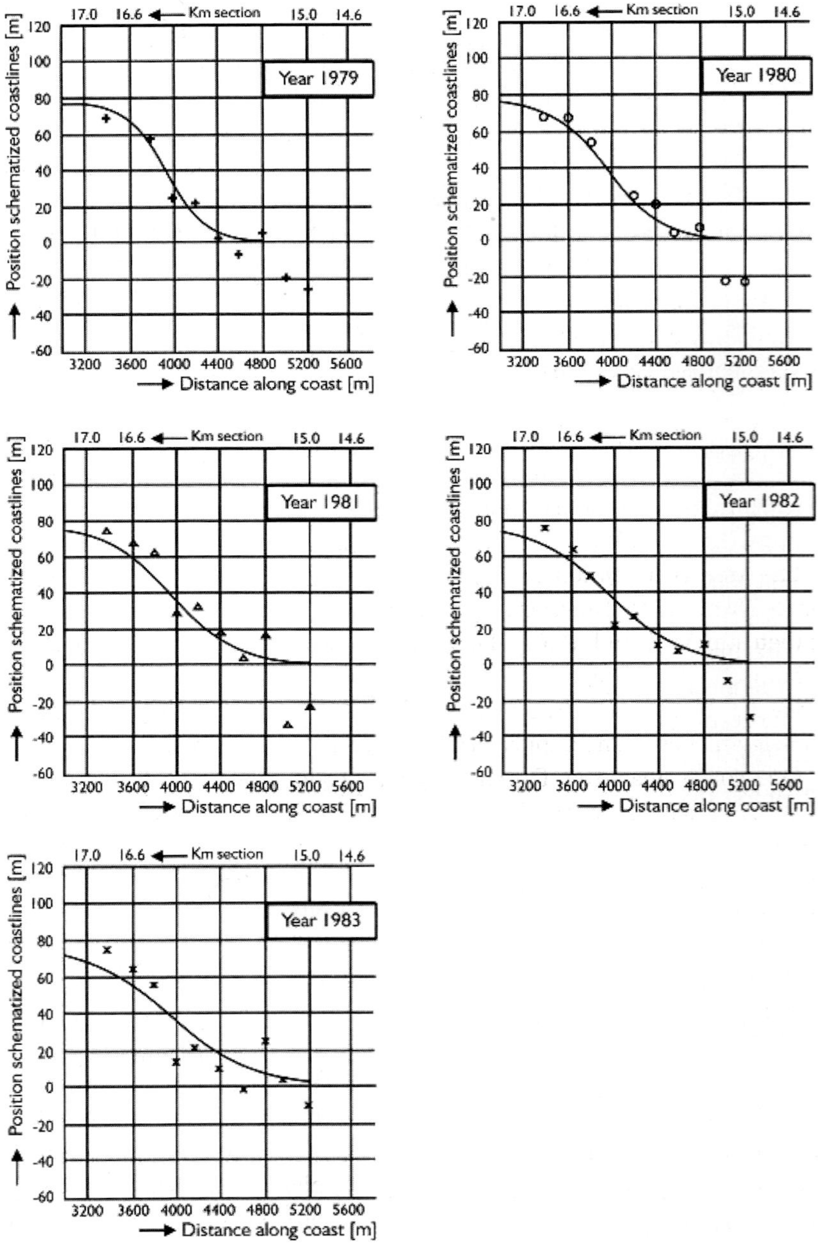

Fig. 5.44 Comparison of measured and predicted coastlines (one line theory).

Computation of beach fill deformation according to the two-line theory; data and analysis. The beach nourishment project at Zeebrugge has been used as well as a second example of the application of the two-line theory. As the two-line theory is more detailed than the one-line theory, it requires more detail with respect to the initial conditions and the boundary conditions. Therefore, in the two-line approach for Zeebrugge, the real situation is approximated better than in the afore-mentioned case of the one-line theory .The attention has been focussed on the east-boundary of the fill, which is located near the Dutch–Belgian border, as shown in Fig. 2.77.

It may be revealed, that a good prediction of the behaviour of the supply, especially at the eastward boundary, was of great importance for the Belgian Government, since expensive additional coastal protection works, such as groynes with a length of 1 km, were taken into consideration to prevent longshore losses in an eastward direction. With the aid of coastline theories, it has been possible to predict, that these groynes were not necessary. After the execution of the nourishment project, it appeared that the losses were even less than predicted.

The area between km 15.6 and 16.8 has been considered in detail (see Figs. 2.73 to 2.77). However, in a more approximative way, also the area between km 14.2 and 15.6, where the ancient, fifteenth century entrance to the harbour of Bruges, the "Zwin" debouches, has also been taken into consideration, as well as the discontinuity at the most easterly groyne along the Belgian coast. The supply originally ended at km 16.2.

The coast near the Dutch–Belgian border consists of two groyne-protected areas, one on the Dutch and one on the Belgian side. These areas are connected by as unprotected area. In the deeper regions, the bottom topography is to a large extent determined by currents in the tidal channel, the so-called Appelzak. The wave influence is especially important at the beach. In the computations, only waves have been taken into account.

The line of the beach has been defined by the region between MSL + 4 m and MSL − 2 m, while the region between MSL − 2 m and MSL − 8 m defines the line of the inshore. The initial situation before nourishment has been modeled by superposition of three stable situations:

The macro-coastline, where the resultant wave-driven longshore transport (Q) in eastward direction is decreased due to the groynes in the protected area, and by the turning of the coast to the direction of the wave crests in the unprotected area (see Fig. 5.45a).

The unprotected part of the coast is modeled by a line of the beach and a line of the inshore between the groynes (see Fig. 5.45b).[1] Just eastward

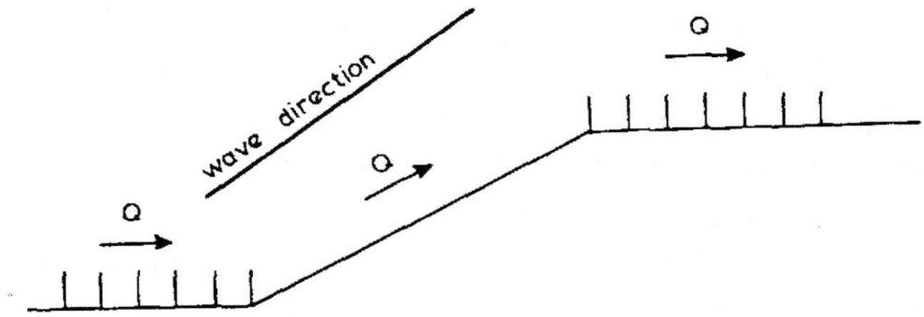

Fig. 5.45a A stable macro-coastline (the transport Q should be everywhere the same).

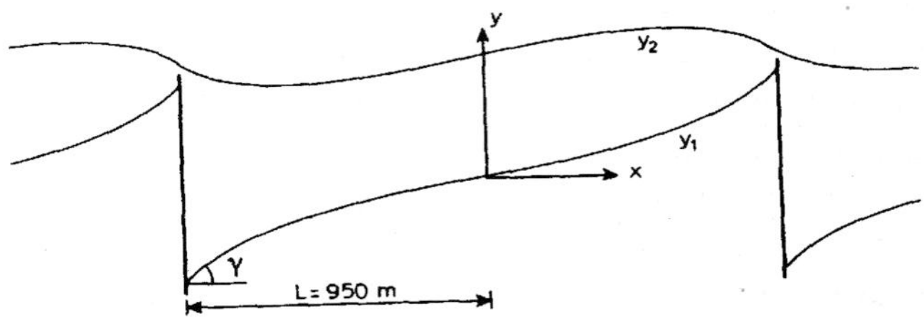

Fig. 5.45b Schematization of unprotected area between the Belgian and Dutch groynes (cf. Fig. 3.33).

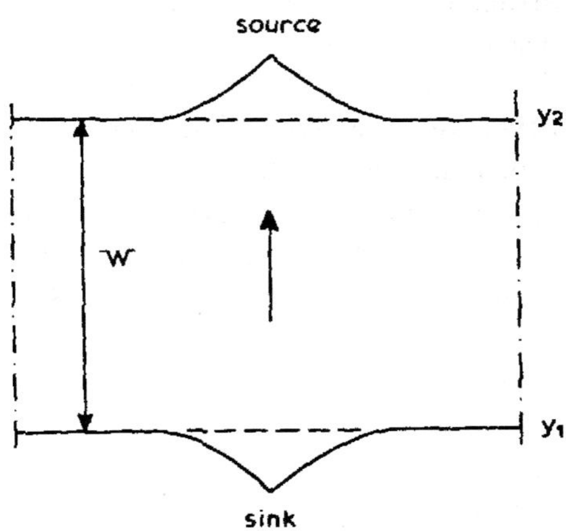

Fig. 5.45c Schematization of the Zwin (cf. Fig. 3.28).

of the last Belgian groyne, sand is transported to the beach, because the profile is flatter than the equilibrium profile. Just westward of the first Dutch groyne, the beach is located more offshore. The profile is here steeper than the equilibrium profile.

The Zwin has been schematized by a sink at the beach and a source at the inshore. Sink and source have the same strength (see Fig. 5.45c).[2] The reason for this schematization is that during storm conditions the sand is dumped into the Zwin channel, where it is transported by tidal currents from the beach to the inshore zone. Since the lines of beach and inshore are located at a larger distance from each other than the equilibrium distance, an onshore transport results, returning the material from the inshore to the beach.

In this way, the coastline of 1977 between km 16.8 and 15.6 (before nourishment), as shown in the upper part of Fig. 5.46, could be reproduced with an accuracy of 12 m. In order to model the fill just after nourishment, the two-line solution for the evolution of a rectangular beach nourishment has been applied (see Background Report),[38] Annex VI).

It has been assumed, that in 1978 the inshore was not yet affected by the fill, while along the beach already a certain deformation of the rectangular shape had taken place.

According to the theory, the deformation of the fill can be described by an error function. The optimal shape of the error function, found from curve fitting, gave an initial inaccuracy of 10.5 m (7.5 m for the beach and 13.5 m for the inshore). Starting from this initial condition and taking the transport constants for the beach and inshore as $s_1 = s_2 = 250{,}000\,\mathrm{m}^3/\mathrm{yr}/\mathrm{rad}$ (thus resulting in $s = 500{,}000\,\mathrm{m}^3/\mathrm{yr}/\mathrm{rad}$, cf. p. 5.48), the lines of beach and inshore of 1980 and 1983 were calculated. The results are shown in Fig. 5.46. An overall accuracy of 17 m was found.

The value of the cross-shore sediment transport constant $s_y = 3\,\mathrm{m}/\mathrm{yr}$ was chose it in such a way, that the behavior of beach and inshore at the western boundary (km 16.8) corresponds with the behavior observed in nature. This value has also been used for the modeling of the stationary line of beach and inshore along the unprotected area and for the modeling of the Zwin.

5.3.5. *Holland coast*

5.3.5.1. *Abstract*

In this subsection the *Coastal Defence Study*, Rijkswaterstaat,[40] as executed in the Netherlands, involving a 100-year prediction for the coastline evolution of specific parts of the Holland coast passes the review. In order to create this prediction, a mathematical two-line model (cf. Section 3.4) was used. In this model the effects of wave driven as well as tidal transports

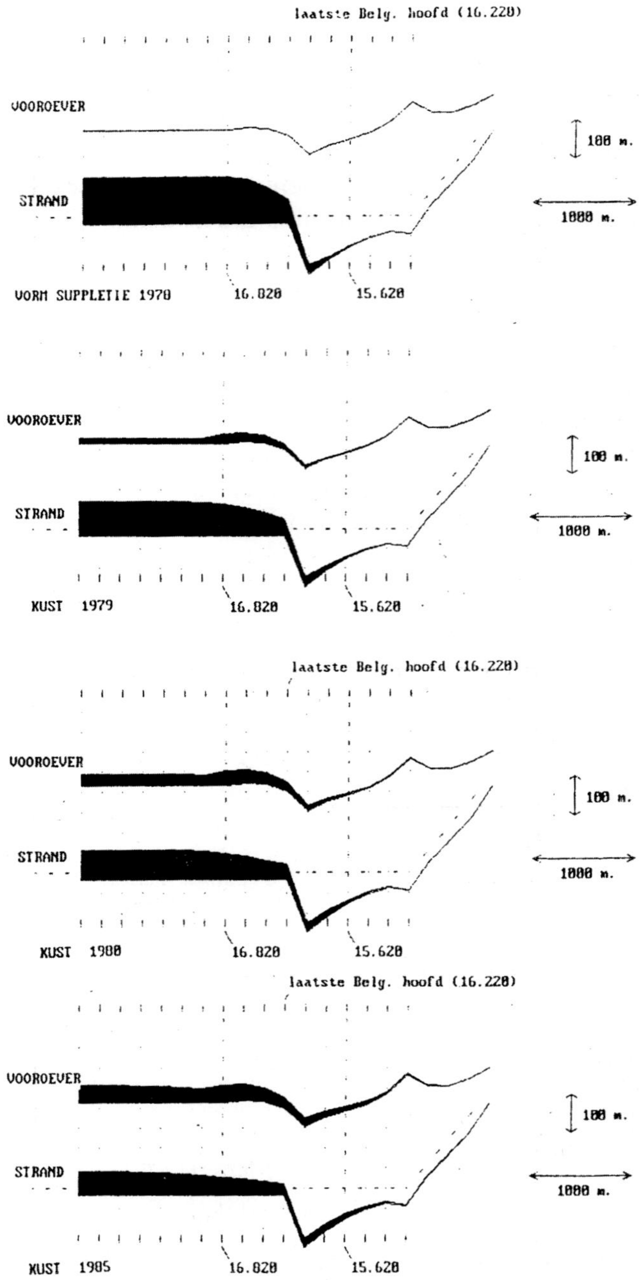

Fig. 5.46 Development of supply in 1978, 1979, 1980 and 1985.

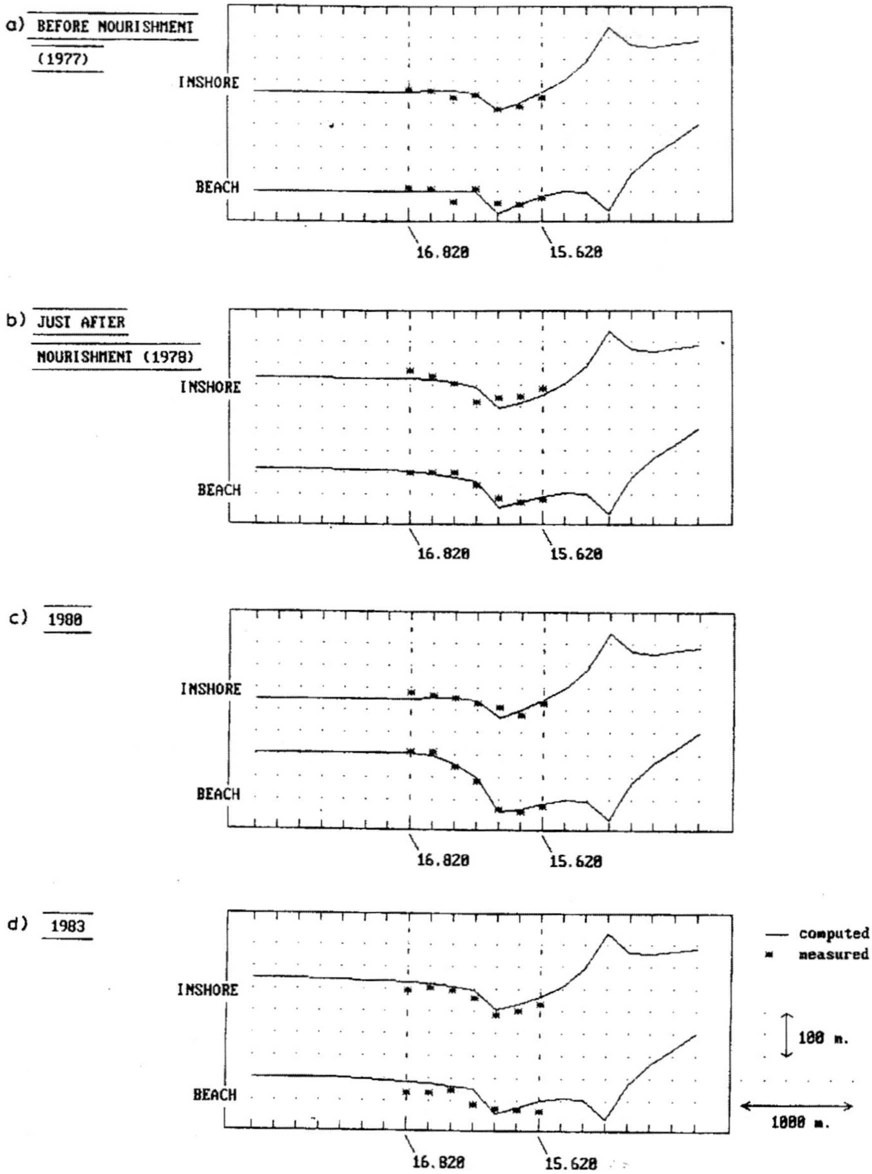

Fig. 5.47 Results of two-line computations for Zeebrugge.

are incorporated, also cross-shore transport is implemented. The model is calibrated using sand balances, derived from measurements of coastal profiles. Qualitatively correspondence is reached, but quantitative differences between model and nature remain.

Refer to Chapter 2 concerning the aspects of the Dutch coast with a phenomenological character.

5.3.5.2. *An introduction concerning the coastal dynamics of the Holland coast*[l]

The mathematical prediction of the coastline evolution of the Holland coast for the Coastal Defence Study is based on a two-line model,[m] as described by Bakker,[1] and in the Manual on Beach Nourishment.[38] The way of prediction is described by Dijkman *et al.*[21]; the present chapter gives a condensed version of this report.

Referring to Chapter 2: the Holland coast is the central one of the three parts (Rhine/Scheldt estuary, uninterrupted Holland coast; Wadden islands) into which the Dutch coast can be divided (see Fig. 2.1, c.f. 5.48). Three harbors (Hook of Holland, Scheveningen and Ijmuiden) partition the Holland coast. The harbor moles of Hook of Holland and IJmuiden have a length of 3 km and therefore have a dominating effect on the coastal behavior

5.3.5.3. *Model boundaries*

For the model the Holland coast was divided into three areas (cf. Fig. 5.48):

- Callantsoog to IJmuiden
- IJmuiden to Scheveningen
- Scheveningen to Hook Holland

These areas all stretch from one harbor mole to the other, except for the most Northern border. Here the model was limited to Callantsoog, because of the influence of the tidal inlet near Den Helder becoming to strong in the more Northern region.

Model parameters

Wave climate. The wave climate used is derived from measurements in the period 1978–1984. These measurements lack direct information on the

---

[l]N.B. for a phenomenological description of the Dutch coast, refer to Chapter 2.
[m]Cf. Chapter 3.

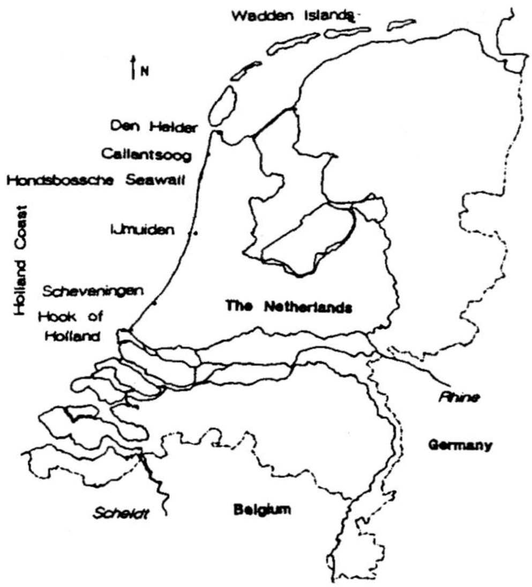

Fig. 5.48 The Holland coast.

direction of wave propagation, only the wind direction is given. To overcome this problem a correction to the directions has been applied based on the correlation as proposed by Hokke and Roskam,[25] who found this correlation between the wind direction and the wave direction from measurements using Wavec-buoys. The so-corrected wave climate significantly differs from the original 'uncorrected' one.

Wave data originating from three different sites (Eierland, IImuiden and LE Goeree) have been used. A comparison between these three sites gives an impression of the variation of the wave climate along the Holland coast, see Fig. 5.48.

Wave data originating from three different sites (Eierland, IImuiden and LE Goeree) have been used. A comparison between these three sites gives an impression of the variation of the wave climate along the Holland coast, see Fig. 5.49.

Furthermore the wave climate as now determined is valid for deep water with a depth of approx. 20–25 m. In order to obtain the required wave climate at shallower water (of approx. 10 m depth) the ENDEC program (Energy DECay, Delft Hydraulics, 1988) has been used. This program incorporates the effects of breaking waves, refraction, shoaling, bottom friction and wind.

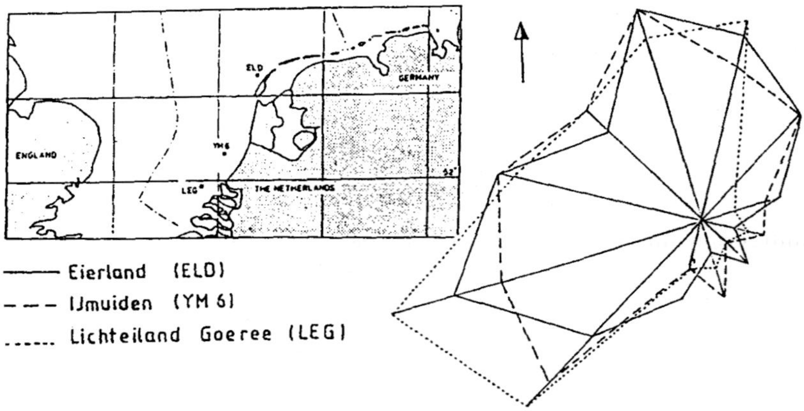

Fig. 5.49 Variation of the wave climate.

Tidal influence on the Holland coast. The tidal data have been taken from the HOKU model (HOllandse KUst, which stands for Holland Coast) (Delft Hydraulics, Hartsuiker, 1988).[24] The non-linear relation between tidal velocities and sediment transports has been compensated by using a morphological tide, which is found by multiplying the average tide with a 'tidal factor' (here taken to be 1.1).

In the direct surroundings of the harbors the influence of the tidal effects can be expected to be much more noticeable than at some distance from those harbors. This as a result of the converging/diverging tidal flows, causing sediment transport gradients and accessory erosion/accretion patterns. Qualitatively the formation of an eddy at the leeside of a harbor mole causes a reversal of the longshore transport in this region. For the opposite tide the longshore transport directed to the mole will remain constant until a much smaller distance from the mole. The resulting effect of a complete tidal sequence will therefore be an accretion at a small distance to the mole, followed by an erosion that at growing distance from the mole becomes less noticeable.

The tidal effects are calculated in the vicinity of the harbors up to the point where the effect of the harbor mole is negligible. In between the harbors a linear interpolation has been used. The tidal velocities at the points in the grid are calculated for an interval of 30 mm (25 in total, covering 12 h). The probability function for the orbital velocities caused by waves is determined for each grid point. For the fraction of time waves are breaking in this point there is assumed to be no contribution to the tidal transport for that particular combination of waterdepth, wave direction and wave height. Then with the Bijker formula, using both the orbital velocity distributions and the tidal velocities, a resulting tidal sediment transport is calculated. This tidal transport is assumed to be time independent. The results are shown in Fig. 5.49.

Diffraction near the harbors. In the direct vicinity of the harbors the effect of diffracting waves around the harbor moles is an important factor. This effect is implemented in the model, however in a somewhat limited manner: the diffraction and refraction effects have been calaculated seperately. For the method used see Ref. [11].

- Coastal constants

These are the constants (constant only in time, but varying along the coast as a result of a variation of the wave conditions along the coast) which determine the longshore wave driven sediment transport ($S_{0,i}$ and S_i in Eq. (3.1)). Calculation of these constants is done using the following assumptions:

- The wave driven longshore transport between two depth contours is proportional to the longshore component of the energy-loss of the driving wave between those depth contours.

The equation used to calculate the longshore transport is a variation of the CERC formula

$$S = A \cdot (n_{\text{br}} H_{\text{br}}^2 c_{\text{br}} \sin(\varphi_{\text{br}}) \cos(\varphi_{\text{br}}) \tag{5.35}$$

where S is the longshore sediment transport (m^3/yr); A, proportionality constant (–); the wave characteristics at the breaker line: n (–), H(m), c(m/s), φ(rad).

For a more detailed description of the method used see Refs. [3,11,44].

The calculations were performed using the KC program (Kust Constanten, which stands for Coastal Constants, by Casteleyn, 1975)[19] which includes diffraction and refraction effects. In order to compensate for the overestimated diffraction effects, caused by using monochromatic waves, a reduction factor for the calculated diffraction effect has been applied, this factor is taken to be 0.75. (The difference between the calculated values in the diffraction zone and the value outside this zone is multiplied by the reduction factor). Originally the KC program gives values for the coastal constants for a one-line model, the necessary two line values are derived from the one-line values by assuming a constant ratio between the values for beach and inshore, respectively. From calculations a ratio (beach:inshore) of 1:4 was found.

The only remaining unknown in the coastal constants is the proportionality constant A (see Eq. (5.42)). This will be determined in the calibration of the model, just like the cross-shore coastal constants s_y and W (see Eq. (2.2)), page 58.

5.3.5.4. *Coastal defence works of the Holland coast*

As described in Chapter 2, there are three different defence methods in use along the Holland coast groynes, the Hondsbossche Seawall and sand supplies

The groynes partially block the longshore sediment transport. The effectiveness of the groynes has been roughly estimated by considering the depth to which they reach, in relation to the sediment transport distribution over the depth as found by Bakker.[3] In the model this has been accounted for by using a reduction factor for the coastal constants. Note that this method implies neglecting the impact the groynes may have on the tidal sediment transports: the tidal transports mainly take place at greater depths than the depths to which the groynes reach.

The Hondsbossche Seawall is a rigid seawall, below MSL-1 m it is covered by sediment. At both sides of the seawall there are several groynes preventing this area from eroding too severely. In front of the Hondsbossche Seawall the beach zone does not really exist, in the model this has been accounted for by taking the coastal constants in this area equal to zero. In the inshore zone there can be some sediment transport. However, at both sides of the Hondsbossche seawall very strong curves in the coastlines occur, which are not varying too much in time. Therefore the coastal constant s_2 must also be taken equal to zero for this area. This assumption can be backed up by the geological consistency of the bottom material in front of the Hondsbossche Seawall: there are some almost baring clay layers present there.[33] This implies there can only be a constant transport along the Hondsbossche Seawall: what goes in on one side must go out the other. The model does not implement the sand supplies, but does take these into account in the calibration of the model, see Section 5.3.5.5.

5.3.5.5. *Calibration and output*

The proportionality constant A, used in the calculation of the coastal constants (see Eq. (5.42)), is used as the instrument to calibrate the model.

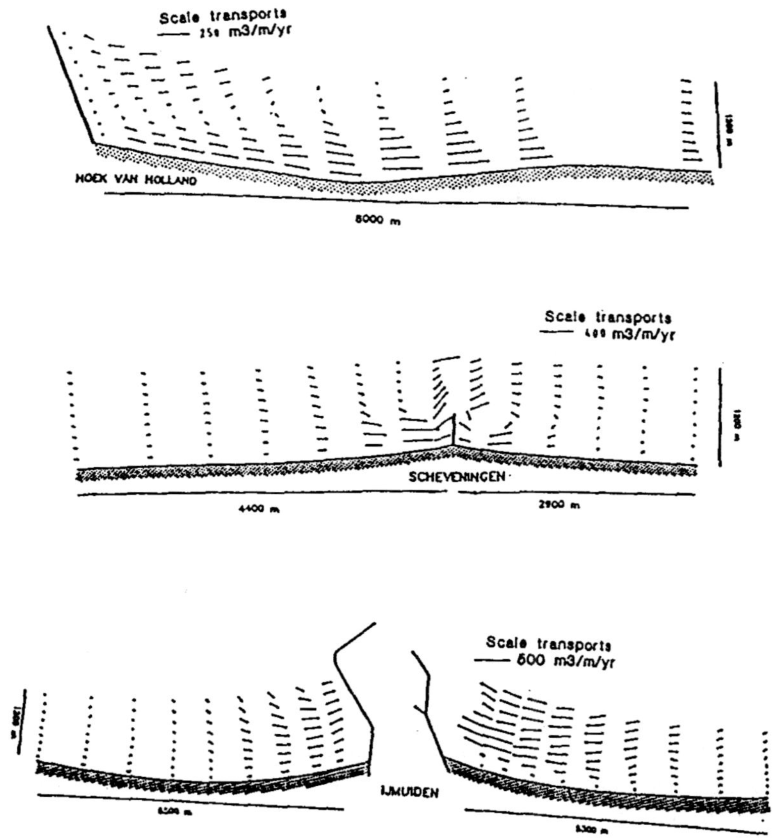

Fig. 5.50 Resulting tidal sediment transports near the harbors.

The calibration is based on the measurements of the coastal profiles of the Dutch Coast on a yearly basis, as stored in the JARKUS database. These measurements are used to create sand balances for the specific parts of the Holland Coast. Some results are shown in Fig. 5.50, showing the relative amount of sand (relative to the first year) for three subareas.

By making some assumptions these sand balances can be used to give an idea of the average sediment transports in the considered areas. The necessary assumptions are:

- No sand leaking around the harbor mole near Hook of Holland.
- No sand from above the MSL−5 m passes from one side of the harbor moles of Scheveningen and IJmuiden to the other.

- Equally distributed sand supply from deeper water over the MSL-5 m line for an area between two harbors.

With these assumptions the following conclusions can be derived:

- Accretion of the area Hook of Holland-IJmuiden of 370.103 m^3/yr, or 6 m^3/m/yr.
- For the area IJmuiden-Callantsoog this accretion is 40.103 m^3/yr or 0.9 m^3/m/yr. This value has been neglected in Eq. (3.6). This relatively low value (relative to the value in the area Hook of Holland-IJmuiden) can be explained by the increase (going Northward from IJmuiden) of the Northward tidal flow in deeper water, consuming' the sediment which would otherwise come to benefit the area above the MSL-5 m level.

The onshore transports found here compare reasonably well to those found by Stive.[43] The value Stive proposes, is a bit larger: approximately 10 m^3/mlyr, almost constant over the entire Holland coast. However this value is derived for an onshore transport over the MSL-8 m line, where in this paper the transport over the MSL-5 line is considered.

- The optimum value for the proportionality constant A is 500, determined for the area Hook of Holland IJmuiden. This value is about a third of the conventional values! With the proportionality constant A now determined, the longshore transports can be calculated, the results are shown in Fig. 5.51.
- Again, comparing the so found longshore transports to those found by Stive,[43] a reasonable agreement is found, see Dijkman *et al.*[21]
- The coastal constant for the cross-shore transport, s_y (at a depth of MSL-1 m) is found to be in the order of 0.05 m/yr (for the region Hook of Holland — Scheveningen) tom/yr (for Scheveningen — IJmuiden). These results were found by using a value of 280 m for the equilibrium distance between beach and inshore line. In the modela, constant value of 0.10 m/yr is used.

Other scenarios, for instance large offshore transport caused by rip-currents and tidal currents in the immediate vicinity of the harbor moles, have been elaborated with respect to their consequences for the coastal development. (The report by Dijkman *et al.*[21] shows results). However, these scenarios had to be rejected because of the coastal behavior according to these scenarios proved to be unrealistic.

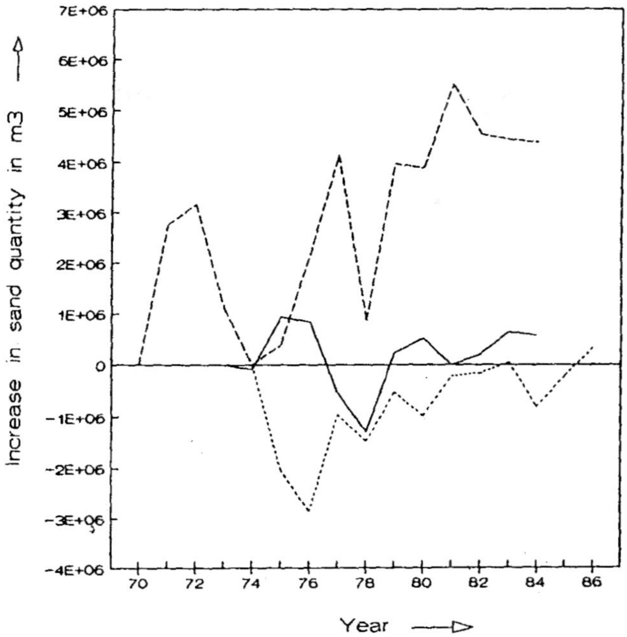

Fig. 5.51 Sand balance for the Dutch coast.

5.3.5.6. *Results of the coastal model of the Holland coast*

Figure 5.52 shows the calibrated coastal development between 1973 and 1984, for the area Callantsoog to IJmuiden. On the left-hand side one observes the effect of the harbor mole of IJmuiden, and between the ranges 20 and 25 the effect of the Hondsbossche Seawall.

Figure 5.53 shows predictions found in this way for the years 2020 and 2090. Finally the same results are presented in a different manner, namely with the reference as the initial situation. Figure 5.56 shows the development of the beach-line, Fig. 5.57 that of the inshore line.

Results for the area IJmuiden to Hook of Holland are not presented here, they can be found in the report by Dijkman *et al.*

Conclusions concerning the Holland coast

- The results (certainly for an area with harbor moles) strongly depend on the wave climate used.[21] Therefore not only the accuracy, with which

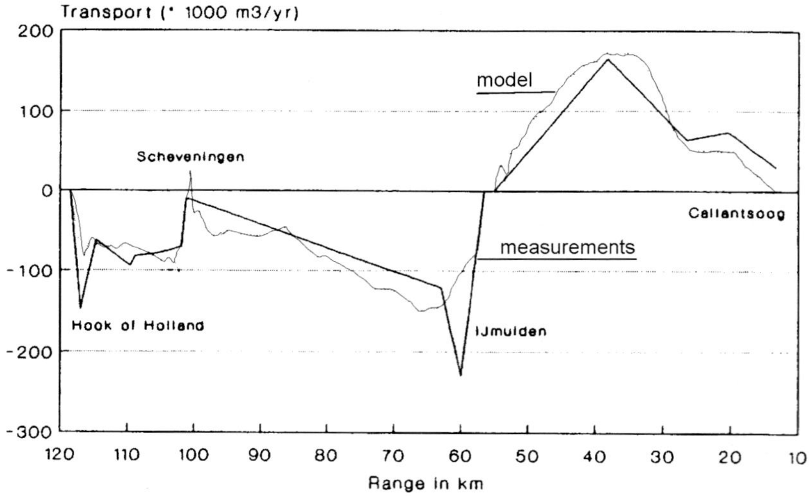

Fig. 5.52 Calibration of the coastal constants.

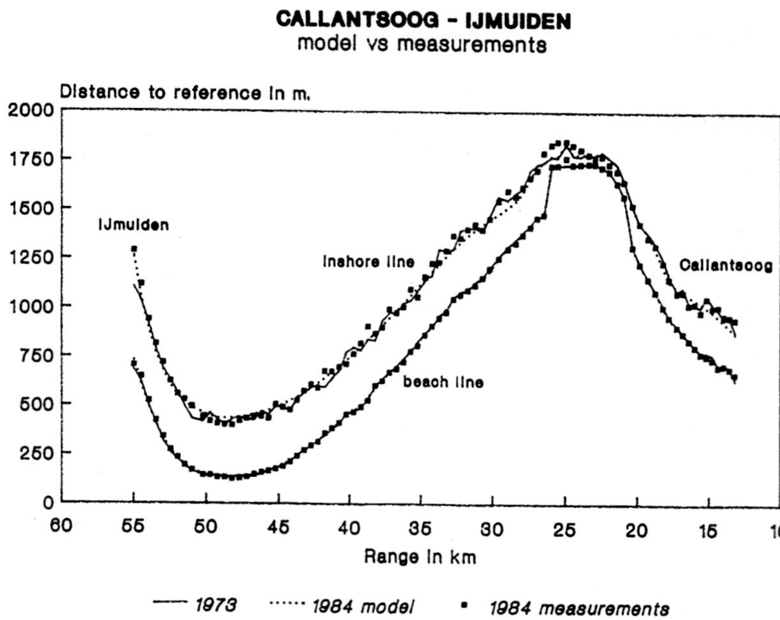

Fig. 5.53 Calibrated coastal development for Callantsoog to IJmuiden.

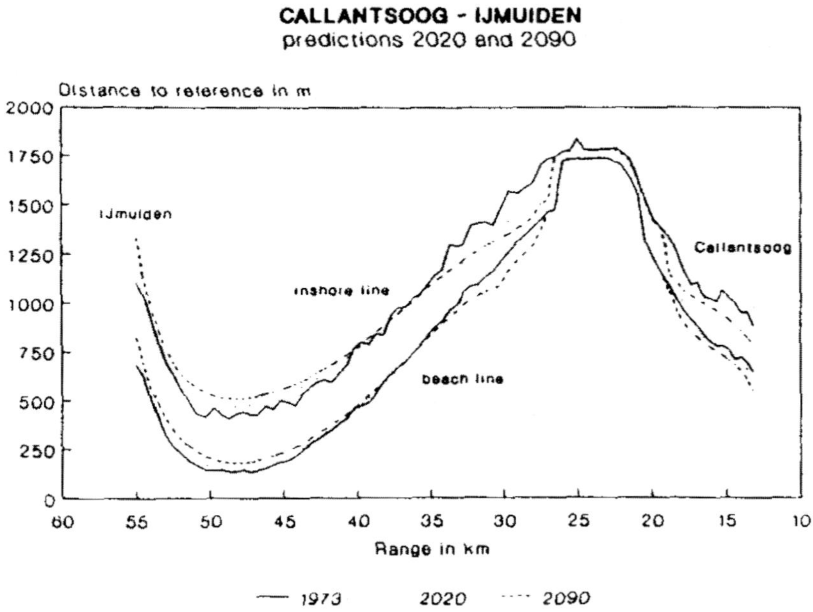

Fig. 5.54 Predictions coastal development for Callantsoog to IJmuiden.

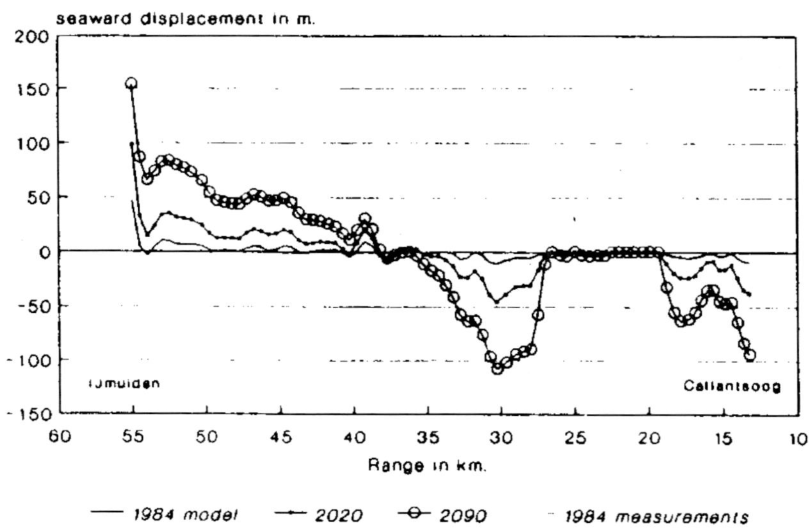

Fig. 5.55 Development of the beach-line.

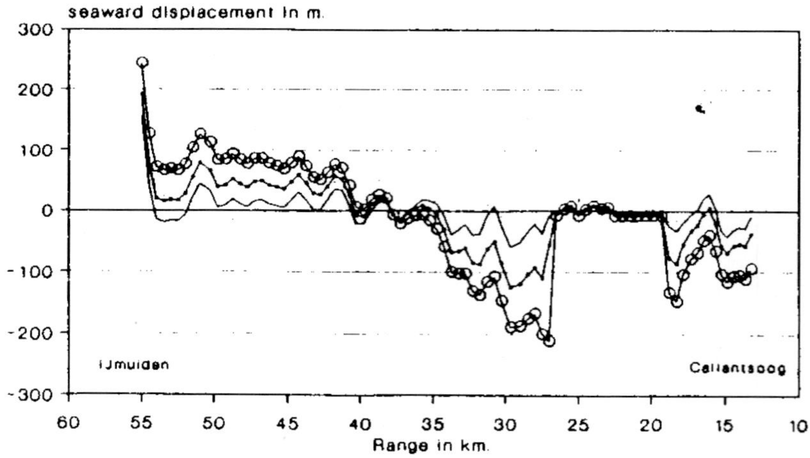

Fig. 5.56 Development of the shore-line.

the present wave climate is known, but also the accuracy, with which the future wave climate can be predicted highly determines the accuracy of a prediction of the coastline evolution. Accurate wave data (especially concerning wave directions) are vital for accurate coastal prediction.

- The proportionality constant A (giving the ratio between the sediment transport and the longshore component of the wave energy flux, see Eq. (5.42) found here, is much lower (about a third) than more conventional values.
- The use of the two-line model as an instrument to create a prediction of a coast-line evolution is in creating a qualitative picture giving eroding and accreting areas, but limited in its quantitative accuracy.

5.3.6. *The NOURTEC cases (Terschelling and Norderney)*

5.3.6.1. *Introduction on Nourtec*

The NOURTEC- project is an EU-research project in the scope of the program for Marine Science and Technology (MAST II). Purpose is Innovative Nourishment Techniques Evaluation. For aim and scope is referred to Mulder *et al.*[32] National coastal authorities in three different countries around the North Sea (Germany, the Netherlands and Denmark) have carried out full-scale experiments with alternative coastal nourishment techniques (i.e. shoreface nourishments and combinations of shoreface and beach nourishments). These experiments are implemented under a range of environmental conditions. Extensive modelling and monitoring programmes form an essential part of the projects.

One kind of models used for this goal is line-models. These will be applied on all selected supply sites in the three participating countries.

The present paper deals with the prediction of the coastal behavior with the aid of those models and comparison with prototype data for the Dutch and German supply: at Terschelling and Norderney (Fig. 5.56) respectively. In the future, also an analysis of the Danish supply will be given. A more detailed description of the investigations is given by Bakker and Kersting[12] and by Bakker *et al.*,[13] respectively.

Terschelling

Situation. An underwater supply of ca. 2 Mm3 has been carried out in spring and summer of 1993 on the Dutch Wadden isle of Terschelling. Figure 5.57 shows the site.

Three breaker bars are gradually moving seaward along the whole stretch; motion is periodic; the mutual distance between the breaker bars is 500 m; seaward propagation velocity is 50 m and period is 10 years.

Depth-averaged (between NAP+3 m and NAP −2.75 m; "NAP" denotes a level, which is about mean sea level) the area under consideration shows a coastal retreat of about 2.75 m/yr.

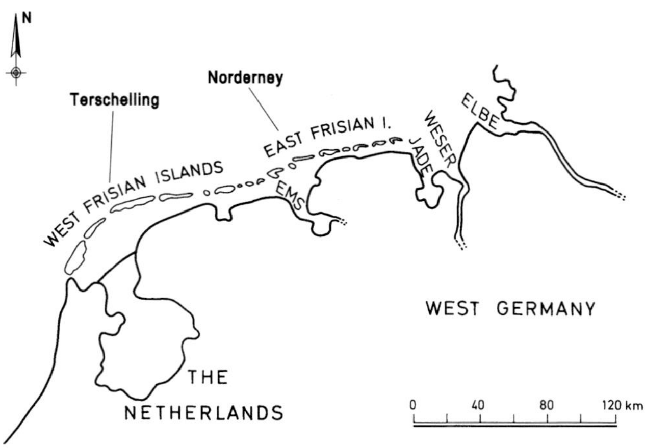

Fig. 5.57 Location Terschelling and Norderney.

Three-line model. For the principles of line modeling, refer to Chapter 3.

In the three-line model used for this case (Subsection 3.4.1) the coastal zone is schematized in three zones (Figs. 3.39 and 5.58), which will be called: "beach", "inshore" and "deeper part".

In each zone it is assumed that all contour lines remain parallel and that the profile in this zone only moves horizontally. It is sufficient to compute only an average line in each zone, because it characterizes this zone. There is assumed to be a horizontal separation plane between the zones. Refer to Section 3.4.

Coastal constants concerning Terschelling

Cross-shore constants Terschelling. From the supply (between two breaker bars; Fig. 5.58) the breaker bars get an "antropogenic distortion". It is assumed, that by a diffusion process the sand, involved in this distortion will diffuse to the neighboring breaker bars. Furthermore, it is assumed, that the velocity of diffusion is related to the time laps between the replacement of one bar by the next landward bar. Per cycle, a certain percentage (determined by "engineering judgement") is retained; the rest of the sand is transported to the two adjacent breaker bars.

The engineering judgement is based upon the solution of Smit[42] for the cross-shore diffusion, reproduced in Subsection 3.4.2.

From his equations it can be derived, that in a period T_{01} (defined after Eq. (3.92)) the process of spreading the supply evenly over the profile is almost 80% completed. It seems logical to assume, that the diffusion process to a similar rate is accomplished during the propagation period of the bars. The periods T_{01} and T_{02} are therefore chosen both equal to

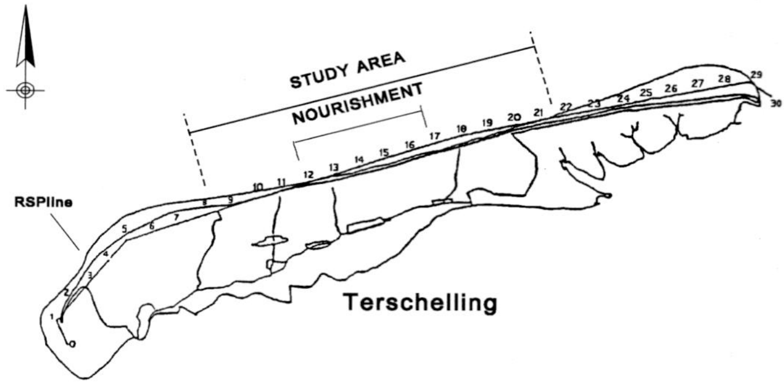

Fig. 5.58 Location study area Terschelling.

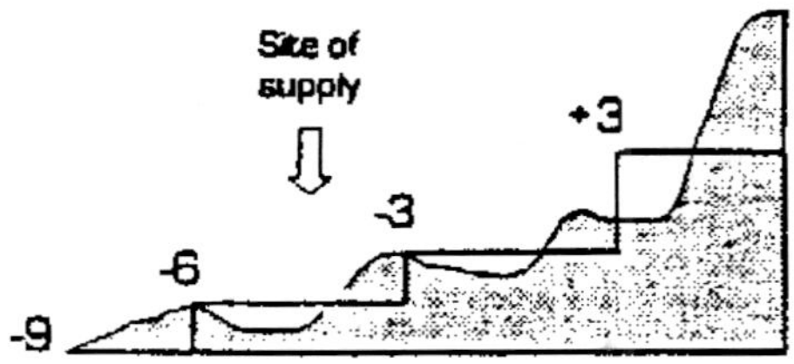

Fig. 5.59 Chematization 3-line theory.

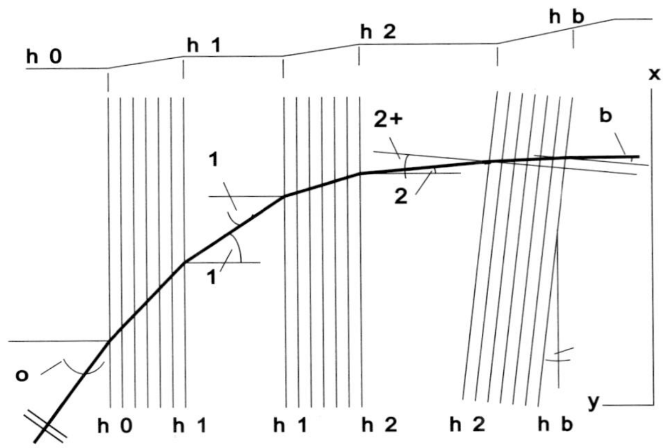

Fig. 5.60 Schematization used for transport computations.

10 years. With values for $h_{1,2,3}$, as shown in Fig. 5.59 (6 m, 3 m, 3 m), values of s_{y1} and s_{y2} of 0.2 and 0.15 m/yr, respectively are found. A higher cross-shore diffusivity at the upper part is found than at the lower part, which seems reasonable.

It is assumed that processes dominated by the diffusion of the supply can be linearly superponed to the natural longshore and cross-shore coastal processes. The latter processes are implemented by assuming an autonomous erosion of 2.75 m/yr for the zone between NAP -9 m and NAP $+3$ m.[48]

In the mathematical model it is assumed that cross-shore transport of supply below a level of NAP -9 m is neglegible (during the time of consideration) compared to the transport above this level. Thus, in the model no period averaged cross-shore transport by the periodical motion of the breaker bars is assumed.

Longshore constants Terschelling. To determine the longshore constants for the line-model use is made of the CERC-formula and the assumptions

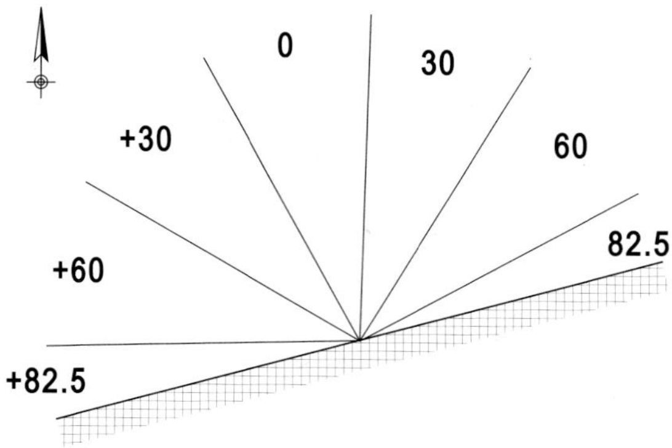

Fig. 5.61 Directions of wave approach.

of Svasek:[44] the local sediment transport between two contour lines is proportional to the longshore component of the decrease of the energy flux between those two lines. For detailed description is referred to Chapter 4, i.e. to.[3,8,11,26,44]

Sloping shores have been assumed between terraces, separating beach, inshore and deeper zone; local wave height is taken proportional to local depth.

Total wave climate has been taken into account, taking discrete wave classes, each defined by a certain wave height H, period T and direction ϕ and a probability of occurrence:

$$S_\phi = S_{H \times T\phi} \times P_r(H, T, \phi) \tag{5.43}$$

Figure 5.61 shows the various classes of wave direction, translated into angles of incidence.

Vertical tide has been taken into account by discretization in a number of levels, combined with probabilities; furthermore it has been taken into account, that high waves from certain directions are linked with water level rise (storm surge height).

The black arrows in Fig. 5.61 give the transport rates when all contours are parallel. Under those circumstances the method results in a total depth integrated longshore transport for a certain (H, T, ϕ) as given by the CERC-formula:

$$S_x = 0.040 \times H^2_{\text{b,sig}} \times n_{\text{b}} \times c_{\text{b}} \times \cos(\varphi_{\text{b}}) \times \sin(\varphi_{\text{b}}) \tag{5.44}$$

The columns in Fig. 5.61 denote the various angles of wave incidence; the "rules" in the figure the three zones. The volumes of the arrows are proportional to the transport (weighted with its probability of occurrence) integrated over all classes of H and T. Probability of occurrence has been derived from SON, a measurement station in the vicinity of Schiermonnikoog. Wave data have been made available by Rijkswaterstaat.[39]

For both upper zones the longshore transport has been computed three times for each wave direction: apart from assuming parallel depth contours also the angle of the beach (inshore) has been turned over an angle δ with respect to the other contour lines. For instance: in Fig. 5.61 the angle of the beach is turned. The dotted arrows in Fig. 5.61 show the effect, where δ has been chosen $\pm 2°$. Table 5.6 shows the numerical results.

The constants s_1 and s_2, which can be shown theoretically[37] to be the derivative of the longshore transport to the coastal direction, are calculated from:

$$s_i = \frac{\sum_{+82.5°}^{-82.5°}[S_i(\varphi_i + \delta) - S_i(\varphi_i - \delta)]}{2 \times \delta}$$

$$s_1 = \frac{804.2 - 551}{2 \times 2 \times (\pi/180)} \times 10^6 = 3.63\ 10^6\ \text{m}^3/\text{yr/rad} \qquad (5.45)$$

$$s_2 = \frac{198.5 - 131.2}{2 \times 2 \times (\pi/180)} \times 10^6 = 0.964\ 10^6\ \text{m}^3/\text{yr/rad}$$

Terschelling: results. For s_1 and s_2 values of 3.63×10^6 and $0.96 \times 10^6\ \text{m}^3/\text{yr/rad}$ were found; the calculation of the total longshore transport gave $0.85 \times 10^6\ \text{m}^3/\text{yr}$ as result. Diffusitivity in longshore direction along the deeper part could be neglected, according to the calculations. Effect of curvature of the inshore on the coastal constant for the beach (the inshore turns more or less in the same way at the same time as the beach) has not been taken into account. Neglecting this effect can give a rather important overestimate of the coastal constant of the beach.[9,11] Another extreme (no longshore diffusion) is given by the formulae of Smit[42] (Subsection 3.4.2).

The left-hand side of Fig. 5.62a,b,c shows (extreme high longshore diffusitivity) a plan view of the expected development of beach, inshore and deeper part (mind the different vertical scales). The autonomous erosion of the coast is depicted on the right-hand side of the figure.

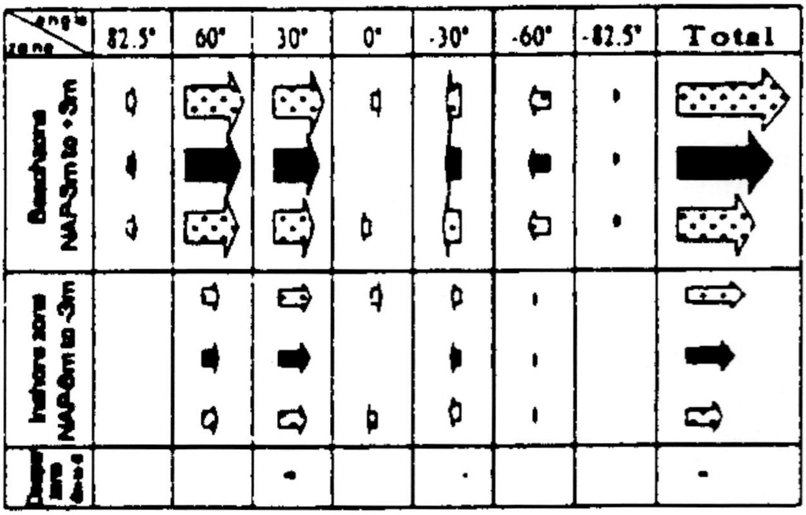

Fig. 5.62 Transport rates along the three zones as a function of the angle of approach of the waves at deep water and the effect of $\delta = \pm 2°$.

Table 5.6 Transport rates (10^3 m^3/yr).

	42.5°	60°	30o	0°	−30°	−60°	−42.5°	ES
−2°	41.42	500.4	431.4	40.53	−114.6	−04.03	−10.14	804.2
0°	40.55	478.3	387.3	0	−128.7	−87.91	−10.25	679.3
−2°	39.46	453.8	341.2	−40.53	−142.1	−90.54	−10.30	551.0
−2°	0.004	68.36	161.9	15.22	−30.16	−0.030	0	190.5
0°	0.004	67.57	148.5	0	−41.90	−0.010	0	165.3
−2°	0.004	66.30	134.4	−15.22	−45.56	−0.732	0	131.2
0°	0	0	5.510	0	−1.420	0	0	4.09

The volumes of the arrows in Fig. 5.64 are proportional to numbers in Table 5.6.

Norderney

Conditions at the site of supply. Near the west point of the German Wadden island of Norderney, 450.000 m^3 of sand has been supplied in 1992 to the beach and the shore-face (Fig. 5.63).

On this supply has been reported by Bakker *et al.*,[13] which publication is reproduced below.[n]

[n] Apart from some typographical changes.

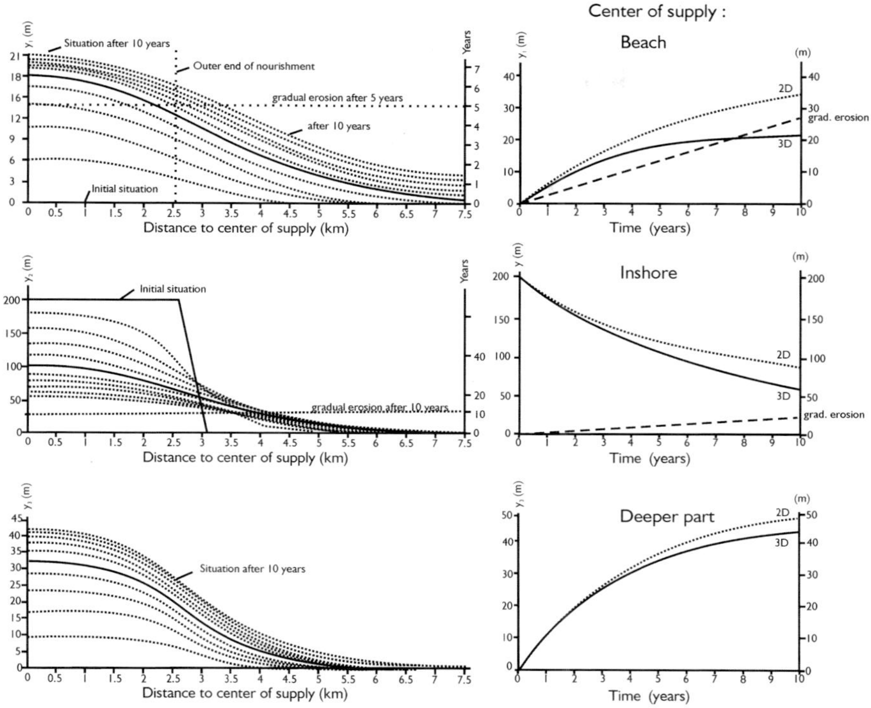

Fig. 5.63 Left-hand side: calculated shapes of the line of the beach, inshore and the deeper part. Right-hand side: site of beach, inshore and the deeper part (center of the supply) in the course of time. Interrupted line (beach and inshore) autonomous erosion. Two extremes are shown: no, resp. extreme longshore diffusion. Furthermore, the gradual erosion of the whole beach and inshore is displayed (interrupted line).

The seaward limit of the nourishment coincides with the 5 m depth contour. The coastline is protected by groins, revetments and sea-walls. For the largest part the area is inside the "Riffbogen" (bars formed by the outer delta). Wave attack is important on the North West side.

At the South West side (Norderneyer Seegat) current dominates.[31,34,35,46]

Since the first artificial nourishment (1.25 mm^3 in 1951/1952) (major) sand supply in this area and its vicinity was repeated 6 times (incl.1992). Figure 5.63 shows the area of coastal nourishment in 1992: an amount of $0.5 \times 10^6\,m^3$, of which 50% supplied above NN-1 m (as NAP the level NN is about mean water level). Ca. 20% was supplied between NN–3 m and NN–5 m.

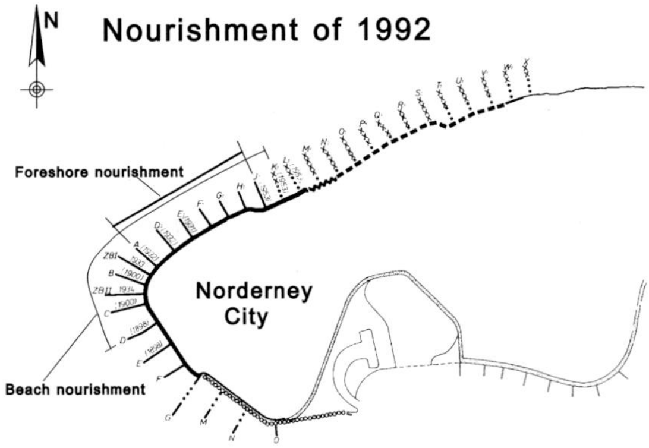

Groins	Date from (ca):
B to E	1900
ZBI & II; A,D1,E1	1933
J1, K1, L1	1953

Fig. 5.64 Coastal protection at the Westend of Norderney and: combined beach and shoreface nourishment.

Wave climate and wind set-up. The outer delta largely shelters the considered area from the deep water wave field. However, even more than at Terschelling, high waves and high surge levels are often linked together. Kersting *et al.*[29] compute the sand transport using the SON deep water wave climate as well for Norderney as well as Terschelling. However, storm surge levels at Norderney are higher than at Terschelling. Thus, if Norderney would have been located near Terschelling, its (present) outer delta would give better protection; it would reduce wave height at the Westpoint shore more than in the present situation. Kersting et al. estimate that the effect of location on the surf height results (for the same wind, i.e. offshore wave height) in surge top water level rises, being 22% higher at Norderney than at Terschelling. They use theoretical considerations[41,51] as well as analysis of the SON wind and wave data.

Sand transport computations. Schematization: investigations concern the coastal area with a length of ca. 5 km between the tidal inlet and the site where the outer delta merges with the island (between points I–IV; Fig. 5.65). The wave and climate on the inshore and the beaches in this area is strongly influenced by the bars of this outer delta. These bars reduce the energy of the incoming waves and also influence the propagation direction of the waves by refraction.

Figure 5.65b,c shows the schematization chosen; contour lines are schematized to sets of parallel straight lines. Furthermore, Fig. 5.65a displays the real contour lines and a refraction diagram, from Niemeyer,[34] which indicates an analogy in similarity of hydraulic response between

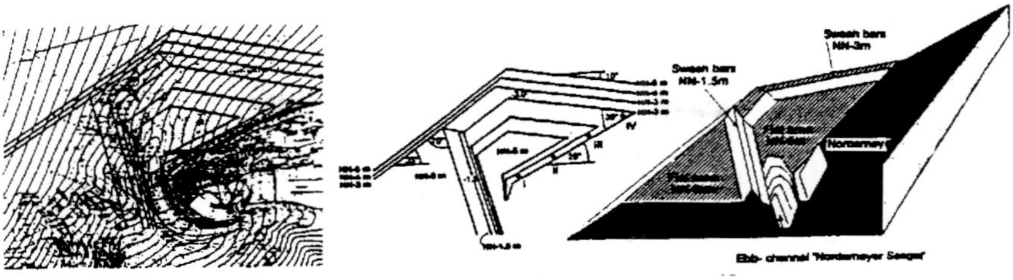

(a) (b) (c)

Fig. 5.65 (a) Refraction pattern. (b) Schematisation. (c) Schematisation (3D).

schematization and prototype. Further research with respect to this aspect is proceeding. Convergence and divergence of wave rays are, in the schematization, partially taken into account: difference in direction of beach and inshore is implemented. However, the effect of the curvature of beach and inshore on the refraction coefficient is not taken into account. By the chosen schematization, the three parts of the outer delta together act as a convex lens, focussing the wave rays. From the refraction diagram of Fig. 5.64a the same picture arises. Because of this, a convex beach, making everywhere (about) the same angle with the wave crests will be stable, where for an offshore topography with totally straight contour lines, parallel to the beach, a straight beach would have been stable. This explains the convex character of the Westend of the island.

Up to now, only the transport along the shore (beach and inshore) of the island has been calculated. The transport along the seaward shoreface of the outer delta will be calculated in the future. Again, the Svasek method of computation has been used (cf. Subsection 4.3.2) Effects of current are not taken into account. Preliminary computations concerning diffusivity constants $s_{1,2}$ have been carried out (for point IV; Fig. 5.64b) by varying the direction of beach or inshore as done in Chapter 4. Apart from this, diffusivity constants for Norderney will be calculated from measured data.

Results. Figure 5.65 shows the calculated transport capacities in the points I–IV , which would occur, if the coastline is assumed straight. It shows that accretion in the coastal area between points I and IV would result (according to the computations). Longshore diffusivity constants s_1 and s_2 in point IV are found, equal to 0.6×10^6 and $2.8 \times 10^6\,\mathrm{m^3/yr/rad}$ for beach and inshore, respectively. From the values calculated one would find, that a convex coast with a difference in coastal direction of ca. 9°–12° between points I and IV (Fig. 5.67b) would give a stable coast. The present difference in coastal direction is ca. 15° (which is greater) thus erosion (as occurs) can be expected according to the computations, refer to.[29] As the accuracy is low, the result has only qualitative importance.

Because of the existing groynes the real transport, along the inshore, will be much less than the transport capacity. Bakker and Joustra,[5] comparing coastal erosion along the dutch coast before and after the construction of groynes, indicate that groynes might reduce coastal transport with a factor of 4. This would reduce the value of s_2 (in case of erosion) in the same way.

Line modeling. Coastal constants; relation between curvature and erosion: Bakker,[2] describes another way to find the coastal constants: from coastal measurements. The data for volumetric change and curvature are related; latter taken as difference of coastal direction $\Delta\phi$ between left-handside and right-handside of the stretch. In order to obtain uniformity of data, a stretch of (only) 890 m (between groynes "A" and "H1") near the Western end of Norderney has been analyzed. Figure 5.66 displays this relationship; from this figure Kersting *et al.*[29] derive values of $s_1 = 1.2 \times 10^6$ and $s_2 = 0.9 \times 10^6\,\mathrm{m^3/yr/rad}$. As for the first observation leading to s_2 the groynes will have been less effective (the supply extended further seaward than the groynes), this observation has been discarded, leading to $s_2 = 0.2 \times 10^6\,\mathrm{m^3/yr/rad}$. Based upon Fig. 5.67d, s_3 has been taken zero. The constants $s_{y1,2}$ (both assumed as 4 m/yr) have been found by estimating T_{01} from the measurements and by iterative approximation, using

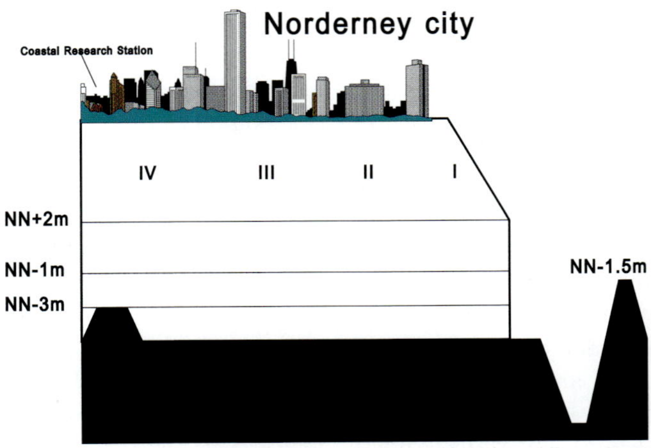

Fig. 5.66 Transport capacity (without groynes), if the coast is assumes straight.

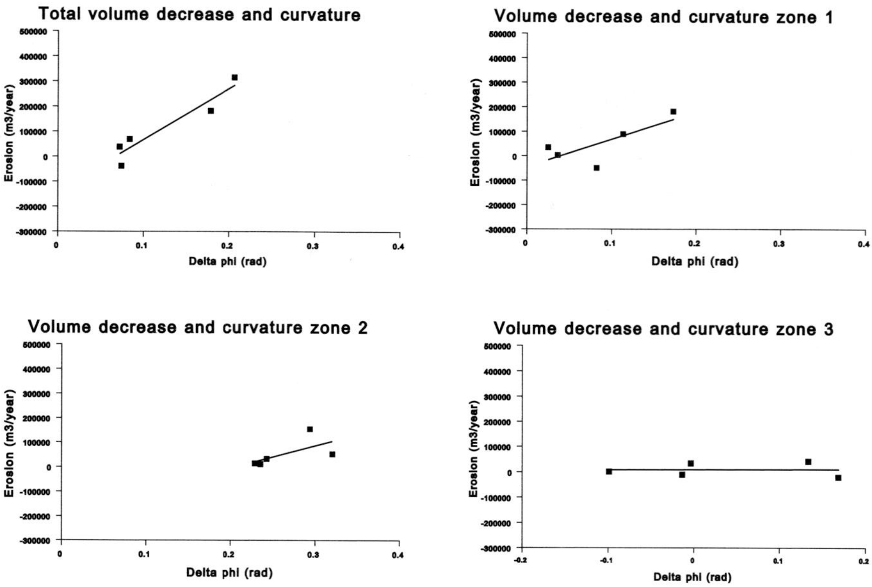

Fig. 5.67 (a–d) Volume decrease & curvature between certain reference levels.

the numerical three-line model (Chapter 3.4). Initial and boundary conditions are imposed by prototype data.

Results. In the Norderney case, "$y_{1,2,3}$" will be defined as the surplus volume remaining after the supply, (expressed in $\mathrm{m}^3/\mathrm{m}'/\mathrm{m}$ depth) between two successive separation planes (see Fig. 5.66). Comparison between measurement and computation gives ($y_{1,2,3} = f(x)$; t as parameter) and Fig. 5.68 ($y_{1,2,3} = f(t)$; x in the center of the field). For y_1 and y_2 good agreement is found; y_3, which in the model can only change because of cross-shore transport, appears to remain too large in the model. This suggests scour by current in the area below the groynes.

The NOURTEC cases: discussion and conclusions. Terschelling: depending on the assumptions (of which the extremes are: "only cross-shore diffusivity" and "cross-shore plus unfavorable high value of longshore diffusion", a "life time" of the supply can be estimated. This varies respectively between 13 and 7 years (estimating the autonomous erosion as $2.75\,\mathrm{m/yr}$).

Norderney:

- Concerning Norderney only preliminary conclusions and some indications can be given. Comparing longshore coastal constants, derived from

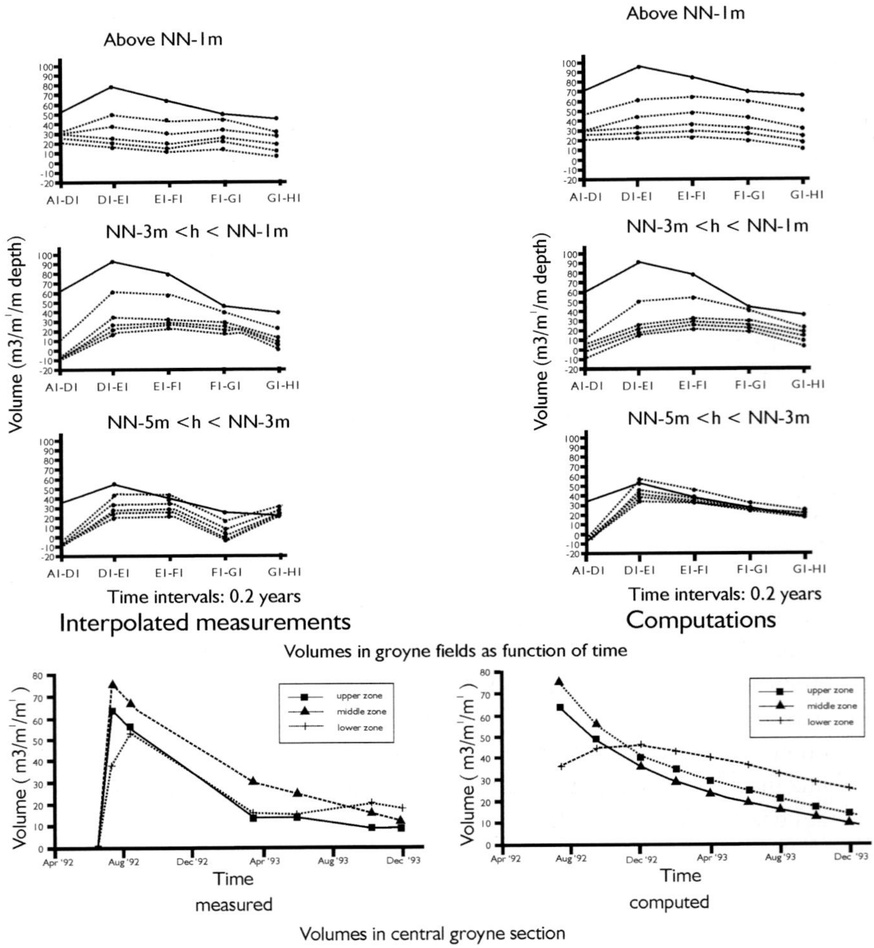

Fig. 5.68 (a–d) Volumes in central groyne section.

transport formulae and from evaluation of topographic field data the s_2-value, found from those field data is lower than expected (even considering the effect of the groynes), where the s_1-value is considerably larger. The total of $s_1 + s_2$ is in the expected range.

- The coastal measurements indicate much larger longshore diffusivity on the beach (above NN − 1 m) than on the inshore (NN − 3 m $< h <$ NN − 1 m).
- In the future effects of currents should be carefully examined, especially concerning the zone seaward of the groynes. Three-line model (which is able to reproduce the development of beach and inshore rather well) underestimates (without current effects) the erosion of the deeper zone.

General:

- Comparison of coastal constants in general for various stretches of coast is given by Bakker *et al.*[11] In that scope, the maximum values of the longshore diffusivity assumed for Terschelling are rather high: values of $1 - 2 \times 10^6$ are more common.
- The cross-shore transport constants $s_{y1,2}$ are (physically) strongly dependent upon the depth of the separation plane, as well on the schematization. Compared to the values, mentioned by Bakker *et al.*,[11] the cross-shore transport constants assumed for Terschelling are somewhat low. For instance, using Kalman filtering, de Vroeg[49] finds $s_y \approx 0.7$ m/yr at NAP.
- 0.72 m ($T_{01} \approx 7$ years) for the Dutch Rijnland-coast. Compared to this, the value of $y_{1,2}$ of 4 m/yr at NAP.
- −3 m ($T_{01} = 0.4$ year and $T_{02} = 0.25$ year) seems extremely high. Up to now, T_0-values of the order of 1 year were the smallest ever found (in Cadzand,[16] $s_y = 3$ m/yr at NAP −2 m). As in Norderney, in Cadzand a tidal channel is near to the coastal area. In the Norderney case, the wave attack is probably larger than in Cadzand.

The difference in character between the supplies of Terschelling and Norderney is remarkable:

- Supplying high on the beach instead of on the inshore would have been favorable for the Terschelling case if the costs of supplying on beach or inshore would have been equal. In Norderney, however, this is different because (most probably) the groynes largely reduce the longshore diffusivity of the inshore (compared to the beach).
- Probably the deeper zone in Norderney will be shown to be more active (current effects near to the Norderneyer Seegatt) than at Terschelling.

References

1. Bakker, W. T., *The dynamics of a coast with a groyne system*, London, 11th Conf. on Coastal Engineering, Ch. 31, 1968.
2. Bakker, W. T., *A mathematical theory about sand waves and its application on the Dutch wadden Isle of VIieland*, Shore and Beach, Oct. 1968b.
3. Bakker, W. T., *Calculation of the longshore wave driven transport using the method of parallel depth contours*, (in Dutch), Rijkswaterstaat (Dutch Ministry of Public Works), Dept. for Coastal Res., Rep. WWK 69-7, 1969.
4. Bakker, W. T., E. H. J. Klein Breteler and A. Roos, *The dynamics of a coast with a groyne system*, Washington, 12th Conf. on Coastal Engineering, Ch. 64, 1970.
5. Bakker, W. T. and D. Sj. Joustra, *The history of the Dutch coast in the last century*, Washington, Proc. 12th Conf. on Coastal Eng., 1970.
6. Bakker, W. T., Computation of the wave-induced littoral drift according to the Svasek method, assuming parallel depth contours, Rijkswaterstaat (Dutch Ministry of Public Works), Dept. for Coastal Res., Study Rep. WWK 71-2, 1972.
7. Bakker, W. T., The influence of longshore variation of the wave height on the littoral current, Rijkswaterstaat, Dir. Water Management and Hydraulic Research, Dept. for Coastal Research, Study, Rep. WWK 71-19, 1971.
8. Bakker W. T., H. G. H. ten Hoopen and G. R. H. Grieve, Berekening van het zandtransport volgens de methods Svasek bij een strand en een vooroever, die een hoek met elkaar maken. "(Computation of the sand transport according to the Svasek method, when beach and inshore make an angle with each other (in dutch), Rijkswaterstaat, Dir. for Water Management and Hydraulic Res., Dept. for Coastal Res. Study Rep. WWK 71-18, 1971.
9. Bakker, W. T. and G. Delver, *Coastal changes, caused by a shallow water sanddam in front of the Delfland coast*, Delft University of Technology; Coastal Engineering Group, 1986.
10. Bakker, W. T. and J. H. de Vroeg, *Coastal modeling and coastal measurements in the Netherlands*, Proc. Seminar SOGREAH: Prototype measurements to validate numencal models of coastal processes, Grenoble, 1987.
11. Bakker, W. T., C. van der Kerk and J. H. de Vroeg, *Determination of coastal constants in mathematical line models*, 2nd European workshop on Coastal Zones, as related to Physical Processes and Coastal Structures, 1988.
12. Bakker, W. T. and N. F. Kersting *Pilot study concerning the behaviour of coastal supply on the Dutch Wadden island of Terschelling*, Delft University of Technology, Coastal Engineering Group, 1994.
13. Bakker, W. T., M. F. Kersting and H. D. Niemeyer, *Line modeling of shoreface nourishment*, Kobe, Proc. 25th Int. Conf. of Coastal Engineering, 1994.
14. Barcelo, J. P., *Experimental study of the hydraulic behaviour of groyne systems*, London, 11th Conf. on Coastal Engineering, Ch. 13, 1968.
15. Byrne, R. J., Field occurrences of induced multiple gravity waves, *J. Geophys. Res.* **74**(10), 2590–2596, May 15, 1969.

16. CUR, Rijkswaterstaat; WL | Delft Hydraulics, *Manual on Arificial Beach Nourishment*, 1986.
17. Van Bochove, G., *Elaborations of a model test conducted by Barcelo*, The Hague, Rijkswaterstaat, Dir. Water Management and Hydraulic Research, Section Coastal Research, Memo 72-2, 1972.
18. Bijker, E. W., Longshore transport computations, *Proc. A.S.C.E.*, **97**(ww 4), 687–701, Nov. 1971.
19. Castelyn, J. A., *Numerical computations of the changes of coastal shapes under the influence of wave attack*, Delft University of Technology, Coastal Engineering Group, Rep, 75-1 (part I, II and III), M.SC. Thesis, 1975.
20. Dorrestein, R., *Wind and Wave Data of The Netherlands Lightvessels*, Royal Dutch Meteorological Institute, Mededelingen (announcements), No. 90, 1964.
21. Dijkman, M. J., W. T. Bakker and J. I. L. de Vroeg, Two-line model for the Holland Coast (in Dutch). Voorspelling ontwikkeling kustlijn 1990–2090, *(Development prediction coast-line 1990–2090)*, Rijkswaterstaat, fase 3, part. Rep. 3.2 (of Tech. Rep. 5), 1989.
22. Galvin, C. J., *Classification of Breaking Waves on Three Laboratory Beaches*, Res. Divn, 1967.
23. Groenewoud, M. D., W. T. Bakker, J. van de Graaff, R. Spanhoft and Chr. Laustrup, *Evaluation of shoreface nourishments by linemodelling*, Orlando, Proc. 25th Int. Conf. of Coastal Engineering, pp. 2857–2870, 1996.
24. Hartsuyker, G., *Tidal Flow Model for the Holland Coast* (In Dutch), WL | Delft Hydraulics, 1988.
25. Hokke, A. W. and A. P. Roskam, *Measured Deep Water Wave Climate* (In Dutch), Rijkswaterstaat, Tidal Waters Division, GWAO 87.291, 1987.
26. Ten Hoopen. H. G. H and W. T. Bakker, *Erosion problem of the Dutch Island of Goeree*, Copenhagen, Proc. 14th Int. Conf. on coastal Eng., 1974.
27. Hulsbergen, C. H., *Origin, effect and suppression of secondary waves*, Copenhagen, 14th Conf. on Coastal Engineering, Ch. 22, 1974.
28. Hulsbergen, C., W. T. Bakker and C. van Bochove, *Experimental verification of groyne theory*, Honolulu, Proc. 15th Int. Conf. of Coastal Engineering, 1976 (Also: WL | Delft Hydraulics, Rep. 176).
29. Kersting, N. F., W. T. Bakker and H. D. Niemeyer, *Evaluation of a Combined Foreshore and Beach Nourishment at the German Wadden Island of Norderney*, Delft University of Technology; Coastal Engineering Group *and* Coastal Research Station, Norderney, 1995.
30. Kroon, A., P. Hoekstra, K. T. Houwman and B. G. Ruessink, *Morphological Monitoring of a shorefce nourisment — NOURTEC experiment at Terschelling, The Netherlands*, Proc. 24th Int. Conf. on Coastal Eng, Kobe, Japan, pp. 2222–2236, 1994.
31. Luck, G., *Inlet changes of the East Frisian Islands*, Honolulu, Proc. 15th Int. Conf. on Coastal Eng., 1977.
32. Mulder, J. P. M., J. van der Kreeke. and P. van Vessem *Experimental shoreface Nourishment. Terschelling (NL)*, Kobe, Proc. 24th Int. Conf. on Coastal Eng., 1994.

33. De Mulder, E. J. F., *Geological history of the Hondsbossche Seawall* (in Dutch), 'Vrienden van de Hondsbossche', 3rd edn., Rijks Geologische Dienst, 1983.
34. Niemeyer, H. D., *Ausbreitung and Dampung des Seegangs im See- und Wattengebied von Norderney (Spreading and attenuation of swell around Sea and Wadden of Norderney, in German)*, 1986.
35. Niemeyer, H. D., *Morphodynamics of tidal inlets "course" Coastal Morphology*, Delft University of Technology, Proc. COMETT-course, 1990.
36. Patzold, U., *Erosion von sandigen Küsten und Kliffs*, Master Thesis, 1980.
37. Pelnard Considere, R., *Essai de theorie a l'evoLution des forms de rivages en plages de sable et de galets*, Paris, Quatrieme Journees de 'I Hydraulique, Les Energies de La Mer Question 3, June 13–15, 1954.
38. Pilarczyk, K. W. and J. van Overeem (Ed.), *Manual on Artificial Beach Nourishment*, Rijkswaterstaat, Delft Hydraulics, CUR, 1986 (N.B.: Annexes!, ca. 500 pages background information.
39. Roskam, A. P., *Golfklimaten langs de Nederlandse kust (Wave climates along the dutch coast; in Dutch)*, Rijkswaterstaat, Dept Getijdewateren, 1988.
40. Rijkswaterstaat (Dutch Ministry of Public Works), *Kustverdediging na 1990 (Coastal Defense after 1990)*, 1989.
41. Schalkwijk, N., *A Contribution of The Study of Storm Surges on The Dutch Coast* Royal Dutch Meteorological Institute, Med. en Verh. SeriesB, part I No. 7, 1947.
42. Smit, E. S. P., *De rol van het sediment dwarstransport in de ontwikkeling van de Hollandse kust (The effect of cross-shore transport for the development of the Dutch coast; in Dutch)*, Delft University of Technology; Coastal Engineering Group, 1987.
43. Stive, M. J. F., J. A. Roelvink and H. J. de Vriend, *Large Scale Coastal Evolution Concept*, 1990.
44. Svasek, J. N. and E. W. Bijker, *Two methods for determination of morphological changes induced by coastal structures*, Paris, XXIInd Int. Nav. congress, Section II, item 4, 1969.
45. Swart, D. H., *A schematization of onshore-offshore transport*, Copenhagen, 14th Conf. on Coastal Engineering, Chapter 51, 1974.
46. Thilo, R. and G. Kurzak, *Die Ursachen der Abbrucherscheinungen am West- und Nordwest-Strand der Insel Norderney (the causes of erosion at the West- und North-west-beach of the island Norderney (in German)*, Die Küste, 1952.
47. Tomlinson, J. H., *Groynes in Coastal Engineering, A Literature Review and Summary of Recommended Practice*, Report No. IT 199, Hydraulics Research Station, Wallingford England.
48. Van Vessem, P., *Morfologische analyse kustvak Terschelling, raai 13 tot 20 (Morphological analysis coastal area Terschelling; range 13 to 20; in Dutch)*, Rijkswaterstaat, Tidal Waters Division, 1992.
49. de Vroeg, J. H., *The use of Kalman filtering for investigation of coastal behaviour*, Delft University of Technology, Coastal Engineering Group, 1987.
50. De Vroeg J. H., E. S. P. Smit and W. T. Bakker, *Coastal genesis*, Malaga, 21st Int. Conf. on Coastal Eng., 1988.

51. Weenink, M. P. H., *A Theory and Method of Calculation of Wind Effect on Sealevels in a Partly Enclosed Sea, with Special Application to The Southern Coast of the North Sea*, Royal Dutch Meteorological Institute, Med. and Verh., No. 73, 1958.
52. WL | Delft Hydraulics, The influence of groynes on the coast (in Dutch), Report M 918 part I (experiments) March 1976.
53. WL | Delft Hydraulics, Comparison of theoretical and experimental coast lines (in Dutch), Report M 918, part lll, 1976.
54. WL | Delft Hydraulics, Berekeningen van de kustlijn ten Oosten van Zeebrugge (Computations of the coastline East of Zeebrugge), Report R 1301, May 1978.

Chapter 6

Evaluation of the One- and Two-line Model

N.B. notation: mind the essential difference in transports S_t, S_b, S_s, S_y on one hand and transport S_1 and S_2 on the other hand. The first-mentioned category of symbols denote (local, specific) transports per unit of width; the second-mentioned category transports integrated over a total zone; like the beach or the inshore. Thus transports of the first-mentioned category have the dimension [m^2/s] and of the last-mentioned category the dimension [m^3/s].

6.1. Introduction

In this evaluation of the one- and two-line model system, three aspects will be highlighted:

(a) Some questions concerning assumptions will be considered (Section 6.2):
(b) The effect of variable wave conditions on coastline development will be treated (Section 6.3):
(c) The two-line theory originally has been developed in order to simulate the behavior of a coast with groins. Independent of the question, whether a groin system is the most feasible way of coastal protection, it is worthwhile to remind the conclusions concerning the behavior of groins, as this deepens the insight in coastal behavior in general (Section 6.4).

Consideration of (a). (Section 6.2): the following questions arise:

(1) Basis of the former chapters has been the validity of the CERC-formula and the Svasek method for the calculation of the longshore transport. How the results would have been changed, if for instance, the Bijker method had been used? Section 6.2.1 deals on this subject.
(2) Is wave refraction near the coast simulated in an acceptable way? Two aspects (at least) ask for consideration:
    - the effect of the inshore on the transport along the beach (Section 6.2.2); it should be reminded, that beach and inshore in reality are not equivalent. Wave motion on a beach has a memory on the inshore, where the inverse is not the case. This element has already been considered in Subsection 4.3.6 ("Effects of schematization"). However, the results will be summarized (in words) as well in Section 6.2.2.
    - the effect of the curvature of the inshore. (Subsection 6.2.3)
(3) Which is the role of radiation stress?(Subsection 6.2.4)
(4) Which is the magnitude of the rip-currents; which is the amount of sand transported by rip currents? (Subsection 6.2.5).
(5) Which is the effect of tides? For the answer on this question, refer to Section 2.4

Under the guidance of Battjes and Bakker the questions "4" and "5" have been investigated in a number of M.Sc. Theses: by Klein Breteler,[11] Opdam,[12,13] and Voogt and Romijn[14]; a further elaboration of the effect of radiation stress is given in [6] and [7]. The first question "(a)" is treated by Bakker.[3]

Consideration of (b): variation of wave conditions (Section 6.3): Variation of wave conditions especially affects the beach shape in the vicinity of a groin or a harbor mole. If wave conditions vary in time, the boundary conditions of the coastal (differential) equation (3.6) vary, thus introducing moving sandwaves (Subsection 6.3.1)

In the case of wave shelter, caused by a harbor mole, wave conditions depend strongly on location. Remind the problem of the Holland coast (Section 5.2). In Section 6.3.2 especial attention will be given to the combination of wave shelter and time-varying wave conditions.

Consideration of (c): Effect of groins: (Section 6.4): in this section special attention will be given to the conclusions of the investigations on groins, reported by van Bochove,[10] Nearly all the elements mentioned at

(a) and (b) (above) found their application here. Results were implemented in the KC/KL program (Section 4.3.4), in the analysis of laboratory experiments mentioned in Section 5.1 and in the site investigations as Cadzand (Section 5.2.4) and the Holland coast (Section 5.2).

6.2. Remarks on the Sand Transport Mechanism Assumed

6.2.1. *Comparison between the Bijker method and the Svasek method*

Another book could be written on sediment transport formulae. The only purpose of this subsection, however is to provide something like a sensibility analysis.

In chapter 5 Bijker, characterizes the (coast-parallel) bottom transport S_{b} per m coastal profile as

$$S_{\mathrm{b}} = 5D \times \frac{V}{C_h} \times \sqrt{g} \times e^{-\left[\frac{0.27\Delta D\rho g}{\mu\tau_{cw}}\right]} \tag{6.1}$$

For *sand* grains under *prototype* circumstances the stirring parameter is very small, and then one can write:

$$S_{\mathrm{b}} \approx 5D \times \frac{V}{C_h} \times \sqrt{g} \tag{6.2}$$

Analogous to Vanoni and Einstein, Bijker formulates the suspended load as:

$$\frac{S_{\mathrm{s}}}{S_{\mathrm{b}}} = f\left(\frac{h}{K}, \frac{w}{v'_*}\right). \tag{6.3}$$

where D is the grain diameter; C_h, Chezy coefficient; S_{s}, suspended sediment transport; S_{b}, bottom sediment transport; w, fall velocity; K, an estimate of bottom roughness; $v*'$, combined shear stress velocity; V current velocity.

The combined shear stress velocity is caused by waves and current and therefore as well a function of the depth h:

$$v^{*\prime} = \left[1 + \frac{1}{2}\left(\xi\frac{u_0}{v}\right)^2\right]^{\frac{1}{2}} v^*. \tag{6.4}$$

Here, v_* is the "normal" shear stress velocity caused by current velocity and the amplitude of the orbital velocity, u_0. The parameter $>$ is given by:

$$\xi = \frac{p\kappa c_r}{\sqrt{g}} \tag{6.5}$$

in which $p = 0.45$ and K is the von Karman constant (0.4).

It is useful to get an impression of the influence of the depth (h) on S_s/S_b and therefore the quantity w/v'_* in Eq. (6.3) has been replaced by a variable, independent of h, i.e. $w/\sqrt{gk}$. Equation (6.5) has been substituted Eq. (6.4), which yields:

$$v^* = \frac{V\sqrt{g}}{C_h}; \quad u_0 \frac{H}{2}\sqrt{\frac{g}{h}} = \frac{\gamma}{2}\sqrt{gh} \tag{6.6}$$

Furthermore, it has been assumed that if $\frac{1}{2}(\xi \frac{u_0}{V})^2$ is much larger than 1 (example: when $u_0 = v$ then $\frac{1}{2}(\xi \frac{u_0}{V})^2 \approx 4.5$) and therefore the term "1" in Eq. (6.4) may be neglected.

One finds:

$$\frac{w}{v^{*\prime}} = \frac{2\sqrt{2}}{\gamma p \kappa} \times \frac{w}{\sqrt{gK}} \times \sqrt{\frac{K}{h}} \tag{6.7}$$

For given values of $\frac{h}{K}$ and $\frac{w}{gK}$ one can find $\frac{w}{v^{*\prime}}$ and thus $\frac{S_s}{S_b}$ from Eq. (6.3). Figure 6.1 shows a dimensionless plot[3]:

$$\frac{S_s}{S_b} = f\left(\frac{h}{K}, \frac{w}{\sqrt{gK}}\right) \tag{6.8}$$

As S_s in the surf zone is large compared to S_b from Eq. (6.8) for the total transport S_t (per definition $S_b + S_s$) can be derived:

$$S_t \cong S_s = S_b \times f\left(\frac{h}{K}, \frac{w}{\sqrt{gK}}\right)$$

and substituting for S_b using Eq. (6.2) yields:

$$S_t = 5\left[D \times \frac{\sqrt{g}}{C_h} \times f\left(\frac{h}{K}, \frac{w}{\sqrt{gK}}\right)\right] \times V \tag{6.9}$$

In the area shown in Fig. (6.1), $f[h/K, w/\sqrt{gk}]$ can be approximated well by $S_s/S_b = (h/K)^m$, in which m is a function of $w//(gK)$.

Under prototype circumstances, m appears to be ca. 1.2–1.65.

Furthermore for V is found Bakker[3]:

$$V = 5.5\gamma C_h h \times h_{br}^{-1/2} \sin\phi_{br} \cos\phi_{br} \tan\alpha \tag{6.10}$$

Under the present assumptions, V is proportional to h and therefore, the total transport S_t should be proportional to $h^{2.2-2.65}$. This does not deviate

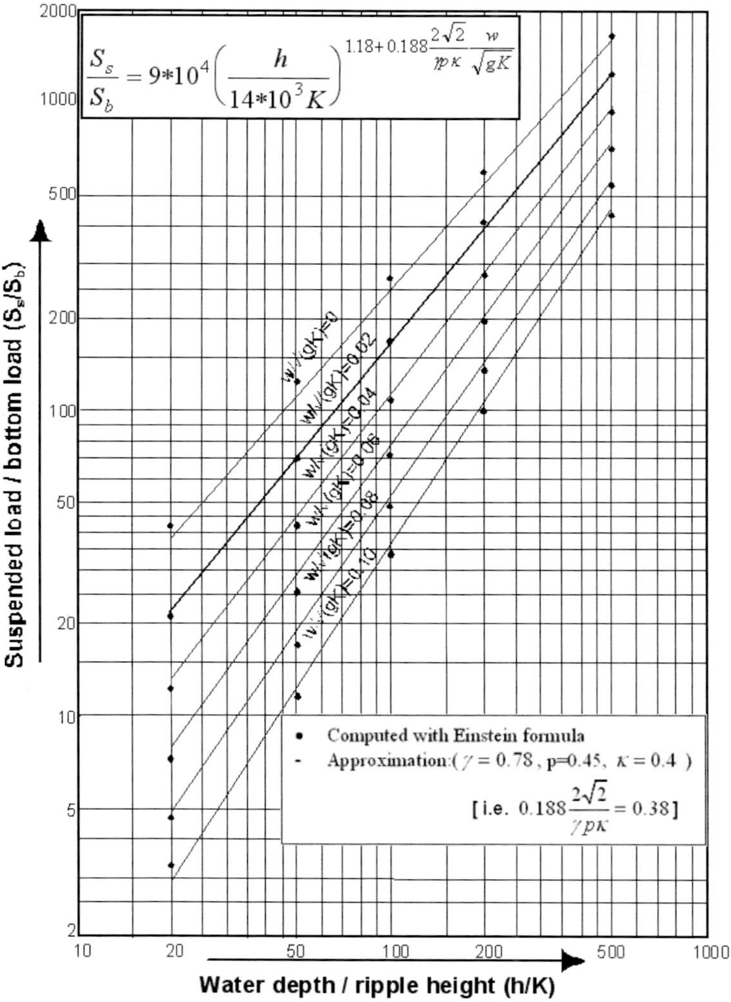

Dimensionless fall velocity $w/\sqrt{gK}$ as a function of grain diameter *d* and ripple height *K*									
D in :	*w m/s.*	*K (m)*	0.01	0.02	0.03	0.04	0.06	0.08	0.10
		$\sqrt{gK}$ m/s	0.314	0.443	0.543	0.627	0.767	0.886	0.991
100	0.007		0.022	0.016	0.013	0.011	0.009	0.008	0.007
150	0.015		0.048	0.034	0.28	0.024	0.020	0.017	0.015
200	0.023		0.073	0.052	0.42	0.037	0.030	0.026	0.023
250	0.032		0.102	0.072	0.59	0.051	0.042	0.036	0.032
300	0.0395		0.126	0.089	0.73	0.063	0.051	0.045	0.040

Fig. 6.1 Dimensionless fall velocity as a function of (d, k).

much from Svašek's assumption, that S should be proportional[a] to h^2. Also the order of magnitude of the transports as calculated via Svašek and Bijker matches rather well Bakker.[3]

A big advantage of the Bijker method compared to the Svašek method is the fact that the longshore current velocity occurs explicitly; this advantage will be shown in Section 6.2.4 concerning radiation stress.

6.2.2. *Effect of refraction on the inshore on the transport along the beach*

The transport S_1 along the beach shown by Eq. (4.26) is a function as well of φ_1 as of φ_2 (the angle the beach, c.q. the inshore makes with the x-axis). In Fig. 4.10 the relative effect of the inshore (in relation to the effect of the beach) is displayed. This "relative effect" is defined as $[\frac{\partial S_1}{\partial \delta_2} \Big/ \frac{\partial S_1}{\partial \delta_1}]$. The ratio depends upon the angle of wave incidence in deep water (in Fig. 4.10 curves are shown for a number of angles of wave incidence, 0°–75°) and of the thickness of the beach layer h_1 compared to the one of the inshore layer h_2. The horizontal scale in Fig. 4.10 represents $h_1/(h_1+h_2)$. Thus the left end of the figure represents "all inshore" (evidently then the influence is maximal) and the right-hand side a zero layer thickness of the inshore (then its influence will be zero as well). Thus the figure demonstrates, that when the relative thickness of the inshore and the angle of wave incidence are not too large, neglecting this aspect is acceptable.

6.2.3. *Effect of the curvature of the beach on the refraction*

In the theory, explained in Sections 2.4 and 4.3.3, parallel depth contours have been assumed and the effect of the refraction for oblique incident waves has been taken into account via Snel's Law.

However, the effect of convergence or divergence of wave rays has been neglected. Starting from a bottom topography, shown in Fig. 6.2, in Battjes[6]

[a]Assume in the surfzone, $\cos\phi \approx 1$; then the local energy flux in the surfzone is proportional to H^2C and the longshore component of it is proportional to $H^2C\ \sin\phi$, or expressed in h, proportional to $\gamma^2 F_r g h^{2.5}.\sin\phi$. Thus the loss of energy flux between two depth contours,. Δh apart from each other is proportional to $h^{1.5}\Delta h \sin\phi$. Assuming parallel depth contours over the considered zone, Snel's Law gives, that sin ϕ is proportional to C, i.e. to $(gh)^{0.5}$. Thus, the longshore component of the energy loss will be proportional to $h^2\Delta h$.

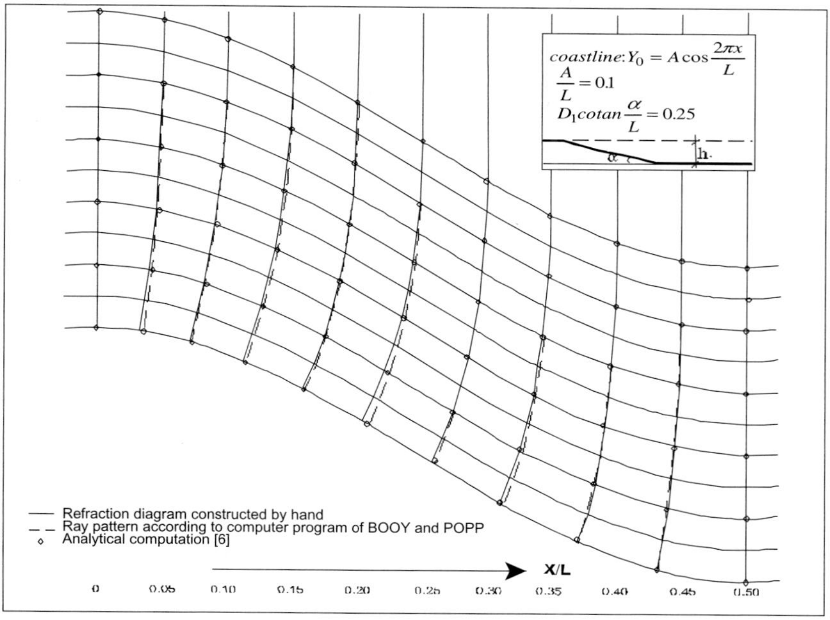

Fig. 6.2 Comparison of analytical calculated ray-pattern and other computations.

an equation for the breaker depth has been derived, analogous to (4.18a), however, taking the effect of curvature of the coast into account.

For this, the shallow water theory has been used ($C = F_r\sqrt{gh}$).

6.2.4. *Effect radiation stress*

An important element of coastal hydraulics, and thus of coastal dynamics, is "radiation stress". Especially in the case of a strong variation of wave height in longshore direction, the effect on the longshore current has to be taken into account. (Figs. 6.3 and 6.4). Instead of equation (6.10) one finds:

$$V = 5.5\gamma C_h\sqrt{h} \times \sqrt{\frac{h}{h_1}} \times \frac{dy_0}{dx} + \left(\frac{d(y_{\mathrm{br}} - y_0)}{dx}\right)\tan\alpha \qquad (6.11)$$

The following assumptions have been made:

(a) Bottom topography as shown in Fig. 6.5[b]

[b]A constant bottom slope between $h = 0$ and $h = h_1$ has been assumed.

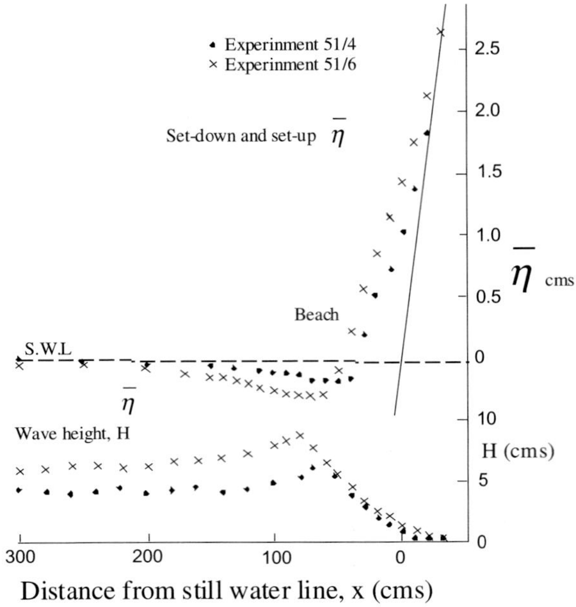

Fig. 6.3 Effect of difference of wave set-up according to Bowen *et al.*[7]

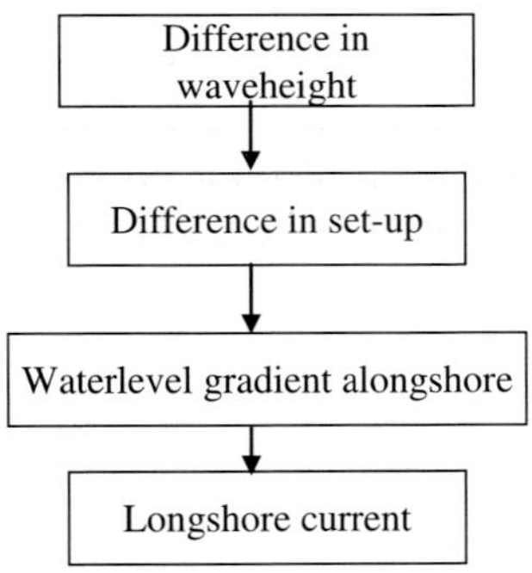

Fig. 6.4 Mechanism longshore current in the case of difference in waveheight.

(b) The coordinate system has been chosen in this way, that the wave crests at the plane at depth h_1 are parallel to the x-axis (Fig. 6.5). If h_1 would be equal to $h_{\rm br}$, the x-axis makes an angle $N_{\rm br}$ with the coast. Then the approximation holds:

$$\tan\phi_{\rm br} = \sin\phi_{\rm br} = -\frac{dy}{dx}.$$

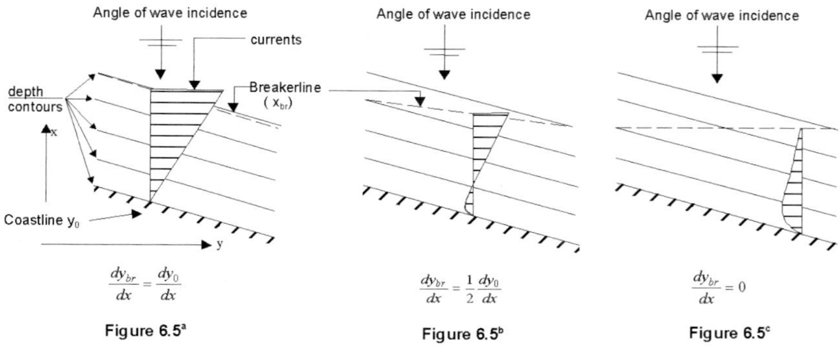

Fig. 6.5 Cross-shore pattern of longshore velocity from [5].

(c) It is assumed, that $\varphi_{\rm br}$ is small ($\cos\varphi_{\rm br} = 1$). If in that case $y_{\rm br} = y_0$, i.e. if the breaker depth along the coast does not change, Eq. (6.11) changes into (6.10).

(d) The value of γ is assumed to be so small, that $3\gamma^2/8$ can be neglected with respect to 1[c]. For a more accurate formula, refer to Battjes.[6]

(e) The effect of turbulence on the velocity field has been neglected.

Figure 6.5 a,b,c depicts the effect of the variation in width of the breaker zone on the velocity distribution in the surf zone: for a strong variation, as occurs for instance in the diffraction zone behind a breakwater the direction of the longshore current even can be opposite to the alongshore component of the propagation velocity of the waves (Fig. 6.5c).

Numerical example: for a coastal slope 1:100 a difference in breaker depth of 1 m between two points at a mutual distance of 200 m alongshore already gives a current reversal.

6.2.5. *The magnitude of the rip currents.*

The basic idea of the present rip current theory originates from Bowen.[7–9] It demonstrates the large effect of only a small variation in wave height, as treated in the former Subsection 6.2.4.

Figure 6.6 shows the pattern of a rip-current according to Bowen,[8] generated by a variation of wave height. Combination of the Bowen theory with the transport formula (6.9) results in a value for the magnitude of

[c]In the theory of Battjes,[6] this assumption has not been made; it only happens here for the sake of transparency of the formulae. Refer to Battjes,[6] for more accurate formulae.

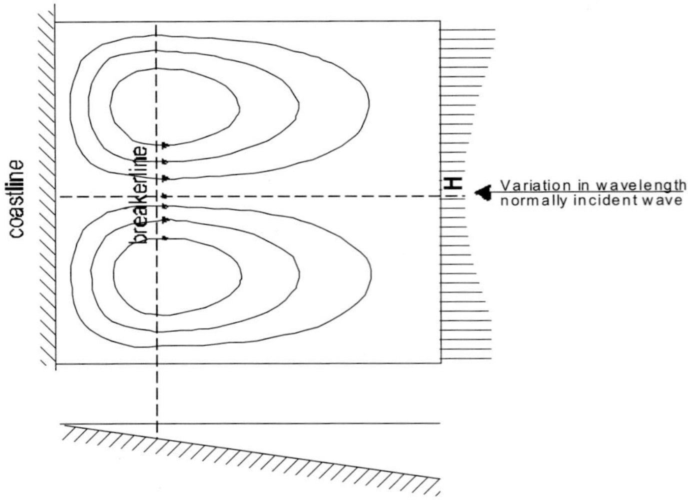

Fig. 6.6 Rip-current; according to Bowen.[8]

S_{rip}.[d] Tentative results of v.Bochove and ten Hoopen[10] show the right order of magnitude, comparing those with results found from laboratory experiments reported in Section 5.2.

6.3. Effect of Variable Wave Conditions

6.3.1. *Effect of variable wave conditions in time*

The mathematical solution of the change of a coastline near a harbor mole, caused by variable wave conditions in time has been given in Subsection 3.3.5.

$$y = \frac{\hat{\phi}\lambda_\varphi}{2\pi\sqrt{2}} \times e^{-k_\phi x} \cos(\omega_\phi t - k_\phi x) \qquad (3.85);\text{repeated}$$

Using (4.26c) and substituting Table 4.3 $A_1 = 2 \times 4 \times 10^{-2}, \gamma = \frac{0.4}{\sqrt{2}}$, $F_{\text{r}} = \sqrt{1.4}$, then one finds:

$$\lambda_\phi = 0.308\, h_{\text{br}}^{3/4} (gT_\phi^2)^{1/4} \qquad (6.12)$$

[d]As mentioned before, this book is based on former publications. Reminding of this paragraph might be of value in order to indicate, how the problem has been solved for the reported practical cases. It gives some physical insight as well. In present times, computer modelling (with implementation of inertia aspects as well) for cases like this should be used at a complementary basis (at least).

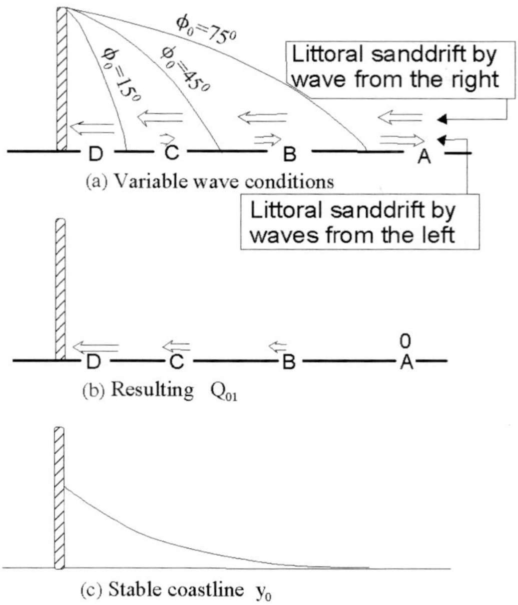

Fig. 6.7 Effect of variable wave conditions on resulting coastal shape.

If for instance one chooses 1 week for T_ϕ, (which is long for Dutch circumstances) and $h_{br} = 1\,m$, then λ equals 550 m.

As the attenuation of this sandwave is very large, it means that at a distance of $\lambda_\phi/2$ practically no effect will be perceivable. This implies, that as long as the features to be investigated are larger than $\lambda_\phi/2$, it is justified to calculate with time-average features.

Thus the effects, to be considered in the next subsection will prevail in most cases.

6.3.2. *Effect of wave shelter; conditions depending on location*

Apart from the effects of periodical change of wave direction (Section. 6.3.1), a second influence of a groin or harbor mole will be the wave-shelter. Assume first, that the sheltered area is large with respect to $1/2\ \lambda_\varphi$ the "constants" S_0 and s_1 become functions of x, called $S_{01}*$ and s_1*.

Now first, the effect of diffraction will be treated, and afterwards the effect of time-varying wave conditions, combined with this shelter effect.

Considering Fig. 6.7. The influence of diffraction will be neglected with respect to the influence of changing wave conditions.

The computation of S_0 and s_1 in area A can be performed as explained in Subsection 4.3.2, using Eq. (4.5). But applying Eq. (4.5) to area B, all wave classes with $\varphi_0 \geq 75°$ must be excluded in the summation, in area C all wave classes with $\varphi_0 \geq 75°$, and so on.

When, for instance, the resulting transport in area (A) would be zero for a coastline parallel to the x-axis (Fig. 6.7), the transport $S_{01}*$ $(\partial y_1/\partial x = 0)$, (denoted with an asterix because it changes in x-direction) will be larger and larger (in negative direction) in the areas (B, C and D) and also s_1* will change.

Now we have returned to the normal KC/KL program, only with other values for y_0 and y' than in the diffraction cases.

The stable coastline y_0 can be found from continuity: for y_0 the transport is everywhere the same:

$$[S_{01}]_{x=\infty} = S_1 \rightarrow \quad [S_{01}]_{x=\infty} = S_{01}^* - s_1 \times \frac{dy_0}{dx} \rightarrow \quad \frac{dy_0}{dx} = \frac{S_{01}^* - [S_{01}]_{x=\infty}}{s_1^*} \tag{6.13}$$

From (6.13) the stable coastline y_0 can be found, which gives the initial value of the unstationary part y_1'.

In case of changing wave conditions, and no resultant drift, the final coastal shape near one single groin will become just the stable shape y_0, because then everywhere the resultant drift is zero. This will give accretion on both sides of the groin, which will be withdrawn from a very long stretch of coast (Fig. 6.8a). in case of a row of groins, the sand for the accretion near the groin is withdrawn from the area in the sides between the groins, and only near the boundary of the groin system some real accretion can be expected (Fig. 6.8b). However, after some time this sand will move to the areas between the groins, and so this shelter effect may give some accretion (in case of no resultant drift), starting from the boundaries of the groin system.

In case $1/2\ \lambda_\varphi$ is not small with respect to the sheltered area, the best way of computation is a kind of "hindcasting", using the one-line or two-line computer program described before, and changing the wave conditions during the program. This has been done at the Coastal Research Department of Rijkswaterstaat.

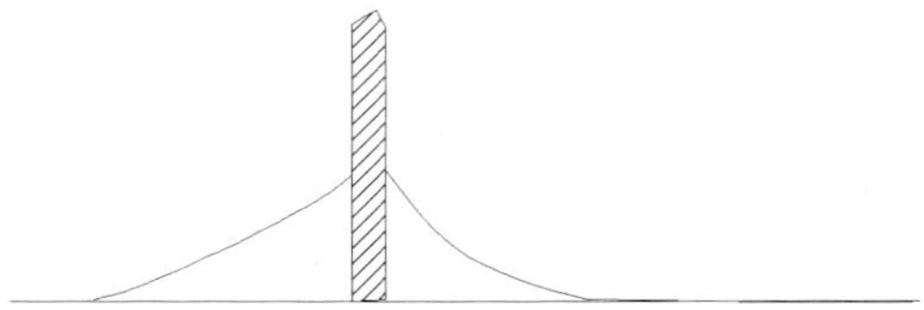

(a)

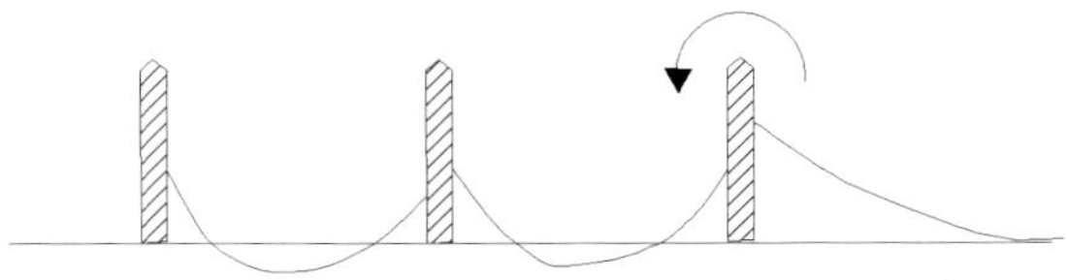

(b)

Fig. 6.8 Stable shape of coastline in case of changing wave conditions. (a) Near a single groin or harbor mole. (b) Near a groin system.

6.4. Effects of Groins as Result of the One- and Two-line Theory

Consider a coast, where over a long stretch groins are constructed at time $t = 0$. What is the influence of the groins on the coast?

Three influences may be distinguished, which can be superponed (for all influences have been assumed linear):

(1) the influence of external causes;
(2) the influence of oblique incidence of the waves (stationary transport);
(3) the influence of the shape of the coast.

At (1) external causes: assume a wave direction, perpendicular to the coast and a straight coastline at $t = 0$ (Fig. 3.8).

Assume, that by one or another external cause (for instance a tidal channel embouching left of the left boundary in Figs. 3.7 or 3.8) the coastline would have become y_0 in Fig. 3.7.

Refer to Section 3.2.5 for dealing with this kind of problem.

At (2) the influence of oblique wave incidence:

At (3) the influence of the shape: according to the one-line theory groins reduce the coastal constant (cf. Section 3.2.5 concerning the effects).

Figure 6.11 shows the results of the erosion of a convex coast according to the two-line theory. It is assumed, that before $t = 0$ beach and inshore

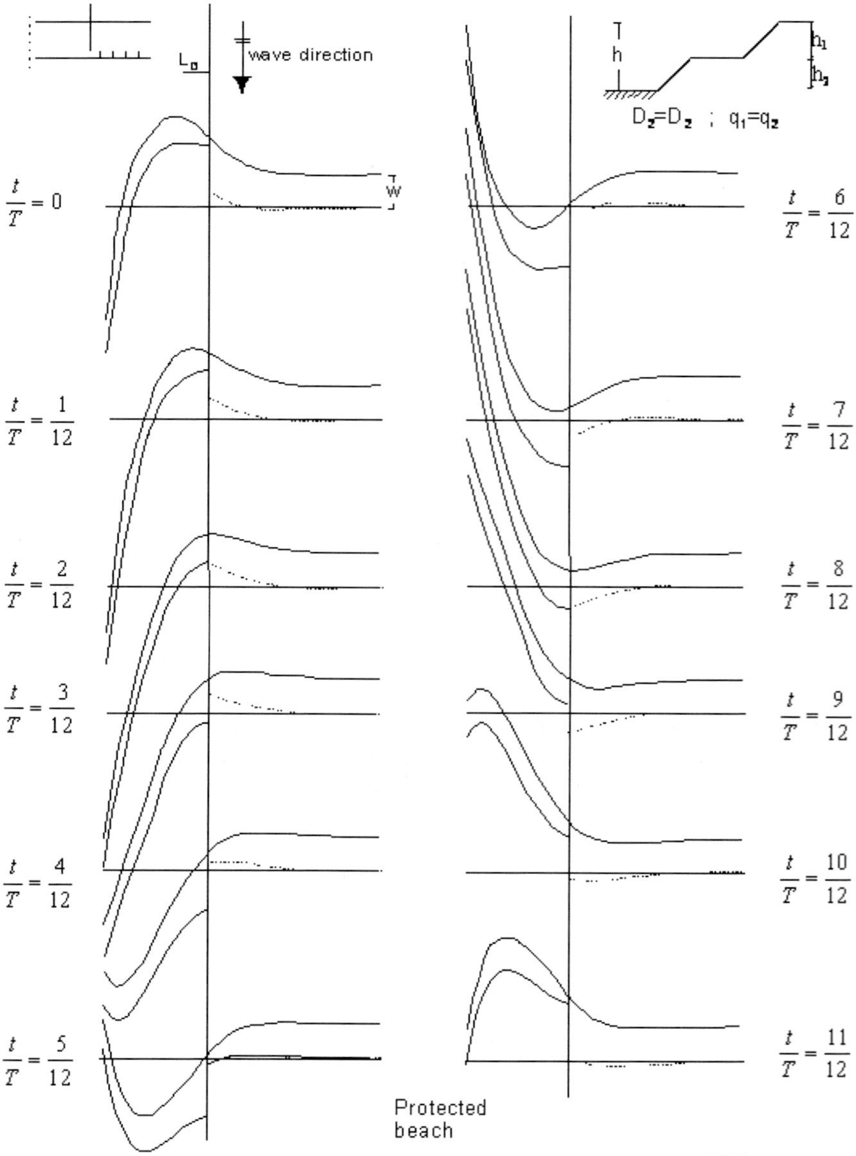

$$T = 4\pi T_0 = 4\pi \frac{h_1 h_2}{S_y h} \qquad 1 = L_0 = \frac{1}{\sqrt{S_y\left(\frac{1}{h1} + \frac{1}{h2}\right)}} = \sqrt{\frac{sT_0}{h}}$$

Fig. 6.9 The solution of sand waves, following from the two-line theory for this kind of cases.[1]

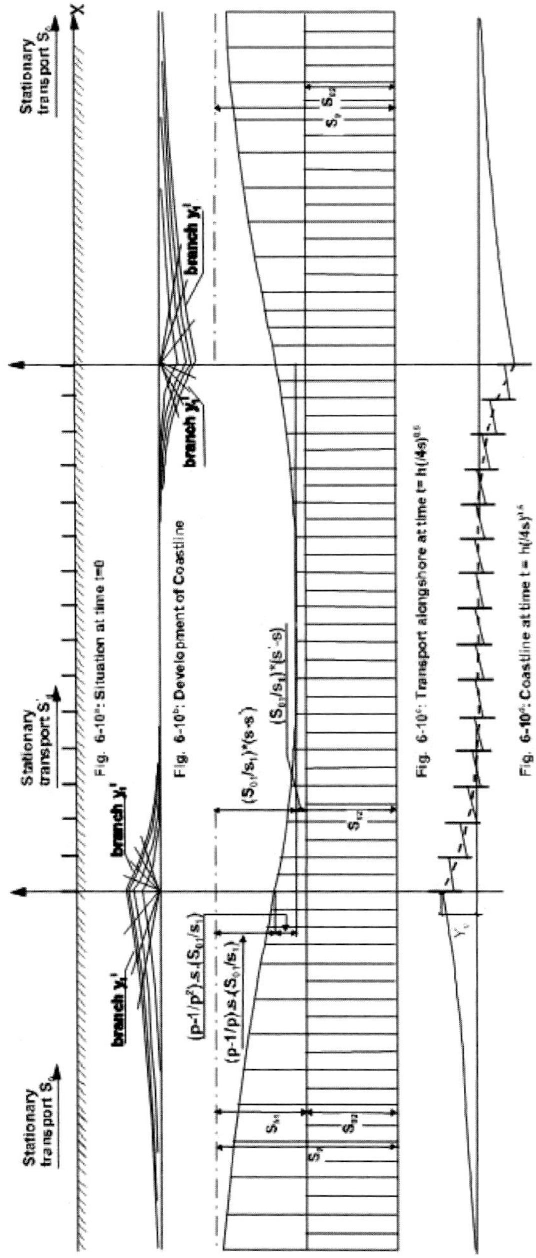

Fig. 6.10 The kind of solution, which is based upon the solution given in Subsection 3.2.5.

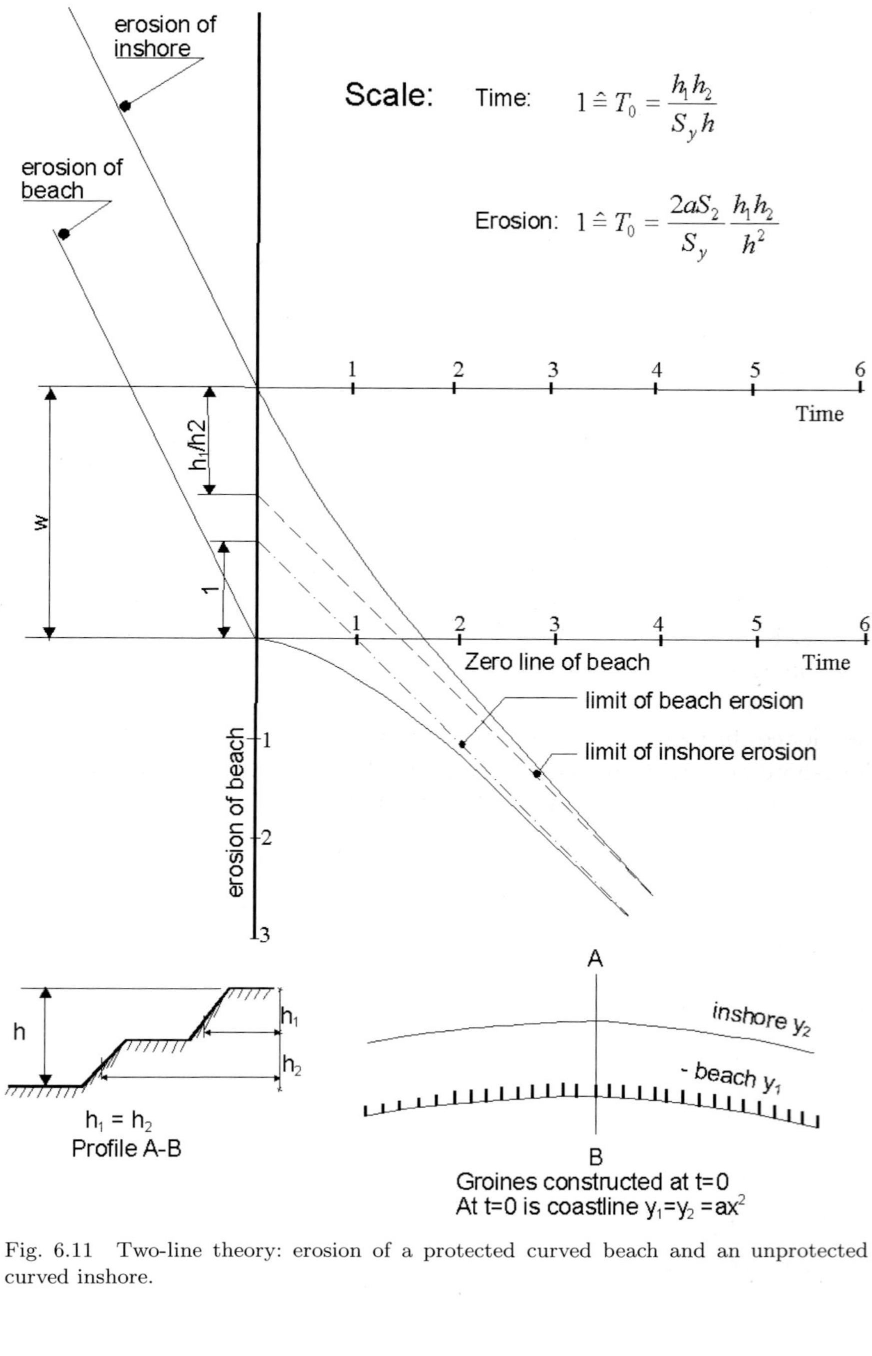

Fig. 6.11 Two-line theory: erosion of a protected curved beach and an unprotected curved inshore.

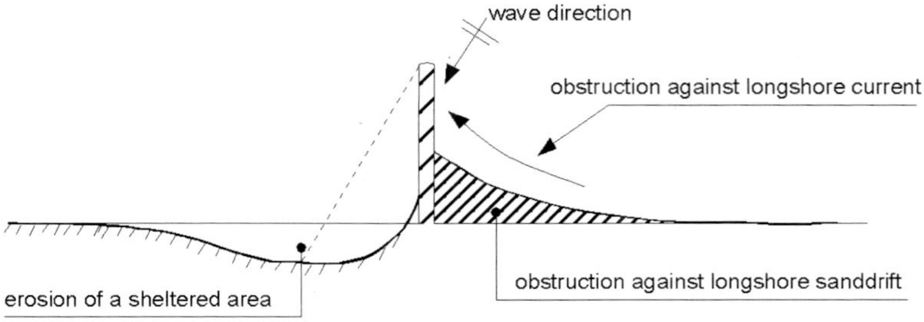

Fig. 6.12 The effects of construction of a groin.

have the same parabolic shape and an equilibrium profile. Both erode with the same velocity. At time $t = 0$ groins are constructed at the beach, near enough to each other to prevent all transport along the beach. The erosion along the inshore is not stopped, however, the profile becomes too steep and sand moves from beach to inshore.

The erosion of the inshore diminishes and the erosion of the beach starts again. Eventually, beach and inshore erode together, but the profile remains steeper than the equilibrium profile; the total rate of erosion is less than before, in accordance with the result of the one-line theory (remind the same kind of effect as found by Smit with more-line theories (Section 3.4).

Looking in a more subtle way to groins, mathematical theory has been developed, bringing apart from the obstruction of longshore sanddrift and longshore current (Subsections 3.3.3 and 3.3.4) as well the formation of a sheltered area into account (Fig. 6.12). Refer to Subsection 6.3.2 for the mathematical simulation of the coastline in the sheltered area and to Fig. 5.15 concerning the kind of (numerical) results.

Near coasts with large tidal influence, also the effect of the scour hole in front of the groin should be taken into account. Up to present, the only way of implementing the scour hole has been the assumption of a large value of s_y (Zeebrugge; Norderney).

References

1. Bakker, W. T., *The dynamics of a coast with a groyne system*, London, Proc. of the 11th Conference on Coastal Engineering, 1968.
2. Bakker, W. T., E. H. J. Klein Breteler and A. Roos, The dynamics of a coast with a groyne system, Rijkswaterstaat, Department of Coastal Research, Rep. W.W.K. 70-13,1970.

3. Bakker, W. T., *Littoral drift in the surf zone*, Rijkswaterstaat, Department of Coastal Research., Rep. W.W.K. 70-16,1970.
4. Bakker, W. T. and H. J. Opdam, Over de invloed van golven en getij op het zandtransport in de brandingszone (The effect of waves and tides on the littoral drift in the surfzone), Rijkswaterstaat, Department of Coastal Research, Rep. W.W.K. 70-9,1970.
5. Bakker, W. T., The influence of longshore variation of the wave height on the littoral current, Rijkswaterstaat, Department of Coastal Research, Rep: W.W.K. 71-19,1971.
6. Battjes, J. A., Radiation stress in short crested waves. *J. Maritime Res.* **30**(1), 1972.
7. Bowen, A. J., D. L. Inman and V. P. Simmons, Wave set-down and set-up. *J. Geophys. Res.* **73**(8), 2569–2577, 1968.
8. Bowen, A. J., Rip-currents. *J. Geophys. Res.* **74**(23), 5467–5490, 1969.
9. Bowen, A. J., The generation of Iongshore currents on a plane beach. *J. Marine Res.* **74**(2), 1969.
10. Van Bochove, G. and H. G. H. ten Hoopen Report model T23 and T23a, Elaboration of the results of model T23, Delft University of Technology, Dept. of Civil Eng., M.Sc. Thesis, 1971.
11. Klein Breteler, E. H. J., Getijberekening voor de kustzone tussen Hoek van Holland en ljmuiden met behuip van het getijmodel van Booy (Tidal computation concerning the coastal area between Hoek van Holland and IJmuiden, using the Booy tidal model), Rijkswaterstaat, Department of Coastal Research, Rep. W.W.K. 70-10, 1970.
12. Opdam, H. J., Een mathematisch model van de kustzone, betreffende de invloed van golven, getij wind en corioliskracht op de stroming langs de kust (A mathemetical model for the current in the littoral zone, concerning the effect of waves, tides, wind and coriolis force), Rijkswaterstaat, Department of Coastal Research, Rep. W.W.K. 70-11, 1970.
13. Opdam, H.J., Experimentele fluoreseine metingen betreffende de invloed van golven en getli op de stroming langs de kust (Experimental fluorescine dye measurements concerning the effect of waves and tides on the littoral current), Rijkswaterstaat, Department of Coastal Research, Rep. W.W.K. 71-5, 1971.
14. Voogt, B. and A. J. Romein, De invloed van een leidam op het zandtransport (The effect of guidance dam on the sand transport to the Marsael Brega harbor), Delft University of Technology, Dept. of Civil Eng., M.Sc. Thesis, 1971 ("Over de haven van Marsa el. Brega").

Chapter 7

Interaction Between Outer Delta's and Tidal Basins

7.1. Introduction

7.1.1. *General overview*

A significant part of the world's coastline consists of systems of barrier islands and tidal inlets. The maintenance of such coasts is notoriously difficult, especially if the inlets are fixed by manmade works (e.g. jetties). Moreover, the basin usually contains a sensitive system of channels and intertidal areas of great ecological value, which is easily disturbed by anthropogenic effects. An accelerated rise of the mean sea level is another threat, because it is still unclear whether the intertidal flats will be able to follow the sea level. If so, the increased sediment demand of the basin may have detrimental effects on the barrier islands, especially if they have to be kept in place (e.g. because of inhabitation).

These are all good reasons to investigate the morphological behavior of these systems, and to try and predict their response to human interference and exogenous changes. This may explain the long history of tidal inlet research, from purely empirical work,[52,53] and classification,[39] via mathematical analyses[31,32,80,81] to highly sophisticated 3-D morphodynamic simulation models.[89,90]

By definition, simulation models work at the time scale of the constituent physical processes, in this case the tide. Hence they are rather time-consuming and difficult to control if they have to cover the time scales of decades or more which are typical for inlet systems. These models yield encouraging results on the channel dynamics, but they have not yet been able to reproduce the empirical equilibrium relationships for the cross-section of the gorge,[21,43] the intertidal area[59,60] or the volume of the outer

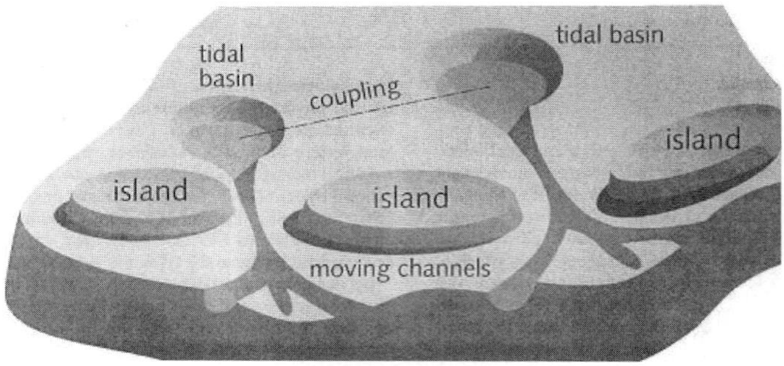

Fig. 7.1 A barrier island coast can be considered as a compound system of tidal basins, inlets, deltas and island coasts. The final system model in mind will comprehend all the elements, shown in this figure.

delta.[88] This has created a demand for a more aggregated type of model, which works at the time and length scales of the relevant morphological evolutions and includes the equilibrium state.

One line of development was initiated in The Netherlands, stimulated by the need to have long-term predictions of the evolution of the Wadden Sea. It starts from the idea that a barrier island coast can be considered as a compound dynamic system of tidal basins, inlets, deltas and island coasts (Fig. 7.1). For each of these subsystems we know a number of empirical parameters of the equilibrium state (also see Ref. [34]), and in some cases we can even estimate the time scale of evolution. Based on this information and physical process knowledge, and with the support of sophisticated simulation models, a set of semi-empirical models was developed for a single-inlet system far from resonance, one for the basin[85] and one for the outer delta combined with the adjacent island coasts.[24]

As yet, these models have not been coupled, which means that they have to be forced artificially at the interface, so in the gorge. Moreover, they apply to non-resonant single-inlet systems only. The latter is not sufficient for the Western Wadden Sea, which is a near-resonant multiple-inlet system. Therefore, a special model concept for this type of system has been developed, at the same time trying to avoid the interfacing problem in the gorge.

The present chapter outlines the concept of this model, which is based on Lorentz's[49] successful linear network model of the tide in the Western Wadden Sea. Without claiming that all quantitatively important processes are included, the basic principle of the morphodynamic evolution model is explained. The concept is shown to have the potential to become a practically useful tool, which can easily be extended with other transport mechanisms and which can be integrated with an outer delta and coastline model.

7.1.2. *Closer analysis; content of this chapter in relation to former chapters*

Consider again Fig. 7.1 the barrier island coast, consisting of a compound system of tidal basins, inlet deltas and island coasts.

Comprehending the morphological interaction between all the elements shown in this picture is the final goal of the theory developed in the present chapter.

It makes sense to simplify the problem first by splitting into elements, for instance by looking first to one of the inlets apart, Fig. 7.40. The principal elements of such a system are shown in Fig. 7.41.

Comprehension of such a system requires knowledge concerning:

a. The morphological behavior of the outer delta and the adjacent coasts;
b. Morphological tidal inlet models

Phenomenology concerning these kinds of systems has been treated in Chapter 2; Section 2.5 in particular dealt on the effect of tides.

Where the *basics* on *outer delta* behavior has been given already in Chapters 3–6, no highlight is thrown up to now on the basics of the morphological behavior of the *tidal inlets*

Caution is required for the coastal engineer, who enters the alien territory of an estuary: a territory with other laws than the one he is accustomed to:

a. "Sediment" is not equivalent to "sand" any more: silt behavior can become dominant.
b. Gradients of salt content become important; those may affect as well the tidal-averaged circulation as the coagulation of silt.
c. Tidal-averaged vertical circulation may result from the long-wave equivalent of "undertow" as occurs in short waves; this phenomenon is called "rain in the drain" in long waves (cf. Section 7.4 below).
d. Inherent to the vertical circulation (density-driven or rain in the drain[a]) in the estuary is upwelling (and downwelling); this may overrule the fall

[a]Mind that density-driven circulation and rain in the drain counteract each other.

velocity of the sediment. Upwelling can cause locally high concentrations of silt, a "silt plug".[b]

e. Horizontal sediment concentration gradients may determine the local momentaneous change in concentration at least as much as vertical concentration gradients.

The present chapter will be the mathematical/physical counterpart of Section 2.5: "phenomenology of estuaries". In its introduction, Subsection 2.5.1, "binary" selections concerning the topic were made (quote):

"Making "binary" selections on the topic "coastal dynamics; tides" instead of splitting in "phenomenology" (Chapter 2) or "mathematics" (Chapter 7) other "dimensions" are:

a. "outer deltas" or "tidal basins", assigned as **"O"(uter)** and **"I"(nner)**;

b. "linear effects" or "quadratic effects".

Altogether, this gives $2^3 = 8$ cases to be treated, of which four in Section 2.5 and the other four in Chapter 3, c.q. 7.

Thus in this Section 2.5, in the various subsections one or more of four sub-selections (O&linear, O&quadratic, etc.) can be treated. In the following paragraph for each of the subsections those four sub-selections are put in an 2*2 array; a "dot" denotes, that the sub-selection is not treated in that subsection." The Introduction of Chapter 8 (Mathematics; interaction between outer delta's and tidal basins) will give similar arrays for the sections of Chapter 7" (unquote). After that, a schedule was displayed, where "O's appeared only in the first column of the arrays; "I's in the second one (see below) (quote):

Subsection	2.5.3	2.5.4	2.5.5
Linear	O.	.I	..
Quadr.	O.	.I	O I

(unquote)

Thus, a similar table as the one in Subsection 2.5.1 now should be made for the sections of Chapter 7: Table 7.1.

However, in the case of a mathematical/physical treatment, it is necessary to treat the motion of sand and water separately. This brings a third[c]

[b] A "slib prop" in Dutch.

[c] Apart from the alternative "phenomenology" versus "mathematics".

Table 7.1 Combinations of binaries (omitted: inner&waves; saltgrad&waves; nearly all outer&no_waves)

	Sed. (3)	Avrgd (2)	Saltgrad (4)	Multichan (5)	Outer (1)	Waves (6)	Section
1	Water	Moment	Clear	Single ch.	Inner	No waves	7.2
3	Water	Moment	Clear	Single ch.	Outer	No waves	2.4 &7.2
4	Water	Moment	Clear	Single ch.	Outer	Waves	2.4
5	Water	Moment	Clear	Multichan	Inner	No waves	7.2
7	Water	Moment	Clear	Multichan	Outer	No waves	2.4 & 7.2
8	Water	Moment	Clear	Multichan	Outer	Waves	no
9	Water	Moment	SaltGrad	Single ch.	Inner	No waves	7.2
13	Water	Moment	SaltGrad	Multichan	Inner	No waves	no
17	Water	Avrgd	Clear	Single ch.	Inner	No waves	7.4
20	Water	Avrgd	Clear	Single ch.	Outer	Waves	4.2.2
21	Water	Avrgd	Clear	Multichan	Inner	No waves	7.6
24	Water	Avrgd	Clear	Multichan	Outer	Waves	4.2.2.
25	Water	Avrgd	SaltGrad	Single ch.	Inner	No waves	7.3
29	Water	Avrgd	SaltGrad	Multichan	Inner	No waves	7.3.2 c
33	Sed.	Moment	Clear	Single ch.	Inner	No waves	7.6.5 c
36	Sed.	Moment	Clear	Single ch.	Outer	Waves	4.2
37	Sed.	Moment	Clear	Multichan	Inner	No waves	No
40	Sed.	Moment	Clear	Multichan	Outer	Waves	4.2
41	Sed.	Moment	SaltGrad	Single ch.	Inner	No waves	NO!
45	Sed.	Moment	SaltGrad	Multichan	Inner	No waves	NO!
49	Sed.	Avrgd	Clear	Single ch.	Inner	No waves	7.6
52	Sed.	Avrgd	Clear	Single ch.	Outer	Waves	7.7
53	Sed.	Avrgd	Clear	Multichan	Inner	No waves	7.8
56	Sed.	Avrgd	Clear	Multichan	Outer	Waves	7.7
57	Sed.	Avrgd	SaltGrad	Single ch.	Inner	No waves	7.9??
61	Sed.	Avrgd	SaltGrad	Multichan	Inner	No waves	no

binary ("Sed;(3)") in the picture: the investigation of

a. "water" (which refers to clear-cut "Hydraulics");
b. "sediment".

Furthermore, "tidal-averaged effects" (called "Q" ("quadratic") in Section 2.5) are not necessarily "quadratic" effects: also a longitudinal density gradient may play a role. This introduces binary (4),"Saltgrad":

c. "salt gradient (y/n)".

Mind, that the shape of the inlet system can be different. It can be schematized as a single-channel or a multiple channel system (binary "multichan" (5)).

Finally, the resulting shape of coast and/or estuary will be different, depending of waves are playing a role in shaping the coast (binary "waves" (6)).

So the considerations above mean a "small" amendment on the statement in Subsection 2.5.1: (quoted before:): "Altogether this gives $2^3 = 8$ cases to be treated, of which 4 in Section 2.5 and 4 in Chapter 3, c.q. Chapter 8" (unquote).

In fact, only the number of combinations in Chapter 8 appears to be already 26 = 64.

However, there is some consolation:

a. Some possibilities are more likely than other ones;
b. This book is not the only one on coastal dynamics: some possibilities will not be treated at all (without denying the importance of these possibilities).

More specificly:

a. Effects of waves internally in the estuary will be disregarded in this book. Mind, that inside an estuary "effect of waves" in fact should imply as well: "effect of wind". And (considering the dynamics of estuaries) "wind-driven current in estuaries" implies as well: "stirring of sediment by wind"; "horizontal and vertical circulation by wind set-up", etc. Logistics more than logic bans the subject from this book.
b. Effects of waves combined with effects of salt gradients will be disregarded.
c. Apart from the subject of linear tidal computations (Chapter 7.2) in the "Outer" area the presence (and effect) of waves is presumed.

Those presumptions reduce the number of possibilities from 64 to the 26, mentioned in Table 7.1.

Remarks:

7.2 linear tidal computations
7.3 residual currents, caused by density effects
7.4 residual currents, caused by non-linear effects
7.6 morphological stability of a tidal channel
7.7 morphological behavior outer delta's
7.8 tidal inlet models Wadden & Western Scheldt

Table 7.2 Content of Chapter 8.

Binary 3:	Water	Water	Water	Sed.	Sed.	Sed.
Binary Section	7.2	7.3&4	7.6	7.6	7.7	7.8
Linear (2)	O I	* *	* *	* *	O *	* *
Secondary: Salt gradient (4)	* *	* I	* *	* *	* *	* *
Secondary: non-linear (2)	* *	* *	* I	* I	O *	* I
Number of channels (5)	1 > 1	1 *	1 *	1 *	* *	* >1

Thus a table concerning the content of Chapter 8 (leaving the sixth binary (waves) out of consideration) similar to the one in Subsection 2.5.1 gets roughly the following shape:[d,e]

A more detailed description is given below; it is illustrated by Table 7.2

It makes sense to treat elements of those basics in three introducing parts:

a. Section 7.2: linear tidal Hydraulics (the Lorentz method);
b. Section 7.3&4: second-order water motion caused by tides and density currents:
   7.3 residual currents caused by density currents
   7.4 residual currents caused by tides
c. Section 7.6: resonance and morphological stability of a prismatic channel.[f]

After this (refer to the key figure 7-14), this chapter consists of the two basic constituents, mentioned above, i.e. respectively:

d. the morphological behavior of the outer delta and the adjacent coasts (Section 7.7);
e. tidal inlet models, in general and its applications for Wadden Sea and Western Scheldt (Section 7.8). Section 7.8 extends Section 7.6 by considering a network of channels instead of a simple channel.
   The chapter ends with a discussion and summary (Section 7.9).

The following final remark should be made in this introduction.

[d]On the horizontal scale is binary (1): "water/sed". Indicated in the table are the binaries "outer" (1), represented by the symbols "O" and "I" and "multichan" (5) (indicated as "1" or "> 1") where a star denotes: "not treated in this section"

[e]Section 7.5 gives a summary and remarks w.r.t. the preceding investigations concerning hydraulics.

[f]As well hydraulic as morphological matters are considered.

The present subject leads from the coast in the estuary or lagoon. Morphodynamics of estuaries and lagoons is essentially different from coastal dynamics, for instance because of:

f. the effect of longitudinal gradients of Salt content,
g. the effect of silt,
h. effects of animal life on sedimentation and
i. the specific effects of meandering in tidal channels (periodic submergence of tidal flats, giving differences in current pattern during ebb and flood, cf. Subsection 2.5.3 concerning the theory of van Veen.[49,55]

Hence, although extending a one-dimensional perception of the morphology of a prismatic channel with sandy bottom (Section 7.3) to a network of channels (Section 7.5) may give some feeling on the dynamics of the delta⇔inlet interaction, the present display of investigations is certainly not sufficient to reach the ultimate insight.

Mind, that the response on the input of higher harmonics generated on or seaward of the outer delta will not be treated in this book; refer for this to Refs. [29,85]. Mind the very relevant remark of Bliek on page 8.12 concerning the opposite effect of Salt gradients to vertical circulation compared to the effect of "rain in the drain", as explained in Section 7.4. With respect to this subject "Salt gradients", refer to Subsection 7.3 concerning the effect of the Salt content on the vertical circulation of the water motion and on the silt content.

Furthermore, refer to the concise annotated summary of *other* important elements in estuary dynamics, given in the afore-going Subsection 7.1.1

Focusing on all these estuary-specific subjects would lead outside the scope of this book.

A challenge for the future is: merging of the various elements which have been investigated apart from each other until now.

7.1.3. *A positive sign*

a. For outgoing sailors and coastal engineers evidently the seaward direction is the positive direction.
b. It is general custom to let a positive axis point from left to right.
c. It is a general custom of cartographers to plot a card with the Westside to the left-handside of the Eastside.

A. However, the Lord, in his Righteousness, decided to situate the Sea to the West of the Netherlands.

Thus, considering examples in the Netherlands of, for instance, salt intrusion in an estuary can lead to contradiction between the rules mentioned above.

Where in general cases the rules "a,b,c" are applied, in examples concerning specificly the Western Scheldt Rule A of the Lord has been followed. Mind, that in Figs. 7.14 and 7.19 flood comes in from the left-hand side: current profiles right of the z-axis are flood current profiles.

Also in Figs. 7.17 and 7.19 sea is on the left-hand side. On the other hand, in the "general" Fig. 7.20 the positive velocities are in *sea*ward direction. Also in the "general" Fig. 7.22 the sea is on the right-hand side.

Sub-subsection (c.) specificly deals on the Western Scheldt and thus in Figs. 7.27 and 7.28 the sea is on the left-hand side. Figure 7.30 (Sub-subsection (d.)) refers again to Fig. 7.23 and has the sea on the right-handside.

In Figs. 7.31–7.33 again Western Scheldt situations are considered (sea on the left-hand side).

In the general graphs of Fig. 7.34a,b again the closed end of the channel is on the left-hand side.

Figures 7.35 and 7.62b,c refer again to the Western Scheldt situation (positive velocities directed to the East).

Also in Fig. 7.36 (Western Scheldt) the sea is on the left-hand side. Upward velocities are positive.

Sea on left-hand side	Sea on right-hand side
7.14, 15, 17–19, 24–31, 33, 34	7.9, 22, 32^{ab}, 35, 36, 37,38

7.2. Momentaneous velocities: Linear Tidal Hydraulics. The Lorentz Method and Considerations on Velocity Profiles

7.2.1. *Introduction to tidal computations in general and the Lorentz method in particular*

This Section deals on momentaneous water motion as caused by tides in estuaries (cf. Table 7.3) page 365.

It is an introduction in (linear) tidal hydraulics as far as necessary to understand the morphodynamics of tidal estuaries. Thus the linear Lorentz method of tidal computation, genial in its simplicity, forms the mean menu of this section (Subsections 7.2.2–7.2.10).

A relevant question is the accuracy of the linearization. Section 7.4 deals about it.

Evidently,[g] for long term morphodynamics use of straight-forward linear tidal computation leads to frustration: nothing should change! That means: unless also the residual current (depth-averaged or not) and its interaction with the first harmonic are considered as well. So the goal of the present section is to get the tools to get a grip on the linear tidal motion, in order to be able (in subsequent sections) to eliminate it by tidal-averaging.

The effect of a longitudinal salt gradient on the momentaneous velocity profile will be considered in Section 7.3.

Thus the section starts by giving an outline of the Lorentz method of tidal computation as introduced by Lorentz[49] and used by the Lorentz Committee in the 1930s. At that time an investigation was required to know the effect on the tide of the closing of the Zuiderzee.[h] Some more details are given below.

The magnitude of both waves (absolutely and relative to each other) depend on the boundary. The linearization of the friction terms in the hydrodynamic equation (7.4) (cf. page 357) is the basis of the Lorentz Method.

The method (linear computation) can be used for single- or multi-connected tidal basins. The method simulates the total tidal communication within assumed local boundaries between the inner delta (tidal basins) and the outer delta (sea or ocean), using only a few complex variables.

Physics (in general) is the art to translate reality into mathematical equations.

The Lorentz method exposes an outstanding ability to translate the topography in the area in which the tidal motion takes place in a simple set of complex numbers and to translate the network of channels (which may have multiple boundaries) in a simple set of linear equations.

For instance, consider the Western Wadden area (Fig. 7.2); topic of the Lorentz report.[49] Mind the dam enclosing the Zuiderzee (area south of this dam in Fig. 7.2). Figure 7.4 shows the network of tidal channels as schematized by the Lorentz Committee; the areas representing the various channels are shown in Fig. 7.2. Mind, that many of the channels consist of

---

[g]Apologies for this section in a book "Coastal Dynamics". It does not deal on sand, so: no "coast" involved. Furthermore, after a wave period, according to the theory in this section, water returns to exactly the same position as a period before, so it is no "dynamics" either. It is like giving critics on a concerto of a Bach fugue by describing the air pump of the organ. This section follows Thijsse's classnotes.

[h]After the enclosure this area was called "IJssel Lake".

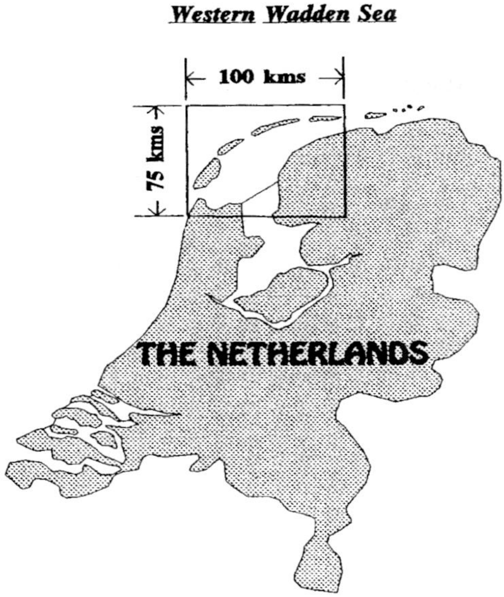

Fig. 7.2 Western Wadden Sea area and enclosure dam, designed by the Lorentz Committee. "Zuiderzee" is the original name of the area South of the enclosure dam under consideration in the Lorentz report.

a link of a series of two or three serial sub-channels; furthermore, each of those sub-channels may consist of three parallel sub-sub channels with arbitrary width and depth.[i] Eventually, the whole hydraulic behavior of this set of 3 times 3 sub-sub-channels is characterized by two complex numbers α and β (Subsection 7.2.8). The vertical tide Z (characterized in each point by a complex number as well) is calculated firstly in the various nodes; continuity and kinematics yield equations, expressing Z in the considered node in Z in the adjacent nodes, where the coefficients in these equations contain the α- and β-factors of the linking channels. When Z in the nodes is known, the Z-values in the transition points between the sub-channels (for instance between sub-channels 2a and 2b; Fig. 7.3) can be interpolated, taking into account, that those sub-channels have various hydraulic characteristics, determined by "own" values of α and β, different from the one of the integrated channel.[j] When the vertical tide at the ends of a sub-channel are known, as well as the horizontal tide along the sub-channel are

[i]The confinement to three serial and three parallels is not essential for the method.

[j]The value of α and β of the integrated channel can be derived from the values of the sub-channel and the configuration (the result for a link "2a–2b–2c" will differ from

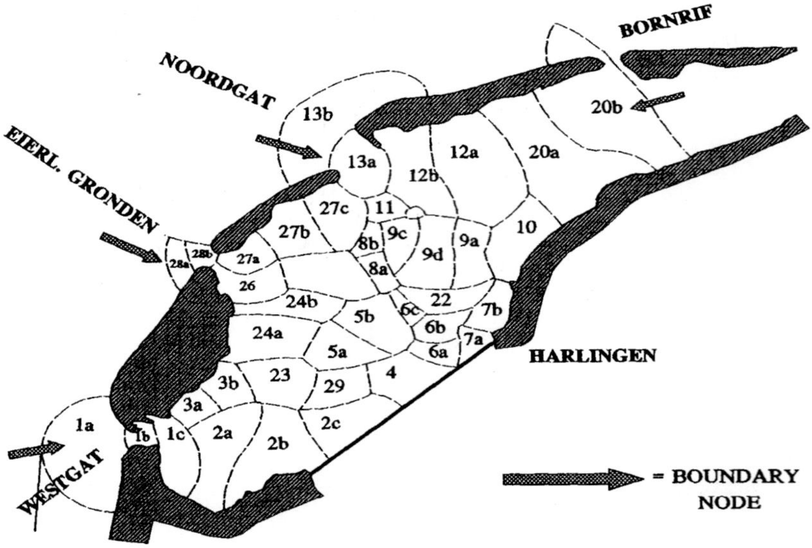

Fig. 7.3 Schematization–tidal channels of the Western Wadden Sea.

known. More precisely, the horizontal tide in the various parallel sub-sub-channels will differ, as the dimensions and friction of those sub-sub-channels will differ. As one might expect, those coefficients α and β (and a third coefficient Δ, related to storage) determine the non-linear motion as well (Section 7.8).

The following subjects concerning the Lorentz method are treated in the present Subsections 7.2.2–7.2.8.

First (Subsection 7.2.2) some background concerning tidal motion in general is given; then the variables playing a role are introduced.

Subsections 7.2.3–7.2.7 treat the linear hydraulics of the most simple case of a uniform straight prismatical channel: the hydraulic equations in 7.2.3, leading to expressions for amplitude and velocity in 7.2.4. The general concept is displayed, that the tidal motion in such a channel consists of two basic components, which can be (in more or less simplified terms) characterised as two propagating and attenuating waves: an "incoming wave" propagating and attenuating in one direction and a "reflected wave"

"2a–2c–2b", when dimensions and friction of each section (2a, 2b and 2c) will remain the same in both cases.)

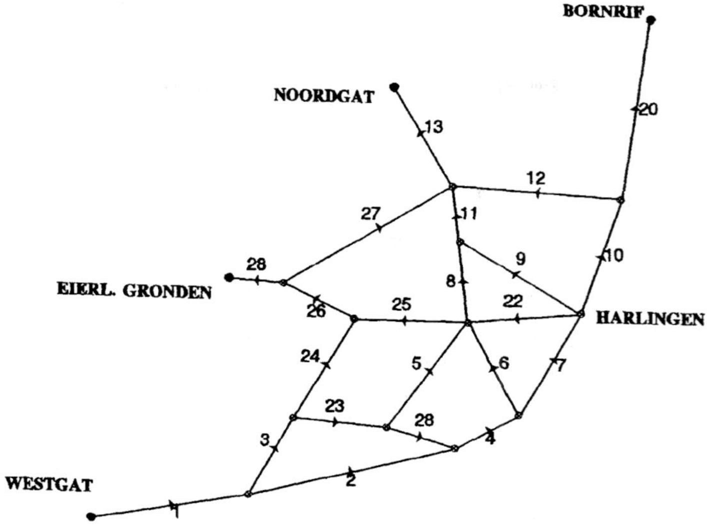

Fig. 7.4 Schematization–network of the Western Wadden sea.

propagating and attenuating in the other direction. Evidently, the conditions: cf. Subsection 7.2.5, especially Table 7.3.[k] (page 365)

Subsections 7.2.6–7.2.8 deal on practical applications. Subsection 7.2.6 displays a graphical method 7.2.6 and 7.2.7 an analytical one. The derivation of the α- and β-coefficients, mentioned above is treated in subsection 7.2.8. A special treatment (Subsection 7.2.7) shall be given to the case of a channel, closed on one side, as this is the basic schematization for the estuaries, treated in Sections 7.3–7.6

As mentioned, in this section, only the propagation of tide and the water motion in one single prismatic channel is investigated. So this is only a part of the story of the preceeding paragraphs; refer to[78] and [79] for the whole story.[l] For the understanding of the morphological implications, most important is the understanding of the physical meaning of the α- and

[k]Subscript "0" stands for the left-hand boundary of the channel; "L" for the right-hand side of the channel; "Z" stands for "vertical tide" and "Y" for "horizontal tide", i.e. the harmonic discharge at the indicated side of the channel

[l]A comprehensive computer version of the Lorentz tidal computation is given by Jallah and Bakker.[42] It also includes checks of the program and considerations on accuracy (Ref. [42] in Chapter 4); irregular cross-sections (Ref. [42] in Chapter 2), influence coefficients (cf. Fig. 7.31 of the present chapter) and computer graphics as polar diagrams and three-dimensional films of the water surface; furthermore the calculation of the coefficients α, β and Δ occurring in (present) Subsections 7.2.8 and 7.2.9.

β-factors, the factor Δ (7.64) and of the variable θ determining the friction of the channel.

7.2.2. *Origin of tides; schematization; variables of interest*

In the study of the characteristics of tides, it is often important to know the impact it will have on, for instance a structure, estuary or area of interest for which the study is requested. Hence, the knowledge of the tidal amplitude (the vertical tide) and the current (the horizontal tide) is significant. Knowledge of the tidal process is important for tidal computations.

Tides are generated by the mutual attraction of earth, moon and sun. More exactly, because of the difference between the mass-averaged attraction force between two of these bodies and the local force at a particular spot of these bodies (for instance: the earth). If it is assumed that these celestial bodies move at a constant distance from the earth (in the plane of the equator), a semi-diurnal-tide is the result of this phenomenon with a period of half the time the earth requires to revolve around its axis relative to the bodies.

The period and frequency, resulting from the effect of resp. moon and sun on the tide are:

Moon
Period: $T_M = 12\,h$ and $25\,min$
Frequency: $\omega = 2\pi/T_M = 1.405 \times 10^{-4}$ radian per seconds
Sun
Period: $T_s = 12\,h$
Frequency: $= 2\pi/T_s = 1.454 \times 10^{-4}$ radian per seconds

Because the sun attracts the water "harder than average" on the side of the earth, facing the sun, this gives a "hump of water" on this side of the earth, facing the sun.

Because the sun attracts the water "less than average" on the side of the earth, opposite to the sun, here the water lags behind: this gives, on that side of the earth a "hump of water" as well!

Evidently, rotation of the earth around the sun with its centripedal forces keeps the sun at a safe distance. So, an earth covered with water would show two humps of water, one directed to the sun and one opposite; the earth would rotate below this layer of water with a 24-h period;

so a local observer experiences a period of the "humps" of 12 hours (not 24!).

The explanation above is VERY simplified "hydraulics":[m] mind that the explanation deals on FORCES, not on MOTION. Water has its own "reply" on forces; this explains the number of pages of the present chapter. However, the explanation on ("half") period holds in the following chapter (apart from influence of higher harmonics. Explanation of T_M is similar to explanation of T_s

Moon and sun act simultaneously, thus causing spring and neap tides.

Thijsse[79] distinguishes:

a. "Tidal prediction" in which the local tide at a certain spot is predicted by former tidal registrations at the same spot. Emphasis is on higher harmonics, of which the periods can be derived from knowledge of astronomy.[n] Keyword(s) is: "Fourier analysis".
b. "Tidal computation" in which (originally)[o] only a harmonic tide was computed in a physical way. Mostly the M_2-tide is the relevant topic, but depending on the local circumstances (resonance) it could be another dominant tide as well. Keyword: "harmonic method".

In the present section, only (the original) tidal *computation* will be treated and will be used for its implications on morpho-dynamics. These "narrow-minded" assumptions (confining to one harmonic) will lead to "narrow-minded results". It is hoped that confining to essentials in tidal hydraulics leads to finding the essentials of morpho-dynamics.

---

[m] Continents obstruct the continuous water wave around the earth. However, this is not the case South of Cape Fear and Cape the Good Hope. Result is a tide-generating zone over there (resonance at 63°–64° Southern latitude) and tidal waves propagate from there through the oceans. After one or two days the "tidal signal" from Cape the Good Hope reaches the Netherlands. Maximum Spring tide in the Netherlands is one or two days after the conjunction of sun and moon (not at the same day).

[n] Resulting in, for instance, effects of ellipticity (orbits of earth and sun and (more so) of moon around the earth are not circles, but ellipses). The distances sun–earth and earth–moon vary periodically, resulting in fluctuation on strength of tides (annual, resp. monthly); "declination": sun and moon are generally at an angle (declination) with the equator; reflection and interference of tidal waves caused by irregularity of bottom and boundaries of oceans, etc.

[o] Scientists like Mazure, Schonfeld (and many others) extended the method to, for instance, higher harmonics. Tragically, later in the twentieth century, the importance of those sophisticated and intelligent, however time consuming and difficult to grasp computational methods was murdered by the brutal force of the digital computers. The original Lorentz method is saved for this fate because of its simplicity.

Confining the picture further to a straight prismatical channel of uniform depth h, leading (Fig. 7.5) from "Here" ($x = 0$) to "There" ($x = L$), the following variables will appear to be important:

x : horizontal coordinate (m)
t : time (second)
ζ : instantaneous water level (elevation compared to tidal-averaged value); function of x and t; in meter
H : tidal-averaged water depth
u : instantaneous tidal velocity, function of x and t
g : gravitational attraction of the earth (taken $9.81\,\text{m/s}^2$)
C_h : the Chezy coefficient in m1/2/s
f_c : the dimensionless friction coefficient, $(8g/C_h^2)$
θ : an angle of internal friction (0 if no friction and 45° if infinite friction; will be specified in Subsection 7.2.4 a).
k_f : linear friction coefficient (will be specified later on).
ω : angular frequency of the tide
$\hat{z}$: amplitude of motion of water level $\zeta(x,t)$
$\hat{u}$: amplitude of the tidal velocity; function of x and t
Z : complex number, indicating the water level

$$(\varsigma = \text{Re}(Ze^{i\omega t}); \quad \hat{z} = |Z|)$$

U : complex number, indicating the tidal velocity $u = \text{Re}(Ue^{i\omega t})$; $\hat{u} = |U|$
φ : phase difference between vertical and horizontal tide

The use of complex variables is quite common for solving problems with harmonic motion. It is an artifice, which allows the variable "time" by replacing it by the phase of the tide (relative to some reference tide) and thus by a constant number. The method will be illustrated below.

7.2.3. *The hydraulic equations*

Equations for the Lorentz approach are based on a 1D linear model. The method shows the relationship between the tidal boundary conditions and the tide in the tidal basin. One relevant possibility is to express the discharge (q_{here}) in the tidal channel by the vertical tide (z_{here} and z_{there}) The expressions "here" and "there" represent beginning and end of the channel (Fig. 7.5).

The approach to the Lorentz computation-method by Thijsse[79] and Bakker and De Vriend[10] (cf. Section 7.6) and the method adopted in

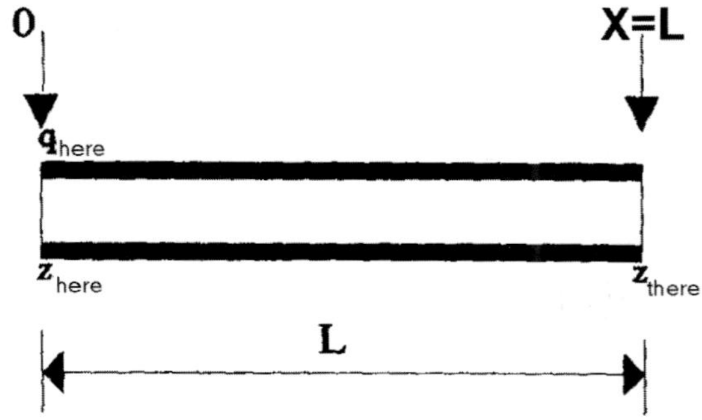

Fig. 7.5 Definition of vertical tide and horizontal tide.

a computer program[42] is discussed. In the first part of the discussion, much emphasis will be given to the procedure of linearization discussed by Thijsse.[79]

In the Lorentz method, tidal waves are assumed to be sinusoidal for which the vertical and horizontal motion of (c.q. in) the waves are given by the water level and currents. These phenomena are described by Eqs. (7.1) and (7.2).

Considering the most simple case, the vertical tide in a certain point is described as:

$$\varsigma = \hat{z}\cos\omega t \tag{7.1}$$

The horizontal tide is described by:

$$u = \hat{u}\cos(\omega t + \varphi) \tag{7.2}$$

The definitions according to (7.1) and (7.2) imply that the vectors in the complex plane Z and U (Subsection 7.2.2) locally make an angle φ with each other.

The idea of linearizing tidal computations is described in the introduction, Section 7.2.1.

Two non-linear equations describe the propagation of those tidal waves:

a. The Continuity Equation

$$\frac{\partial\zeta}{\partial t} = -(h+\zeta)\frac{\partial u}{\partial x} - u\frac{\partial\zeta}{\partial x} \tag{7.3}$$

b. The Dynamic Equation:

$$\frac{\partial u}{\partial t} + u\frac{\partial u}{\partial x} = -g\frac{\partial\zeta}{\partial x} - \frac{f_c u|u|}{8(h+\zeta)} \tag{7.4}$$

The two equations are non-linear and therefore a sinusoidal wave does not give a proper solution. Because of the distortion waves undergo due to the non-linear terms, the continuity and dynamic equations are simplified to contain only linear terms. The linearization of the friction term in the dynamic equation, introduced by Prof. H.A Lorentz is achieved as follows.

If the water elevation ζ is much smaller than the depth h, i.e $\zeta \ll h$, then the last term in the equation of continuity becomes negligible. Thus,

$$u\frac{\partial \zeta}{\partial x} \approx 0$$

Also, in the dynamic equation, the second term is small with respect to the first. Thus,

$$u\frac{\partial u}{\partial x} \ll \frac{\partial u}{\partial t}$$

Therefore, the second term on the left-hand side of (7.4) will be neglected.

The last term of (7.4), the friction term, which caused damping is linearized upon the assumption that the momentaneous friction force F is proportional to u and not to u^2

Therefore, the last term in Eq. (7.4) can be written as:

$$\frac{f_c u|u|}{8(h+\zeta)} = k_f u \tag{7.5}$$

k_f being a friction factor. As f_c is dimensionless, the dimension of k_f is [s^{-1}]. Physically, one would prefer to replace k_f as well by a dimensionless coefficient. In anticipation of the derivations given in the next Subsection 7.2.4, it is stated, that a nice way of definition (leading to physical grip on the matter and to dimensionless "rating" of friction) turns out to result from the introduction of a kind of "friction angle"[p] called θ, in the following way:

$$k_f = \omega \tan(2\theta) \tag{7.6}$$

One may observe from (7.6), that $\theta = 0$ results in frictionless fluid, where $\theta = \pi/4$ indicates an infinitely high friction.

In this way the governing equations (7.3) and (7.4) reduce to:

[p]Admitting that the assumption looks far-fetched at first sight and that θ cannot be assigned to an angle as occurring in nature, like the friction angle occurring when a block slides along a slope.

The Continuity Equation

$$\frac{\partial \zeta}{\partial t} + h\frac{\partial u}{\partial x} = 0 \tag{7.7}$$

The Dynamic Equation:

$$\frac{\partial u}{\partial t} + g\frac{\partial \zeta}{\partial x} + (\omega \tan 2\theta)u = 0 \tag{7.8}$$

7.2.4. *Derivation of expressions for amplitude and velocity*

a.) *Deriving an expression for the angle of friction, θ*

The objective of this subsection is to introduce the value of θ in such a way, that the error made by linearization of the friction term is minimized. Friction causes damping of waves due to energy loss. The procedure is described below.

The work dW done by friction in a time interval dt by a current velocity u, as found from the last term of Eq. (7.4) is:

$$dW = udt \times \frac{f_c u^2}{8h} \tag{7.9}$$

where f_c is the bottom friction coefficient, as defined in Subsection 7.2.3.

Equation (7.4) representing the velocity is substituted[q] in Eq. (7.9). Thus,

$$dW = \frac{f_c}{8h}(\hat{u}\cos\omega t)^3 dt \tag{7.10}$$

Integrating over a quarter tide period yields the expression for the real total work done as:

$$W = \frac{f_c}{8h}\int_0^{T/4}(\hat{u}\cos\omega t)^3 dt = \frac{f_c}{12\omega}\frac{\hat{u}^3}{h} \tag{7.11}$$

Remind, that in the derivation of (7.11) no assumptions about linearizing friction have been made.

Similarly, an expression for work-done is derived, using the expression of the last term in equation (7.8.), i.e. starting from friction linearization. Then the result is:

$$dW = udt^* k_f u = k_f u^2 dt \tag{7.12}$$

[q]Omitting φ, which is irrelevant in the present context (changing the start "$t = 0$" in an appropriate way)

The value of k_f in practice is unknown; in fact, it is the goal of the present exercise to express this in the known quantities like f_c, occurring in (7.11). Comparison of the expression last found with (7.9) is difficult because k_f is not dimensionless like f_c; thus substitution of (7.6) is appropriate.

Substituting for k_f and integrating over a quarter tide period, gives an expression for total work-done in the case of linearization. Thus,

$$W_r = (\omega \tan 2\theta) \int_0^{T/4} (\hat{u} \cos \omega t)^2 dt = \frac{\pi \tan 2\theta}{4} \hat{u}^2 \tag{7.13}$$

Equating "resistance without linearization" (Eq. (7.11) to resistance with linearization (Eq. (7.13)) yields an expression for the friction angle, expressed in known physical quantities:

$$\frac{f_c}{12\omega} \frac{\hat{u}^3}{h} = \frac{\pi \tan 2\theta}{4} \hat{u}^2 \tag{7.14}$$

to be elaborated as:

$$\tan(2\theta) = \frac{f_c}{3\pi\omega} \times \frac{\hat{u}}{h} \tag{7.15}$$

Equation (7.15) contains the term $\hat{u}$, which has to be calculated by iteration. The procedure for calculating the velocity is explained in Subsection 7.2.4.c.), later on.

b) *Derivation of expressions for the vertical tide*

Expressions for the water elevation ζ and velocity[r] u are derived using Eqs. (7.7) and (7.8). The expression for ζ results from differentiating the continuity and dynamic equations to eliminate the velocity u; more precisely: the terms $h\partial^2 u/\partial x \partial t$ and $h \partial u/\partial x$.

Differentiate (7.7) with respect to t,

$$\frac{\partial}{\partial t}\left[\frac{\partial \zeta}{\partial t} + h\frac{\partial u}{\partial x}\right] = 0 \tag{7.16}$$

Also differentiating (7.8) with respect to x, multiplying with h yields:

$$h\frac{\partial}{\partial x}\left[\frac{\partial u}{\partial t} + g\frac{\partial \zeta}{\partial x} + (\omega \tan 2\theta) u\right] = 0 \tag{7.17}$$

[r]For u refer to the next Subsection 8.2.4 c.

Thus an equation for the water level ζ results

$$\frac{\partial^2\zeta}{\partial t^2} - gh\frac{\partial^2\zeta}{\partial x^2} + (\omega \tan 2\theta)\frac{\partial\zeta}{\partial t} = 0 \tag{7.18}$$

Transferring ζ to its complex "counterpart" Z (Subsection 7.2.2), a primary solution of Eq. (7.18) is found as (only the real part of the complex variable has to be retained):

$$Z = Z_0 e^{kx} \tag{7.19}$$

Resuming the meaning of the symbols in (7.18) and (7.19):

$\omega = 2\pi/\text{Ttide}$ period: sinusoidal wave frequency

ζ is the water elevation $(\text{Re}(Ze^{i\omega t})$

Z_0 is the complex presentation of the vertical tidal motion at $x = 0$

x is the coordinate in channel direction

k is the *complex* wave number, considered below

The exponent k has a real and an imaginary part and for the former, the formula expresses logarithmic damping. For the latter, it shows that Z has a periodic character in x-direction.

The factor k can be derived by substituting Eq. (7.19) in Eq. (7.18) reminding that

$d(ae^{bx})/dx = b(ae^{bx})$:

$$\{-\omega^2 - ghk^2 + i\omega^2 \tan 2\theta\}Z = 0 \tag{7.20}$$

from which results[s] for k:

$$k = \pm\frac{\omega(i + \tan\theta)}{\sqrt{gh(1 - \tan^2\theta)}} \tag{7.21}$$

The phase of the tide is determined by the imaginary part of $Z(x)e^{i\omega t}$. The propagation velocity C of the tidal wave ς (represented by Z) is defined as the velocity, with which the point with a certain, given phase ϕ_1 propagates in x-direction. Considering the elementary solution (7.19), C is a constant: from the definition of phase it shows that $[\text{Im}\{\ln(Z)\}] + \omega t$ should be constant, i.e. (referring to (7.19)):

[s]The easiest way to check this result is to consider $k\sqrt{gh/\omega}$. Equation (7.20) states the module of its square is $1/\cos(2\theta)$ and its argument is $\pi - 2\theta$. Equation (7.21) states that its module is $(1/\cos\theta)/\{\sqrt{\cos(2\theta)}/\cos\theta\} = 1/\sqrt{\cos(2\theta)}$ and its argument is $(\pi/2) - \theta$.

$\mathrm{Im}(kx) + \omega t$, i.e. $\mathrm{Im}(kCt) + \omega t$ should be constant

$$C = -\omega/\{\mathrm{Im}(k)\} \tag{7.22}$$

From Eq. (7.21), one may derive an expression for the propagation velocity C of an elementary harmonic tidal wave as given by (7.19):

$$C = \pm\sqrt{gh(1 - \tan^2\theta)} \tag{7.23}$$

If $\theta = 0$, meaning no friction, this celerity becomes $C = \sqrt{gh}$; when $\theta = 45°$ (infinite friction; $f_c = \infty$; Eq. (7.15)) zero celerity is the result.

With substitution of C, Eq. (7.21) is written as:

$$k = \pm\frac{\omega}{C}(i + \tan\theta) \tag{7.24}$$

In the following, "k" will be defined as the "+" solution. Thus the notation for the "−" will be "$-k$". Furthermore, the symbol "C" will be reserved for the positive value of the right-hand side of Eq. (7.24). This detaches it from the general meaning of propagation velocity C, i.e. the velocity with which a certain phase of a wave propagates through the medium. If one would start from the original definition, it will show in the following, that even in the harmonic case combination of solutions of Eq. (7.18) may lead to propagation velocities, which are quite dependent of x. Some insight in this feature follows from the notion, that in frictionless water motion two equal propagating waves with propagation velocity C and $-C$, respectively, give a standing wave with undetermined propagation velocity.

As well the solution "Z" of (7.18) with positive and the one with negative k are relevant. Those will be called "Z_+" and "Z_-", respectively, and because (7.18) is linear, a linear combination of solutions is a solution as well. Therefore, in the following, the more general solution Z will be considered:

$$Z = Z_- + Z_+ \tag{7.25}$$

The value of "Z_0" at $x = 0$ for the primary solution in (7.19) may be different[t] for the "+" and "−" solution in the combined solution as given in (7.25); call Z_{0-} and Z_{0+}. In the rest of this chapter, "Z0" will be defined as the (complex) sum of Z_{0-} and Z_{0+} (at $x = 0$, evidently).

[t]However, not in the case the basin is closed at $x = 0$.

Summarizing the solution $\varsigma(x, t)$ in the "real" world is the real part of a combination of a wave Z_-, propagating and attenuating in positive direction and of a similar wave Z_+ propagating and attenuating in negative direction where:

$$\zeta(x.t) = \text{Re}\{(Z_- + Z_+)e^{i\omega t}\} \tag{7.26}$$

where (cf. (7.24) and (7.23)) :

$$\begin{aligned} Z_- &= Z_{0-}e^{-\frac{\omega x}{c}(i+\tan\theta)} \\ Z_+ &= Z_{0+}e^{\frac{\omega x}{c}(i+\tan\theta)} \end{aligned} \tag{7.27}$$

The factors in Eqs. (7.26) and (7.27) are explained as follows:

$$e^{i\omega t}:$$

a. Equation (7.26) shows that the phenomenon is periodic with period $T = 2\pi/\omega$.

$$(\omega x/C)i :$$

b. The factor $\omega x/C$ depicts the site on the x-coordinate of channel direction related to the wave length $2\pi C/\omega$. Remember from (7.23), that the wavelength reduces in case of larger friction (first effect of θ). Mind, that the phase of the waves, indicated by $\pm i\omega x/C$ changes an angle 2π over a distance CT and that the Z_--wave moves in positive x-direction[u], where the Z_+-wave moves in negative x-direction.

$$(\omega x/C)\tan\theta :$$

c. As $\tan(\theta)$ is positive, it shows, that the Z_--solution in (7.26) attenuates logarithmically in positive x-direction, where the Z_+ - solution attenuates in negative x-direction. The larger θ, (i.e. the larger friction) the stronger the attenuation, as to be expected.

Summarizing (7.26) and (7.27) show that the vertical tide consists of a wave Z_+ propagating and attenuating in negative x-direction plus a wave Z_- propagating and attenuating in positive x-direction.

Remark. A considerable part of this chapter[v] will be focused on semi-infinite closed (prismatic) channels. Then , the whole solution will be normalized to the vertical tide Z_0 at the closed end of the channel. Refer for

[u] same phase when $x - Ct$ remains the same; cf. (7.26) and (7.27).
[v] 7.2.10; ch 7.3–7.6

instance Fig. 7.9a, depicting the amplitude of the vertical tide in such a channel; the value at the open end is one. It will be shown in the next sub-subsection c.), that in a case like that the values of Z_{0-} and Z_{0+} should be exactly the same. Thus, in the normalized case of Fig. 7.9a those values should be 1/2 and in the general case $Z_0/2$.

c) *Derivation of expressions for the horizontal tide*

In this sub-subsection, a complex expression[w] $U(x)$ for the horizontal tide will be derived. This implies: the values of $\hat{u}$ and φ will be expressed into:

a. the characteristics of the Z_+- and Z_--wave,
b. into the water depth h,
c. the propagation velocity C (as given by (7.23)) and:
d. into the friction angle θ.

The complex function $U(x)$ thus has to be related to the other complex functions: Z, Z_-, Z_+, etc., appearing in (b) The relation simply follows from (7.7), reading in complex terms (after replacing Z by Z_- and Z_+):

$$\frac{\partial U}{\partial x}e^{i\omega t} = -\frac{1}{h}\frac{\partial}{\partial t}\{(Z_- + Z_+)e^{i\omega t}\} \tag{7.28}$$

$$\frac{\partial U}{\partial x} = \frac{i\omega}{h}(Z_- + Z_+) \tag{7.29}$$

From (7.27) it is shown that:

$$\int Z_{+,-}dx = \pm\frac{C}{\omega(i + \tan\theta)Z_{+,-}} \tag{7.30}$$

and thus, after multiplication of numerator and denominator of (7.30) with $ie^{i\theta} = \cos\theta(i - \tan\theta)$ from (7.29) and (7.30) the result is found:

$$U = \frac{C\cos\theta}{h}e^{i\theta}(Z_- - Z_+) \tag{7.31}$$

Complex solutions can be visualized easily in a complex plane. It can be useful for easy understanding, to plot velocities and waterlevels in the same picture; however, then for comparison the units should be the same. For this purpose, Thijsse introduced the "velocity length" Y; a vector (i.e. a

[w]For the definition and its relation to "real world" velocity $u(x,t)$, refer Chapter 7.2.2 and Eq. (7.2).

complex number with a length scale, just as the complex presentation of the water level Z), with the definition:

$$Y = \frac{h}{C\cos\theta}U \tag{7.32}$$

In the following, Y will be called: the "velocity length". Introduction of it expresses (7.32) in the simple shape:

$$Y = e^{i\theta}(Z_- - Z_+) \tag{7.33}$$

where Z_- and Z_+ are defined in Eq. (7.27).

It is logical to split Y in its constituants:

$$Y = Y_- + Y_+ \tag{7.34}$$

where:

$$\begin{aligned} Y_- &= e^{i\theta}Z_- \\ Y_+ &= e^{i\theta}Z_+ \end{aligned} \tag{7.35}$$

7.2.5. *Boundary conditions of channel system*

Equations (7.26) and (7.31) determine the vertical tide ζ and current u at any time (t) and place (x) in a channel if the two elementary wave components Z_- and Z_+ are known.

There are various kinds of boundary conditions from which Z_- and Z_+ are to be determined. Thijsse[79] distinguishes 5 cases, shown in Table 7.3.

Case 4 symbolizes the elementary solution (7.27) (Z_--solution), where Y follows from (7.32), taking Z_+ equal to zero.

Table 7.3 Boundary conditions (from[79]).

	Water level and current			
Case	Z_0	Y_0	Z_L	Y_L
1	X	?	?	0
2	X	?	X	?
3	X	?	0	?
4	X	?	–	–
5	X	X	?	?

x, Known; ?, Unknown; –, no reflected wave.

Thijsse calls case 5 "the standard case" in his class notes:[79] vertical and horizontal tide known at one side of the channel; wanted are the same characteristics at the other side. A graphical solution which will elucidate the physics is treated in the next Subsection 7.2.6.

Case 1 is important for the application, treated in Section 7.3, where it delivers the hydraulics, necessary for grasping the morphodynamic stability of a tidal channel, closed on one side. Case 1 will be considered in Subsection 7.2.7. Contrarily to the notation of Thijsse[79] (cf. Table 7.3) in Subsection 7.2.7 the value of Y_0 will be taken zero, where requested are Z_l and Y_l in relation to Z_0. The case of those subsections therefore is a combination of "case 1" and "case 5" of Thysse.[79]

Case 2[x] (Table 7.3) is important for the calculation of the tide in a network of channels, like for instance the Waddensea (Fig. 7.54). It gives the possibility to translate the continuity equations in the nodes of a channel system (a matter of discharges) in equations, having only the vertical tide in its variables (cf. for instance Eq. (7.178) in Subsection 7.6.3). This case is treated in Subsection 7.2.8. In lastmentioned subsection also the inverse variant of case 2[y] (known discharges on both sides; requested the vertical tides) will be treated.

Evidently, Case 3 is just a particular solution of Case 2.

7.2.6. *Graphical presentation of the tidal wave in the complex plane: a "polar diagram"*

a) *Introduction to 7.2.6; graphical presentation of the "Z -wave"*

As Thijsse[79] points out, the presentation of a physical tidal wave can be easily depicted[z] in the complex plane.

In the present Subsection 7.2.6, an explanation will be given how the constituents of the "Z"-wave (Eq. (7.25) can be graphically displayed in a polar diagram. Its constituents are defined in Eq. (7.27) as "Z_-" and Z_+". In the *present* subsection a.) will be focused on Z_-. In the same polar diagram the horizontal tide, linked with Z_-. (i.e. $Y-$, (7.35)) can be presented (Subsection ad.b). Graphical presentation of Z_+ and $Y+$ is treated in subsection ad. c.). Finally (subsection d.), it will also be explained how the constituents Z_- and Z_+ can be graphically derived from the given

[x]Indicated by Thijsse,[79] as: "channel connecting two oceans"
[y]Not mentioned by Thijsse[79]
[z]His Fig. 1, reproduced here in Fig. 7.6.

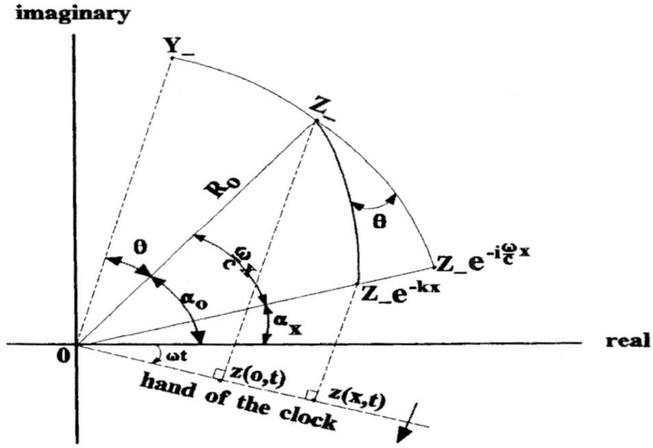

Fig. 7.6 Polar diagram depicting $Z_{-}(x)$ and $Y_{-}(0)$. From [79].

vertical and horizontal boundary conditions at the origin ($x = 0$). Thus, this deals on the solution of the standard Case 5 (Table 7.3; Subsection 7.1.2).

Thus, in this subsection the vertical tide will be considered. Figure 7.6 depicts the Z_{-}-wave; cf. Eq. (7.27); for the moment the $Z+$ -wave is left out of consideration.

Let the argument[aa] of $Z0-$ be α_0 and denote by R_0 the value of $|Z_{0-}|$. From (7.26) it shows, that the momentaneous water level at $x = 0$ is depicted by the projection of the Z_0- -vector on the hand of a clock, which makes a total revolution in a tidal period T, i.e. rotates with an angular speed of ω.

In Fig. 7.6 also the vertical tide at an arbitrary point $x = x$ has been graphically displayed.

The arc of a circle has been drawn through the point Z_0-. Furthermore, the complex value of the vertical tide Z_{-}, as given by Eq. (7.27) is depicted by the thick curved line in Fig. 7.6, ending in the point $Z_{0-}e^{-kx}$ Here "k" is given in Eq. (7.24).

From (7.24) it also shows, that the curve would end up on the circle (the point "$Z_{-}e^{-i\omega x/C}$"), if friction would have been neglected ($\theta = 0$). Following the endpoint of the vector $Z-(x)$ for increasing values of x, one finds a logarithmic spiral (the line between Z_{-} and $Z_{-}e^{-kx}$) of which the tangent makes every where the angle θ with the circle between $Z-$ and $Z_{-}e^{-i\omega x/C}$.

[aa]For shortness the value $|Z_{0-}|$ is abbreviated as Z_{-} in Fig. 7.6.

(b) *Graphical presentation of the Y-wave*

The way to find Z_- at $x = L$ has been illustrated in Fig. 7.6. Consider now Y_-, the vector presenting the horizontal tide according to (7.35). Again, for clarity sake, first the effect of Z_+, occurring in (7.31) will be omitted. Under this condition, from (7.35) a phase shift of θ between $Z_-(x)$ and $Y_-(x)$ can be derived, independent of the value of x; cf. Fig. 7.6. From (7.33) it shows, that because of the scaling of Y (7.32) the modules of Y_- and Z_- are the same. Again the magnitude of the local and momentaneous horizontal tide can be found from projection of the respective vectors on the hands of a clock. In this case, the value of φ in (7.2) thus is permanently equal to θ. Mind, that by the assumption of a propagating wave the positive u-direction, for which this result holds, is automatically the direction of propagation of the wave.

This remark will become important in Subsection 7.2.10, in which a prismatical channel, closed on one side is considered. There, the origin is defined at the closed end of the basin and the x-axis points in seaward direction, i.e. flood current is negative and ebb current is positive. The tidal wave, intruding from sea, propagates and attenuates in landward direction. Thus, the main feature considered in Subsection 7.2.10 is the Z_+-wave (not the Z_--wave as in the present sub-subsection). It is shown from Eq. (7.27) that in that case the phase lag between horizontal and vertical tide is $\pi - \theta$ instead of θ. Refer the lowest figure of Fig. 7.9. In last-mentioned figure as well incoming as reflected wave are considered, i.e. the phase of $Z_+ + Z_-$, in which evidently the incoming wave Z_+ dominates the reflected wave Z_-.

Summarizing: the velocity length in $x = 0$ is represented in Fig. 7.6 by:

$$Y_-(x=0) = R_0 e^{i(\theta+\alpha_0)} \tag{7.36}$$

where R_0 is the module of $Z-$. From there in positive x-direction the trace of the velocity–length curve (for clarity not depicted in Fig. 7.6) is a logarithmic spiral with the same rate of decay as the one for $Z-$.

(c) *Graphical presentation of the Z+ and Y+ -wave*

In Fig. 7.7[bb], an illustration is shown on how the Z_+-wave can be graphically represented in a similar way as the Z_--wave. In this case, the calculation of Z_+ and Y_+ at $x = x$ are required. Again, let Z_+ in Fig. 7.7

[bb]Derived from Thijsse's class notes as well.

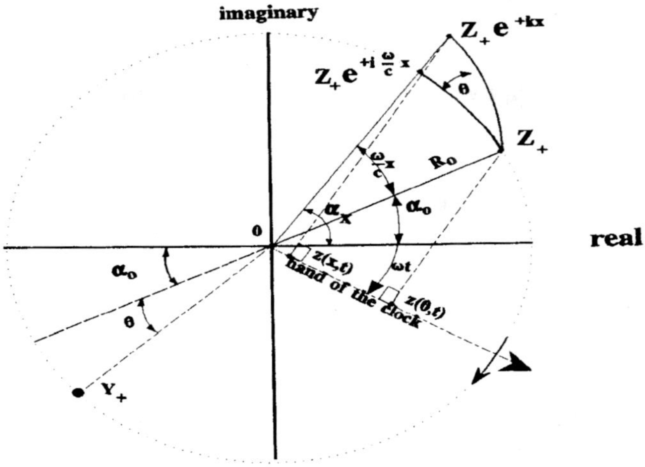

Fig. 7.7 Polar diagram of the Z_+-wave and of $Y_+(0)$. From [79].

represent the water level at $x = 0$. The point Z_+ is a short form of $Z_+(0)$, i.e. Z_{0+}. The effect of the imaginary part of k on the Z_+- vector is found by turning the vector $Z_+(0)$ over aen angle $\omega x/C$ in positive direction of α (i.e. against the hand of the clock).

According to Eq. (7.27), for the Z_+-wave the decrease of amplitude at $x = 0$, compared to its value at $x = x$, caused by the logarithmic damping of the wave amounts a factor $e^{-(\omega x/C)\tan\theta}$. The end result is the value Z_{0+}. Therefore, the amplitude of the Z_+-wave at $x = x$ should be $Z_{0+}e^{(\omega x/C)\tan\theta}$. The logarithmic spiral between Z_{0+} and $Z_{0+}e^{(\omega x/C)\tan\theta}$ is formed by increasing the value of x. The tangent to the spiral makes everywhere an angle θ with the circle with the origin as center and the vector to Z_+ as radius (Fig. 7.7). In this figure, the wave Z_+ at $x = 0$ is represented by:

$$Z_{0+} = R_0 e^{i\alpha_0} \tag{7.37}$$

As for the Z_-wave (cf. Figs. 7.6 and 7.7) projecting Z_{0+}on the hand of the clock, one finds the value of ζ_+ at $x = 0$, being:

$$\zeta_+(0,t) = R_0\cos(\omega t + \alpha_0) \tag{7.38}$$

At a point x, the value of Z_+ becomes:

$$Z_+(x) = Z_{0+}e^{kx} \tag{7.39}$$

This value, again, has to be projected on the hand of a clock to provide $\zeta_+(x,t)$. Evidently, in practice one will be more interested in the value $\zeta(x,t)$ of the sum of $\zeta_+(x,t)$ and $\zeta_-(x,t)$ and add $Z_+(x)$ and $Z_-(x)$ vectorially before projecting on the hand of the clock.

In the same way, the velocity length Y_+, which describe the horizontal tide is found by turning the logarithmic spiral over an angle θ. In this case Y_+ results from (7.33) (omitting Z_-) as:

$$Y_+(x) = -z_{0+}e^{i\theta+kx} \tag{7.40}$$

As given by (7.35) for $x = 0$ a velocity length $Y_+(0)$ belonging to the incoming wave $Z_+(x)$ is found, being $-Z_{0+}e^{i\theta}$. Thus, if $x = 0$ is defined as a reflection point of the channel (Case 1 of Table 7.1: no resultant discharge; $Y_+(0)+Y_-(0) = 0$) the consequence of (7.36), (7.37) and (7.33) is[cc]: $Z_{0+} = Z_{0-}$

d) *Graphical solution of the standard case: finding* $Z(x)$ *from* $Y(0)$ *and* $Z(0)$

After consideration of both components Z_- and Z_+ apart, now it will be explained how the components $Z_-(x)$ and $Z_+(x)$ can be graphically derived from the given vertical and horizontal boundary conditions (Z_0 and Y_0) in the point $x = 0$. For in the general case, tide and current will be the composite effect of both elementary waves, propagating and attenuating in opposite directions.

Solving Z_{0-} from (7.25) and (7.33) by elimination Z_{0+} for $x = 0$ the relations can be derived:

$$\begin{aligned} Z_{0-} &= (Z_0 + Y_0e^{-i\theta})/2 \\ Z_{0+} &= (Z_0 + Y_0e^{-i\theta})/2 \end{aligned} \tag{7.41}$$

So, turning Y_0 clockswise over an angle θ one can construct a parellellogram, from which half the diagonals are Z_{0-} and Z_{0+} respectively (Fig. 7.8). From those values, the vertical tide at an arbitrary place $x = x$ results from (7.27).

In the next Subsection 7.2.7 , the graphical polar representation given here will be transformed in its analytical shape.

[cc]This statement is no more than a visual presentation of a result, which directly could have been stated from (7.35). Compare Figs. 7.6 and 7.7, copied from.[79] The vector length "R_0" denotes the length of Z_{0-} in Fig. 7.6 and the value of Z_{0+} in Fig. 7.7. Only in the case that the channel is closed in $x = 0$ these lengths have to be the same.

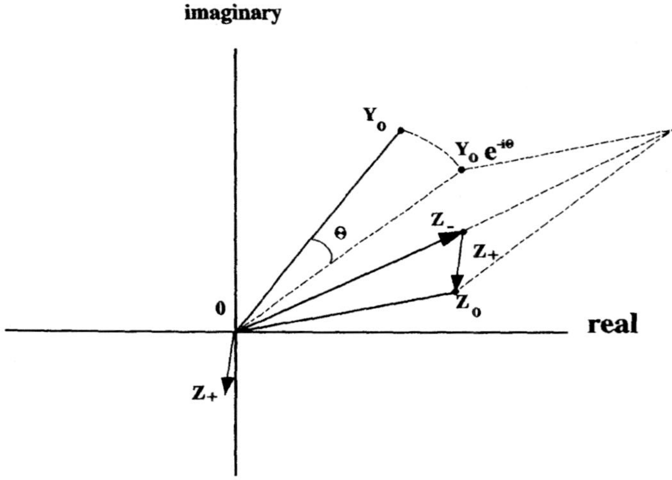

Fig. 7.8 Graphical construction of Z_{0-} and Z_{0+} in a polar diagram, given Z_0 and Y_0. From those values, $Z(x)$ and $Y(x)$ can be easily derived (Figs. 7.6 and 7.7).

7.2.7. *Analytical solution*

In the present Subsection 7.2.7 an *analytical* solution will be given, for the problem *graphically* treated in the former Subsection 7.2.6. There will be focused on the Cases 1 and 5 of Subsection 7.2.5 (Table 7.3): the problem of a semi-infinite prismatic channel.

In later Subsections (7.2.8 and 7.2.9) special attention will be given to network calculations, as necessary for solving vertical and horizontal tide in for instance the Wadden Sea (Fig. 7.3) and for calculation of influence coefficients, as displayed in Subsection 7.8.4. Background is given for solving of problems, hinted at by schematical formulae like (7.40) in Section 7.8.4.

In network calculations, continuity in nodal points will play an important role. As the channels involved do not have equal dimensions, in general, discharges in nodal points can be more important than local velocities. Therefore, formulae will be transferred from velocity length Y to local discharge Q.

Then (Subsection 7.2.8), the solution will be given in terms of discharges Q instead of velocity lengths Y. The background of Eq. (7.40) (Subsection 7.2.6) will be displayed.

Then (Subsection 7.2.9) the solution for a network will be indicated.

The following analytic relation concerning the vertical tide can be given between the complex value $Z(x)$, (representing the vertical tide ζ at $x = x$)

and $Z(0)$ at $x = 0$, given that the horizontal tide at $x = 0$ is presented by velocity length $Y_0(= Y(0))$, as defined by (7.32).

Start from (7.25) to (7.27), which can be written as (using k from (7.24)):

$$Z(Kx) = Z_{0-}e^{-kx} + Z_{0+}e^{+kx} \tag{7.42}$$

where substitution of (7.41) leads to:

$$Z(Kx) = Z_0 \cosh kx - Y_0 e^{-i\theta} \sinh kx \tag{7.43}$$

where k is complex[dd]. Still, Eq. (7.43) only yields a partial solution of the problem posed, as "k", derived in (7.21) is a complex number. A solution showing directly the characteristics of Z (amplitude and phase) for a given site x, Eq. (7.43) expressed in the unit $T\sqrt{gh}$ would make life much easier.

However, as (7.21)–(7.23) shows, the solution is a function of θ as well.

For the case that $Y_0 = 0$, i.e. the "Cases 1 & 5"-combination, mentioned in Subsection 7.2.5, this simple presentation is given below.

Assuming (as mentioned) $Y_0 = 0$, Eq. (7.43) will be written in the shape:

$$\hat{z}(kx)/\hat{z}_0 => \text{Fctn}_1(x/T\sqrt{gh}, \theta) \tag{7.44}$$

Because of the chosen condition: $Y_0 = 0$, evidently an analogous display of relative velocity lengths in the shape:

$$\hat{y}(kx)/\hat{z}_0 => \text{Fctn}(x/T\sqrt{gh}, \theta)$$

is not viable. In the equation above, the module of Y has been denoted by $\hat{y}$.

It is more logical to relate $\hat{y'}(kx)$ to $\hat{z}_0$ and the relation displayed will have a shape like:

$$\hat{y}(kx)/\hat{z}_0 => \text{Fctn}(x/T\sqrt{gh}, \theta)$$

Despite of introducing different dimensions, a more transparent picture would result, if the velocity *length* (represented by $\hat{y}$ at the left-hand side of this equation) would be replaced by the velocity amplitude $\hat{u}$ itself, as in

[dd]In that case, real and imaginary parts of cosh kx and sinh kx can be found, for instance from,[1] Eqs. (4.4.49) and (4.4.50). Real and imaginary parts of cos kx and sin kx can be found from,[1] Eqs. (4.4.55) and (4.4.56). If friction is negligible, k is pure imaginary and cosh kx is cos ikx, where sinh kx is sin $(\omega x/\sqrt{gh})$; cf. (7.21).

the present way a quite irrelevant effect of θ is introduced on the left-hand side of the equation above: from (7.32) and (7.23) it shows:

$$\frac{\hat{u}(kx)}{\hat{y}(kx)} = \sqrt{\frac{g}{h}\cos(2\theta)} \tag{7.45}$$

Thus, for depicting velocity amplitudes as function of site and bottom friction (i.e. θ) $\hat{y}(kx)$ gives a distorted picture. It is more logical to plot:

$$\frac{\hat{u}(kx)}{\hat{z}_0\sqrt{g/h}} = \frac{\hat{y}(kx)}{\hat{z}_0}\sqrt{\cos(2\theta)} \tag{7.46}$$

Resuming: there will be calculated:

$$\frac{\hat{u}(kx)}{\hat{z}_0\sqrt{g/h}} => \mathrm{Fctn}_2(x/T\sqrt{gh}, \theta) \tag{7.47}$$

The phase between vertical and horizontal tide[ee] will be expressed as:

$$\arg Y(kx) - \arg Z(kx) = \mathrm{Fctn}_3(x/T\sqrt{gh}, \theta) \tag{7.48}$$

Anticipating on the results: those are displayed in Fig. 7.9; the relative tidal amplitudes $\hat{z}(kx)/\hat{z}_0$ and relative velocity amplitudes $\{\hat{y}(kx)\sqrt{\cos(2\theta)}/\hat{z}_0\}$ are denoted as $\hat{z}'(kx)$ and $u'(kx)$ respectively.[ff]

For the calculation of the relations given in Fig. 7.9, use is made of Eqs. (4.3.55)–(4.3.57) from[1] concerning the real and imaginary parts of a circular function with complex argument $\psi(= \psi_r + i\psi_i)$:

$$\begin{aligned} \sin\psi &= \sin\psi_r \cosh\psi_i + i\cos\psi_r \sinh\psi_i \\ \cos\psi &= \cos\psi_r \cosh\psi_i - i\sin\psi_r \sinh\psi_i \end{aligned} \tag{7.49}$$

With respect to the module, the following relations hold (Eqs. (4.5.54) and (4.5.56) in[1])

$$\begin{aligned} |\sinh\psi| &= \sqrt{\sinh^2\psi_r + \sin^2\psi_i} \\ |\cosh\psi| &= \sqrt{\sinh^2\psi_r + \cos^2\psi_i} \\ \arg(\tanh\psi) &= \arctan\left\{\frac{\sin(2\psi_i)}{\sinh(2\psi_r)}\right\} \end{aligned} \tag{7.50}$$

[ee] Mind that the *phases* of $Y(kx)$ and velocity $U(kx)$ are the same (cf. (7.32)).

[ff] In this Chapter 8, vertical tidal amplitudes will be made dimensionless as well with as $\hat{z}_0$ well with the waterdepth h, depending on the physical problem tackled. Symbols assigned to the dimensionless quantities are $\hat{z}'$ and $\hat{z}^*$, respectively.

Given kx in (7.43) (to be written as: $k_r x + ik_i x$), the belonging value of $x/T\sqrt{gh}$ results from the relationship (7.24):

$$\begin{aligned} k_r &= \frac{\omega}{\sqrt{gh}} \frac{\sin\theta}{\sqrt{\cos(2\theta)}} \\ k_i &= \frac{\omega}{\sqrt{gh}} \frac{\cos\theta}{\sqrt{\cos(2\theta)}} \end{aligned} \tag{7.51}$$

where the module of k is:

$$|k| = \frac{\omega}{\sqrt{gh}} \frac{1}{\sqrt{\cos(2\theta)}} \tag{7.52}$$

Thus, the relative amplitude $\hat{z}'(kx)$ at a certain site (compared to the amplitude $\hat{z}_0$ at the landward side of the basin) is equal to:

$$\hat{z}'(kx) = |\cosh kx| \tag{7.53}$$

according to (7.43), given that velocity (length) Y_0 has been taken zero.

Its value can be derived for given values of $T/\sqrt{gh}$ and θ from (7.50), where the constituants k_r and k_i of k result from (7.51). This relative amplitude is displayed in Fig. 7.9a.

The horizontal tide at $x = x$ can be depicted in a similar way:

Start from (7.33), which can be written as:

$$Y(kx) = e^{i\theta}(Z_{0-}e^{-kx} - Z_{0+}e^{+kx}) \tag{7.54}$$

Substituting Z_{0_-} and Z_{0_+} from (7.41) yields:

$$Y(kx) = \frac{1}{2}e^{i\theta}[(Z_0 + Y_0 e^{-i\theta})e^{-kx} - (Z_0 - Y_0 e^{-i\theta})e^{kx}]$$

which can be written as:

$$Y(kx) = -Z_0 e^{i\theta} \sinh kx + Y_0 \cosh kx \tag{7.55}$$

As mentioned, there is focused on the case $Y_0 = 0$: the case of a closed semi-infinite prismatic channel. For this case, in Fig. 7.9b the relative velocity amplitude $u'(kx)$ is displayed, equal to: $\{\hat{y}(kx)\sqrt{\cos(2\theta)}\}/\hat{z}_0$. Remind from (7.46), that the reference velocity amplitude is: $u_{\text{ref}} = \hat{z}_0\sqrt{g/h}$. The calculation of it is similar to the one of $\hat{z}'(kx)$ above. According to (7.55), for $Y_0 = 0$ the value of $\hat{y}(kx)/\hat{z}_0$ equals:

$$\hat{y}(kx)/\hat{z}_0 = |\sinh kx| \tag{7.56}$$

$$\frac{\hat{u}(kx)}{\hat{z}_0\sqrt{g/h}} = |\sinh(kx)|\sqrt{\cos(2\theta)} \tag{7.57}$$

The value on the right-hand side can be expressed in its real-valued constituents by using (7.50). Again, those constituants k_r and k_i of k result from (7.51).

In Fig. 7.9b those results are displayed.

In the lowest part of Fig. 7.9 the difference in phase between high tide and maximum current velocity is shown. Its value can be derived from:

$$\arg\{Z(kx)\} - \arg\{Y(kx)\} = \arg\{Z(kx)/Y(kx)\}$$

Substitution of (7.43) and (7.55) respectively, with $Y_0 = 0$, gives:

$$\arg\{Z(kx)\} - \arg\{Y(kx)\} = \arg\{-e^{i\theta}/\tanh(kx)\}$$

$$\arg\{Z(kx)\} - \arg\{Y(kx)\} = -\arg\{\tanh(kx)\} + \pi - \theta$$

Applying (7.50) yields the following result.

$$\arg\{Z(kx)\} - \arg\{Y(kx)\} = -\arctan\left(\frac{\sin(2k_i x)}{\sinh(2k_r x)}\right) + \pi - \theta \qquad (7.58)$$

For small values of x, Eq. (7.58) can be approximated ($\sin \approx \sinh$), looking at (7.51):

$$\arg\{Z(kx)\} - \arg\{Y(kx)\} \approx -\frac{\pi}{2} + \theta + \pi \approx \frac{\pi}{2} \text{ (for small } kx) \qquad (7.59)$$

For large values of x the "arctan" in (7.58) tends to zero, i.e. under those circumstances:

e. without friction vertical and horizontal tide will have a phase shift of 180°.
f. for very large friction the phase shift will be 135°.

In Fig. 7.9c, phase differences between vertical and horizontal tide are depicted: the plots are valid for values of θ, indicated on the right-hand side of this figure.

As shows from Fig. 7.9, if friction is small, there will be a large variation of phases (between approximately 90 and 225) in the most landward wavelength of the channel, due to the effect of the reflected wave, which attenuates in seaward direction. In case of small friction, horizontal and vertical tides are approximately in phase; the phase difference of 180 is due to the fact, that the wave propagates in *negative* x-direction.

In order to give a physical explanation of the "tops and troughs" in Fig. 7.9, refer to Fig. 7.10. This shows the reduction of the "physical wave length" λ, caused by bottom friction.

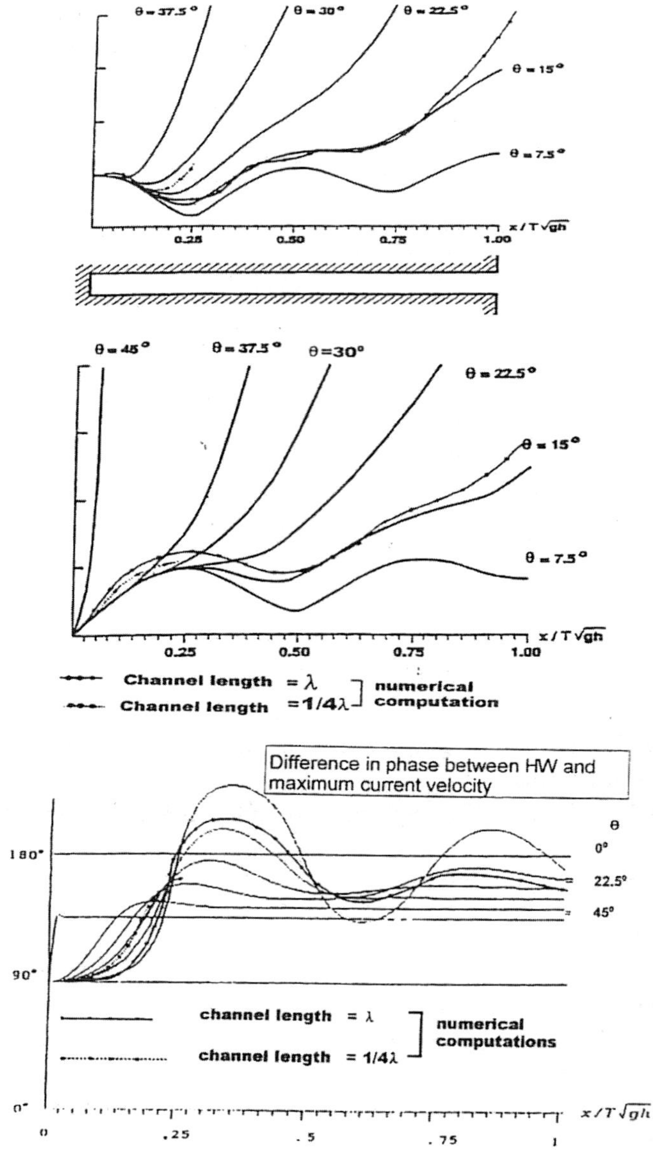

Fig. 7.9 (a–d) Amplitude normalized vertical tide $\hat{z}'$, normalized horizontal tide $\hat{u}'$ and phase difference between vertical tide and horizontal tide; comparison numerical and analytic computation (cf. Subsection 7.2.10). (a) Vertical tide; (b) plan view channel; (c) horizontal tide; (d) phase difference. Mind, that Fig. 7.9a shows the amplitude of the wave, *necessary* at $x = x$ in order to obtain an amplitude of unity at the landward side of the channel. Evidently, in practice the amplitude at the closed end of the channel will be determined by the boundary condition at the seaside.

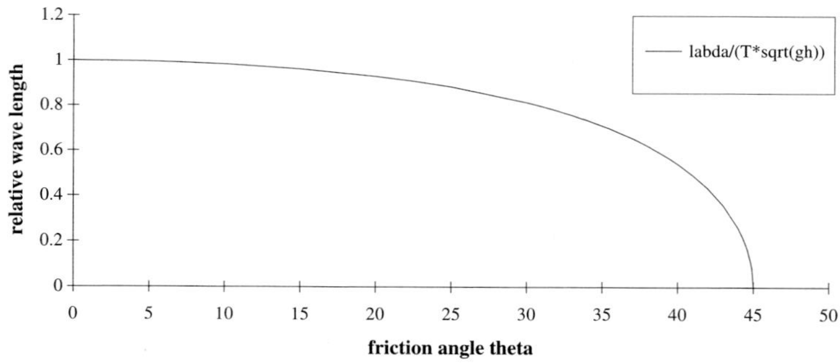

Fig. 7.10 Relative way length $\lambda/T\sqrt{gh}$ as a function of friction angle θ.

This might be defined as the distance, over which the phase of the constituant "Z_--wave", c.q. "Z_+-wave" (sub-subsection a) and c)), change an angle 2π. From (7.39) this distance is shown to be:

$$\lambda = 2\pi/\mathrm{Im}(k) \tag{7.60}$$

by using (7.51):

$$\frac{\lambda}{T\sqrt{gh}} = \frac{\sqrt{\cos(2\theta)}}{\cos\theta} \tag{7.61}$$

However, because of the linear theory (approximation) the presentation in Fig. 7.9 allows that one graph represents an infinite number of channels of all kinds of lengths.

Figure 7.10 shows $\lambda/T\sqrt{gh}$ as function of θ; its relation with the nodes and antinodes in Fig. 7.9 is clear: the nodes in Fig. 7.9 approach more and more the origin when θ increases.

7.2.8. *From channel to network*

In this Subsection 7.2.8 the treatise concerning the way of computation of tides in straight channels will be pursued by considering network of channels. The topic of combination of channels (parallel or serial) to one hydraulicly equivalent single channel (cf. Subsection 7.2.1) will not be treated at all in this book. Refer to the Lorentz report,[49] to Thijsse[79] or to "Lorentz on PC" Jallah and Bakker;[42] a short description concerning "Lorentz on PC",[42] is given in the next Subsection 7.2.9.

The present Subsection 7.2.8 reveals the background of remarks made in Subsection 7.8.4,[gg] leading to Eqs. (7.197) and (7.198) of Subsection 7.8.6.

[gg]So the present treatise is in anticipation of this Subsection 7.8.4.

Consider first "Case 2", pointed at in Subsection 7.2.5: in a prismatical channel the *vertical* tide at both sides of the section is known; wanted is the *horizontal* tide. This matter is of importance: if one links channels together in a tidal computation, evidently continuity in nodal points[hh] should be obeyed.

Thus, consider such a network. Start from an unknown vertical tide, denoted by a complex number Z_k in a nodal point k in a network,[ii] and by Z_{k+1} in an adjacent nodal point $k+1$ of the network. Let those nodal points be linked with a channel. Then "Case 2", as denoted before, will facilitate to find the velocities at the ends of the considered channel. Thus, also the (harmonical) discharges $Q_{k,i}$ and $Q_{k+1,i}$ (where the first subscript, k, stands for a certain nodal point and the second one, i, for one of the channels of the network) will be known.

These kind of relations among Z_k, Z_{k+1}, Q_k and Q_{k+1} will result in a number of linear equations (with unknowns in the shape of complex numbers), from which those unknowns can be solved.

In other words, these thoughts are verbalized in Subsection 7.8.4 (*quote*):

"Via the harmonic solution of the harmonic equations for a single channel, the tide at one end of each section is linked to that at the other end. In a network model, we have to close the mathematical system via another set of relationships between the channels which meet at each nodal point. Per nodal point these relationships can be found from the continuity requirement (the sum of all flow rates toward a nodal point equals zero) and the compatibility requirement (the water levels at the ends of all branches in one node are the same).

Thus we find a system of N equations with N unknown complex amplitudes $Z \dots$ (*unquote*) ... which can be solved.

The considerations above demonstrate, that for the present purpose (where in the nodal points channels with various cross-sections meet) the (harmonic) discharge Q at the end of a channel is more important than the velocity length Y. Thus, in the present section, a complex variable Q, linked to the discharge, will be defined, similar to Y:

$$Q = bhU \tag{7.62}$$

where b is the width of the channel (assumed prismatical) and the complex variable U (linked to the velocity) is defined in Subsection 7.2.2. Evidently, Qhas the same phase as U.

[hh] Displayed for instance in Fig. 7.3 in Subsection 7.2.1.

[ii] Quite similar to $Z(kx)$ in, for instance, (7.43).

From (7.31) it shows:

$$Q = Y \times bC \cos\theta \tag{7.63}$$

In Ref. [12] the notation Δ is assigned to:

$$\Delta = bCe^{i\theta}\cos\theta \tag{7.64}$$

The constant Δ may be seen as a profile-related characteristic of the channel.

From (7.63) and (7.64) it is shown that:

$$\frac{Q}{\Delta} = Ye^{-i\theta} \tag{7.65}$$

This facilitates a concise notation of (7.43):

$$\frac{Q_0}{\Delta} = \frac{Z_0}{\tanh kx} - \frac{Z_x}{\sinh kx} \tag{7.66}$$

or still more specific, referring to a channel length l from "here" to "there":

$$\frac{Q_{\text{here}}}{\Delta} = \beta Z_{\text{here}} - \alpha Z_{\text{there}} \tag{7.67}$$

with:

$$\begin{aligned} \alpha &= \frac{1}{\sinh(kl)} \\ \beta &= \frac{1}{\tanh(kl)} \end{aligned} \tag{7.68}$$

Mind[jj] that (7.67) has the same structure as Table 7.3 of Subsection 7.2.5. "Z_0" and "Z_L" in Table 7.3 are similar to "Z_{here}" and "Z_{there}" in (7.67). However, where Table 7.3 gives question marks, Eq. (7.67) yields answers. The important answer is, that Y(or Q) is a linear combination of the values of the boundary conditions (no surprise after having conveniently erased all non-linear terms in the equations).

Mind furthermore, that one may swap the direction of the x-axis in (7.67). Then also the direction of positive Q swaps; "Z_{here}" becomes "Z_{there}" vice versa, but one gets an answer on the second question mark in "Case 2" (Table 7.3) as well. Generally, the whole Table 7.3 shrinks to the problem of solving two linear equations with two variables and two unknowns.

[jj]Mind furthermore, that for small values of kl (7.67) looks like a linearized Chezy law (discharge proportional to the slope of the water level).

The coefficient Δ incorporates the cross-sectional characteristics of the channel, the coefficients α and β the longitudinal characteristics; the bottom friction angle θ plays a role, as well in Δ as in the (α, β) combination.

The subjective choice of the direction of the x-axis in (7.67) condenses the two equations "left => right" and "right => left" to one equation. If one abstains from that advantage, vector analysis gives another (and still better) possibility for condense notations.

Requested is to formalize the concept, that vertical tide Z and horizontal tide (Q/Δ) are[kk] linearly related,[ll] with "a" proportionality matrix $\overline{\overline{D}}$ ("dimensions") depending on channel characteristics.

This concept is developed here for a single channel; however it is "just a step" (one channel has two ends and a netwerk of channels has more ends) to extend the concept to a network.

The concept of proportionality[mm] "$Q = DZ$", being extended and refined already by a formulation in complex numbers, now will be refined and extended once more by incorporating that Q "on both sides of the channel" depends "linearly" on Z "on both sides of the channel". Finally, it will be be extended to the idea that Q "in the nodal points in the network" depends "linearly" on Z "in the nodal points of the network". Mathematical translation of this procedure is vectorization of Z, D and Q, i.e.:

$$\frac{\vec{Q}}{\Delta} = \overline{\overline{D}}\vec{Z} \tag{7.69}$$

So, first, focus on the "one-channel case" again. Let $\vec{Z}$ be a two-dimensional "vertical tide vector" as (Z_1, Z_2) denotes the vertical tide at the ends of the channel. Let the discharge vector $\overrightarrow{Q}$ denote (Q_1, Q_2), i.e. the discharges $(Q_{\text{here}}, Q_{\text{there}})$. Both vectors $\vec{Z}$ and $\overrightarrow{Q}$ have complex constituents. The complex "dimension matrix" $\overline{\overline{D}}$, relates $\overrightarrow{Q}$ and $\vec{Z}$ according to (7.69) (in

[kk]Given our wishful thinking! Refer to Subsection 7.2.10 to become convinced, that this is not *complete* nonsense.

[ll]Mind the subtility of the "linearity": phase differences are incorporated by the use of complex "proportionality coefficients"; for the mathematical (and physical!) concept and treatment the fact of using either complex or real numbers is not essential! The Netherlands should be grateful for the mental activity performed by Lorentz and Thijsse in the ("roaring") 1920s. In his class concerning "Tides" Thijsse related concerning reproaches (in the coach) at the funeral of Lorentz, because of keeping Lorentz from his work (relativity theory).

[mm]Discharge relative to vertical tide, depending on dimensions of the channel(s).

"two directions": from "left to right" and from "right to left", as indicated above).

Thus, according to (7.67), the two incorporated relations (taking "here" successively on both sides of the channel) now can be reformulated and one may write $\overline{\overline{D}}$ as:

$$\overline{\overline{D}} = \frac{1}{\sinh(kl)} \begin{pmatrix} \cosh(kl) & -1 \\ 1 & -\cosh(kl) \end{pmatrix} \tag{7.70}$$

One thus may conclude directly, that the inverse relationship ($\vec{Z}$ as function of $\overrightarrow{Q}$) can be found by taking the inverse of $\overline{\overline{D}}$.

Defining a two-dimensional matrix $\overline{\overline{A}}$ as:

$$\overline{\overline{A}} = \begin{pmatrix} a_{11} & a_{12} \\ a_{21} & a_{22} \end{pmatrix} \tag{7.71}$$

According to well-known matrices theory, the inverse of $\overline{\overline{A}}^{-1}$ is:

$$\overline{\overline{A}}^{-1} = \frac{1}{|A|} \begin{pmatrix} a_{22} & -a_{12} \\ -a_{21} & a_{11} \end{pmatrix} \tag{7.72}$$

in which $|A|$ is the determinant of $\overline{\overline{A}}$.

Thus, converting (7.69) yields:

$$\vec{Z} = \overline{\overline{D}}^{-1} \frac{\overrightarrow{Q}}{\Delta} \tag{7.73}$$

as $|D|$, the determinant of $\overline{\overline{D}}$ equals $-\sinh(kl)$. Application of (7.72) shows, that $-\overline{\overline{D}}^{-1}$ equals $\overline{\overline{D}}$:

$$\vec{Z} = \overline{\overline{D}}\overrightarrow{Q}/\Delta \tag{7.74}$$

Thus the two simple Eqs. (7.69) and (7.74), with $\overline{\overline{D}}$ given in (7.70) give the solution concerning the problems 2 and 3 of Thijsse[79] mentioned in Subsection 7.2.5, as well as the solution of the inverse problem.

7.2.9. *Lorentz on PC*

a) *Introduction to 7.2.9*

In 1994, Jallah and Bakker[42] published a description of a computer program "Lorentz on PC", which was meant to carry out tidal computations according to the Lorentz method. It has been used for the computations, described in Section 7.8; an abstract is given in the present Subsection 7.2.9.

It is the idea, to publish it on the Web as free software in the future; extensions (in a described format) would be welcomed It consists of the following constituents:

(a) Introduction
(b) Summary of the Lorentz method of tidal computation
(c) The computer program
(d) Checks of the program: Waterway Hook of Holland to Rotterdam
(e) Program application: Western Wadden Sea
(f) Conclusions and recommendations

The investigation was carried out as a part of the Coastal Genesis Project. It was financed partly by the Ministry of Public Works "Rijkswaterstaat" and partly by the Delft University of Technology.

The program enables the following:

(a) Making and visually showing a configuration for a prismatic channel-in-series or network in channels.
(b) Replacing a channel with an irregular cross-section by a prismatic channel with equivalent hydraulic properties.
(c) Calculation of coefficients according to the Lorentz method.
(d) Calculation of representative tidal constituents (vertical, horizontal amplitudes and phases).
(e) Three-dimensional simulation of vertical tide and discharges over a tidal cycle.

A sample of results (influence coefficients) is shown in Fig. 7.55 in Subsection 7.8.5.

The present review will focus mainly on the computer program itself.

(b) *Summary of the Lorentz method of tidal computation*
Apart from the subjects treated in Subsection 7.2.8, Section 2.2 of Jallah and Bakker[42] focuses on:

a. Channel schematization: the calculation of a representative depth h_0, width b_0 and friction coefficient θ_0 for a channel with an irregular cross-section (Fig. 7.11). Here, the total width of the channel equals the sum of the widths of the separate channels; the same water level slopes along all the sections and the discharge q_0 of the original channel equals the discharge of the schematized channel. Refer to,[79] pp. 84–87.

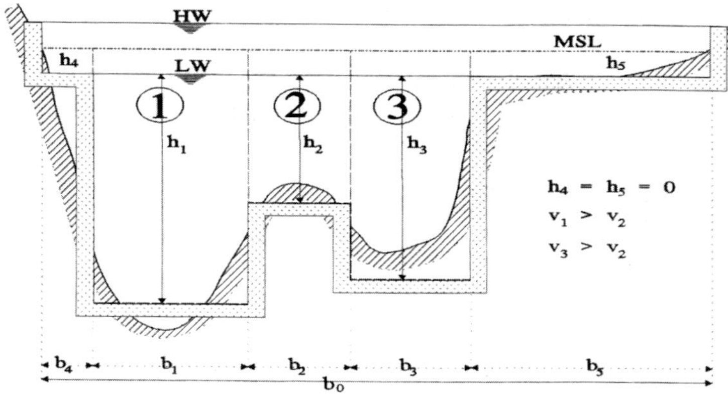

Fig. 7.11 The irregular cross-section (firstly schematized as a series of parallel channels $b_1 \times h_1 \cdots b_2 \times h_2 \cdots b_n \times h_n$) is replaced by a rectangular cross-section $b_0 \times h_0$.

b. Replacing of a series coupled channels with different cross-sections to one representative channel. For example: replacing channel $2^{\mathrm{a}} - 2^{\mathrm{b}} - 2^{\mathrm{c}}$ in Fig. 7.3 by one channel 2.

As shown in Subsection 7.2.8, the essential characteristics of a channel for tidal computations are the coefficients α and β in Eq. (7.67). Using those equations for two (serial) linked channels, one gets two equations for the discharge in the point where the two channels are linked, from which this value can be extrapolated.

Thus, if one links two sections a and b with coefficients $\alpha_{\mathrm{a}}, \alpha_{\mathrm{b}}, \beta_{\mathrm{a}}$ and β_{b} respectively, for the combined channel one gets an equation like (7.67), i.e.

$$Q_{\text{here}} = \beta_{\text{ab}} Z_{\text{here}} - \alpha_{\text{ab}} Z_{\text{there}} \tag{7.75}$$

where the new point "there" is at the end of the second channel.

One finds:

$$\begin{aligned} \alpha_{\text{ab}} &= \frac{\alpha_{\text{a}}\alpha_{\text{b}}}{\beta_{\text{a}} + \beta_{\text{b}}} \\ \beta_{\text{ab}} &= \frac{\beta_{\text{a}}\beta_{\text{b}} + \Delta_{\text{a}}^2}{\beta_{\text{a}} + \beta_{\text{b}}} \end{aligned} \tag{7.76}$$

Note, that α does not change irrespective of the notation "here" and "there". However, β does change. Its value will change if one swops from "here" to "there" (channels a and b. In Eq. (7.75), the value of Q_{b} will be different as channel b is taken as "here" and channel a is taken as "there".

For a series of more than two channels eq. (7.76) is applied repetitively: the combination of the first two channels is combined with the third, etc.

The (important) subject of dealing with channels with irregular cross-section does not come into focus in the present Chapter 7.2; however, it is treated in[42] in accordance to Thijsse.[79] The method has been used for the calculation of the influence coefficients, depicted in Fig. 7.55.

Considering a network of channels (cf. for instance Fig. 7.4), one can distinguish: "junction points", where three[nn] or more channels meet, and "boundary points" where the tide enters the basin considered. In Fig. 7.4 the four points "Westgat", "Eierlandse Gronden", "Noordgat" and "Bornrif" are the boundary points. All the other dots in Fig. 7.4 refer to junction points.

Consider first a junction point, only connected with other junction points, i.e. having no direct connection to a boundary point.

Such a junction point "here" has a number of connections with the other junction points "there". Continuity "here" requires, that the sum of the discharges "here" is zero. Thus, continuity "here" can be formulated by addition of the successive Eq. (7.75) for the various channels involved, which results into:

$$Z_{\text{here}} \sum \beta_{\text{channels}} - \sum \alpha_{\text{channels}} Z_{\text{there}} = 0 \tag{7.77}$$

Consider now the case, in which the junction point has a connection with the boundary point.

So let there be K boundary points and focus on boundary point $k(k \leq K)$, being connected to a junction point, denoted by k as well. Let this junction point have J_k connecting channels, from which the *first* one links the junction point to the boundary point.

Call:

$$\beta_{\text{tot},k} = \sum_{j=1}^{J_k} \beta_{j,k} \tag{7.78}$$

where $\beta_{j,k}$ is the β-factor of the jth channel in the junction point k. Similarly, $\alpha_{j,k}$ is the α-factor in this channel. Thus, one may write Eq. (7.75) for junction point k:

$$\beta_{\text{tot},k} Z_k - \sum_{j=2}^{J_k} \alpha_{j,k} Z_j = \alpha_1 Z_{\text{b}} k \tag{7.79}$$

[nn]If only two channels meet, the two channels can be combined to one, using Eq. (7.76).

where Z_{bk} denotes the complex presentation of the vertical tide in the kth boundary point.

In a similar way, one can treat a case, in which a junction point is connected to two (or more) boundary points. For the case of two connections with boundary points, let the α-factors of those channels (i.e. inlets) be α_1 and α_2 and the vertical tide in those boundary points Z_{b1k} and Z_{b2k}. Instead of Eq. (7.80) one should find

$$\beta_{\text{tot},k} Z_k - \sum_{j=3}^{J_k} \alpha_{j,k} Z_j = \alpha_1 Z_{b1k} + \alpha_2 Z_{b2k} \tag{7.80}$$

Tides at the boundaries are known values; thus one may assign a (complex) value of ς_2 to the ratio between Z_{b1k} and Z_{b2k}; dividing Eq. (7.80) by $1 + \varsigma_2$ reduces (7.80) to the shape of (7.79).

Thus, for N points of a network, N equations can be composed of the kind of either (7.77) or (7.79), resulting in a matrix:

$$\overline{\overline{D_{\text{tot}}}}\vec{Z} = \vec{Z}_{\text{b}} \tag{7.81}$$

in which the elements $\vec{Z}_{\text{b}}$ are either zero, either represent a boundary condition.

The solution is:

$$\vec{Z} = \overline{\overline{D_{\text{tot}}}}^{-1}\vec{Z}_{\text{b}} \tag{7.82}$$

In words, by inverting the matrix $\overline{\overline{D_{\text{tot}}}}$ in (7.81) the vertical tide in a point of a network can be composed by linear[oo] superposition of the vectors, representing the vertical tide in the various boundary conditions.

Thus, for each of the boundary conditions one may compose influence coefficients for every point of the network. Refer to Subsection 7.8.5, Fig. 7.55, in which the modules of those amplitudes of the Waddensea network (Fig. 7.3) are depicted.

(c) *Practical aspects of the program "Lorentz on PC"*
The computer program "Lorentz on PC" contains the following constituants:

a. *Input of data* concerning the (plan view of) configuration: coordinates of boundary points, junction points and intermediate points (points in a

[oo]If reality should be linear.

tidal channel, where the cross-section changes). The program can show a two-dimensional or a three-dimensional view of the configuration; this last aspect is of importance when combined with a three-dimensional view of the water surface (cf. Fig. 7.55[pp]).

b. *Channel characteristics* (widths, depths and lengths); those lead to the calculation of the α- and β- coefficients;
c. *Calculation of the influence coefficients* (Subsection 7.2.9 b.)) and subsequently of the vertical tide in the nodal points, given the boundary conditions;

Additional to the description of the program, also some checks have been given. Choice of well-matched Chezy coefficients is important.

One of the displays shows a film of the vertical tide by assigning to each nodal point of the network a height equal to the momentaneous water level.

7.2.10. *Effect of linearization*

The present Subsection 7.2.10 covers a subject, which is only "at the border" of the overall subject of Section 7.2, which is: "linear tidal hydraulics".

For Subsection 7.2.10 deals on the *accuracy* of the linearization (in the case of a semi-infinite channel). Its content is based on a report of Bult.[22] Analytical computations were checked by Bult,[22] with the non-linear numerical DUFLOW program of SPAANS *et al.*[70]

Refer to Fig. 7.9a showing the dimensionless amplitude $\hat{z}'$ of ς as function of x^*. Results of Lorentz computations are given for various values of θ. In the same way, Fig. 7.9b shows the dimensionless velocity amplitude $\hat{u}'$. For zero friction, curves would have the shape of a $|\cos|$- and a $|\sin|$-function, respectively.

Checking the values with a non-linear computation is difficult, as–because of non-linear friction–a constant Chezy-friction value[qq] does not result in a constant θ-value[rr] over the channel. Furthermore, under normal conditions of real estuaries, for friction values which are physically real the reflected wave practically always will be attenuated after half a wave length

[pp]Figure 7.55 shows a three-dimensional picture of influence coefficients. In the same layout, "Lorentz on PC" allows the display of momentaneous waterlevels, or shows a tidal simulation ("movie"); last rule of this paragraph.

[qq]Occurring in non-linear computation.

[rr]Occurring in Lorentz computation.

or less: values of θ of 7.5° or 15° are not very likely in practice. For testing, however, longer channel/wave length ratios are illustrative.

In order to check the mathematics, some tests with physically unrealistic values of the Chezy-value have been performed. Two cases have been checked, in which the channel length was L_0 and $L_0/4$ respectively. The channel length in either case was equal (86.4 km), but the depth h was different: 0.408 m, resp. $16 \times 0.408 = 6.41$ m. The tidal wave period T was taken as 12 hours. Chezy-values C_h were: 120, resp. $60\sqrt{m/s}$ and the amplitude $\hat{z}_s$ at the seaward boundary was taken $h/10$. When relating C_h c.q. f_c, to θ the velocity amplitude $\hat{u}$ enters the formula (see Subsection 7.2.3). In the case of a semi-infinite channel with a closed end, one finds a channel-averaged value of $\hat{u}$ (if one neglects friction):

$$\hat{u} = \frac{2}{\pi} \times \frac{\hat{z}_0}{h}\sqrt{gh} \tag{7.83}$$

Here $\hat{z}_0$ denotes the vertical wave amplitude in the antinodes. Equating the period-integrated friction loss using a linear friction approach ($\tan(2\theta)$) to the same entity, however calculated starting from the assumption of quadratic friction loss (cf. Subsection 7.2.4 ad. a) a relation between θ and Chézy similar to (7.15) can be derived C_h and θ can be derived.

$$\tan(2\theta) = \frac{8}{3\pi^3}\frac{\hat{z}_0}{h}\frac{L_0}{h}\frac{g}{C_h^2} \tag{7.84}$$

For the $L_0/4$-channel a value $\theta = 25.44°$ is found. For the L_0-channel, the fact that friction is neglected in the derivation of (7.83) gives some ambiguity: in fact $\hat{u}$ in the most landward node (at $x \approx L_0/4$ will be much less than the one in the most seaward node ($x \approx 3L_0/4$). If for z_0/h in (7.84) the seaward boundary value of 0.1 is taken, the same value of θ results as for the $L_0/4$-channel as (7.84) shows.[ss] However, if one takes for z_0/h the value at the landward side of the channel, which is 0.01377/0.408, a value of 11.37° is found. Thus, for the L_0-channel θ should be in the range $11.4° < \theta < 25.4°$.

Figure 7.9a–d shows (between the given ranges of θ) the correspondence for the amplitudes of the vertical and horizontal tide, as well as for the phase difference.

[ss] For the long channel z_0/h remains the same; L_0 is divided by 4 and h is divided by 16; C_h is multiplied by 2.

7.3. Tidal Hydraulics: Resultant Currents, Caused by Longitudinal Density Gradients

7.3.1. *Introduction; kind of estuaries to be modeled; considered topography; assumptions concerning the circulation mechanism and the horizontal density gradient*

The former Section 7.2 dealt on *momentaneous* velocities. As mentioned in the introduction of that section, for the present goal the subject "momentaneous velocities" is just a tool for leading to conclusions concerning *tidal-averaged* currents, and, still more important, concerning tidal-averaged sediment motion and tidal-averaged accretion and tidal-averaged erosion. This kind of bottom response has its feed-back on the water motion. Knowledge on this counteraction finally leads to a vision on long-term change of topography.

In the present Section 7.3 (and two others following, i.e. Sections 7.4 and 7.5) the focus remains on tidal hydraulics, i.e. on the *water* motion; the effect on the bottom of the estuary and on delta formation will be treated later on (Section 7.6–7.8).

Sections 7.3 and 7.4 consider, respectively, two important reasons[tt] for tidal circulation:

7.3: density gradients

7.4 "rain in the drain": undertow in *tidal* waves; it is caused by non-linear aspects of water motion.

It will be made soon clear, that those effects counteract each other. Thus a very relevant question is (put by BLIEK, actually, cf. Section 8.1) which one of the two is the most important. Response on this question will be given in Section 7.5.

A general schedule of Section 7.3 (also in relation to Section 7.4) is depicted in Fig. 7.12.

Refer to the Introduction 7.1, Table 7.1 (p 345) (Cases 25 and 29) concerning the variables involved:

a. The present Section 7.3 only deals on *water* motion (where, by the way, the effect of surface waves is neglected).
b. The estuary is schematized as a single closed prismatic channel (although some excursions will be held to prototype circumstances (Wester Scheldt; multi-channel inlets)).

[tt] Often, the most important reasons.

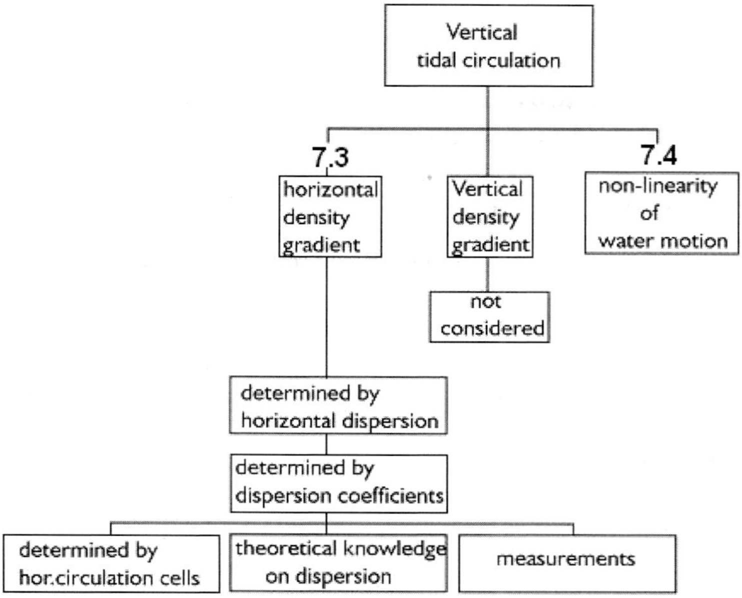

Fig. 7.12 Vertical tidal-averaged circulation: mechanisms to be treated in Sections 7.3 and 7.4.

Thus, focus (in this Section 7.3) on the *density currents.*

The schematization used below is by no means a passe-partout for all estuaries.

Important confinements are:

* vertical stratification is neglected i.e. the difference in salinity between bottom and surface is neglected
* the (tidal- and depth)-averaged velocity is neglegible with respect to the tidal-averaged surface velocity.

The text *(in italics)* below puts the investigation in a more general scope.

Van de Kreeke and Zimmermann[84] (following Hansen and Rattray[38] characterize estuaries by two parameters (resp. on the horizontal and the vertical axis in Fig. 7.13):

- The ratio between tidal-mean surface velocity "u_s":, related to the depth-averaged fresh-water velocity "u_f".

As the latter is assumed neglegible in the present Section 7.3, this ratio thus should be very (infinitely) large:

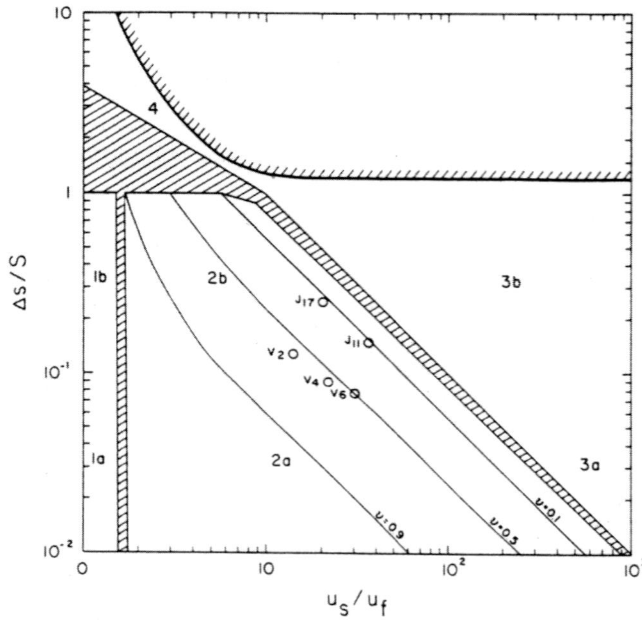

Fig. 7.13 Circulation-stratification diagram (from [84]).

- The difference in salinity "Δs" between bottom and surface, related to the mean density S; as the former is assumed neglegible, this ratio should be very small.

They distinguish four different estuary types (Fig. 7.13):

Type 1: no flow reversals.

Type 2: relative moderate flow reversals, caused by gravitational circulation.

Type 3: gravitational circulation completely controls the upestuary salt flux.

Type 4: Strong stratification, hardly any density circulation.

It is suggested, Fig. 7.13, that for small vertical density gradients, a linear relationship consists between the two above-mentioned parameters for a given estuary, depending on a parameter "ν", being the ratio of the upestuary diffusive flux and the downestuary flux. In the schematization used below, therefore ν is schematized as 1.

The kind of estuary considered in the following investigation ($\Delta s \sim 0$; $u_f \sim 0$) is thus situated at the right-hand side, down-under in the diagram.

Thus the analytic model used is a "type 2" model[uu] ("flow reversals") in the terminology of V.d. Kreeke & Zimmermann,[84] i.e. a model, driven by gravitational circulation only. With respect to circulation, the fresh water discharge is completely neglected, as is the salt concentration gradient in vertical direction. Thus the model displayed below represents the right downward corner in a circulation–stratification diagram (cf. Fig. 7.13, being Fig. 7.2 of:[84] The Western Scheldt estuary (Neth.,)) is thought as an example of this kind of estuary.

Thus the following hydraulic features are considered as important (Figs. 7.14 and 7.19):

(a) Gravitational circulation.
(b) Diffusion: determined by a longitudinal dispersion coefficient D_x. There will be focused on (what is considered as) the main cause of this dispersion, in which the third dimension of the estuary plays an important role: the horizontal circulation in circulation cells.

Thus the concept reads (Figs. 7.14 and 7.15):

(a) The local salt gradient determines the gravitational circulation.
If a (theoretical) shape of density distribution can be assumed, currents can be calculated with the Prandtl theory. This is a subject in its own, and it will be treated first (Subsection 7.3.2), as it leads directly to the core of the matter:
Given the longitudinal density gradient (for which a certain kind of longitudinal distribution will be a priori assumed), what is the magnitude of the vertical circulation? In Fig. 7.15, the investigation "ad a" highlights the relation between the blocs: "vertical circulation" and "horizontal density gradient".

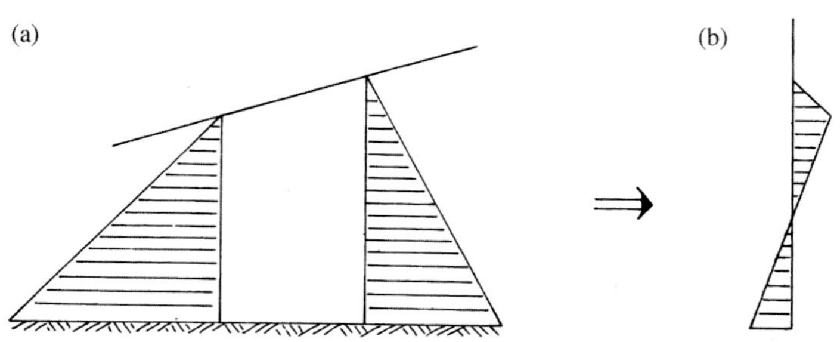

Fig. 7.14 (a) pressure forces acting on a slice of water in the case of a longitudinal density gradient. (b) resulting forces, compensated by internal shear stress.

[uu]Tending to "type 3".

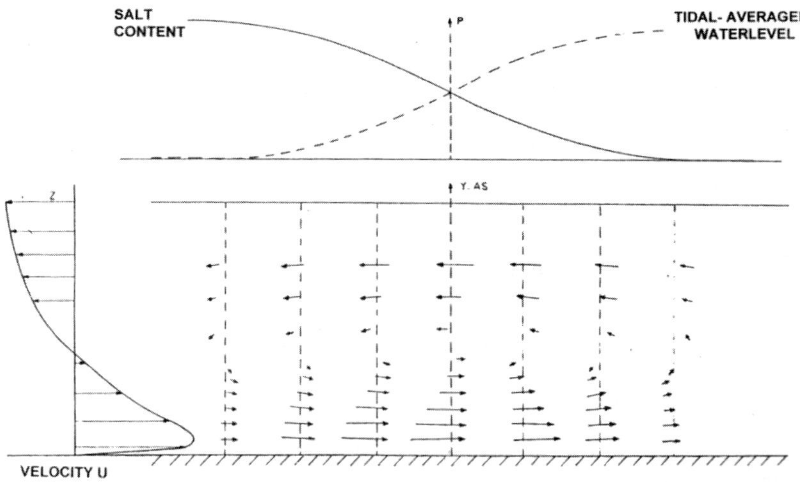

Fig. 7.15 Above: horizontal density gradient, compensated by a (tidal-averaged) water-level gradient. Below: circulation pattern, induced by the resulting internal shear stress.

Afterwards (Subsection 7.3.3) it will be shown, that such an assumption on density distribution makes sense in practical conditions.[vv]

(b) The magnitude of the longitudinal salt gradients is determined by longitudinal dispersion, in its turn determined by horizontal circulation in circulation cells.

Thus the longitudinal dispersion in an estuary (with the characteristics of the Western Scheldt) will be an important subject of investigation in this chapter.

It will be treated in Subsection 7.3.3. Thus, Subsection 7.3.3 covers (Fig. 7.12) the relation between the bloc "horizontal density gradient" and all the connected lower blocs.

More specific (cf. Fig. 7.15):

(c) The relation between vertical tidal circulation and an assumed formulation of longitudinal distribution of horizontal density yields a formula for vertical tidal circulation: cf. Subsection 7.3.2.

(d) The longitudinal distribution of horizontal water density can be measured.

The (management) problem remains, how this distribution depends on changeable circumstances (which ones?). Therefore, knowledge on the

[vv] At least, that there are practical conditions in which it makes sense.

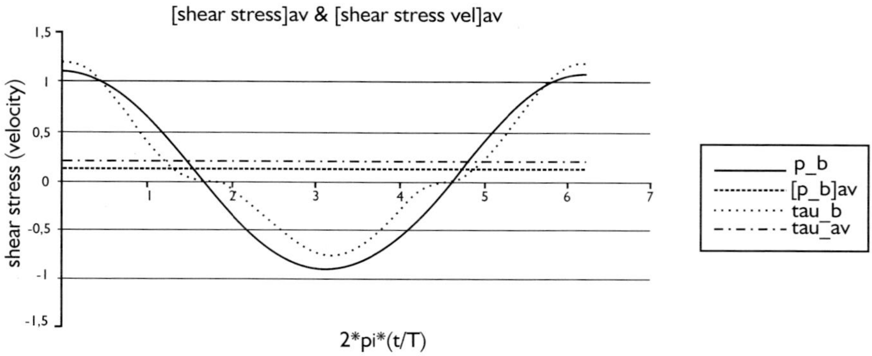

Fig. 7.16 Stress and shear stress velocity.

reason behind the measured density distribution is quite relevant as well. This is the subject of Subsection 7.3.3, the lower blocs in the schedule below "7.3" in Fig. 7.12.

Advantage of the model is simplicity and lack of fuzzy and unknown factors. The salt-driven circulation zone in the estuary has a beginning and an end (cf. Fig. 7.15).

The model is used only for estimation of orders of magnitude of circulation: on the one hand the density circulation, on the other hand the"rain in the drain" (Section 7.4).

Starting from tidal-averaged steady state, an equilibrium equation is considered, in which the local longitudinal pressure gradient (caused by water surface gradient and density gradient) is balanced by the internal turbulent shear stress gradient. Van de Kreeke and Robaczewska[83] state that those terms in a practical case like the considered one are at least a factor 2 larger than other relevant terms.

The tidal variation of the shear stress is implemented in the calculation of the mean shear, which results in a somewhat different formulation of the tidal-averaged turbulent shear stress than by "$\kappa_z \partial^2 \bar{u}/\partial z^2$" given by Pritchard[57] and Rigter.[64,ww] Compare Eq. (7.91) with (7.95). Starting from the Prandtl hypothesis, the dependence of κ_z of z has been implemented.

[ww]Equation (2) of.[83] Retain only the terms of order 10^{-5}, and assume a (negative) proportionality between the surface gradient to the density gradient. Integration to z of this equation yields directly an equation for the tidal-averaged shear stress, which however (because of non-linearity of the shear stress) will not only be a function of the tidal-averaged velocity, but also of the first harmonic of the velocity.

Where the effect of river flow is neglected with respect to the calculation of the velocity profile as function of z, it implicitly plays a role by the assumption of the longitudinal (stationary) density gradient (Fig. 7.14 a,b and Eq. (7.108) below), which (in the way it is defined) implies some current and a fast increasing longitudinal diffusivity[xx] in seaward direction.

7.3.2. *Effect of longitudinal density gradients on the momentaneous velocity profile*

(a) *Introduction: Problem considered; Assumptions*
The present introduction is an interlude to Subsection 7.3.2, in which tidal-averaged motion will be considered. One of the elements in that section will be: the problem of tidal-averaged circulation by density currents.

Before, however, some thoughts may be given to the *instationary* effects of those longitudinal pressure gradients, generated by density gradients. Thus, this subject falls under the scope of the present subsection concerning momentaneous velocities.

The Lorentz computation, treated in the preceeding subsections, yields momentaneous horizontal velocities in a basin. However, it starts from uniformity in vertical sense of the horizontal velocities: remember, that u, defined in Subsection 7.2.2 was a function of x and t: thus unsaid, theory allows slip along the bottom, although taking bottom friction into account. Such a computation yields velocities $u_{\mathrm{Lor}}(x,t)$; the index "Lor" stands for "Lorentz". The variable u_{Lor} replaces u in Subsection 7.2.3, i.e. it stands for "real" and "vertically-averaged" velocities.

In practice and in detail, there are (at least) two reasons for deviations from the Lorentz theory, which together leads to a correction term $u_d(x,z,t)$ of the real velocity compared to $u_{\mathrm{Lor}}(x,t)$:

a. The no-slip condition at the bottom. In the following, the consequences of the Prandtl hypothesis will be considered. The deviation from $u_{\mathrm{Lor}}(x,t)$ will be a periodic function: the correction term u_{d} will be harmonic at the bottom boundary as $u_{\mathrm{Lor}}(x,t)$, to be provided with this correction term, is harmonic.
b. A horizontal density gradient $d\rho/dx$.

Simplifications:

- In the approximation below, variation of the density gradient during the tide is not accounted for.

[xx]Ref. [7].

- The correction term u_{d} will be related to the *local* value of $u_{\mathrm{Lor}}(x,t)$, i.e. it is assumed that the vertical velocity profile is not affected by the longitudinal variation of $u_{\mathrm{Lor}}(x,t)$.
- Thus the term added to the Lorentz computations consist of a periodic one plus a deviatory part $u_{\mathrm{d}}(x,z,t)$:

$$u = u_{\mathrm{Lor}}(x,t) + u_{\mathrm{d}}(x,z,t) \tag{7.85}$$

- Let the tidal-averaged value of the deviatory velocity u_{d} be: $u_{\mathrm{dc}}(x,z)$.
- Continuity of $u_{\mathrm{dc}}(x,z)$ requires tidal-averaged vertical velocities $v_{\mathrm{dc}}(x,z)$. However, in the *present* pilot study (Subsection 7.3.2, concerning the vertical distribution of momentaneous horizontal velocity $u(z,t)$ in a well-mixed estuary with a salt gradient) longitudinal velocity gradients will be neglected.

Thus, apart from the (assumed) stationary horizontal pressure gradient, there remains a one-dimensional problem with the z-direction as variable.

For coastal dynamics, the present subsection is important for calculating the stirring of sediment; however, this (microscale) subject as such will not be treated in this book. Refer to papers elsewhere, for instance.[15]

The subject, however is important as well to reveal (after tidal-averaging) the tidal-averaged shear between horizontal layers of water; knowledge necessary to calculate the tidal-averaged vertical circulation (cf. next Section 7.3).

Thus, in the present subsection attention will be focused on the horizontal velocity.

For this goal, in the next part of this Subsection b.) the way of calculating this momentaneous velocity is displayed, making use of the equation for the internal shear stress velocity. Examples of resulting velocity distributions are shown in Fig. 7.17a and b.

Furthermore as a preparation on Section 7.3, an equation for the tidal-averaged horizontal velocity will be derived. Use of this equation leads in Subsection 7.3.3 to the tidal-averaged circulation, caused by density differences.

(b) *Momentaneous equations*

As mentioned above, in this pilot study concerning the vertical distribution of momentaneous horizontal velocity u in a well-mixed estuary with a salt gradient, longitudinal velocity gradients are neglected. Thus u is assumed to be a function of vertical coordinate z (positive upward; zero at the **bottom**) and of time t. Assume, that the water in horizontal sense (x-direction) has a density gradient $d\rho/dx$; assumed stationary and variable in longitudinal direction only. Accurate up to first order, the horizontal momentaneous velocity $u_{\mathrm{Lor}} + u_{\mathrm{d}}$ is then determined by the equation of motion:

$$\frac{\partial(u_{\mathrm{Lor}} + u_{\mathrm{d}})}{\partial t} = -\frac{1}{\rho}\frac{\partial}{\partial x}\{\rho g(h + \varsigma - z)\} + \frac{1}{\rho}\frac{\partial \tau}{\partial z} \tag{7.86}$$

Shear stress τ is positive, working in positive direction from the upper layer to the lower layer; u_{Lor} is the harmonic Lorentz solution (obeying continuity) determined by the linearized continuity and dynamic equation, (7.85) and (7.86), respectively. For the present goal, the dynamic equation of the Lorentz solution may be written as:

$$\frac{\partial u_{\mathrm{Lor}}}{\partial t} = -g\frac{\partial \varsigma}{\partial x} - \tau_{\mathrm{b}}/\rho h \tag{7.87}$$

in which τ_{b} is assumed harmonic. Subtracting (7.87) from (7.86) yields after tidal-averaging (mind that $\partial(\bar{u}_{\mathrm{Lor}} + \overline{u_{\mathrm{d}}})/\partial t = 0$; periodic features):

$$0 = -g(h + \bar{\zeta} - z)\frac{1}{\rho}\frac{d\rho}{dx} - g\frac{dh}{dx} + \frac{1}{\rho}\frac{d\bar{\tau}}{dz} + \frac{\bar{\tau}_{\mathrm{b}}}{\rho h} \tag{7.88}$$

where the overbar denotes tidal averaging. As mentioned above, variation of the density gradient with the tide (and thus its correlation with ς) has been neglected.

Reminding the harmonic character of ζ and τ_{b}, (7.88) leads to the equation for tidal-averaged variables:

$$\frac{1}{\rho}\frac{d\bar{\tau}}{dz} = g(h - z)\frac{1}{\rho}\frac{d\rho}{dx} + g\frac{dh}{dx} \tag{7.89}$$

which is, after multiplication with ρ, integrated between $z = z$ and $z = h$ to ($\tau = 0$ for $z = h$):

$$\bar{\tau} = -\frac{1}{2}g(h - z)^2\frac{d\rho}{dx} - \rho g(h - z)\frac{dh}{dx} \tag{7.90}$$

For the local turbulent shear stress τ in (7.86), following Prandtl the assumption is used:

$$\tau = \rho l^2 \frac{\partial u}{\partial z} \times \left|\frac{\partial u}{\partial z}\right| \tag{7.91}$$

where for the mixing length l is assumed:

$$l = \kappa z\sqrt{1 - \frac{z}{h}} \tag{7.92}$$

where κ is the von Karman constant.

Introduce the (momentaneous) internal shear stress velocity p:

$$p = l\frac{\partial(u_{\mathrm{Lor}} + u_{\mathrm{d}})}{\partial z} \tag{7.93}$$

as u_{Lor} does not depend on z, the above equation may be written as:

$$p = l\frac{\partial u_{\mathrm{d}}}{\partial z} \tag{7.94}$$

Differentiation of Eq. (7.86) with respect to z gives, using (7.94):

$$\frac{1}{l}\frac{\partial p}{\partial t} = \frac{g}{\rho}\frac{d\rho}{dx} + \frac{\partial^2(p|p|)}{\partial z^2} \tag{7.95}$$

Similar equations[yy] have been used considering the bottom boundary layer in the short wave theory.[15] Concerning a solution for tidal waves, refer to[13]. The problem is a specific case of the diffusion equation. It easily shows that in a semi-finite medium, with $p = 0$ at infinity the z-scale of a periodical solution is proportional to the square root of the time scale, given the bottom-boundary conditions. Thus a "semi-infinite" medium ("bottom boundary layer") is much smaller for short waves with periods of a few seconds than it would be for tidal waves. In the last-mentioned case, apart from the bottom boundary condition the boundary condition at the water surface ($p = 0$ for $z = h$) determines the solution of (7.95) and of the resulting flow pattern.

Equation (7. 95) has to be solved starting from a boundary condition at bottom (p_{b}), c.q. surface (p_{s}), respectively:

$$p_{\mathrm{b}} = \bar{p}_{\mathrm{b}} + \hat{p}_{\mathrm{b}1}\cos\omega t; \quad p_{\mathrm{s}} \equiv 0 \tag{7.96}$$

Figure 7.16 depicts schematically such a bottom shear stress velocity as a function of time.

First problem (A) is to relate the values of tidal-average shear stress velocity $\bar{p}_{\mathrm{b}}$ and amplitude of its first harmonic $\hat{p}_{\mathrm{b}1}$ to a "given" tidal-average bottom shear stress $\bar{\tau}_{\mathrm{b}}$ and amplitude $\hat{\tau}_{\mathrm{b}}$ of the shear stress (depicted in Fig. 7.16 as well).

One then still may have a question (B) how to find the magnitude of afore-mentioned $\bar{\tau}_{\mathrm{b}}$ and $\hat{\tau}_{\mathrm{b}}$ from the watersurface elevation $\zeta(x,t)$, i.e. from tidal amplitude and the gradient of the surface slope, caused by the density gradient.

[yy]Without the first term on the right-hand side.

Focus on one part of question B first:

In a following Subsection 7.33 extensively will be treated, how to find the vertical distribution of the *tidal-averaged* shear stress $\bar{\tau}(z)$; so, in the present context the tidal-averaged bottom shear stress $\bar{\tau}_{\mathrm{b}}$ may be assumed to be known.

With respect to $\overline{p_{\mathrm{b}}^2}(= \bar{\tau}_{\mathrm{b}}/\rho)$, refer to (7.90), stating that:

$$\frac{\bar{\tau}_{\mathrm{b}}}{\rho} = -gh\left\{\frac{1}{2}\frac{h}{\rho}\frac{d\rho}{dx} + \frac{dh}{dx}\right\} \tag{7.97}$$

where the gradient of h is caused by the large-scale tidal-averaged density gradient, but also (Subsection 7.3.3) is affected by the tidal-averaged vertical circulation.

Returning to question A:

The magnitude, in (7.96), of tidal-average shear stress velocity $\bar{p}_{\mathrm{b}}$ and the amplitude of its first harmonic $\hat{p}_{\mathrm{b}1}$ are determined by approximation from the given bottom boundary non-slip condition, in the following way.

Consider $\bar{p}_{\mathrm{b}}$ first (Fig. 7.16), to be related to $\bar{\tau}_{\mathrm{b}}/\rho$ and $\hat{\tau}_{\mathrm{b}}/\rho$.

Squaring bottom shear stress velocity p_{b} from (7.96), Fig. 7.16, and tidal averaging (taking the sign of the shear stress into account) yields $\bar{\tau}_{\mathrm{b}}/\rho$. This evidently depends on the ratio $\bar{p}_{\mathrm{b}}/\hat{p}_{\mathrm{b}1}$. Thus, inversely, the first step finding $\bar{p}_{\mathrm{b}}$ is to find out how the ratio $\overline{p_{\mathrm{b}}^2}/\bar{p}_{\mathrm{b}}^2$ depends on the ratio $\bar{p}_{\mathrm{b}}/\hat{p}_{\mathrm{b}1}$.

Refer to BIJKER,[17] for this subject, who in his Ph.D. thesis calculated the integral:

$$B(a) = \frac{2}{T}\int_{-T/4}^{T/4}(1 + a\sin\omega t)|1 + a\sin\omega t|dt \tag{7.98}$$

In the present context, "a" stands for the ratio $\hat{p}_{\mathrm{b}1}/\bar{p}_{\mathrm{b}}$.

For $a < 1$ the value of this integral is $1 + a^2/2$; for large values of a it approaches the value of $4a/\pi$.

Remains the question on the magnitude of $\hat{\tau}_{\mathrm{b}}/\rho$:

Neglecting instationary effects, the amplitude of the shear stress $\hat{\tau}_{\mathrm{b}}$ at the bottom is $\rho gh\hat{I}$, where $\hat{I}$ is the "amplitude" of the slope of the water level.

From (7.96) one finds, assuming $\bar{p}_{\mathrm{b}}$ smaller than $\hat{p}_{\mathrm{b}1}$, an amplitude, equal to:

$$(\tau_{\mathrm{b\,max}} - \tau_{\mathrm{b\,min}})/(2\rho) = \bar{p}_{\mathrm{b}}^2 + \hat{p}_{\mathrm{b}}^2 \tag{7.99}$$

This leads to:

$$\bar{p}_{\text{b}}^2 + \hat{p}_{\text{b}}^2 \sim gh\hat{I} \tag{7.100}$$

Given the boundary condition at the bottom and surface, Eq. (7.95) can be easily solved numerically. Bakker and Schaart[13] calculate a case for the Eastern part of the Western Scheldt with the characteristics.

Water depth $h = 15.4\,\text{m}$; ripple height $r = 0.03\,\text{m}$; tidal velocity amplitude[zz] $\hat{U} = 0.30\,\text{m/s}$.

Figure 7.17 shows the velocities as function of depth and time at a logarithmic scale. In the plot (on a logarithmic vertical scale), the calculated plotted values have been linearly extrapolated to the vertical axis (line of zero velocity). Starting from the Nikuradze approach of calculating velocity profiles, this would have lead to crossings at a level of $z = r/33$. Due to the salt gradient, calculations concerning the flood would lead to "thicker" velocity profile (smaller apparent roughness) during flood tide than during ebb tide; a feature measured in prototype as well (Fig. 7.18, from[16]).

In the present context, with its focus on *tidal-averaged* circulation, the solution of (7.95) for periodic motion is less relevant than its tidal average.

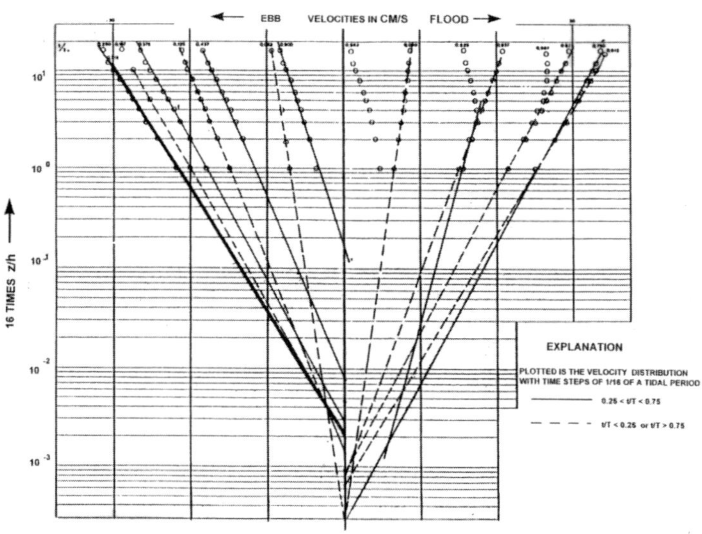

Fig. 7.17a Velocity profiles as numerically calculated by Bakker and Schaart.[13] The circles depict the numerically calculated values. The profiles have been logarithmically extrapolated.

[zz]Mind, that the site is near the closed landward side of the estuary.

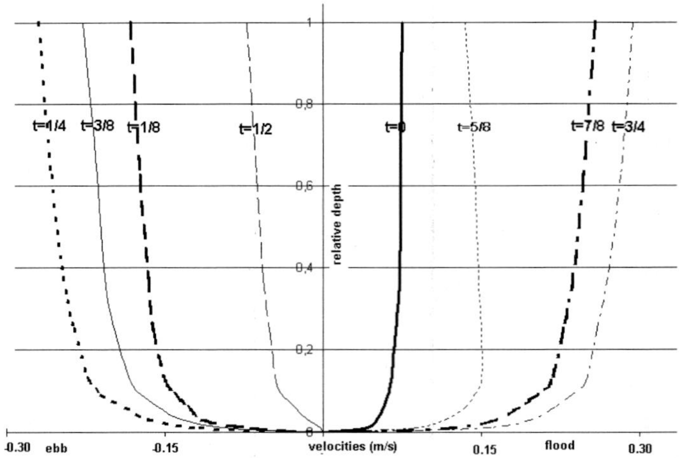

Fig. 7.17b Velocity profiles as numerically calculated by Bakker and Schaart.[13] Depicted are the logarithmically extrapolated profiles, as shown in Fig. 7.17a. A longitudinal dam along the "Nauw van Bath" (cutting off its northern bend) has been supposed in this particular investigation.

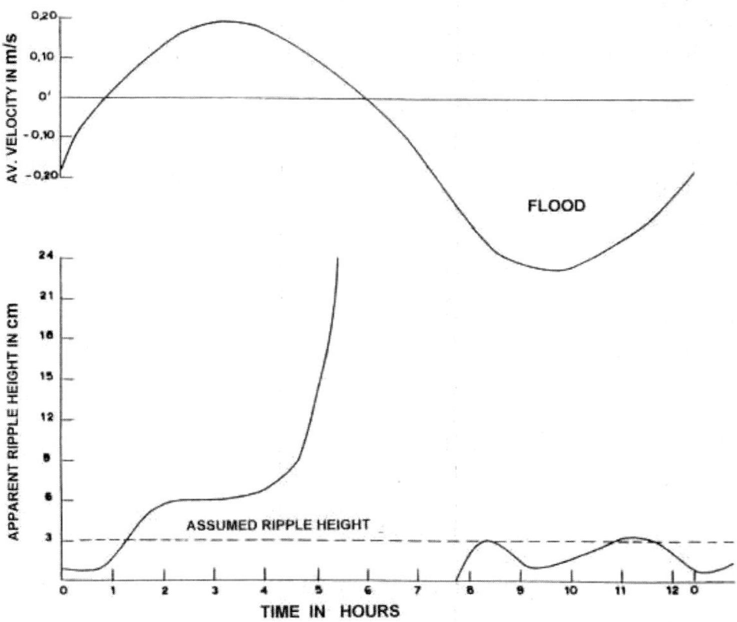

Fig. 7.17c Apparent roughness as function of time.

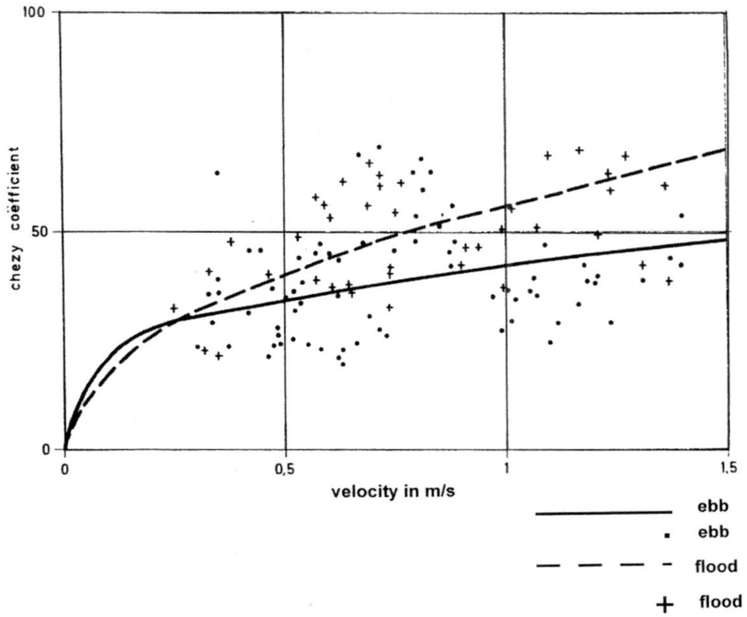

Fig. 7.18 Chezy coefficients as function of depth-averaged velocity as found from Western Scheldt measurements by extrapolation of the logarithmic velocity profile, found above the level of 50 cm above the bottom. From [16] Dots, measurements during ebb tide; crosses, measurements during flood tide; drawn line, best fit through the dots (ebb); interrupted line, best fit through the crosses (flood).

For if a tidal-averaged solution for p of (7.95) would be known, tidal-averaging (7.94) learns: integrating of this equation to z leads to $u_{\rm dc}$, the tidal-averaged u, equal to the tidal mean of $u + U$ as tidal-average $u_{\rm Lor}$ in Eq. (7.93) is zero.

Thus the way for finding the horizontal component $u_{\rm dc}$ of the vertical tidal-averaged density circulation in a prismatic channel gets the following shape:

a. Assume bottom and surface shear stress velocities $p_{\rm b}$, resp. $p_{\rm s}$ in the following way:

$$p_{\rm b} = \bar{p}_{\rm b} + \hat{p}_{\rm b1} \cos \omega t; \quad p_s \equiv 0 \tag{7.101}$$

b. Find the tidal-averaged solution $\bar{p}(z)$ for these boundary conditions.
c. Integrate $\bar{p}(z)$ to $u_{dc}(z)$ using (7.94).

Only the zeroth and first harmonic of (7.95) will be considered

7.3.3. *Mathematical calculations on tidal averaged circulation by longitudinal density gradients*

(a) *Way of solution; assumptions; dimensionless equations*
The following way of approach will be applied:

a. Specify the longitudinal density function in (7.89); put the problem in a dimensionless shape (Eqs. (7.102) to (7.105));
b. Express the (dimensionless) tidal-averaged shear stress $\bar{\tau}^*$ in the (dimensionless) density gradient and water level gradient; cf. (7.106);
c. Split the two-dimensional problem by calculating resp. the (dimensionless) horizontal and vertical tidal-averaged velocity separately. The process of splitting is carried out below ad. b.);
d. Calculate a relation between the dimensionless tidal-averaged shear stress $\bar{\tau}^*$ and the dimensionless tidal-averaged shear stress velocity $\bar{p}^*$; cf ad. c.); page 405.
e. Calculate the dimensionless horizontal tidal-averaged velocity u^*_{dc} by tidal-averaging Eq. (7.94) and integrating it: cf ad d; p. 407.
f. Calculate dimensionless vertical tidal-averaged velocity $\mathrm{v}^*_{\mathrm{dc}}$, using continuity: Sub-subsection e.)

As mentioned in Subsection 7.3.1, the longitudinal density distribution will be assumed a priori according to an error-function (see Fig. 7.19a). Thus, the longitudinal density gradient will vary according to a Gaussian curve.[aaa] Let the standard deviation of this curve be σ.

In order to make the following considerations (and the proportionality constants) as universal as possible, there will be switched to dimensionless constants. Those will be marked with an asterix. Thus call

$$
\begin{aligned}
x^* &= x/\sigma \\
z^* &= z/h
\end{aligned}
\tag{7.102}
$$

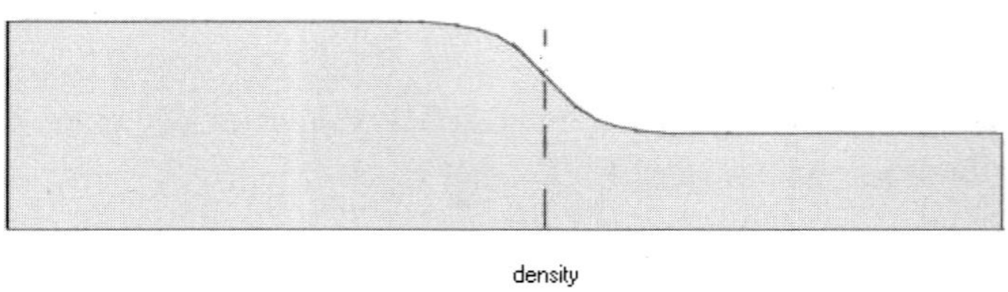

Fig. 7.19a Salt content along the estuary.

[aaa]Remind: the physical reality of this assumption will be treated in Subsection 7.3.4. ad.c.

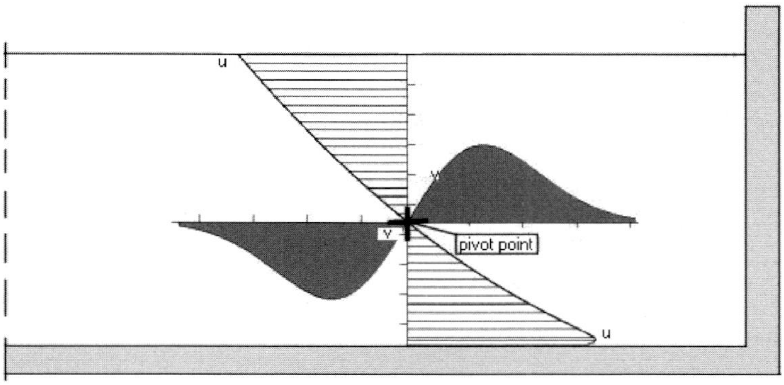

Fig. 7.19b Tidal averaged current velocities in the case the fresh water flux can be neglected (from [9]).

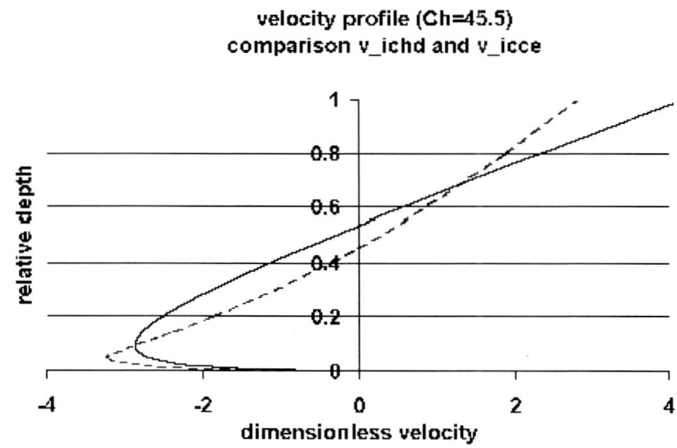

Fig. 7.20 Comparison velocity u_{dc} profile as calculated by Bakker and Barber[9] (---) and in present book (definition mixing length slightly changed)(——) Factor on the left of "[ ]" in (7.141) is omitted.

$\Delta\rho^*$ = density difference between seawater (ρ_{s}) and water in the river (ρ_{r}), related to the average density of the water in the river

$$\Delta\rho^* = \frac{\rho_{\mathrm{s}} - \rho_{\mathrm{r}}}{(\rho_{\mathrm{s}} + \rho_{\mathrm{r}})/2} \tag{7.103}$$

$$\rho^* = \frac{\rho - \rho_{\mathrm{r}}}{\rho_{\mathrm{s}} - \rho_{\mathrm{r}}} \tag{7.104}$$

This implies that ρ^* varies between 0 (riverside) and 1 (sea); dimensional and dimensionless density gradient are related according to:

$$\frac{d\rho^*}{dx^*} = \frac{\sigma}{\Delta\rho}\frac{d\rho}{dx} \tag{7.105}$$

b. Assume a Gauss function for the density gradient:

$$\frac{d\rho^*}{dx^*} = \sqrt{\frac{1}{2\pi}}e^{-x^{*2}/2} \tag{7.106}$$

One may relate the tidal-averaged gradient of the water level linearly to the density gradient

$$\frac{dh^*}{dx^*} = -\beta_{\mathrm{dl}} h_0^* \Delta\rho^* \frac{d\rho^*}{dx^*} \tag{7.107}$$

which is even more transparent in dimensional shape:

$$\frac{dh}{dx} = -\beta_{\mathrm{dl}} \frac{h}{\rho}\frac{d\rho}{dx} \tag{7.108}$$

where β_{dl} is a dimensionless constant to be focused on and in which h_0^* is a dimensionless (mean) channel depth:

$$h_0^* = h_0/\sigma \tag{7.109}$$

The value of β_{dl} will be determined (in the following) by the fact that the depth-averaged resultant current has been assumed negligible (i.e. "zero").

Consider now the dimensionless shear stress. In the present subsection, the effect of vertical tide (treated in extense when considering the "rain in the drain"; cf. Section 7.4) will be overlooked.

Define the dimensionless shear stress

$$\tau^* = \tau/\rho g h \tag{7.110}$$

This puts Eq. (7.90) in the dimensionless shape ($h* = h/\sigma$):

$$\bar{\tau}^* = -\frac{1}{2}h^*(1-z^*)^2\Delta\rho^*\frac{d\rho^*}{dx^*} - (1-z^*)\frac{dh^*}{dx^*} \tag{7.111}$$

As h^* in (7.104) varies hardly in x-direction (in comparison to $d\rho^*/dx^*$) it may be considered equal to h_0^*.

Substitution of (7.99) and (7.100) into (7.104) leads to:

$$\bar{\tau}^* = -\frac{\Delta\rho^* h_0^*}{\sqrt{2\pi}}(1-z^*)\left\{\frac{1}{2}(1-z^*) - \beta_{\mathrm{dl}}\right\}e^{-x^{*2}/2} \tag{7.112}$$

If $\beta_{\mathrm{dl}} > 0.5$, then the tide-averaged shear stress $\overline{\tau}^*$ has the same direction over the whole depth. As vertical tidal circulation is to be simulated, this implies $\beta_{\mathrm{dl}} < 0.5$

(b) *Splitting between horizontal and vertical velocity functions*
One can split $\overline{\tau}^*$ in a function of x^* and a function of z^*:

$$|\overline{\tau}^*| = f_1(z^*)g_1(z^*)f_2^2(x^*) \tag{7.113}$$

in which:

$$\begin{aligned} f_1(z^*)g_1(z^*) &= \left| -(1-z^*)\left\{\frac{1}{2}(1-z^*) - \beta_{dl}\right\}\right| \\ \mathrm{sign}(f_1(z^*)) &= \mathrm{sign}\ (\overline{\tau}^*) \\ f_2^2(x^*) &= \frac{\Delta\rho^* h_0^*}{\sqrt{2\pi}} e^{-\frac{1}{2}x^{*2}} \end{aligned} \tag{7.114}$$

The hidden logic of denoting one function of z* by the product of two functions $f1(z^*)$ and $g1(z^*)$ will appeal in Subsection d.).

A nice perspective develops: the above equations suggest one kind of velocity profile, of which only the scale changes in x^*-direction. Only β_{dl} should be determined in a proper way to match continuity. One hurdle has to be taken: where in stationary flow there is a direct link between shear stress τ and shear stress velocity p, the link is less clear when both variables are tidal-averaged. For (7.114) requires a formula concerning the tidal-averaged p in order to be able to integrate it to tidal-averaged u (call: u_{dc}).

(c) *Tidal-averaged shear stress and shear stress velocity*
Thus the next step in the process[bbb] will be to calculate the distribution of mean shear stress velocity $\overline{p}^*$ from the value of mean shear stress $\overline{\tau}^*$ as given in (7.105). As (7.108) gives some general relation concerning the dependence in x-direction, it is only necessary to concentrate on (7.107) concerning the dependence in z-direction. A special Sub-subsection d.) will be devoted to this problem and the subsequent steps "e" and "f" as displayed in the Introduction a.) to the present sub-subsection.

Which mathematical expression(s) can be given for the vertical distribution of the density-driven circulation current?

In the following, a (reasonable) approximation will be given (valid when the tidal-averaged flow is small compared to the tidal flow).

[bbb]cf. The introduction 7.3.2.a). to the present Sub-subsection.

First, a closer look to (7.106) and (7.107) gives insight in the physical meaning of β.

From (7.114) it shows, that $f_1(z^*)g_1(z^*)$ is zero when $z^* = 1 - 2\beta_{\rm dl}$; thus $\bar{\tau}^*$ is zero, according to (7.113); this means, that the mean velocity u should have an extreme at that level, in other words.: $z^* = 1 - 2\beta_{\rm dl}$ is the level of minimum mean velocity (i.e. of maximum current reversal).

This level will be called $z^*_{\rm min}$:

$$z^*_{\rm min} = 1 - 2\beta_{\rm dl} \Rightarrow \beta_{\rm dl} = \frac{1}{2}\left(1 - z^*_{\rm min}\right) \tag{7.115}$$

When the tidal-averaged flow is small compared to the tidal flow, simple arithmetic gives as a tidal-average of the combination p of a stationary shear stress velocity $\bar{p}$ and a harmonic stress velocity $\widehat{p}\cos\omega t$ a tidal-averaged p^2 (i.e. $\langle\tau\rangle/\rho$), equal to:

$$\text{sign}(\tau)\cdot\overline{p^2}(=\bar{\tau}/\rho) \approx \frac{4}{\pi}\bar{p}\hat{p} \tag{7.116}$$

This can be made clear in the following way.

Consider first the time that p acts in the direction of $\bar{p}$. Neglecting $(\bar{p}/\hat{p})^2$, the time-averaged value of p during this time is $\hat{p}^2$ times {the time-average of $2(\bar{p}/\hat{p})\sin\omega t$ plus the time-average of $\sin^2\omega t$}. The value between accolades is: $(4/\pi)$ times $(\bar{p}/\hat{p})$ plus $1/2$.

Consider now the time that p counteracts the direction of $\bar{p}$. In that case the value between accolades would be $(4/\pi)$ times $(\bar{p}/\hat{p})$ minus $1/2$.

Thus the time-average over the total period is $\hat{p}^2$ times $(4/\pi)(\bar{p}/\hat{p})$ which results in the value of (7.110).

$$p^* = p/\sqrt{gh} \tag{7.117}$$

Equation (7.105) will be applied at the bottom, where the momentaneous shear stress can be determined as $\rho g h I$, where I is the (momentaneous) water level gradient. Neglecting the tidal-averaged water level gradient with respect to its extremes, this results[ccc] into $\hat{p}_{\rm b}^2 \approx gh\hat{I}$, which–made dimensionless with (7.112)–leads to

$$\hat{p}_{\rm b}^* \approx \sqrt{\hat{I}} \tag{7.118}$$

From (7.105) one finds for the dimensionless bottom shear stress at $x = 0$:

$$[\bar{\tau}_{\rm b}^*]_{x=0} = -\frac{\Delta\rho^* h_0^*}{\sqrt{2\pi}}\left(\frac{1}{2} - \beta_{\rm dl}\right) \tag{7.119}$$

[ccc]The "≈" sign keeps the tidal-averaged current in mind.

or, introducing $z^*_{\rm min}$ from (7.109):

$$[\bar{\tau}^*_{\rm b}]_{x=0} = -\frac{\Delta\rho^* h^*_0}{2\sqrt{2\pi}} z^*_{\rm min} \tag{7.120}$$

Now first express (approximately) the dimensionless top shear stress at the bottom $\hat{p}^*_{\rm b}$ in the characteristics of the water surface ζ. For a relatively long basin:

$$\zeta = \hat{\zeta}\sin(\omega t + kx) \tag{7.121}$$

$$\frac{\partial\varsigma}{\partial x} = \hat{\zeta}k\cos(\omega t + kx) \tag{7.122}$$

Thus the maximum slope $\hat{I}$ of the water surface equals:

$$\hat{I} = \hat{\zeta}k \tag{7.123}$$

where the wave number k equals $2\pi/T\sqrt{gh}$, or $\omega/\sqrt{gh}$ and thus, (7.118) can relate the amplitude of the (dimensionless) shear stress velocity $\hat{p}^*_{\rm b}$ to tidal amplitude and astronomical data:

$$\hat{p}^*_{\rm b} = \sqrt{\hat{\zeta}\omega/\sqrt{gh}} \tag{7.124}$$

Very short basins require adaption of (7.115), etc. Substitution of $\bar{\tau}^*_{\rm b}$ into (7.115) according to (7.114), $z^*_{\rm min}$ according to (7.109) and $\hat{p}^*_{\rm b}$ according to (7.118) gives for the local maximum of the tidal-averaged shear stress velocity $[\bar{p}^*_{\rm b}]_{x=0}$ at the bottom:

$$[\bar{p}^*_{\rm b}]_{x=0} \approx -\frac{\pi\Delta\rho^* h}{8\sigma\sqrt{2\pi}} \times \frac{\sqrt[4]{gh}}{\sqrt{\hat{\zeta}\omega}} z^*_{\rm min} \tag{7.125}$$

(d) *Vertical distribution of the tidal-averaged horizontal velocity*
From (7.91) and (7.94) one finds the following relation between instantaneous velocity $u(x, z, t)$ and shear stress velocity p:

$$u(x, z, t) = \int_{z=z_0}^{z=z} \frac{p(x, z, t)}{l} dz \tag{7.126}$$

Time-averaging gives for the mean tidal velocity $u_{\rm dc}(x, z)$, as caused by density currents:

$$u_{\rm dc}(x, z) = \int_{z=z_0}^{z=z} \frac{\bar{p}(x, z)}{l} dz \tag{7.127}$$

or, dimensionless:

$$u_{\mathrm{dc}}^*(x^*, z^*) = \int_{z^*=z_0^*}^{z^*=z^*} \frac{\bar{p}^*(x^*, z^*)}{l^*} dz^* \tag{7.128}$$

In the following it will be assumed, that $\hat{p}$ is large compared to $\bar{p}$ and that the vertical distribution of $\hat{p}$ is not significantly affected by the instationary motion. Specifically, it is assumed that $\hat{p}^2(z)$ has the vertical distribution of an ordinary shear stress and is proportional to $h - z$. Furthermore, assuming a general validity of (7.110) in z-direction, it is logical to allot the functions $f_1(z)$ and $g_1(z)$ in (7.106) to the functions $\bar{p}$ and $\hat{p}$, respectively (apart of a constant).

$$l^* = l/h \tag{7.129}$$

This makes:

$$g_1(z) = \sqrt{h - z} \tag{7.130}$$

or dimensionless:

$$g_1(z^*) = \sqrt{1 - z^*} \tag{7.131}$$

Then (7.106) makes for $f_1(z^*)$ (if one replaces as well β_{dl} by z_{min}^*, using (7.109)):

$$f_1(z^*) = \frac{1}{2} |z_{\mathrm{min}}^* - z^*| \sqrt{(1 - z^*)}$$

For tidal waves the influence of instationary effects on the vertical current distribution may be assumed small.

The above-mentioned assumption, together with (7.104) and (7.110) leads to:

$$\bar{p}^*(x^*, z^*) = [\bar{p}_{\mathrm{b}}^*]_{x=0} \frac{f_1(z^*)}{f_1(0)} \frac{f_2(x^*)}{f_2(0)} \tag{7.132}$$

Assuming a mixing length l in (7.120) and (7.121) as given in (7.92), as vertical gradient of the tide-averaged velocity is found from (7.94) and (7.127):

$$\frac{du_{\mathrm{dc}}^*(x^*, z^*)}{dz^*} = [\bar{p}_{\mathrm{b}}^*]_{x=0} \frac{f_1(z^*)}{l^* \times f_1(0)} \frac{f_2(x^*)}{f_2(0)} \tag{7.133}$$

by using (7.107) and (7.109):

$$f_1(z^*) g_1(z^*) / f_1(0) g_1(0) = \left| (1 - z^*) \left(1 - \frac{z^*}{z_{\mathrm{min}}^*} \right) \right| \tag{7.134}$$

where, according to (7.125):

$$g_1(z^*)/g_1(0) = \sqrt{1 - z*} \tag{7.135}$$

Elimination of $g_1(z^*)$ from (7.129) by substituting (7.130) and substitution of the dimensionless mixing length l^* in (7.128) from (7.92) leads to:

$$\frac{du^*_{\text{dc}}(0^*, z^*)}{dz^*} = [\bar{p}^*_{\text{b}}]_{x=0}\frac{1}{kz^*}\left|1 - \frac{z^*}{z^*_{\min}}\right| \tag{7.136}$$

integrating to (by substituting the dimensionless bottom shear stress velocity $\bar{p}^*_{\text{b}}$; (7.137)).

$$u^*_{\text{dc}}(0, z^*) = -\frac{\pi\Delta\rho^* h}{8\kappa\sigma\sqrt{2\pi}} \times \frac{\sqrt[4]{gh}}{\sqrt{\hat{\zeta}\omega}}\int_{z^*_0}^{z^*}\frac{z^*_{\min}}{z^*}\left(1 - \frac{z^*}{z^*_{\min}}\right)d(z^*/z^*_{\min}) \tag{7.138}$$

$$u^*_{\text{dc}}\left(0, \frac{z}{h}\right) = \frac{\pi\Delta\rho^* h}{8\kappa\sigma\sqrt{2\pi}} \times \frac{\sqrt[4]{gh}}{\sqrt{\hat{\zeta}\omega}}\left[\frac{z}{z_{\min}} - \ln\left(\frac{z}{z_{\min}}\right)\right]_{z_0}^{z} \tag{7.139}$$

$$u^*_{\text{dc}}\left(0, \frac{z}{h}\right) = \frac{\pi\Delta\rho^* h}{8\kappa\sigma\sqrt{2\pi}} \times \frac{\sqrt[4]{gh}}{\sqrt{\hat{\zeta}\omega}}\left[\frac{z}{z_{\min}} - \ln\left(\frac{z}{z_{\min}}\right)\right] - u^*_{\text{dc}0} \tag{7.140}$$

where the constant u^*_{dc0} on the right-hand side of (7.134) is per definition the first term after substitution of $z = z_0$.

So the dimensionless total specific discharge between the bottom ($z/h = z_0/h$) and the level z/h equals the integral of (7.134):

$$\frac{1}{z_{\min}}\int_{z_0}^{z} u^*_{\text{dc}}\left(0, \frac{z}{h}\right)dz = \frac{\pi\Delta\rho^* h}{8\kappa\sigma\sqrt{2\pi}} \times \frac{\sqrt[4]{gh}}{\sqrt{\hat{\zeta}\omega}}\left[\frac{z^2}{2z^2_{\min}} + \frac{z}{z_{\min}} - \frac{z}{z_{\min}}\right.$$
$$\left.\times \ln\left(\frac{z}{z_{\min}}\right)\right]_{z=z_0}^{z=z} - \left(\frac{z}{z_{\min}} - \frac{z_0}{z_{\min}}\right)u^*_{\text{dc}0} \tag{7.141}$$

Because of continuity in longitudinal direction (N.B. the term $f_2(x^*)$ in (7.127)) the value of the integral (7.135) should be zero for $z = h$. Thus by iteratively[ddd] varying the value of z in the right-hand side of (7.135) the value of $h/z_{\min}(= 1/z^*_{\min})$ can be determined for a given value of $z^*_0/z^*_{\min}$. This yields $z^*_0/z^*_{\min}$, i.e. z^*_0, related to the Chezy coefficient C_h by:

$$C_h = 18^*\{^{10}\log(12/(33z^*_0))\} \tag{7.142}$$

[ddd] An Excel file can be very useful in this context.

This results in a relation between Chezy coefficient and relative level of minimum velocity.[eee] Formulae given here deviate somewhat from the ones in,[9] Eq. (5.5). Reason is a slightly different definition of the mixing length in both cases. The present formulae (apart from being simpler) are thought to be physically better. For a given level of z_0^*, the value of $u_{\mathrm{dc}}^*(z_{\mathrm{min}}^*)$ is about the same, however with the present formulae the level z_{min}^* is somewhat higher. Refer to Fig. 5.1 of[9] for a picture of the calculated circulation pattern. Figure 7.20 compares the present calculated velocity profile at $x = 0$ is with the one of Bakker and Barber.[9]

(e) *Horizontal distribution of the tidal-averaged vertical velocity*
The vertical tidal-averaged velocity v_{dc} (caused by density currents) can be derived from continuity equation:

$$\frac{\partial u_{\mathrm{dc}}}{\partial x} + \frac{\partial \mathrm{v}_{\mathrm{dc}}}{\partial z} = 0 \tag{7.143}$$

7.3.4. *Dispersion theory; application concerning the Western Scheldt*

(a) *Introduction to Subsection 7.3.4.*
In the former Subsection 7.3.3, a longitudinal density distribution in the (prismatic) channel was presumed. In the present Subsection 7.3.4, this assumption will be substantiated by considering data of one specific estuary: the Western Scheldt estuary in the Netherlands (Fig. 7.21).

First (ad.b.)), this subsection will memorize elementary stationary diffusion theory.

Next (ad.c.)) some data on concentration of salt and pollutants in the Western Scheldt estuary are displayed. It will show, that for the Western Scheldt estuary a longitudinal concentration distribution as proposed[fff] in Subsection 7.3.3) covers better the reality than the theoretical results, given in Subsection 7.3.3 ad.b.).

Thus in (ad. d.)), the process is inverted: dispersion coefficients are calculated (varying in longitudinal direction) leading to the concentration distribution according to Bakker and Barber.[9] Finally (ad. e.)) a "theoretical" curve in longitudinal direction for "conservative pollutant"- concentration

[eee] Maximum current reversal.
[fff] Eqn. (7.106) in Subsection 7.3.3.

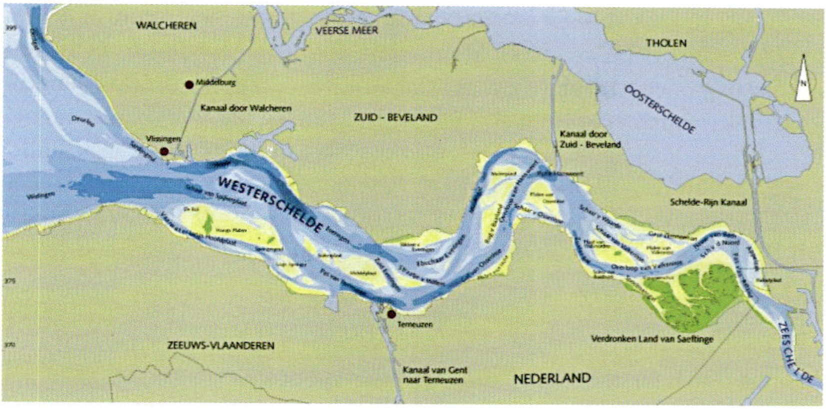

Fig. 7.21 The Western Scheldt estuary in the Netherlands.

is derived,[ggg] which is somewhere in the middle of the distributions found in practice.[hhh]

(b) *Diffusion theory: basics*

Consider a river with uniform cross-section and a steady uniform flow with current velocity u. Concentration $c(x)$ of salt or pollutants (where x is the direction of the current) is stationary as well. For this described case the one-dimensional diffusion equation is:

$$dS/dx = 0 \tag{7.144}$$

where S is the transport of diffusant per unit of cross-section:

$$S = uc - D_x \frac{dc}{dx} \tag{7.145}$$

where D_x is the longitudinal diffusion coefficient, assumed, in this section, independent of x.

Equation (7.146) still can be generalized. Nearly always D_x will be described as a function (for instance an exponential function or an error function) with a certain length scale; this scale will be denoted by σ in the following. This introduces the dimensionless coordinate $x^* = x/\sigma$.

Denote a dimensionless D_x^* as:

$$D_x^* = D_x/u\sigma \tag{7.146}$$

[ggg] Annex 8.

[hhh] The conservancy of those pollutants is questionable.

The theory in this sub-subsection (presuming steady flow) does not imply that D_x is independent of u. Mind, that the same holds for D_x^*! Define the "basic concentration" c_0 as the stationary uniform concentration ($dc/dx = 0$) which results in S, i.e.

$$c_0 = S/u \tag{7.147}$$

Thus one finds for the dimensionless shape of (7.140):

$$c_0 = c - D_x^* \frac{dc}{dx^*} \tag{7.148}$$

Solutions differ for the "salt" case and the "pollutant" case.

Let the river flow into sea at $x^* = l^*$. In the "salt"-case c will be zero at $x = -\infty$ (where dc/dx is zero); thus c_0 will be zero. At sea, $x^* = l^*$, the salt concentration may be assumed a constant, called c_s.

In the "pollutant" case an evident boundary condition is: $c = 0$ at $x = l$, where far upwards the river the concentration will not be influenced by the sea and thus be c_0.

Those boundary solutions lead to the solutions:

$$c/c_s = e^{(x^* - l^*)/D_x^*} \tag{7.149}$$

for the "salt case" and:

$$c/c_0 = 1 - e^{(x^* - l^*)/D_x^*} \tag{7.150}$$

for the "pollutant case". The concentration distributions (7.150 and 7.151) are displayed in Fig. 7.22a,b, respectively.

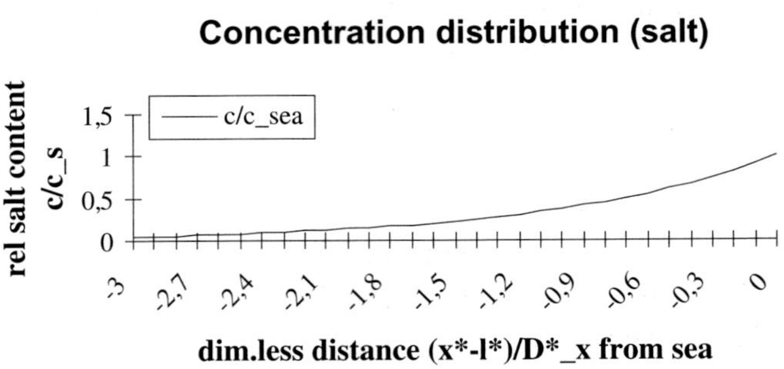

Fig. 7.22a Salt concentration in a river (sea at rightside).

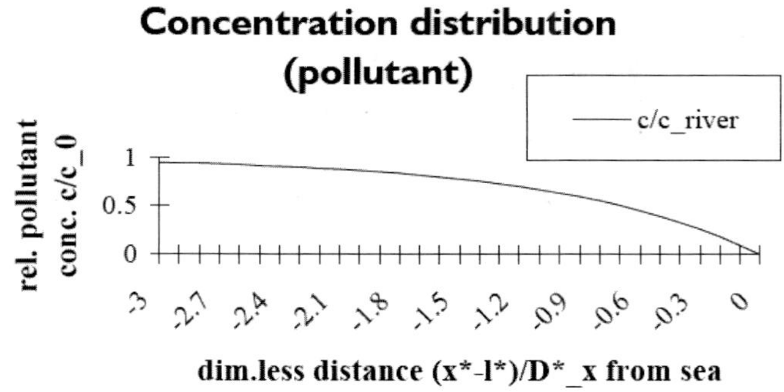

Fig. 7.22b Concentration of pollutant in a river (sea at right side).

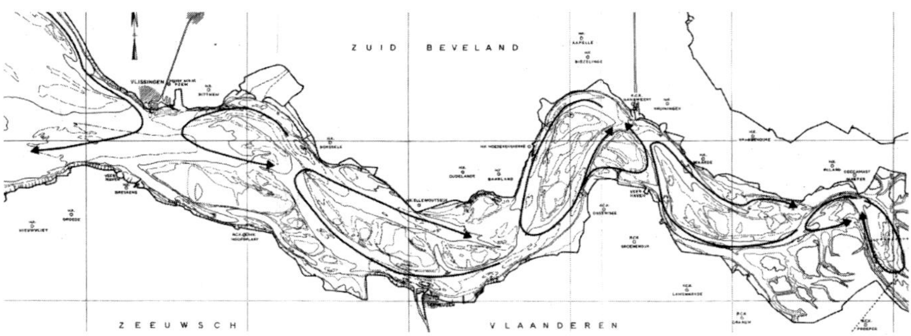

Fig. 7.23 Western Scheldt: horizontal circulation (measurements); Courtesy: Rijkswaterstaat.

(c) *Considerations on the longitudinal dispersion coefficients in the Western Scheldt*

Comparing theory with nature it has to be realized, that in nature the "diffusion coefficient" defined in Subsection 7.3.4 ad.b.) will be a "dustbin coefficient". Thus, the value of this coefficient depends on the accuracy of the schematization: hydraulic circulation in a "cell" (cf . Fig. 7.23, showing these kinds of cells as occurring in the Western Scheldt).

Diffusion at a larger scale can be interpreted as convection at a smaller scale (Subsection 7.3.4 ad. e.).

For this reason, talking about values as found in nature, the word "dispersion" will be used instead of "diffusion" to discriminate between "physics" and "engineering".

With respect to *salt concentration*, an impression of reality, compared to the theory of Sub-subsection b.) can be found from:

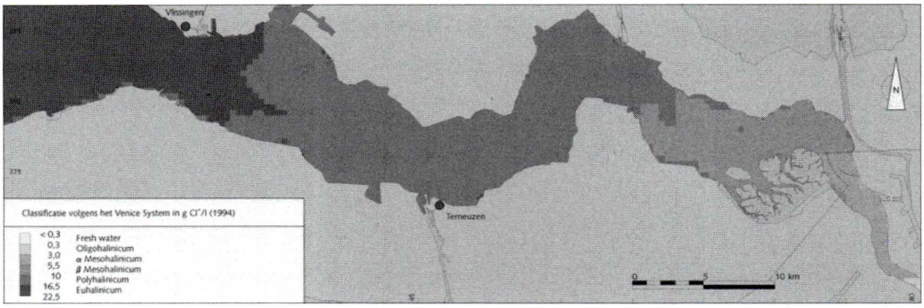

Fig. 7.24 Western Scheldt estuary: density distribution. From: Schelde Informatie Centrum; Data:[40].

a. Figure 7.24: data from the Study Department at Flushing[iii] of the Rijkswaterstaat concerning the embouchure of the Western Scheldt.
b. Figure 7.24: a map found in the "Atlas of the Western Scheldt".[65]
c. A point near Antwerp has been derived from the "Schelde Nieuwsbrief".[7]

From those figures mentioned above Fig. 7.25 has been derived, depicting the salt concentration along the Western Scheldt as function of the distance from Flushing[jjj]. Instead of a concave curve, as in Fig. 7.22a, a convex curve is found. It is much more similar to the one, assumed by

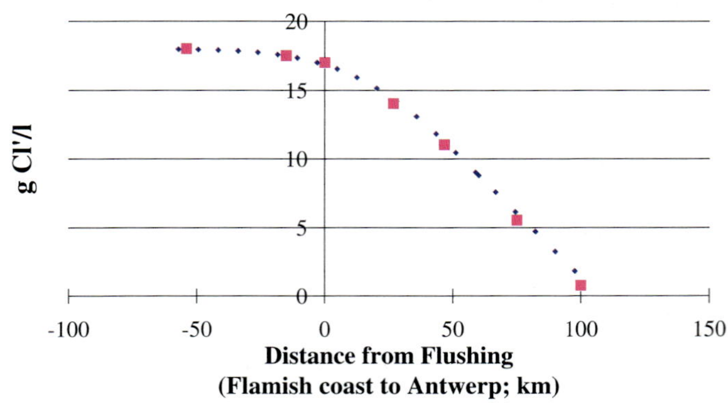

Fig. 7.25 Density distribution along the Western Scheldt as a function of the landward distance from Flushing. The squares indicate data; the diamonds a Gauss function (Eq. (7.152)).

[iii]Drawing 75.198.

[jjj]The distance has been measured along the axis of the river (mouth).

Bakker:[8,9]

$$c = c_s P(x^*) \tag{7.151}$$

$$P(x) = \frac{1}{\sqrt{2\pi}} \int_{-\infty}^{x} e^{-t^2/2} dt \tag{7.152}$$

In Eq. (7.152) x^* means fx/σ and $P(x)$ (Fig. 7.25) is the Gauss function,[1] Eq. (6.22).

In Fig. 7.25, the Gauss-function and the data mentioned above are compared with each other, where the origin of the Gaussian curve has been chosen: 58.8 km landward of Flushing; the value of σ was: 38.7 km. The most landward data point (Antwerp) was derived[kkk] from a report of the RIZA. Furthermore, c_s (Eq. (7.152)) was found to be 18 mg/l. Comparing with Fig. 7.22a, mind, that the seaward part of the curves in prototype (Western Scheldt) are convex instead of concave as in Fig. 7.22a.

Consider now the behavior of the pollutants transported through the Western Scheldt.

Figure 7.26, from RIZA, shows the concentration of three different pollutants (NH_4, O-POP-P and Hg); the first one and the last one are monitored in 1975; the second one in 1975 and 1984.

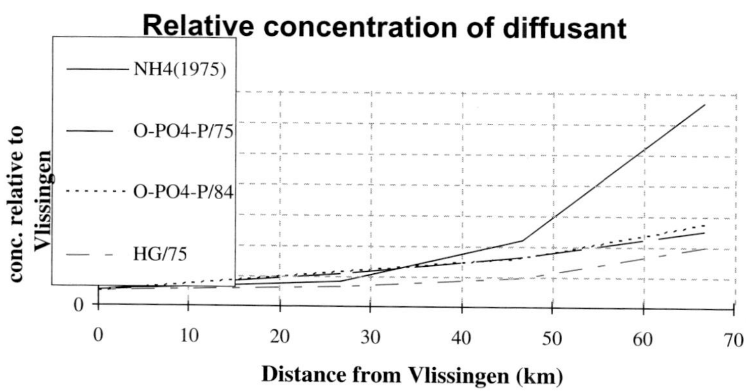

Fig. 7.26 Relative concentration of diffusant as function of the landward distance from Flushing. Compare with Fig. 7.23b:[lll] curves are convex (Fig. 7.28), where these are concave in Fig. 7.23b.

[kkk] "mg salt/l" has been reduced to "mg Cl'/l" by multiplying with 19/34.

[lll] Mind in Fig. 7.22 the sea is on the right-hand side of the pag, where it is on the left-hand side in Fig. 7.28.

As in the case of the salt measurements the measuring data show a curve, opposite of the theoretical ones (Fig. 7.22b): concave instead of convex. In the case of pollutants, evidently also the question of conservancy of material plays a role. However, the main reason for the "unexpected" behavior is the strong increase of the dispersion coefficient in the Western Scheldt in seaward direction, of which the physics will be explained under d.) below.

(d) *Order of magnitude of the diffusion coefficient in the Western Scheldt*

From the data displayed ad c.) above it shows that the assumption of a constant diffusivity constant over the whole Western Scheldt region gives solutions, which are essentially wrong: in order to match reality, anyway a dispersion coefficient, increasing very strongly in seaward direction should be applied. Although this concept has been elaborated in much more detail by Bakker,[7] the result did not show to be essential (up to now) for the investigation of the morphological consequences for the Western Scheldt. Important however is the notion, that the assumption of an error function as longitudinal density distribution in an estuary (as the Western Scheldt) has physical reality (despite simplified diffusion theory), see Fig. 7.25, and some notion why (i.e. under what circumstances) such an increment of dispersion coefficient in seaward direction can happen.

This is the goal of the present sub-subsection: showing the translation of *horizontal* tidal circulation in a dispersion coefficient.

Consider an estuary with a channel, splitting up (at A) in to two branches with equal cross-section (half the one of the channel upstream of A; as the dimension is irrelevant in the following exposure, it will be assumed: "1/2") and joining together again in B (Fig. 7.28). The length of the branches is l.

Let the concentration of some pollutant in A, resp. B be c_A, resp. zero. Let there be a (circulating) current u_{c} in the direction of the arrows in Fig. 7.28 and furthermore a relatively small resultant current u_{r} in both channels in the direction AB; the total currents (direction AB) in channel AB and BA thus are $u_{\mathrm{c}} + u_{\mathrm{r}}$ and $-u_{\mathrm{c}} + u_{\mathrm{r}}$, respectively. In both branches the same value of D_x and σ is assumed. Let the distance between A and B be so large, that the concentration near A in channel AB is practically c_0.

Rephrasing the ideas ad. b.), in the present case a direction x_1 will be defined from A to B, which coincides with the x-direction (cf b above)) for channel AB, but is inverse with respect to the x-direction for channel BA, moving in the direction of the arrows (Fig. 7.27)

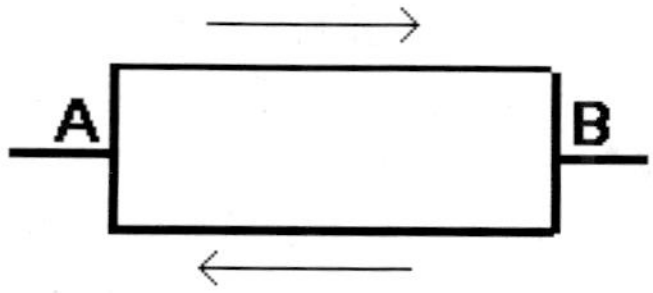

Fig. 7.27 A double channel with circulation.

Referring to Eqs. (7.153) and (7.154) (Fig. 7.23a,b), the concentrations in channels AB and BA will be similar to Fig. 7.22b in channel AB (the "pollution" case) and to Fig. 7.22a (the "salt" case) in channel BA.

$$D'_x \text{ equals } D_x/u_r\sigma$$

The resultant transport of "pollution" will be negligible in channel BA, where it will be about $c_A u_c$ in channel AB. The resultant transport S per unit of (total) cross-section would thus be $c_A u_c/2$. If the concentration in B would have been c_A as well, S would have been $c_A u_r$. Thus from (7.154) it is shown, that $c_0 = c_A$.

With a bird's eye view one could consider the double channel AB+BA as one channel with velocity u_r and concentration gradient $-c_A/l$.

At large scale, a concentration gradient is found:

$$dc/dx = -c_A/l \tag{7.153}$$

Then, large-scale (7.140) would read:

$$c_A u_c/2 = c_A u_r + D_x c_A/l \tag{7.154}$$

from which one finds for D_x:

$$D_x = l\left(\frac{u_c}{2} - u_r\right) \tag{7.155}$$

As u_c is much larger than u_r, this shows that D_x depends on u_c rather than on u_r (mind the remark following Eq. (7.147))!

The treated example (Fig. 7.28) is a specific case, illustrating the character of D_x. Generally, its meaning can be derived from the general equation for the transport

$$S = \overline{cu} + \overline{c'u'} \tag{7.156}$$

where c' and u' are the deviations of c and u from its mean and the overbars denote space averages. Here c' mostly will be proportional to dc/dx, and

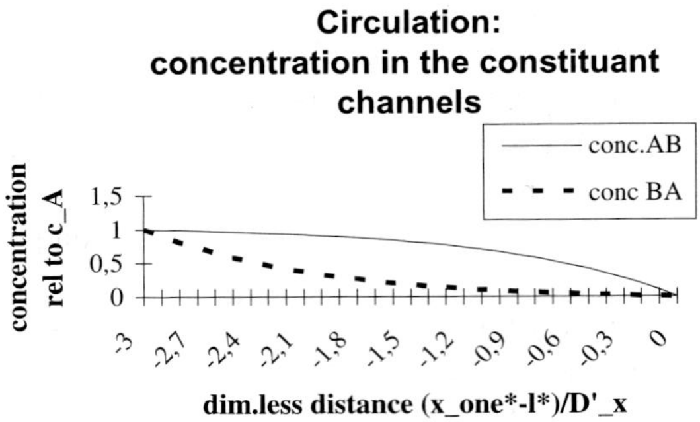

Fig. 7.28 Longitudinal concentration distribution in channels AB and BA.

thus, writing $\overline{c'u'}$ as $D\ dc/dx$ makes sense. In the case of Fig. 7.28 the proportionality of c' with dc/dx becomes even more evident and thus an explicit expression for D can be derived.[mmm]

Looking to Fig. 7.23, showing circulation cells in the Western Scheldt with order of magnitude of 10 km long, with differences of ebb and flood velocities in the ebb -and flood channels of the order of 10 cm/s and a tidal-averaged resultant velocity of 2–5 mm/s, from (7.150) a value of D_x of the order of 500 m^3/s is found, which appears physically reasonable.

Keeping the mechanism sketched above in mind, and realizing, that there has not been focused yet on the funnel-shape of the estuary (larger values of l and u_r at the seaward side than at the landward side of the estuary; shallower inner estuary than outer estuary) one may conclude, that an increase of D_x at the seaward side of the basin is very likely.

7.3.5. *Comparison with results in nature: v.d. Kreeke and Robacewska*[83]

From Fig. 5 of[83] (abbreviated: "KR") maximum surface velocities of the order of 18–20 cm/s are found in the Volkerak on August 16 and 17, 1977; large compared with the depth averaged velocity "$\langle\bar{u}\rangle$" of 0.5 cm/s (tidal discharge Qf = 50 m^3/s; cross-section 104 m^2). These are caused by wind-circulation and gravitational circulation; KR contribute a velocity of the order of 5–7 cm/s to gravitational circulation (their Fig. 9). The basin considered has a length of 35 km (Chapter 1 of their paper).

[mmm]Acknowledgement: the author gratefully acknowledges this interesting comment of Ir. L. Postma concerning (7.150).

With a tidal wave length of ca. 450 km, the length is ca. 0.067 of the wave length. Bakker and Barber[9] consider a stretch of the Western Scheldt with a length of 60 km (and the same assumed depth). As the amplitude of the horizontal tide is about proportional with the length, the top shear stress velocity (proportional to $\hat{u}$) in the Western Scheldt will be of the order of $60/35 = 1.7$ times the value in the Volkerak. Comparing Fig. 3.1 of Bakker and Barber[9] with Table 2 of KR reveals, that longitudinal salt gradients are of the same magnitude. Thus the tidal-averaged shear stress will be of the same magnitude in both cases. From (7.3.16) one could conclude, that in the Western Scheldt case thus a tidal-averaged shear stress velocity $\bar{p}$ would result in the order of $1/1.7 = 0.6$ times the value in the Volkerak (more turbulence in the Western Scheldt, i.e. smaller tidal averaged vertical gradients; Eq. (7.117). So this would result in density-driven tidal-averaged surface velocities of the order of 3 cm/s in the Western Scheldt. The factor in (7.142) on the left-hand side of the square brackets, omitted in the graph of Fig. 7.20 is for the Western Scheldt of the order 0.35E−3. The dimensionless velocity $u^*_{\mathrm{dc}}(0, z)$ has a reference velocity $\sqrt{gh}$: order of 10 m/s.

Figure 7.20 shows a surface value of the order 4. Thus, computation results in a tidal-averaged surface velocity of 1.4 cm/s for the Western Scheldt. Adding the depth-averaged river discharge (order 0.5 cm/s) in the Western Scheldt, neglected in Chapter 4, computations (~2 cm/s) result in values of the order 33% smaller than the result found above. Given the goal (orders of magnitude), the result suffices. Mind, that (apart from the von Karman constant) no matching coefficients have been applied.

7.4. Non-linear Tidal Hydraulics: Vertical Circulation, Caused by "Rain in the Drain"

7.4.1. *Introduction; assumptions concerning the circulation mechanism*

Where in Subsection 7.3.3 the tidal-averaged circulation by longitudinal density gradients has been considered, the present section deals on its counterpart:

the concept of "rain in the drain": introduced by Bakker and de Vriend.[10,nnn]

[nnn] V.d.Kreeke & Ianuzzi,[82] (their Eq. (29) and Fig. 5), give analytical solutions as well.

It is the mechanism of "undertow" (a non-linear feature known from short waves) but physically as well relevant for *tidal* waves. It is caused by an Eulerian tidal-averaged fluid motion in landward direction above the LW-line (discharge Q): the rain in the drain is[ooo] the compensatory flow in each vertical cross-section of an equal quantity of tidal-averaged discharge Q in seaward direction below the LW-line. Thus the mechanism works in opposite direction (seaward near the bottom) as the longitudinal density gradient: a relative comparison of both mechanisms is quite relevant.

As in Section 7.3, in the following the Western Scheldt estuary (refer to Figs. 7.23–7.25) will be considered as an example.

Bakker & de Vriend[10] and Bakker[6] considered the relevance of the rain in the drain for bottom changes and investigated the morphological feedback (under which circumstances the bottom change will boost a larger change and when the change will counteract itself; cf. Section 7.9 of this book).

In the present Section 7.4 analytic computations are given (Subsection 7.4.3) starting from data concerning the Western Scheldt estuary, displayed in Subsection 7.4.2.

Results are compared with the ones from a numeric approach performed with Delft-3D-FLOW (cf.,[9] Section 7 and 8, reproduced here in Subsections 7.4.4 and 7.4.5).

7.4.2. *Schematization; channel characteristics; boundary conditions*

(a) *Boundary conditions and channel characteristics*

The basis of the schematized channel is laid by Gerritsen & Dekker:[37] a one-dimensional schematization of the Western Scheldt.

Thus the following channel characteristics are used in the present section, conform Bakker & Barber.[8,9]

The assumption of a semi-infinite channel ($60000 \times 4500 \times 10$ m and friction and seaside amplitude as given below) is sufficient for the analytical[8] and numerical[19] calculation of the rain in the drain, if the effect of density differences is neglected. In this case, the "closed" end really is closed (no resultant (fresh-)water-discharge).

In the following specific discharges are calculated, i.e. the width of the channel can be assumed equal to 1 m.

[ooo]Concentrating on its definition for a closed prismatic tidal channel.

As vertical tidal amplitude[ppp] at the seaside is assumed: 1.86 m and as tidal period T: 12 h = 43,200 s.

Those characteristics lead to the following values for θ (Section 7.2) and wave length L_0, defined as $T\sqrt{gh}$ (cf. (7.2.6 ad.a)):

* $L_0 = 427.876$ km, i.e. the channel length is 0.1402 times L_0;
* The "friction angle"[qqq] θ depends upon the assumed velocity amplitude $\hat{u}$, which evidently varies over the length of the channel.

b. if $\hat{u}$ is taken 1 m/s (about midpoint):

$$\tan(2\vartheta) = \frac{8g}{C_h^2}\frac{1}{3\pi}\frac{\hat{u}T}{2\pi h} = \frac{8 \times 9.81}{3000} \times \frac{1}{3\pi} \times \frac{1 \times 43,200}{2\pi \times 10} = 1.908$$

$$\theta = 0.544 \text{ rad} \;= 31.17^\circ$$

c. if $\hat{u}$ is taken 1.6 m/s (seaside): $\tan(2\vartheta) = \frac{8g}{C_h^2} \times \frac{1}{3\pi} \times \frac{\hat{u}T}{2\pi h} = \frac{8\times 9.81}{3000} \times \frac{1}{3\pi} \times \frac{1.6\times 43,200}{2\pi\times 10} = 3.053$;

$$\theta = 0.627 \text{ rad} \;= 35.90^\circ$$

Because of the non-linearity of the friction (which has to be averaged over the whole channel), there is reason to give the higher friction at the seaside a higher weight than the friction in the midpoint and use the curves for $\hat{z}$ and $\hat{u}$, valid[rrr] for 330.

(b) *Tide: the harmonic model*
Referring to Section 7.2.7 in general and (7.2.4) in particular ("Y0 = 0"); the linear solution (Fig. 7.9) for the vertical tide ζ (water level w.r.t. S.W.L.) is ($\omega =$ ang. vel. of the tide):

$$\varsigma = \text{Re}[Z_0(\cos kx)e^{i\omega t}] \tag{7.157}$$

ζ being the surface elevation; x is zero at the closed end of the channel; Z_0 (dimension: [m]) is a real or complex value, characterising the tidal motion at the closed end of the channel; k equals:

$$k = \frac{\omega}{C}(1 - i\tan\theta) \tag{7.158}$$

where θ (chosen 330) is an angle determining the friction in such a way, that the period-integrated linear friction equals the period-integrated quadratic

[ppp]Only relevant in present subsection because of chosen friction angle.
[qqq]cf. Sub-subsection 7.2.4a, p. 7–14.
[rrr]Conform Gerritsen and Dekker[37] (taking 33.5°, their p. 6).

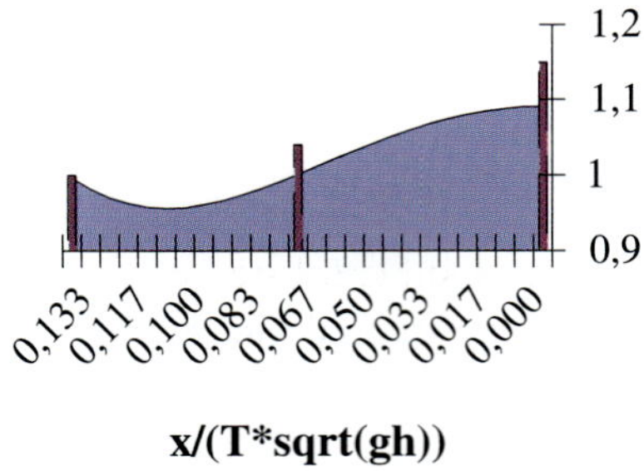

Fig. 7.29 Amplitude vertical tide in the basin, compared to the one at the seaside (on leftside).

friction, assuming a Chezy coefficient according $C_h^2 = 3000\,\mathrm{m/s^2}$. Propagation velocity C is:

$$C = \sqrt{gh(1 - \tan^2\theta)} \tag{7.159}$$

The solution $u(x,t)$ for the horizontal tide is (for $\widehat{u}$: cf. Fig. 7.9b):

$$u(x,t) = \sqrt{(g/h)\cos(2\theta)}\mathrm{Re}\{-iZ_0 e^{i\omega t+\theta}\sin kx.\} \tag{7.160}$$

The gray area in Fig. 7.29 shows the amplitude of the vertical tide (relative to the vertical tide at the seaside); Fig. 7.30 shows the amplitude of the horizontal tide (specific discharge in $\mathrm{m^2/s}$) assuming a vertical tide at the entrance of 1 m; the result will be linear w.r.t. this tide at the entrance (linear computation!).

The thick vertical bars in Figs. 7.29 and 7.30 denote the results of the Delft3D-computation (cf. Subsection 7.4.4) compared with the analytic solution.

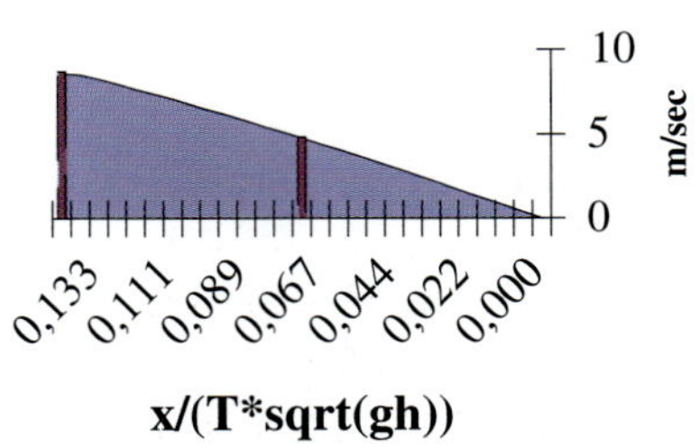

Fig. 7.30 Amplitude horizontal specific discharge in the basin (per meter vertical tide at the landward side; so the unit is m/s).

7.4.3. *Circulation by "rain in the drain"*

The solution for the circulation by the "rain in the drain" can be expressed as follows.

The x-coordinate being at the mean water level, the water velocity $u(x,t)$ is assumed zero when the considered level z is higher than the water level $\zeta(t)$; the tidal-averaged specific discharge $-q_{\mathrm{rd}}$ above the LW-line equals per definition:

$$-q_{\mathrm{rd}} = \frac{1}{T}\int_0^T \int_{-\hat{z}}^{\zeta(t)} u \; dzdt$$

In a closed channel, because of continuity, this specific discharge must be equal to and of opposite sign as the tidal-averaged specific discharge $q_{\mathrm{rd}} = u_{\mathrm{dr}}(h - \hat{z})$ below the LW-line; this defines "undertow" u_{dr}.

Starting from linear wave theory in shallow water, and approximating the LW-level as $z = -\widehat{z}_0$, one finds for q_{rd}:

$$q_{\mathrm{rd}} = -\frac{\hat{u}(x) \times \hat{z}(x)}{2} \cos\varphi(x) \tag{7.161}$$

where φ is the local phase difference between vertical and horizontal tides.

This results in the dimensionless relation:

$$\frac{u_{\mathrm{dr}}}{\hat{u}} = -\frac{(\hat{z}/\hat{z}_0)\cos\varphi}{2\{(h/\hat{z}_0) - 1\}} \tag{7.162}$$

Thus the dimensionless tidal-averaged velocity below LW can be expressed in the dimensionless horizontal and vertical tidal amplitudes (shown as function of the friction angle θ and of the dimensionless distance $x/T\sqrt{gh}$ from the closed end of the channel (Fig. 7.9a,b,c)) and in the phase difference φ, depicted in a similar way in Fig. 7.9d:

$$\frac{h}{\hat{z}_0}\left(\frac{h}{\hat{z}_0} - 1\right)\frac{u_{\mathrm{dr}}}{\sqrt{gh}} = -\frac{1}{2}\frac{\hat{u}}{\sqrt{g/h}} \times \frac{\hat{z}}{\hat{z}_0}\cos\varphi \tag{7.163}$$

The result is shown in Fig. 7.32a,b, respectively: for various values of θ, and (Fig. 7.32b) for a value of θ, equal to 330 (Western Scheldt).

In (7.161)–(7.163) $\hat{u}(x)$,$\hat{z}(x)$ are the amplitudes of horizontal, resp. vertical tide and $\varphi(x)$ the local phase difference between $u(x,t)$ and $\hat{z}(x)$; all as determined from (7.157) and (7.160). The distribution of tidal-averaged horizontal velocity u_{dr} below the Low-Water Line has been assumed uniform, where the one above LW can be calculated as parabolic. The last one is

Fig. 7.31 Tidal-averaged circulation (in a vertical plane) by rain in the drain. Vertical scale of arrows (below LW) largely (factor 2667) exaggerated. Compare Fig. 7.32b. N. B.: positive x to the left (seaward)).

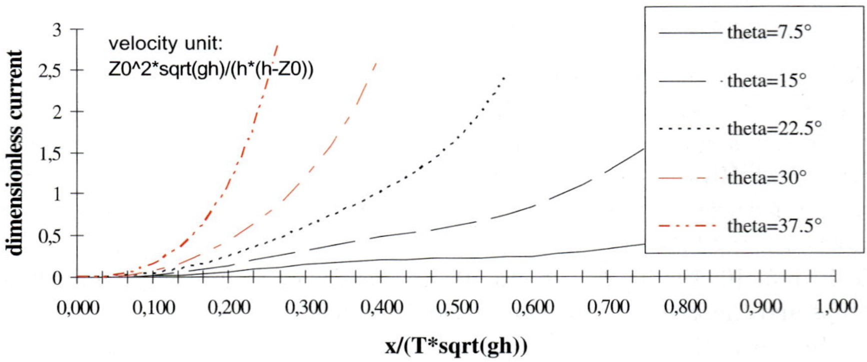

Fig. 7.32a Resultant tidal-averaged horizontal current: dimensionless return current as function of site along the channel (closed at $x = 0$). Various values for θ. Unit on vertical scale: $\frac{\hat{z}_0^2\sqrt{gh}}{h(h-\hat{z}_0)}$.

defined as the integral of the Eulerian velocity during the part of the tide the point under consideration is submerged. Near the level of HW this time lapse considered tends to zero (as does the local $u_{\rm dr}$); near the level of LW positive and negative velocities annulate each other. The distribution of vertical tidal-averaged velocity below LW results from continuity. In Fig. 7.31 the vertical scale has been distorted w.r.t. the horizontal scale, in such a way, that the upper-left arrow is under an angle of 45.

Results of the computation are incorporated in Fig. 7.34 of the next Section 7.5 (comparison numerical and analytical computations): it shows

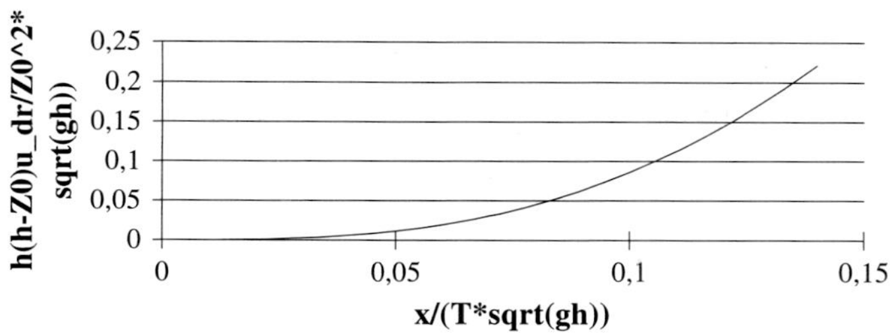

Fig. 7.32b Dimensionless return current as a function of site along the channel (closed at $x = 0$) Unit: $\frac{\hat{z}_0^2\sqrt{gh}}{h(h-\hat{z}_0)}\theta = 33°$.

the downwelling at half depth, caused by the rain in the drain, relatively to the downwelling at Flushing.

7.4.4. *Description numerical model*

A three-dimensional numerical model of the Western Scheldt estuary was created with the rectilinear grid feature of Delft3D. The geometrically simplified rectangular grid approximated the estuary as 60 km long, 4.5 km wide, and 10 m deep. The river discharge equals $100\,\mathrm{m}^3/\mathrm{s}$. Longitudinal and transverse grid cells were spaced at 500 m. Vertically, the model was divided into 10 layers with the following percent thickness (bottom to top): 5, 5, 5, 5, 10, 10, 20, 20, 10 and 10. The model was run using 1-min time steps and the k-epsilon turbulence closure model.

In the initial run, the model assumed a seaside amplitude of 1 m, a seaside salinity boundary condition of 33.1 psu, a longitudinal diffusion coefficient of $342.5\,\mathrm{m}^2/\mathrm{s}$, a 15.7 psu boundary condition at the closed end, and no surface water inflow. The model was run for a three month time period with an initial salinity condition of 25 psu. Only the result from the final tidal cycle (final 25 h) was stored for analysis. As a result of this computation procedure, the thickness of each layer changes as the wave amplitude in the model varies. Model results are linearly interpolated to fixed depth grid positions so that direct comparisons could be made with the analytical results shown in Fig. 8.33.

At the seaside, as prescribed by the boundary condition, water depths varied between 9 and 11 m.

In a subsequent run, a gradient of longitudinal diffusion coefficients was created. The gradient ranged from $200\,\mathrm{m}^2/\mathrm{s}$ landside to $525\,\mathrm{m}^2/\mathrm{s}$ seaside. All of the other model parameters were kept constant.

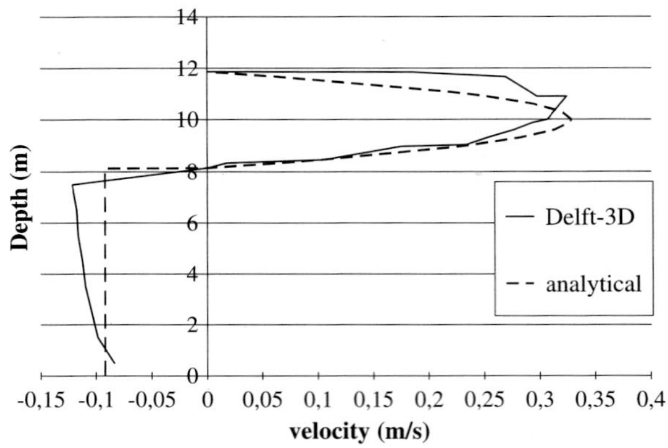

Fig. 7.33 Comparison: numerically and analytically calculated tidal-averaged horizontal velocities. (seaside of basin: "rain in the drain"). N.B.: in this figure landward is positive.

7.4.5. *Comparison between analytical and numerical circulation computations*

At the moment of publication of this book, results of computations with density gradients in the basin are not yet available.

Figure 7.33 shows the comparison of the analytical "rain in the drain" computations with the numerical computations. Depicted are the horizontal tidal-averaged velocities u_{dr} at the seaward end of the basin. For an easy comparison with Fig. 7.31, landward velocities are taken positive in Fig. 7.33.

The correspondence is satisfactory

7.5. Summary Concerning Tidal-averaged Vertical Circulation; Concluding Remarks

7.5.1. *Total solution of the tidal-averaged vertical circulation in the Western Scheldt. Summary concerning considered non-linear effects and rain in the drain*

In this Chapter 7, two mechanisms of vertical circulation have been treated:

a. on the one hand (Section 7.3) "circulation because of a longitudinal density gradient" and
b. on the other hand "rain in the drain" (i.e. undertow in *tidal* waves; Section 7.4).

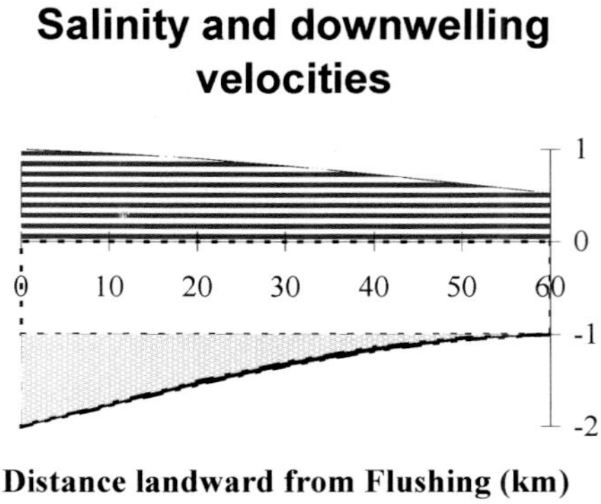

Fig. 7.34 Salinity and downwelling velocities. Upper part: salinity (rel. to salinity at seaside); lower part: downwelling, caused (gray area) by rain in drain and (black area) by density currents rel. to total downwelling at seaside. $x = 0$: Flushing; $x = 60$ km: Belgian border.

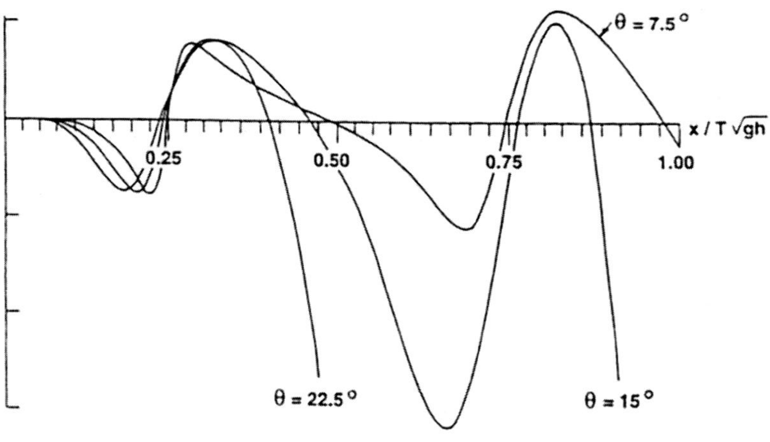

Fig. 7.35 The relation between accretion velocity, site in a prismatical channel (closed on left side) and bottom friction angle θ. Cf Eq. (7.180).

c. There has been focused on the Dutch part of the Western Scheldt (Figs. 7.21, 7.23 and 7.24).

In the present Section 7.5 those two mechanisms are compared (Subsection 7.5.2). The calculations suggest (Fig. 7.34 below), that in the considered part[sss] of such a basin as well "rain in the drain" as salt gradients

[sss]That is, the Dutch part.

cause tidal averaged downwelling. If the landward part would have been extended (including "the Belgian part", Fig. 7.21) at that site "rain in the drain" could have been neglected, where the salt gradients would cause a certain (small) upwelling (conform the right-hand side of Fig. 7.19b).

a. Analytic and Delft-3D calculations concerning the "rain in the drain" feature confirm each other satisfactory (Fig. 7.33).
b. Further checks concerning the analytic computations on the effect of a longitudinal density gradient will be published in the future.

7.5.2. *Comparison between the order of magnitude of "rain in the drain" and the longitudinal density gradient*

The upper part of Fig. 7.34 shows (analogeous to Fig. 7.25) the density distribution of salinity, relative to Flushing for an area of 60 km landward of Flushing.

With respect to the values applied, refer to Fig. 7.25 and Subsubsection c.) of Subsection 7.3.3.

The lower part of Fig. 7.34 shows the magnitude of upward currents (downward is negative); absolute values are relative to the total value at Flushing. The figure suggests a large dominance of "rain in the drain" (green or gray area) compared to the density gradient circulation (black area). This can be substantiated in the following way.

With respect to the *density circulation*: consider first the dimensionless term in square brackets in Eq. (7.141).

From Fig. 7.19b a mean dimensionless velocity over the upper part of the profile of about 2 can be derived. The velocity scale is $\sqrt{gh}$ (ca. 10 m/s) times the magnitude of the dimensionless constant outside the square brackets in Eq. (7.3.41), which magnitude is omitted in the scale of Fig. 7.21. This magnitude is of the order of 4×10^{-4}. For $\Delta\rho^* \approx 0.02$:[ttt] THIJSSE[78] and Fig. 7.27; $\hat{z}_0/h \approx 0.2$, i.e. 4 m tidal difference (cf., for instance[56]); $\sigma \approx 80/\sqrt{2\pi} \approx 30$ km; cf. Fig. 7.25. Thus, the tidal averaged velocity of water above the "pivot point" (which is at about half depth; Fig. 7.20) can be estimated as $2 \times 4 \times 10^{-4} \times 10 = 8$ mm/s. Thus, per unit of width, the estimate for the tidal averaged flux of water, caused by the salt-gradient is estimated (over 5 m depth) as $0.04\,\mathrm{m}^2/\mathrm{s}$.

Compare this with the flux of water involved in the rain in the drain (Fig. 7.32b).

[ttt] Thijsse states: Hook of Holland, 0.016; Waddensea, 0.008–0.023; North Sea, 0.024.

Consider the most seaward point: Flushing. As the deep water wave length $L_0 = T\sqrt{gh}$ equals 427.9 km (cf. Sub-sub-section a.), the considered stretch (60 km,≈ Flushing to Belgian Border) will be $60/427.9 = 0.1402$ times L_0. From Fig. 7.34b a dimensionless return current of 0.22 can be derived.

Considering the reference unit of this current, being $(\hat{z}_0^2\sqrt{gh})/\{h(h-\hat{z}_0)\}$, it shows, that the reference unit for the specific discharge below the line of LW is: $(\hat{z}_0^2\sqrt{g/h})$, being in the present case $4\sqrt{0.981}$, i.e. of the order of $4\,\mathrm{m}^2/\mathrm{s}$. So this is a factor of 100 larger than the effect of the salt gradient!

Evidently, the differences in pivot points, as well in horizontal sense as in vertical sense, should be taken into account.

7.5.3. *Some remarks, concerning effects, not contained in the present model: vertical circulation, caused by wind; overestimate of vertical-plane turbulence at the landside of an estuary; measurements by V.d. Kreeke and Robaciewska; consideration on other turbulence models*

It should be repeated: this book deals on coastal dynamics and the present investigation nearly let us forget the taste of salt.

This (hydraulic part of) chapter should have given the notion that formation of outer deltas (i.e. a sand source to the beach, c.q. inshore; (Section 2.5 and Chapter 3) is present even without sediment-loaded rivers: the "rain in the drain" delivers this source for the beach and inshore in its own.

Furthermore, it should be understood, that siltation at the landward side of this basin is stimulated by the upward flux of the density-gradient driven circulation: it may balance the still-water fall velocity and thus may lift the silt to the marshes.

In Chapter 7.9, a more general discussion will be given. Mind however, aspects which have been neglected in the present schematization:

a. Wind-induced circulation: refer to the discussion by V.d. Kreeke and Robazeweska;[83] they estimate its effect of the same order as the density-driven cirtculation in an area like the Volkerak area (i.e. an area like the landward side of the Western Scheldt).
b. The vertical exchange of momentum at the landward side of the basin is over-estimated by using Eq. (7.108) in combination with (7.103) for

the top-shear stress at the bottom; therefore, circulation at this site will be under-estimated.

c. Evidently, the use of the Prandtl theory for the vertical exchange of turbulence couples the advantage of straight-forwardness to the disadvantage of rigidness and thus abstaining from the possibility of matching theoretical investigations to prototype circumstances by introducing more degrees of freedom. Compare for instance the turbulence model of HANSEN & RATTRAY.[38]

Finally, it may be kept in mind, that analytic and Delft-3D calculations concerning the "rain in the drain" feature confirm each other satisfactory (Fig. 7.33).

7.6. Resonance and Morphological Stability of a Prismatic Channel

7.6.1. *Introduction: method*

In this Section 7.6, the morphological consequences of harmonic tide, occurring in a prismatical channel with tidal-averaged depth h, closed on one side (Fig. 7.9) will be analyzed. The investigation consists of a morphological computation and a stability analysis. The last one reveals, how hydraulic conditions change, when the water depth changes and whether this can be a reason for enhanced erosion.

Rate of change of depth as function of depth (Eqs. (7.175) and (7.181)) renders a differential equation, resulting in depth as function of time. Thus (in)stability can be determined.

As insight is the primary goal, a simple transport formula will be assumed:

sand transport is assumed proportional to the third power of the velocity, the latter being characterized by its zero and first harmonic. Analytically the tidal-averaged gradient of the sand transport (=erosion/accretion velocity) is calculated, as function of those (local) zero and first harmonics and the gradients in x-direction. The gradients in the tidal-averaged currents are caused by the rain in the drain.

Dimensionless presentation is possible: variables are the dimensionless length of the channel (expressed in $T\sqrt{gh}$) and the friction angle θ.

The effect of salt-gradients (and its effect on the siltation of the basin) is neglected.

Subsection 7.6.2 results in a formula for the local erosion rate, expressed in local characteristics in the channel, as: local "rain in the drain", vertical and horizontal tidal amplitude; furthermore containing evidently sediment characteristics and waterdepth.

In Subsection 7.6.3 those local characteristics are translated into more basic characteristics, described below.

Erosion, c.q. accretion rates have been expressed as a function of the ratio $\hat{z}_0/h(\hat{z}_0$ being the tidal amplitude at the closed end of the channel), multiplied by a function f, depicted in Fig. 7.37 and having as independent variables:

a. the site x in the channel (relative to the closed end, expressed as the dimensionless $x/T\sqrt{gh}$) and
b. the friction angle θ.

With the aid of the relations, mentioned above, morphogical instability of channels can be traced. Here instability will be defined as an acceleration of erosion. To this problem will be focused on in Subsection 7.6.4. (with a summary in Subsection 7.6.5). Criteria for this stability will shown to be:

c. the dimensionless length of the channel: $l/T\sqrt{gh}$, where l is its length;
d. the increase of the tidal amplitude at the landward side of the channel, when the channel becomes a little bit deeper;
e. the "tidal friction angle" θ.

Roughly speaking, for "normal" angles of θ only channels with a larger length than 1/4 of the wave length $T\sqrt{gh}$ will show to be unstable (under circumstances; cf. Subsection 7.6.4).

7.6.2. *Calculation of accretion/erosion rate: method*

Let S_1 be (in magnitude) the transport (per unit of width) when the current velocity u is 1 m/s. Write, in the same units, for momentaneous transport S:

$$S = S_1 u^3 \tag{7.164}$$

The appropriate unit for S_1 is $[\mathrm{s}^2/\mathrm{m}]$ since the transport is expressed in$[\mathrm{m}^2/\mathrm{s}]$.

The calculated transport rate is expressed in terms of deposited sediment volume. In order to obtain the solid volume, one should multiply by 1–the pore content of the bed.

Denoting the tidal-averaged value of u by $u_{\rm dr}$, the third-power sand transport formula yields for the erosion rate dS/dx:

$$\frac{\partial S}{\partial x} = 3S_1(\tilde{u} + u_{\rm dr})^2\left(\frac{\partial \tilde{u}}{\partial x} + \frac{du_{\rm dr}}{dx}\right) \tag{7.165}$$

where $\tilde{u}$ denotes the first harmonic of u.

Define δ as the ratio between $u_{\rm dr}$ and $\hat{u}$, the amplitude of u:

$$\delta = u_{\rm dr}/\hat{u} \tag{7.166}$$

Per definition, "the rain in the drain"[uuu] $v_{\rm dr}$ will be the vertical tidal-averaged influx through the LW-plane per unit of length; from continuity it shows, that:

$$v_{\rm dr} = (h - \hat{z})\frac{du_{\rm dr}}{dx} \tag{7.167}$$

From the definitions (7.167) and (7.168), $v_{\rm dr}$ can be expressed as:

$$v_{\rm dr} = (h - \hat{z})\left(\hat{u}\frac{d\delta}{dx} + \delta\frac{d\hat{u}}{dx}\right) \tag{7.168}$$

Denoting the tide-averaged values by an upper bar, the third-power sand transport formula yields for the tide-averaged erosion rate,

$$\frac{d\bar{h}}{dt} = \frac{3}{2}\frac{S_1\hat{u}^2}{h}(v_{\rm dr} + 2\delta\omega\hat{z}\sin\phi) \tag{7.169}$$

This can be derived in the following way.

Choose the time "$t = 0$" such, that $\tilde{u}$ can be expressed as:

$$\tilde{u} = \hat{u}\sin\omega t \tag{7.170}$$

Consider furthermore the linearized continuity equation for water:

$$-h\frac{\partial \tilde{u}}{\partial x} = \frac{\partial \zeta}{\partial t} \tag{7.171}$$

According to (7.172) $\partial\tilde{u}/\partial x$ is proportional to $\partial\zeta/\partial t$, of which[vvv] the phase is 900 ahead with respect to ζ. Denoting the phase difference between $\tilde{u}$ and ζ by φ, the phase difference between $\tilde{u}$ and $\partial\tilde{u}/\partial x$ will be $900 - \varphi$.

[uuu] Cf. definition in Chapter 7.4.1; $v_{\rm dr}$ is positive in downward direction.
[vvv] $\mathrm{Re}(i\omega\varsigma)$.

Thus sand continuity transfers[www] (7.166) into:

$$\frac{\partial \bar{h}}{\partial t} = 3\frac{S_1 \hat{u}^2}{h}\overline{(\sin \omega t + \delta)^2 \left\{ \omega \hat{z} \sin \left(\omega t + \frac{\pi}{2} - \phi \right) + v_{\mathrm{dr}} \right\}} \tag{7.172}$$

Neglecting of δ^2-terms leads to (7.170), as all odd powers of $\sin \omega t$ and combinations of $\sin \omega t$ and $\cos \omega t$ cancel during the time-averaging process.

Remembering, that gS_1 is dimensionless, one may write (7.173) in a dimensionless shape (cf. Subsection 7.6.3):

$$\frac{dh^*}{d(\omega t)} = -\frac{gS_1}{(h^{*2})(h^* - 1)} f(x^*, \theta) \tag{7.173}$$

x^* being:

$$x^* = x/(T\sqrt{gh}) \tag{7.174}$$

and[xxx]

$$h^* : h^* = \bar{h}/\hat{z}_0 \tag{7.175}$$

where $\bar{h}$ is the tidal-averaged value of h. Cf. Subsection 7.6.3 for an expression for $f(x^*, \theta)$.

7.6.3. *Calculation of accretion/erosion rate: details*

Purpose of this Subsection 7.6.3 is to express δ (rate of resultant flow compared to amplitude of oscilatory flow) and v_{dr}, which together determine (Eq. (7.163)) the local sedimentation, into:

a. the location in the channel; the site of the location will be expressed in a dimensionless x-coordinate. Reference length will be: $L_0 = T\sqrt{gh}$
b. the tidal amplitude-to-depth ratio; as standard for the amplitude of the vertical tide, the amplitude $\hat{z}_0$ at the closed side of the channel will be chosen;
c. the bottom friction: this is determined by friction angle θ (Subsection 7.2.3).

[www] Neglecting $\hat{z}$ w.r.t. h in (7.166).

[xxx] Mind, that h^* has another reference than h_0^* in Section 7.3 (cf. for instance (7.109)). No confusion should result.

In other words, purpose of this subsection is to determine the function $f(x^*, \theta)$ in Eq. (7.174).

The local erosion according to Eq. (7.170) can be derived from the erosion Eq. (7.173).

The equation has been made dimensionless by dividing by $-\omega\hat{z}_0$; thus emerges the left-hand side of (7.174).

Call $\hat{z}^*$ the dimensionless local amplitude of ζ (i.e. related to $\hat{z}_0$) and φ the phase difference between vertical and horizontal tide.

Combine (7.167) with (7.163) and introduce a more concise variable δ^*:

$$\delta^* = \frac{\hat{z}^*}{2}\cos\varphi, \quad \delta = \frac{\hat{z}^*(\cos\varphi)}{2(h^* - 1)} \tag{7.176}$$

and thus, according to (7.177), $\delta \approx \delta^*/(h^* - 1)$. Thus (7.168) reads, in a dimensionless way:

$$\frac{v_{\mathrm{dr}}}{\omega\hat{z}_0} = \frac{1}{2\pi(h^* - 1)}\left(\hat{u}^*\frac{\partial\delta^*}{\partial x^*} + \delta^*\frac{\partial\hat{u}^*}{\partial x^*}\right) \tag{7.177}$$

Substitution of (7.177) into (7.178) gives, in a dimensionless shape:

$$\frac{\partial h^*}{\partial(\omega t)} = \frac{gS_1}{h^{*2}(h^* - 1)}\frac{3}{2}\hat{u}^{*2}\left[\frac{1}{2\pi}\left(\hat{u}^*\frac{\partial\delta^*}{\partial x^*} + \delta^*\frac{\partial^*\hat{u}^*}{\partial x^*}\right) - 2\delta^*\hat{z}^*\sin\varphi\right] \tag{7.178}$$

Using (7.174) and (7. 177), and after substitution of (7.176), one can express the function $f(x^*, \theta)$, (7.174) in values, to be derived from the Lorentz theory (Section 7.2):

$$\begin{aligned} f(x^*, \theta) = \frac{3}{2}\hat{u}^{*2}\Bigg[\frac{-1}{4\pi}\Bigg(&\hat{u}^*\hat{z}^*\sin\varphi\frac{d\varphi}{dx^*} + \hat{u}^*\cos\varphi\frac{d\hat{z}^*}{dx^*} \\ &+ \hat{z}^*\cos\varphi\frac{d\hat{u}^*}{dx^*}\Bigg) + \hat{z}^{*2}\cos\varphi\sin\varphi\Bigg] \end{aligned} \tag{7.179}$$

This function $f(x^*, \theta)$ has been depicted in Fig. 7.37. For its derivation, the equations of the Lorentz theory have been used (cf. Subsection 7.2.4). Details are given in Bakker.[5]

7.6.4. *Morphodynamic stability analysis*

Returning to Eq. (7.174), in this subsection the stability of this equation will be considered. Instability will be defined as acceleration of erosion.

Mechanism will be:

increased h gives increased tidal wave length; thus the ratio "channel length/wave length" becomes less. When originally the channel was somewhat larger than 1/4 of a wave length, a subsequent decrease of the relative channel length enhances the wave amplitude $\hat{z}_0$ at the closed end of the channel. This may enhance the erosion process in the channel, etc.

In the following, some simplifying assumptions will be made: the effect of changes in friction (in angle θ and thus in $f(x^*, \theta)$) will be neglected. As θ tends to decrease during the erosion process described, the probability of instability will be underestimated.

Let $x_s^* = x_s/T\sqrt{gh}$ be the dimensionless channel length and let $\bar{f}$ be the mean value over the channel of $f(x^*, \theta)$. From (7.174) it is shown that no accretion of depth has to be expected, if $\bar{f} > 0$. If $\bar{f} < 0$, take as time scale t_0:

$$t_0 = -\bar{f}gS_1T/(2\pi) \tag{7.180}$$

Thus, t_0 will be larger than 0.

Now, first a conversion has to be made, as the variable h^* (equal to $\bar{h}/\hat{z}_0$; Eq. (7.176)) occurring in Eq. (7.174), depends not only on the tidal-averaged depth, but also on the value of $\hat{z}_0$ as well, which will not be a constant during the erosion (c.q. accretion) process.

Thus a new variable h' will be introduced:

$$h' = h/\hat{z}_s \tag{7.181}$$

where $\hat{z}_s$ is the amplitude at the seaward boundary

Denoting t/t_0 by t' and $\hat{z}_s/\hat{z}_0$ by $\hat{z}_s'$, the non-dimensional depth-evaluation equation (7.174) takes the shape of:

$$\frac{\partial h'}{\partial t'} = [(h'\hat{z}_s')^2(h'\hat{z}_s' - 1)]^{-1} \tag{7.182}$$

where $h'\hat{z}_s'$ equal to $h/\hat{z}_0$ can be assumed much larger than 1. Instability occurs, if the right-hand side of Eq. (7.183) increases with h'. This implies, that $h/\hat{z}_0$ should decrease as h' (or h) increases. In other words, instability occurs when $\partial(h'\hat{z}_s')/\partial h'$ is negative. Thus the stability criterion is:

$$\hat{z}_s' + h\frac{\partial \hat{z}_s'}{\partial h} > 0 \Rightarrow \text{stable} \tag{7.183}$$

In order to implement the condition (7.184) in a figure like Fig. 7.36 it is worthwhile to replace the variable "h" in (7.184) by the variable

$x^* = x/T\sqrt{gh}$; differentiation of $x^* = x/T\sqrt{gh}$ with respect to h gives $-x^*/(2h)$ as result.

$$1 + h\frac{\partial(\ln \hat{z}_s')}{dh} > 0 \Rightarrow \text{stable} \tag{7.184}$$

Here one finds $\partial(\ln \hat{z}_s')/\partial h$ as:

$$\frac{\partial(\ln \hat{z}_s')}{\partial h} = \frac{\partial(\ln \hat{z}_s')}{\partial x^*}\frac{\partial x^*}{\partial h} \tag{7.185}$$

Thus the stability criterion becomes:

$$\frac{\partial(\ln \hat{z}_s')}{\partial x^*} < \frac{2}{x^*} \Rightarrow \text{stable} \tag{7.186}$$

This happens for large friction angle θ or for large channel lengths.

In (7.187) x^* denotes the channel length, expressed in $T\sqrt{gh}$. In Fig. 7.36, both $\partial(\ln \hat{z}_s')/\partial x^*$ and $2/x^*$ are plotted. All the points above the line $2/x^*$ (which corresponds with a certain dimensionless channel length and a certain friction angle θ) refer to potential unstable channels, provided that $\bar{f} < 0$ (Fig. 7.36).

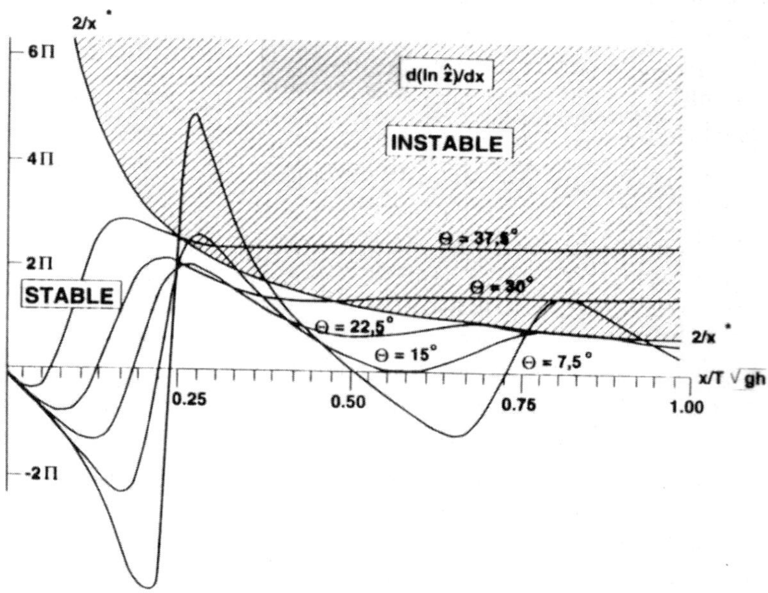

Fig. 7.36 Instability occurs, when the longitudinal gradient of the tidal amplitude exceeds $2/x^*$.

7.6.5. *Summary: results*

Stability of a closed prismatic channel depends on the ratio $L^* = L/(T\sqrt{gh})$, where L is length of the channel and $T\sqrt{gh}$ would be the tidal wave length, if friction would be neglected. If friction is very low ($\theta = 7.5°$) instability would occur for values of $1/4 < L^* < 3/8$ and $3/4 < L^* < 7/8$ (standing wave effects). If friction is higher, near the entrance no effect of reflection is found: here the tidal wave is a propagating, attenuating wave. After some initial deepening reflection may reach the entrance; this will increase the tidal motion and trigger further deepening. This kind of instability occurs for $L^* > 1/2(\theta = 30°)$ or $L^* > 1/4(\theta = 37.5°)$.

7.7. A Morphological Behavior Model for the Outer Delta of Tidal Inlets

7.7.1. *Introduction to Section 7.7*

The theory, displayed in the preceeding sections of this Chapter 7 concerning tidal *basins* now will be extended to the main topic of the present chapter, i.e. the *interaction* of those basins with the outer deltas.

Refer to the "O" in column "7.7" of Table 7.2 of the introduction Section 7.1: finally, this means a return to *coasts*, the subject of this book.

After presenting, in this Introduction 7.7.1, a general overview concerning inlets, a conceptual large-scale model for a single inlet will be developed (Subsection 7.7.2).

The next sections will concentrate on the subsystems "outer delta" and "adjacent coasts". The physical processes, the equilibrium state, the autonomous transient behavior and the response to external forcing will be analysed. Indications will be given how this knowledge can be incorporated in a large-scale behavior model.

More detailed theory concerning the outer delta model is displayed in Subsection 7.7.3, thus returning in sense to Chapter 2, but extending the considerations of that chapter in a broader sense. On the validation of the model, displayed in Subsection 7.7.3 deals Subsection 7.7.4, after which the subject returns (again) to the interaction with the flood basin (Subsection 7.7.5) and ends with discussion (Subsection 7.7.6) and conclusions (Subsection 7.7.7).

Thus, start with the general overview.

Shallow tidal bays which are sheltered from the open sea by barrier islands and connected with that sea by narrow inlet gorges are found on many of the world's coasts. The intertidal zone inside such bays forms an important element in the ecosystem, as feeding and breeding grounds for many species. The deeper channels through the gorge are often used as natural shipping channels, and the barrier islands form a natural protection of the bay and the bordering land. If the islands are inhabited, fixing their position is a coastal defence issue. The bay can act as a source or sink of large amounts of coastal sediment,[76] which can have important implications for the evolution of the adjacent coast.

These functions will be affected by changes in the principal controlling factors of tidal inlet evolution, viz. mean sea level, tide and wave climate. Human interference with these controls or with the morphology can also have adverse and sometimes unforeseen effects on these functions. These are good reasons to investigate the morphodynamic behavior of tidal inlet systems, and to develop tools for its quantitive prediction.

There is a wide variety of tidal inlet types, as it becomes evident from the various classifications (see[72] for a review). Here we will concentrate on the type of inlets which constitute the Dutch Wadden Sea (Fig. 7.37), with a tidal range of typically 2 m and a yearly average significant wave height of less than 1 m (see Refs. [30,68] for more detailed information). In the hydrodynamic classification proposed by Hayes,[39] most of these inlets are of the "mixed-energy, tide-dominant" type, with large outer deltas, wide inlets and drumstick-shape barrier islands.

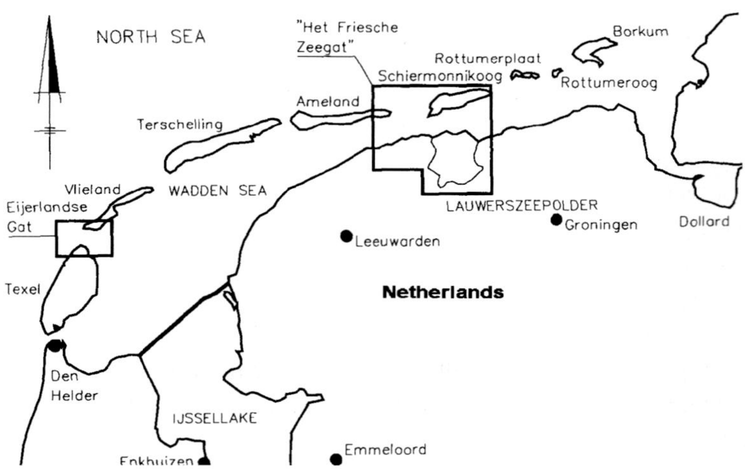

Fig. 7.37 The Dutch Wadden Sea.

The morphology of the Wadden Sea inlets has recently been the subject of a number of extensive studies (for instance, see Ref. [34]), which give access to extensive databases and valuable new information. There is a vast descriptive literature on the behavior of tidal inlets and the adjacent coasts (for instance, see Ref. [2]). The literature on modelling mostly focuses on:

a. the equilibrium state of the gorge,[52,53]
b. the equilibrium state of the delta,[88]
c. the stability of inlets,[20]
d. the short-term response of waves, currents and sediment transports to e.g. man-made works.[89]

The transient response at scales of several kilometres and decades has received much less attention. Semi-empirical behavior models are only available for the basin.[85]

In the present Section 7.7, a behavior model for the outer delta and the adjacent coast of a single-inlet system will be proposed, starting from the two-line model concept given in Section 3.3. This concept is modified in such a way, that the principal transport mechanisms near the inlet are included. The viability of the concept is shown for a number of hypothetical cases, and the validity is tested for the well-documented response of the Friese Zeegat, one of the inlets of the Western Wadden Sea, to a major human interference.

7.7.2. *A conceptual large-scale model for a single inlet*

Figure 7.38 shows an example of a Wadden Sea tidal inlet: the Eijerlandse Gat, between the islands Texel and Vlieland. The principal elements of the system are shown, viz.:

a. the basin, with a distinct channel/shoal system and without a clearly distinguishable flood delta,
b. the outer delta, also with a distinct channel/shoal system, but of a different nature than in the basin,
c. the gorge, in which the channels of the outer delta and the basin meet,
d. the updrift coast of Texel, which is losing large amounts of sand to the inlet,
e. the downdrift coast of Vlieland, with a vast beach plane due to the abundant sand supply from the outer delta.

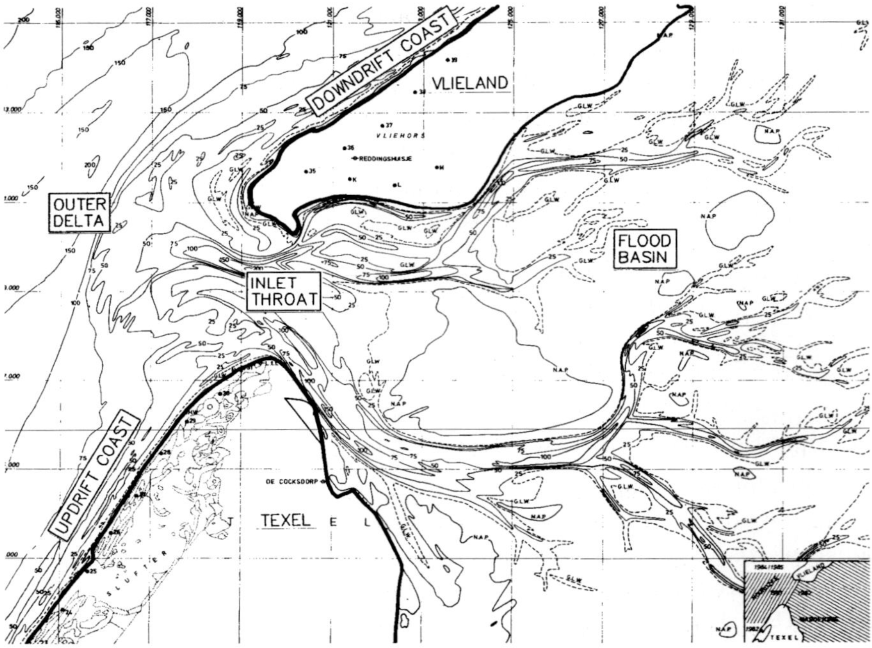

Fig. 7.38 The Eijerlandse Gat, one of the Wadden Sea inlets, cf. Fig. 7.3 ("Eyerlandse Gronden").

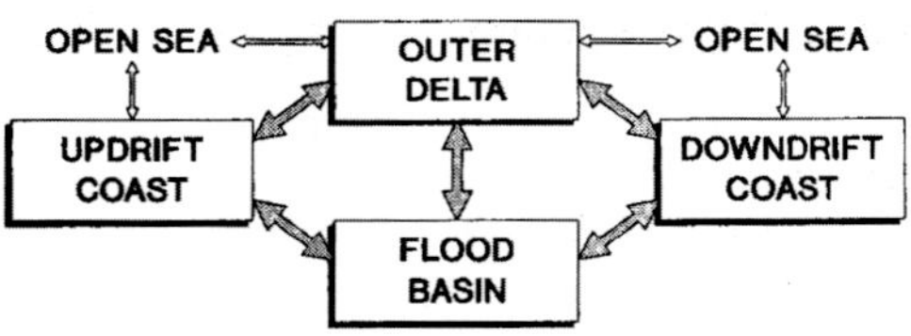

Fig. 7.39 Aggregate flow chart of a tidal inlet system.

These principal elements are linked by "sediment conveyors", three of which go through the gorge (Fig. 7.41).

The morphological state of each of the elements can be described by a small number of aggregate state variables (e.g. the total sand volume of the outer delta, or the total water volume below MSL and the area of the intertidal zone in the basin). Conservation of sediment requires sediment exchange with other elements, or with the open sea.

If we focus on the time-scales of transient evolutions of the principal elements, we can leave the very slow evolutions of the system in its quasi-equilibrium state out of consideration. There is ample empirical evidence that many morphological parameters in the equilibrium state are uniquely related to characteristics of the hydrodynamic forcing.[33]

Assuming these hydrodynamic characteristics (MSL, tide, wave climate) to be independent of the state of the system, we can postulate that:

f. if an element would not be influenced by the others, and the extrinsic conditions (MSL, tide, wave climate) were fixed, that element would exhibit its own inherent behavior and tend to an equilibrium state, and
g. if the extrinsic conditions would remain unaltered, the system as a whole would also tend to a state of (quasi-) equilibrium, i.e. evolution at a much larger time scale than the inherent behavior of each of the elements.

Apart from this inherent tendency towards an equilibrium state, each element responds to external forcings, from other elements, from changes in extrinsic conditions and from human interference. Hence it exhibits a combination of inherent and forced behavior.

The above postulations rule out the type of inlet instability which was first identified by Escoffier,[31] and investigated later on by many others (for instance, Bruun and Gerritsen,[21] Bruun,[20] and Van de Kreeke[80,81]). This instability rests essentially upon the interaction of the hydrodynamic forcing and the morphological state. Most of the Wadden Sea inlets are quite stable,[72] so ignoring this interaction is justified here.

7.7.3. *The outer delta model*

(a) *Processes*

The processes at the outer delta are not known very well, if it were only because of the adverse environment for field measurements. Observed morphological changes form the principal source of information, although remote sensing and exploratory numerical model studies are increasingly important.

The principal elements of the Wadden Sea outer deltas are

a. the main ebb channel, often bifurcated towards the delta margin,
b. a number of smaller flood-dominated channels, often near the heads of the adjacent barrier islands,

c. a shallow terminal lobe, covered by a series of shoals which tend to move slowly downdrift along the delta margin.

Depending on the net sediment drift and the orientation of the main tidal channels, the outer delta can have a smooth transition to the adjacent coast, or it can be clearly separated from it. Figure 7.13 shows that either type occurs near the Eijerlandse Gat: a smooth transition to the northerly island (Vlieland) and a distinct separation from the souterly island (Texel).

The sediment circulation around a delta should be thought of as an intermittent rather than a continuous process. During periods of moderate waves, the tidal transport mechanisms dominate and sand is picked up from the island heads, conveyed through the flood channels and the ebb channel, and deposited at the delta edge. There it remains until the next storm, when waves are breaking on the delta edge and give rise to transport along the edge towards the island heads and shorewards across the edge (bulldozer effect).

Figure 7.40, from an extensive numerical model study of the Eijerlandse Gat[61,86] illustrates this intermittent character. It shows the residual sediment transport vectors for moderate and for high wave conditions. For moderate wave conditions there is a lot of sediment transport on the shoals at the outer delta and in the deep channels, and all the sediment which reaches the delta margin is deposited there. Only if the waves are high enough to break at about MSL -5 m, this sediment is removed by wave-driven currents along and across the margin.

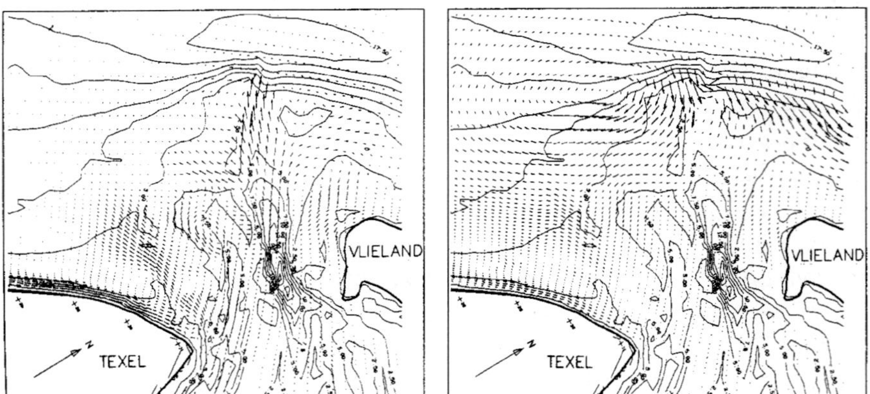

Fig. 7.40 Drift between Texel and Vlieland.

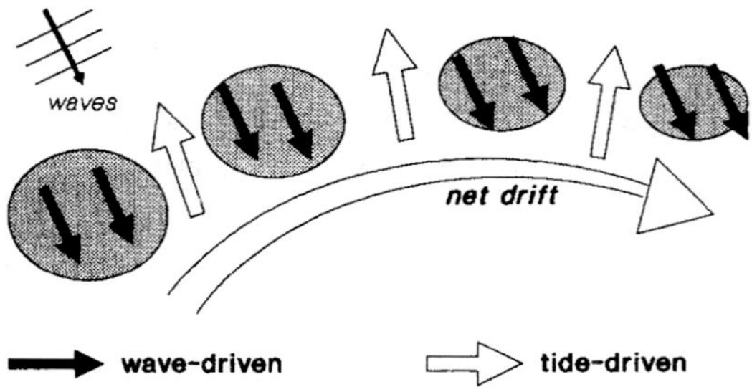

Fig. 7.41 Residual drift mechanism ("bar bypassing") via the shoals on the terminal lobe of the outer delta.

Via this effect, high waves tend to move large amounts of sediment shorewards. This is the most important mechanism of delta reformation if, for any external reason, the tidal motion through the inlet throat is weakened.[74] If the delta is in equilibrium, this "bulldozering" effect of the waves is compensated by the tidal motion in periods of calm weather.

If the wave climate is such, that there is a longshore drift along the island coasts, the long-term residual sediment paths near the shoals are probably not closed, but rather spiraling or even zigzagging (Fig. 7.41; also see Ref. [30]).

The net effect is a sediment drift along the terminal lobe, from the updrift to the downdrift island (bar bypassing), accompanied by a downdrift displacement of the shoals (although these two phenomena are not necessarily linked).

Although this is all rather speculative and more research is definitely needed, we assume for the time being that the long-term residual sediment fluxes can be represented schematically as in Fig. 7.44.

Figure 7.44 may also be an over-simplification of reality, because it takes the main ebb and flood channels as fixed. In reality, these channels may migrate more or less systematically, and when they reach an unfavorable position they vanish and another channel takes over (cf. Section 2.6 in Ref. [87] and Ref.[11]. For the Wadden Sea inlets, the lifetime of a channel is typically some decades to a century, i.e. of the order the time scales which we would like to cover with our long-term dynamic model. Therefore, we can probably not ignore channel migration indefinitely. Yet, it is ignored for the time being, because we think it is not vital to the basic understanding of the outer delta evolution.

Making use of the available empirical information, we will try to set up a semi-empirical long-term behavior model for the outer delta and the adjacent island coasts. Thus we will try to circumvent the indefinite and error-prone time-integration associated with physics based models. For a detailed description of the model, refer to.[18]

In order to reduce the system to its essentials, start from the two-line model concept for uninterrupted coasts.[3,yyy] The basic idea of this concept is that the morphologically active part of the coastal profile is divided into two sections of fixed height (h_1, h_2; see Fig. 7.43), but variable volume. Each section is represented by the mean cross-shore location of its seaward face. The cross-shore positions of these points (y_1, y_2) are the dependent variables. Thus the model produces the time evolution of two lines, called the beach line and the inshore line, which represent the state of the coast (location, steepness). The sediment balance determines the cross-shore displacement of the lines, and includes the wave-induced longshore sediment transport along each line (S_1, S_2) and the sediment exchange between the lines (S_y). Dynamic feedback comes in via the dependence of the longshore transport on the angle of wave incidence relative to the orientation of each line, and via the dependence of the cross-shore transport on the distance between the lines. In a linear approximation, this becomes

$$d(S_1) \therefore d\left(\frac{\partial y_1}{\partial x}\right); \quad d(S_2) \therefore d\left(\frac{\partial y_2}{\partial x}\right); \quad S_y \therefore (W - y_2^{'} + y_1) \qquad (7.187)$$

in which S_y denotes the transport from the beach line to the inshore line and W is the equilibrium distance between the lines. Substitution of Eq. (7.7.1) into the sediment balance yields the following constituting equations of the two-line model (cf. Subsection 3.3.5):

$$h_1\left(\frac{\partial y_1}{\partial t}\right) = s_1\frac{\partial^2 y_1}{\partial x^2} - s_y(W - y_2^{'} + y_1) \qquad (7.188)$$

$$h_2\left(\frac{\partial y_2}{\partial t}\right) = s_2\frac{\partial^2 y_2}{\partial x^2} + s_y(W - y_2^{'} + y_1) \qquad (7.189)$$

where

$y_1; y_2^{'}$ is the position of the lines with respect to a fixed shore-parallel reference line; t, time; x, longshore coordinate; $h_{1,2}$, height of the beach or inshore zone; $s_{1,2}$, coastal constant concerning beach, c.q. inshore (longshore); s_y, cross-shore transport constant; W, equilibrium value of the distance between the beach line and the inshore line.

[yyy] Cf. Section 3.3.

The suffix "1" refers to the beach line, and the suffix "2" to the inshore line. The constants s_1, s_2, and W are related to the wave climate. Bakker *et al.*[14,zzz] (cf. Sections 3.3 and 3.4) describe how a two-line schematization should be made and how these constants can be determined.

This concept can be extended to a situation where the coast is interrupted by a tidal inlet.

Figure 7.44 shows how a concentrated transfer of sediment from the beach to the inshore can make this model produce a delta-type feature (also see Ref. [69]).

In the case of a single inlet, we have to distinguish three zones in this model: the updrift coast, the delta and the downdrift coast. The model for the two island coasts is identical to the above two-line model (Eqs. (7.190) and (7.191)). The delta model has to be different, because:

- the distance between the lines is no longer a measure for the rate of sediment transfer between them; although there are many cross-shore transport mechanisms at work, including the "classic" wave-borne transport,[75] their effects are not always a function of the cross-shore dimension of the delta,
- an important part of the sediment transfer between the lines is concentrated in tidal channels (cf. Fig. 7.44),
- the beach line does not exist in the gorge, nor is the sediment transport near the island heads exclusively wave-driven (cf. Fig. 7.42), and

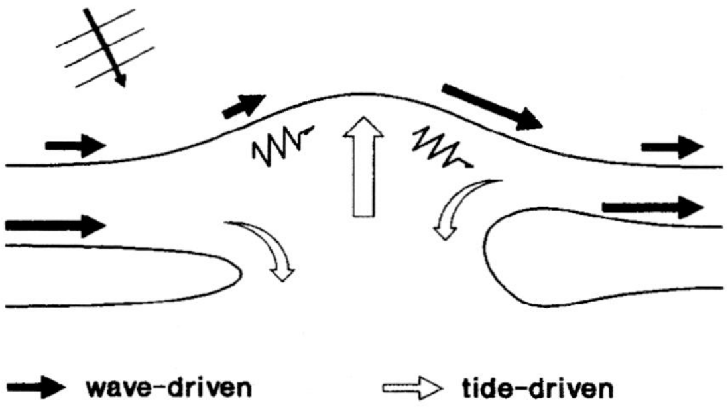

Fig. 7.42 Schematic picture of the long-term residual transport pattern near the outer delta.

[zzz] Cf. Subsections 7.7.4 and 7.7.5.

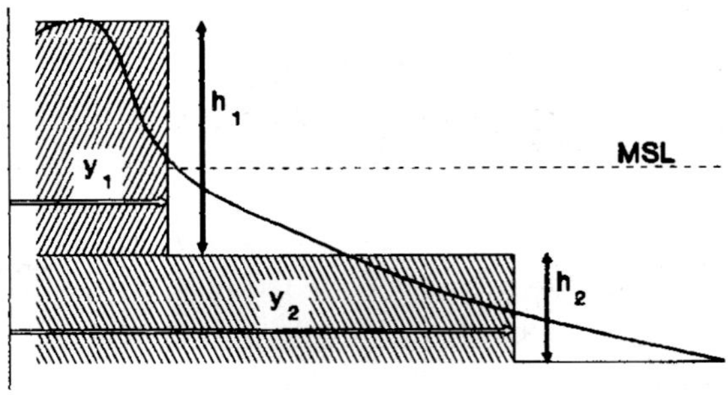

Fig. 7.43 The two-line model concept (cf. Chapter 3.3).

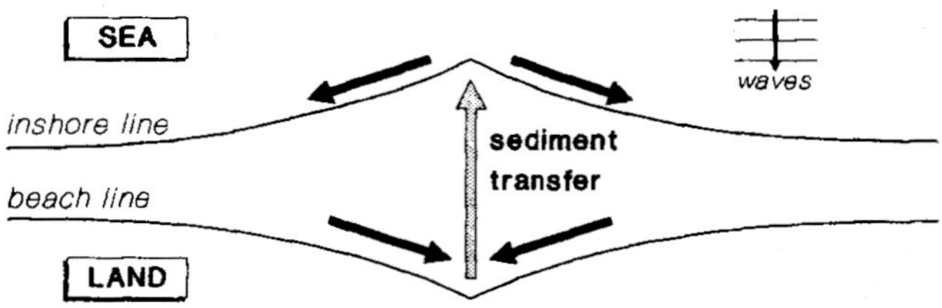

Fig. 7.44 Two-line model concept with a concentrated sediment tranfer from the beach line to the inshore line.

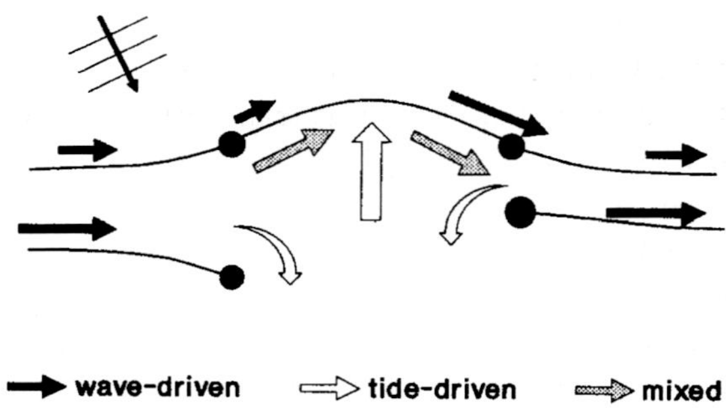

Fig. 7.45 Schematic picture of the long-term residual transport pattern near the outer delta.

- the transport along the inshore line is not exclusively driven by longshore currents, and hence it does not depend exclusively on the orientation of the line relative to the incoming wave direction.

Together with the process analysis in the previous section, this leads to the following model concept (also see Fig. 7.46).

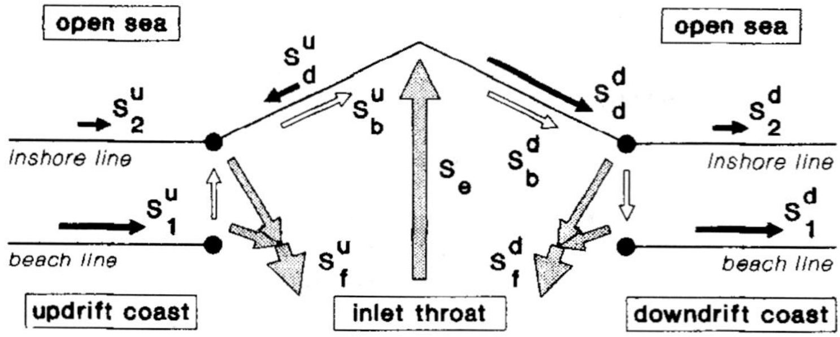

Fig. 7.46 Schematic representation of the outer delta in its equilibrium state.

a. The margin of the outer delta is treated as a continuation of the inshore line of the adjacent coast, i.e.

$$h_2\left(\frac{\partial y_2}{\partial t}\right) = s_2\frac{\partial^2 y_2}{\partial x^2} + S_y(x) \tag{7.190}$$

b. where $S_y(x)$ is a distributed cross-shore sediment flux. The cross-shore sediment flux term is a vehicle to include the wave-driven "bulldozer effect", as far as this is not balanced by the tide-driven radial expansion tendency of the delta.[54]

c. The transport via the shoals at the terminal lobe is assumed to be proportional to the angle between the wave-induced and the tide-induced currents, which is approximately equal to the angle of wave incidence relative to the inshore line. Hence this transport is proportional to (but not necessarily equal to) the longshore transport along the inshore line. This introduces another coastal constant for the transport along the edge of the delta.

d. The transport through the updrift flood channel, which is fed both from the inshore line and from the beach line, is a function of the tidal motion and possibly the wave climate, but in any case independent of the state of the delta. For the sake of simplicity, we assume this transport to be proportional to the tidal prism. The constant of proportionality can be determined from the equilibrium state (see the next section).

e. The same goes for the transport through the downdrift flood channel. In view of the likely dependence on the orientation (via the "tidal motive force"; see;[11,87] i.e. Sections 2.5 and 2.6), we must allow for a difference between the updrift and the downdrift flood channel transports. This introduces another calibration constant, which can also be determined from the equilibrium state.

f. The transport along the inshore line towards the updrift nodal point is split into two parts, one continuing along the delta margin and the other feeding the updrift flood channel. The ratio between these parts is a control parameter of the model.
g. The same goes for the distribution of the transport towards the inshore nodal point at the downdrift coast. In principle, the ratio can be different from that for the updrift nodal point.
h. The transport through the main ebb channel is considered as an input quantity, as it is likely to depend on the state of the basin and the inlet throat, rather than the state of the outer delta.
i. This is supported by the observation that the volume of the delta responds to a sediment demand in the flood basin,[58] and that it depends on the aspect ratio of the throat.[50]

(c) *Equilibrium state*
If the whole system is in equilibrium, the net transport through the gorge (flood and ebb channels together) equals zero.

If we disregard cross-shore transports, the inshore line in the equilibrium state situation consists of four straight sections (Fig. 7.46), and the system is described by 15 sediment fluxes and the position of six nodal points (the four indicated in Fig. 7.46, plus the tip of the delta and the throat of the inlet). In each of these points sediment continuity is required.

Four out of these 15 fluxes, viz. those along the island coasts, depend only on the wave climate, and not on the morphology of the delta. The relevant coastal constants can be determined according to the usual practice in two-line modelling (see e.g. Chapter 4).[14] Four other fluxes, viz. those along the margin and on the terminal lobe of the delta, can be related to the wave climate and the orientation of the inshore line (and hence to the protrusion of the delta). Since they have the same controlling parameters, the transports along the delta margin and over the terminal lobe are lumped into one sediment flux of the wave-driven type, with a coastal constant a.

The seven remaining fluxes are those through the main ebb channel, the "feeders" of the flood channels at either side, and the connections between the beach line and the bypassing terminal lobe. They can be solved from five continuity equations and two expressions for the flood channel transports in terms of the tidal prism.

As a result of these operations, all fluxes which are relevant to the delta morphology are expressed in terms of the "delta aspect ratio" (the maximum protrusion divided by the longshore extent), which is the only remaining unknown. Empirical information on outer deltas in equilibrium should provide an estimate of this quantity.

Recent analyses of the Wadden Sea outer deltas have shown that they have returned to an equilibrium state, after a long period of disturbance due to the closure of the Zuiderzee in 1932[58] (see the next section). Their present sand volumes are related to the tidal prism and the wave climate via.[34,88]

$$V_\Delta = c_\Delta P^{1.23} \tag{7.191}$$

where V_Δ is the sand volume of the outer delta, taking the hypothetical coastline without the inlet as a reference; c_Δ, a constant, which corresponds with Walton and Adams[88] value for highly exposed coasts; and P, tidal prism.

If the delta dimensions maintain their mutual proportion, the maximum protrusion of the inshore line would be given by

$$\lambda_\Delta = c_\lambda P^{0.41} \tag{7.192}$$

where λ_Δ is the maximum protrusion with respect to the islands' inshore lines; and c_Δ, a constant, which probably depends on the wave climate.

Sha and Van den Berg,[68] found for a large number of deltas at the Dutch coast, the U.S. Atlantic coast and the Gulf coast that the protrusion ratio, defined as the delta protrusion divided by the width of the gorge, is primarily a function of the tidal range, and much less so of the incoming wave energy. As the wave climate effects on the delta protrusion and the gorge width may compensate each other, this provides no evidence on the validity of Eq. (7.194).

Upon closer inspection of Sha's[66,67] data on the Wadden Sea deltas, it turns out that Eq. (7.194) should be modified into (see Fig. 7.47):

$$\lambda_\Delta = c_\lambda P^{0.6} \tag{7.193}$$

in which the best fit is obtained for $c_\lambda = 0.044$. This is the empirical relationship for the equilibrium protrusion which we will use in the model.

The longshore extent of the delta is assumed to be invariant to changes of the protrusion. In that case, the longshore extent can just be determined from the initial condition, i.e. from a map. Taking a constant longshore extent is the best we can do on the basis of the material available. Further research into this topic is definitely needed.

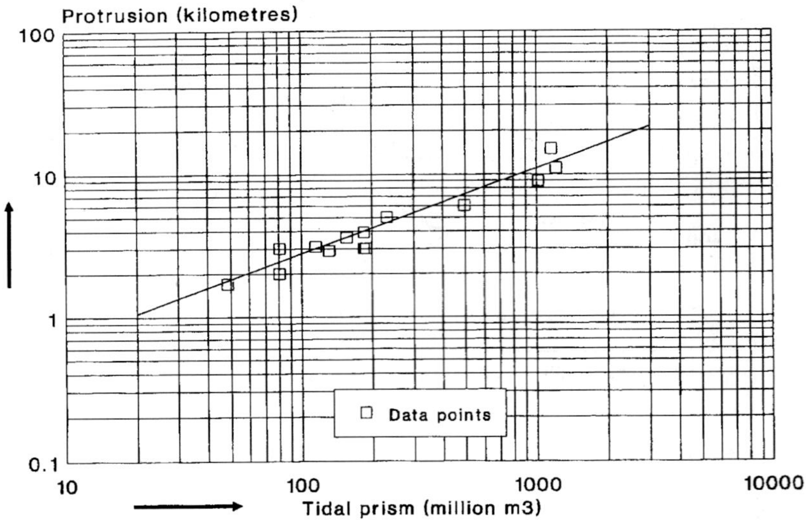

Fig. 7.47 Maximum delta protrusion vs. tidal prism for the inlets of the Western and Eastern Wadden Sea (based on data from [66]).

We can now calculate the sediment fluxes in Fig. 7.46 and have a picture on the intensity of the sediment circulation on the delta if the system is in equilibrium.

If the tide is assumed to be independent of the state of the delta (which rules out the possibility of long-term instability via the feedback of the outer delta on the tidal prism), the sediment fluxes through the ebb and flood channels are independent of that state. The only sediment fluxes which do depend on it are those along the delta margin and over the terminal lobe.

If the delta is out of equilibrium, we need additional information on the inherent transient behavior and on the sediment distribution mechanisms. This will be discussed in the next section.

(d) *Transient evolution of the outer delta*

The inherent behavior of the outer delta and adjacent coast subsystems is included in Eq. (7.189) for the delta margin and in Eqs. (7.190) and (7.191) for the adjacent coasts.

The system will exhibit only its inherent behavior if the "autonomous" sediment fluxes (flood channels, main ebb channel) are given in the right proportion, such that there is no interaction with the flood basin (zero flux through the gorge) or the open sea. Otherwise, forced behavior will also come into play.

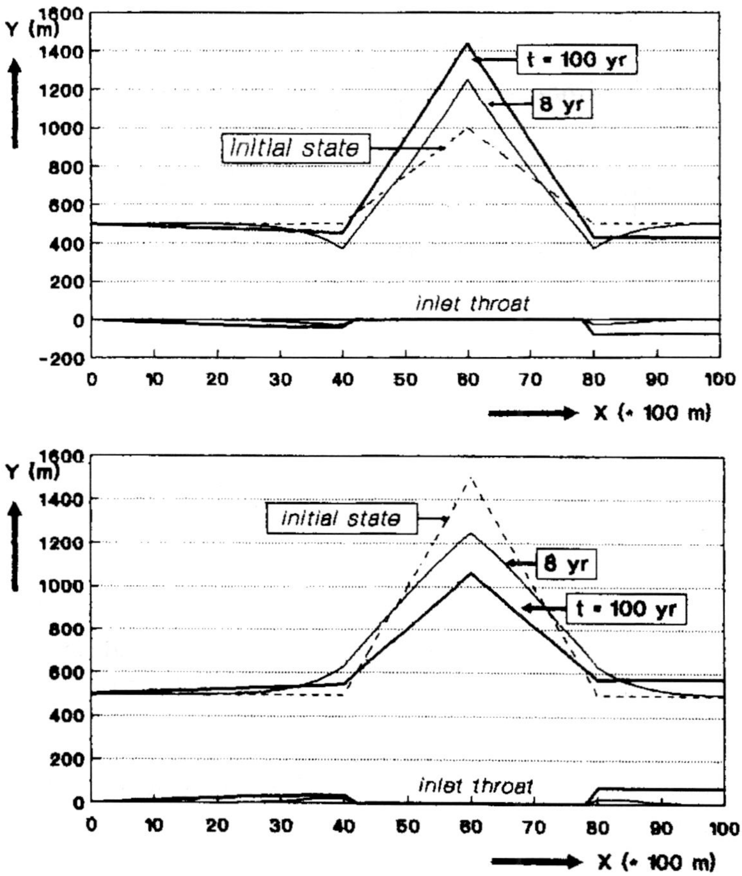

Fig. 7.48 Fictitious example of inherent delta behavior, if the delta is too small (upper figure) and it is too big (lower figure).

Since we have assumed the flood channel transport independent of the morphological state of the delta, the sediment fluxes in the flood channels must be equal to those in the equilibrium state. This means that, in order to show the inherent behavior, the transport through the main ebb channel must be taken equal to the sum of the flood channel transports.

Figure 7.48a,b shows the results for two such cases, both concerning the response to a non-equilibrium initial condition. If the initial delta is smaller than the equilibrium size which corresponds with the tidal prism (Fig. 7.48a), it borrows the sand needed for its growth from the adjacent islands, first from the tips, but after some time from the entire island coasts.

If the delta is initially too large (Fig. 7.48b), the opposite occurs, as must be expected from a linear model. Globally, the deviation from the equilibrium state will tend to decay more or less exponentially, but locally the system may exhibit a more complex behavior (cf. the island tips in Fig. 7.48a). Since the size of the tidal channels or the shoals on the delta is not figuring as such in the model, its possible role as a temporary source of sediment is not taken into account.

In order to model the forced behavior, we have to add a distribution algorithm for surplus (+ or −) sediment fluxes.

Part of this algorithm is included implicitly in the two-line model, but at the following points, additional information, has to be provided (also see Fig. 7.48):

a. How much of the beach transport along the updrift island is bypassed via the shoals on the terminal lobe? The transfer mechanism is probably very complicated, with sediment piling up at the tip of the island, pushing away the flood channel until this has reached such an unfavorable position, that a new channel forms and cuts the piled-up sand off the island.
b. How are the sediment fluxes due to bypassing and longshore transport along the delta margin distributed between the inshore line, the beach line and the flood channel at the downdrift island? The observed behavior of the island coasts (cf.[4]) suggests that the transfer mechanism here is a combination of a more or less continuous sediment flux and an intermittent process of erosion and accretion due to the shoreward migration of the channel and shoal system.

Research on these aspects is still in an early stage. It will be based on studies of observed delta behavior, with physics-based numerical models as a useful aid. For the time being, these distributions are controlled via user-supplied model parameters.

7.7.4. *Model validation*

(a) *Viability of the concept*

In order to assess the viability of the concept, i.e. the potential to qualitatively reproduce the system's behavior, we have to run a number of hypothetical test cases, which were derived from documented situations in the Wadden Sea.[18]

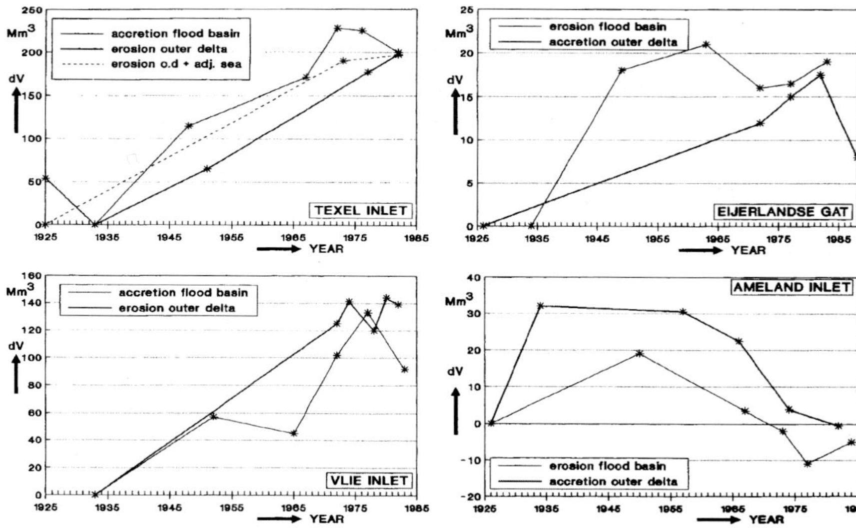

Fig. 7.49 Observed response of the deltas of various Wadden Sea inlets to the closure of the Zuiderzee in 1932 (after[58]).

For instance, the model responds to a sudden increase of the tidal prism by gradually enlarging the outer delta, and to a decrease of the prism by gradually shrinking the delta. This is in line with the observed response of the deltas of various Wadden Sea inlets to the closure of the Zuiderzee in 1932 (see Fig. 7.49; the prisms of the Texel and Vlie inlets were reduced by the closure, the one of the Eijerlandse Gat was increased).

When sediment is continuously taken away from the delta, the model predicts a retreat of the delta and a reorientation of the adjacent coastlines, until these deliver the removed amount of sand per unit time (Fig. 7.50). Also this reponse is as expected.

When, for any reason, the basin is continuously importing sand (e.g. due to continual sand mining or land reclamation), the delta and the adjacent coasts will retreat until the system reaches a new quasi-equilibrium state, in which the adjacent coasts can deliver the required amount of sand. This behavior has been observed in various inlets, also in areas other than the Wadden Sea.

On the basis of these indications, we conclude that the model concept is viable, at least as far as the overall evolution of the delta and the adjacent coasts is concerned. The model has not been tested for its potential to

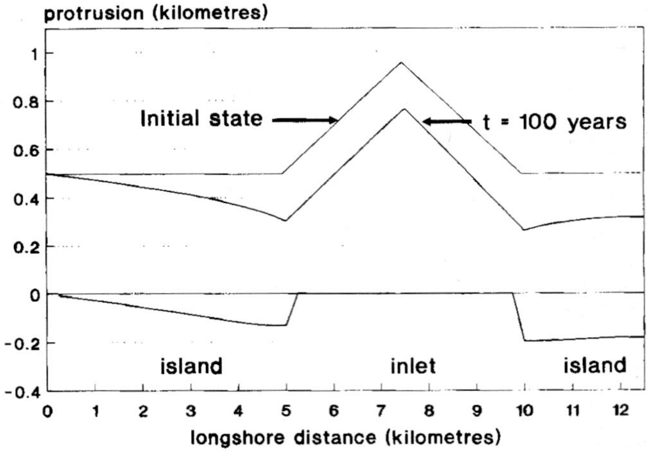

Fig. 7.50 Response to a continuous removal of sediment from the delta.

describe the details of delta evolution in space and time, and it is not claimed to have this potential.

(b) *Validation for response of the "Friese Zeegat"*
The quantitative validity of the model for the inlet system to the reclamation of a large part of its flood basin in 1969. This led to a reduction of the tidal prism from 300 to 200 km^3. According to Hayes[39] classification, this inlet is of the "mixed-energy with tidal dominance" type, with a large outer delta and drumstick-shaped barrier islands.

The two-line model for this inlet, in the equilibrium state before reclamation, is outlined in Fig. 7.51.

The observed changes in the volume of the basin after 1969 were used to define the sediment demand via the ebb-channel transport as a function of time. All parameters were established beforehand:

a. the coastal constants were calculated according to Bakker *et al.*[14,aaaa] and Chapter 4, to yield:

$$h_1 = 9.25\,\text{m}; \quad h_2 = 6.25\,\text{m}$$
$$s_1/h_1 = 1.1 \cdot 10^5\,\text{m}^2/\text{yr}; \quad s_2/h_2 = 1.6 \cdot 10^5\,\text{m}^2/\text{yr};$$
$$s_y/h_1 = 0.054/\text{yr}; \quad s_y/h_2 = 0.08/\text{yr}; \quad 1000\,\text{m}$$

[aaaa] cf. Subsection 4.2.3

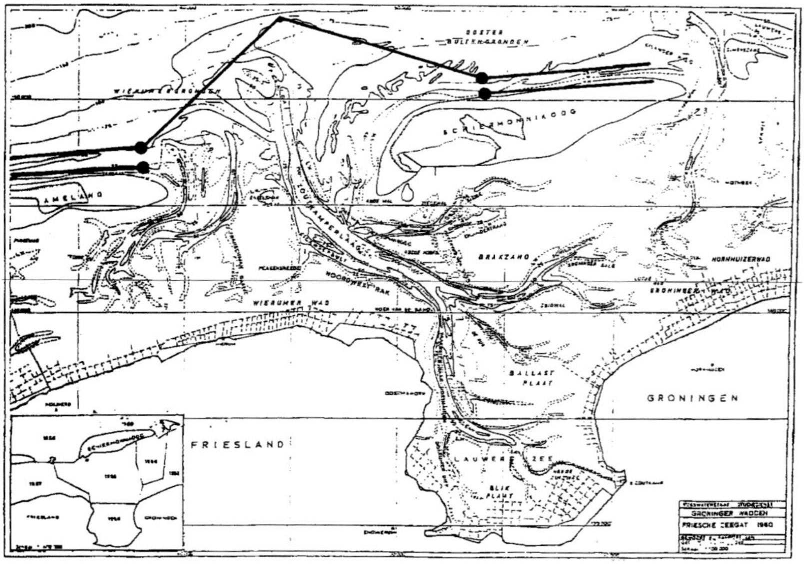

Fig. 7.51 Two-line model of the "Friese Zeegat". See Fig. 7.39 for the situation.

b. the distribution ratio of the updrift sediment transport along the inshore line was taken as 0.75, i.e. 75% of the supplied sediment is fed into the updrift flood channel and 25% goes on along the delta margin and over the terminal lobe,
c. the distribution ratio of the sediment transport along the margin and over the lobe towards the inshore line of the downdrift island was also taken as 0.75, i.e. 75% is fed into the flood channel and 25% goes on along the inshore line of the island.

Only the coastal constant of the terminal lobe was used as a calibration parameter (in this case $s_3/h_2 = 3.0 \times 10\, m^2/yr$). Thus a reasonable agreement of computed and observed changes was achieved, as is shown in Fig. 7.52 for the protrusion. As the shape of the delta as a whole did not change dramatically during the observation period, this protrusion is a good measure for the behavior of the whole delta. Also the observed slow accretion of the islands was reproduced by the model.

These results seem to justify the conclusion that the model is valid for the Wadden Sea, although further verification for other Wadden Sea inlets is certainly necessary. This will be one of the top priorities in the follow-up of this work.

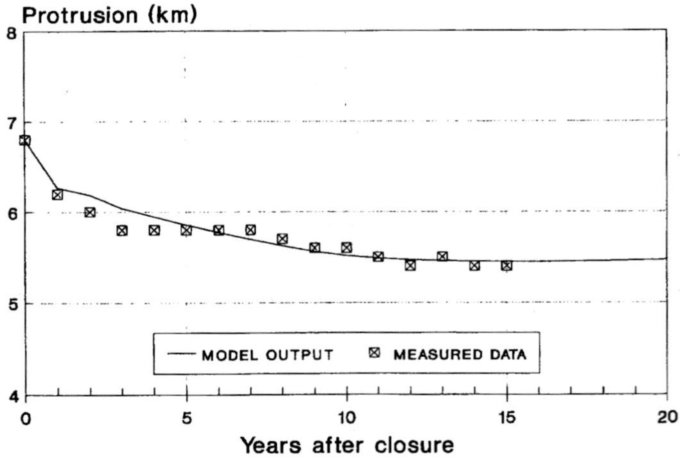

Fig. 7.52 Protrusion of the Friese Zeegat outer delta after reclamation of part of the flood basin.

(c) *Application*

The model has been applied to predict the impact of a mean sea level rise of 0.85 m per century, which represents a pessimistic, but not impossible scenario for the Dutch coast.

Figure 7.53 shows the results if the basin topography rises along with the sea level, which implies a sediment demand of about 1 km^3 per year.

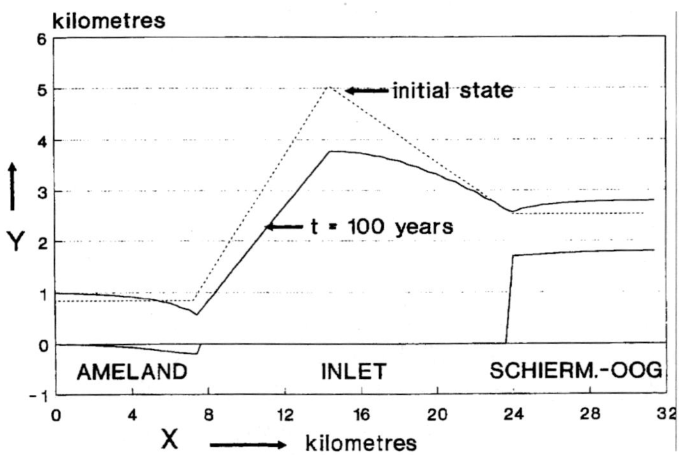

Fig. 7.53 Results if the basin topography rises along with the sea level.

The model shows that the delta tends to respond at a much shorter time scale than the island coasts. In the long run, the system tends towards an equilibrium state in which the size of the delta corresponds with the tidal prism, whereas the delta is located further shorewards and the island coasts take such an orientation that they can supply the required flux of sediment (except for the probably small amounts which come from deeper water).

Note that the delta (like the islands) supplies sand to the basin via the flood channels, not via the ebb channel. As the tip of the delta retreats faster than the island heads, it will become gradually more difficult for the waves to bring the sediment from the delta tip to the flood channels. As a consequence, the delta gradually reduces its sediment supply to the basin and the island coasts take over. Hence the island heads retreat and the island coasts start to change their orientation (via longshore transport processes). The delta tip lags behind the island heads now, whence the protrusion increases again. This process goes on until the new equilibrium state is reached. The interpretation that the delta acts as a sediment buffer for the basin is therefore only partly correct, certainly in the longer run.

7.7.5. *Interaction with the flood basin*

So far, we have considered the basin and the outer delta plus the island coasts as mutually independent subsystems, but in reality they will interact.

This is readily illustrated by the response of the inlets in the Western Wadden Sea to the closure of the Zuidersee in 1932. This affected the tidal prisms in the basins of the Western Wadden Sea,[28] whence the inlets started to tend towards a new equilibrium state. Figure 7.49 shows the adjustment process of the flood basin and outer delta volumes for each of these inlets.[58] The picture is very consistent: if the flood basin has a sand deficit (surplus) and imports (exports) sediment, the outer delta erodes (accretes), but, as long as the new equilibrium is not reached, the amounts of sediment involved do not match.

This implies that the delta is a sand buffer for the basin, but not the only one. Sediment is also borrowed from or lent to the adjacent coasts and the North Sea (see Fig. 7.49a), the latter probably to a lesser extent. List *et al.*[48] showed for the Louisiana coast, that there must be a significant exchange of sediment between the barrier islands and the inlets, presumably via the inshore. Also the lateral migration of inlets, which implies erosion of the downdrift island head, makes sediment available to be trapped in the basin. However, as long as the inlets tend to maintain their size, the amount of erosion tends to be compensated by deposition at the tip of the updrift island. In other words, inlet migration does not affect the overall sediment balance.

In the present model, we have assumed that the long-term residual sediment exchange through the gorge is governed by the morphological state of the flood basin. This is in line with the assumption that the tidal prism for the basin is a given quantity, which does not depend on the morphological state of the system. It implies that the basin evolution always leads the delta evolution, and that the system always tends towards an equilibrium state which is a unique function of the prescribed tidal prism.

Reality can be more complicated, for instance because:

a. the tide inside a basin is influenced by energy losses in the gorge, which are a function of geometry;
b. the tide in a basin is nearly resonant (e.g. if the basin length is close to a quarter of the tidal wave length), whence small changes in the delta or gorge geometry can have large effects on the tidal prism;
c. the tide in the gorge of an inlet is influenced by the geometry of the outer delta;
d. the transport through the flood channels of an inlet is influenced by the geometry of the outer delta.

In order to take these aspects into account, we have to know more of the tide in the system and its response to morphological changes.

The inclusion of these phenomena can have dramatic effects on the model behavior. The relatively simple inclusion of the effects of energy losses in the gorge, for instance, introduces the possibility of destabilization of the whole system.[31,80,81] Therefore, the present concept with a prescribed tidal prism cannot be more than a first step on the way to more sophisticated inlet models.

7.7.6. *Discussion*

The model concept decribed herein seems to have the potential of becoming a useful tool in the prediction of the dynamic large-scale response of the outer deltas of tidal inlets like those along the Dutch and German Wadden Sea coast. A prerequisite is the availability of sufficient calibration data from the inlet to be investigated, or a sufficient similarity with another inlet for which the model has been or can be calibrated.

Although the limits of applicability remain to be specified clearly, and calibration will always be necessary, there is no reason why the concept should not be applicable to other inlets than the Friesche Zeegat, or even to other areas than the Wadden Sea. This can only be proved, however, by verifying the model and its submodels against data from other inlets. Besides, further verification against new data from the Friesche Zeegat is recommendable (e.g. the response to the 18.6 yearly nodal cycle; see;[18] recent data have been provided by Oost *et al.*,[55]).

The concept is based on a number of assumptions which need further substantiation, e.g:

a. the residual sediment transport pattern around the delta, including the underlying mechanisms,
b. the effects of storm waves and surges, especially in non-equilibrium situations,
c. the net transport through the flood channels, as a function of tide and wave climate,
d. shape parameters (e.g. protrusion, longshore extent, skewness) of deltas in equilibrium, as functions of the extrinsic conditions,
e. the role of the location of ebb and flood channels, and
f. the mechanisms and the effects of channel migration.

The essence of the present concept, and one of its most challenging aspects, is that the incorporation of empirical morphological information enables us to go around two of the weakest points of physics-based models, viz.

g. the computation of the sediment transport rate as a function of instantaneous hydrodynamic parameters; not only the complexity of the sediment transport mechanisms makes this an
h. error-prone element, but also the high demands on the description of the tidal motion, since we need an accurate value of the residual transport;
i. the selection of the conditions for which the instantaneous water motion has to be computed in order to obtain a good estimate of the long-term residual transport pattern.

Whether this approach can be extended from the present storage-basin model to models for more complex situations remains an open question, but such an extension, maybe even to ID network models and 2D depth-averaged models, seems worth pursuing.

This does not dispute the value of physics-based modelling. Short-term models of this type have proven to be quite useful to analyse the flow and sediment transport processes in complex systems such as tidal inlets.[73,86] Recent advances in medium-term morphodynamic modelling[89] show that physics-based models also provide promising tools for medium-term dynamic simulations.

Process-based mathematical models are also quite suited for the analysis of long-term processes, and to substantiate semi-empirical models like the present one (e.g. as an aid to establish a relationship between the flood channel transport and tide and wave climate).

Attempts to formally integrate real-time simulation models over the smaller timescales up to the timescales which are of interest to morphologists lead to quite promising results.[23,25]

Together with the possibility of a significant cross-fertilization, these are good arguments for the parallel development of two approaches, viz. "down-scaling" from observations and "upscaling" from first physical principles.[77]

7.7.7. *Conclusions concerning Section 7.7*

The large-scale dynamic model concept which is presented herein describes the equilibrium state and the transient morphological evolution of the outer delta and the adjacent coast of mixed-energy tidal inlets of the same type as those of the Dutch and German Wadden Sea. Although it is possible to solve the two-line equations with a very high spatial resolution, its results should be considered at a rather high aggregation level, e.g. in terms of the protrusion or the sand volume of the delta as a whole. The concept does not allow for a more detailed interpretation, such as the migration of channels and shoals.

The model, which can be run on a simple PC, is an easily accessible and versatile tool, which elegantly integrates the outer delta and the adjacent coasts. However, it has to be fitted to the class of inlets to be considered. To that end, information is needed on the equilibrium state of the deltas in the area, and on the transient evolution of some of them.

The viability of the model concept has been tested for a number of hypothetical situations derived from the Wadden Sea. In all these cases, the model behaved as expected and in line with observations. Hence it is concluded that the concept is viable and worth further investigation.

The validity of the model was tested on the extensively documented response of one of the Wadden Sea inlets to a major human interference. The agreement between model results and measured data is satisfactory.

The overall conclusion from the foregoing is, that the modified two-line concept is a useful approach for inlets like those of the Wadden Sea. One of its principal shortcomings as a predictive tool is the necessity to prescribe the sediment flux through the gorge. A logical next step in the development of long-term tidal inlet models would therefore be to couple the outer delta model with a flood basin model.[27,85]

The next Section 7.8 will attribute further to this subject: tidal inlet models in general and applications for Wadden Sea and Western Scheldt.

7.8. Morphology of Multiple-channel Tidal Inlet Systems, like Wadden Sea and Western Scheldt

7.8.1. *Introduction to Section 7.8*

The view, directed to the seaside in Section 7.7, once again will return to the landside. However, the subject "hydraulics" treated extensively in Sections 7.2–7.6, now will be just used as stepping stone for considerations concerning the final goal: "erosion; sedimentation; morphological stability". This was the subject of Section 7.6; however, now the multiple-channel tidal inlet systems will be considered instead of the single channel (cf. Table 7.2 in Subsection 7.1.2).

First (Subsection 7.8.2) the approach to the subject "multi-channel system" will be explained.

The kind of morphodynamic equations is indicated in Subsection 7.8.3. The effect of the characteristics of the Lorentz tidal computation on the format of the results with respect to currents are highlighted in Subsection 7.8.4 as far as the basin is concerned and with respect to the boundary conditions (Subsection 7.8.5) Evidently, the "rain in the drain" (cf. Section 7.4) will appear to be important, not only for a single channel (Section 7.4) but also for a network of channels: see Subsection 7.8.6. Finally, morphodynamic consequences are considered specificly in Subsection 7.8.7.

7.8.2. *Approach*

As common in this chapter, the geometrical schematization of the basin area is a channel network. See Fig. 7.54 (Western Waddensea[bbbb]) and Fig. 7.62 (Western Scheldt). Such a model domain is divided into a number of elements (sections). Each channel cross-section consists of one or more

[bbbb]Fig. 7.3/7.4 in another display.

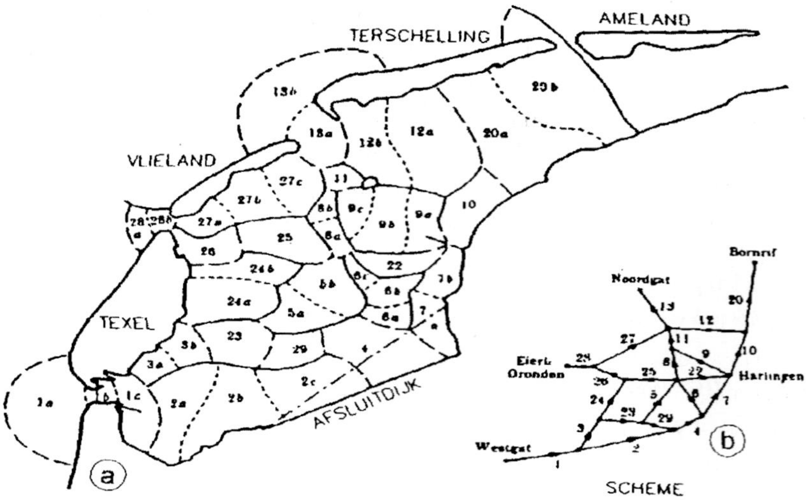

Fig. 7.54 Schematization of the Western Wadden according to Lorentz *et al.*[49] (a) Topography; (b) schematization.

deep and one or more shallow parts.[cccc] The former represents the water and sediment conveying channels in the relevant section, the latter the intertidal flats, which are all assumed to lie at mean water level. Transfer of water and sediment between sections is concentrated in the nodal points, so there is no direct transfer between shoal areas. By letting some network branches stick through the inlets into the sea, the undisturbed tide can be imposed at the seaward ends of these branches. Thus the interfacing problem in the gorge can be avoided.

The tide in this network is described[dddd] with a numerical version of Lorentz's[49] linear tidal model.[42] This linearity is instrumental to the concept, so non-linearity effects have to be included via perturbation methods, effective stress models or otherwise. In order to keep complexity within bounds, the inclusion of these elements is postponed till a later stage, irrespective of their quantitative importance. Only the topographical rectification effect of flooding and drying is taken into account.

The sediment transport rate is described by a simple power-law formula, which relates the transport rate in an element to the local flow velocity.

[cccc]Cf. Fig. 7.11.
[dddd]Subsection 7.2.9

The key element of the model is the morphological evolution equation per section, which is based on the sediment balance and elaborated using the hydrodynamic and sediment transport equations. The bed level in the channels is the only remaining dependent variable. This is also the only morphological variable in the model, since vertical accretion or erosion of the shoals is precluded by fixing them all at the mean water level. Horizontal extension of the shoals is precluded as long as the channel and shoal widths are fixed.

The model consists of the morphological evolution equations for all channels in the network. They form a coupled mathematical system, in which the coefficients are derived from a tidal model run on the initial bed topography. In the next sections, the various parts of this evolution model will be discussed in further detail.

7.8.3. *Transport module and its consequence for dynamic equations*

Logically, the transport equation used will be the same as in Section 7.7, i.e., Eq. (7.190).

When combined with the sediment balance equation, locally equations like (7.195) are found.

This equation can be elaborated to a tide-averaged evolution equation. For small perturbations to a given morphological this can be written as

$$\frac{\partial \Delta h}{\partial t} = \text{fnct}(\Delta h) \tag{7.194}$$

Where Δh is a perturbation to the water depth below mean water level in the deep part of the channel. The most important assumptions underlying this equation are:

a. the bottom of the deep part of the channel is the only morphologically active part of cross-section, i.e. shoals remain of the same size and at the same level,
b. the time scale of the morphological evolution is large compared with the tidal period.

Equation (7.196) describes the stability of the channel morphology against perturbations. Depending on whether the right-hand part is decreasing or increasing as Δh increases, the system is morphologically stable or unstable.

7.8.4. *Network model: tidal module*

Via the harmonic solution of the hydrodynamic equations for a single channel, the tide at the one end of each section is linked to that at the other end. In a network model, we have to close the mathematical system via another set of relationships, between the channels which meet at each nodal point. Per nodal point, these relationships can be found from the continuity requirement (the sum of all flow rates toward a node equals zero) and the compatibility requirement (the water levels at the ends of all branches in one node are the same).

This results in a system of N equations with N unknown complex amplitudes z, which can formally be written as (cf. 7.81):

$$\overline{\overline{D_{\text{tot}}}}\vec{Z} = \vec{z}_{\text{b}} \tag{7.195}$$

$\overline{\overline{D}}_{\text{tot}}$ = the N × N system matrix
$\vec{Z}$ = the vector of unknown amplitudes
$\vec{z}_{\text{b}}$ = the K-dimensional source vector, which contains the seaward boundary conditions, K being the number of seaward boundary points.

The solution for the vector of nodal points can be written as:

$$\vec{z} = \overline{\overline{D_{\text{tot}}}}^{-1}\vec{z_{\text{b}}} \tag{7.196}$$

Equation (7.198) shows that the tide in any nodal point is a linear combination of the tides at the seaward boundaries (cf. 7.55). For a multiple-inlet system this means that the contribution to the tidal amplitude can mapped for each of the inlets separately. As an example, Fig. 7.59 shows these influence functions for each of the inlets of the Western Wadden Sea, as computed with the network model shown in Fig. 7.30.

7.8.5. *Van Veen's "motoric power" translated in Lorentz' values*

It is tempting to analyze the "motoric power" of Van Veen (Subsection 2.5.3; Figs. 2.36 and 2.45, [2.4]) into Lorentz' terms. This excursion should be seen as a contribution to understanding rather than as a contribution to accuracy: essential higher harmonics incorporated in Van Veen's approach are neglected in this way. Furthermore, the Lorentz schematization has not been extended, up to now, to comprise multi-channel outlets.

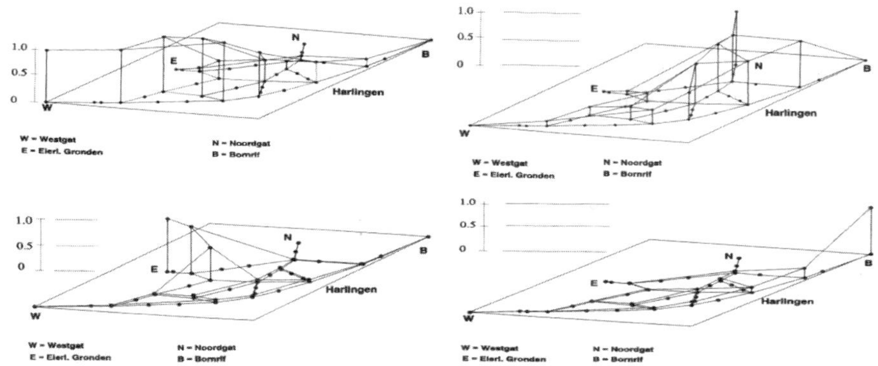

Fig. 7.55 Influence coefficients for the various inlets of the Western Wadden Shoals. (a) The Inlet Westgat; (b) the inlet Eierland Gronden; (c) the inlet Noordgat; (d) the inlet Bornrif.

However, by considering the variation of the M_2-tide along the outer boundary along the seaward side of the outer delta, one could get an indication of the maximum and minimum "motoric power", depending on the direction of "the" channel through the gorge. This could lead to pictures like the two figures in the center part of Fig. 2.45.

7.8.6. *Topography-induced rectification*

Tidal asymmetry is an important transport agent in the channel/shoal system inside the basin. The topography-induced generation of overtides (*hypsometry effect*;[19,71]) is one important effect, which probably plays a role in the equilibrium state as a balancing agent against the sediment importing shallow-water asymmetry at the seaward end of the model.[85]

Another topography-induced effect is rectification, due to which there is a residual circulation between the shoals and the channels.

For the time being, overtides will be left out of consideration (see "Discussion"). The rectification effect is taken into account by treating shoals and channels separately. As the shoals are assumed to lie at mean water level, they fall dry during part of the tide. The net flux of water (mass flux) over a shoal of width B_{sh} during the other part of the tide is given by:

$$\langle Q_{\text{sh}} \rangle = \frac{1}{4} B_{\text{sh}} \hat{z} \hat{u}_{\text{sh}} \cos \phi_{\text{sh}} \tag{7.197}$$

where $\hat{u}_{\text{sh}}$ is the amplitude of the tidal velocity on the shoal; ϕ_{sh}, the phase difference between the vertical and the horizontal tide.

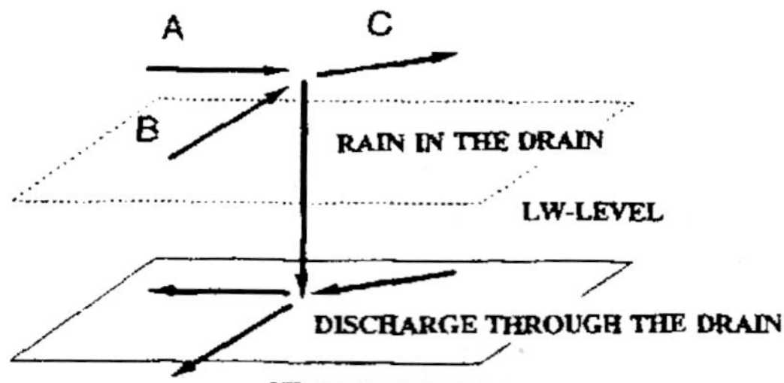

Fig. 7.56 The tide-averaged flux of water above the LW-line should be transferred into "rain in the drain" and subseqeuntly discharged through the parts of the channel below LW.

Note that if this phase difference is 90°, as in the case of a deep-water tide, the net flux goes to zero.

In the channels, the residual mass flux above LW, due to the phase relationship between water level and velocity, follows from:

$$\langle Q_{\rm ch} \rangle = \frac{1}{2} B_{\rm ch} \hat{z} \hat{u}_{\rm ch} \cos \phi_{\rm ch} \tag{7.198}$$

where the suffix ch^+ refers to the tide-induced mass flux in the channel.[eeee]

In order to satisfy the overall mass balance, there has to be another residual current, which is assumed to be concentrated entirely in the channels. From this point of view the channels function as "drains" of the basin. In the same vein, the spatial mass flux gradients, e.g. due to a gradient in the phase difference $\phi_{\rm ch}$, can be considered as producing "rain" which flows off via the "drains" (Fig. 7.56).

Figure 7.57 gives an impression of the distribution of the mass-flux induced "rain" and the corresponding run-off flow through the network. The former follows directly from the mass flux gradients, the latter is the result of a steady-flow computation with a non-linear numerical model with the "rain" as a source term. The above tidal model was used to compute

[eeee]Lorentz[49] and Thijsse[79] implement composite profiles of tidal channels (Fig. 7.11) in their tidal computation. These are first schematized as a set of parallel channels and then replaced by a hydraulically equivalent rectangular channel. The equivalent channel is characterized by its width, depth and friction factor f. The width equals the sum of the widths of its constituent channels. The depth and friction factor are determined by its discharge, equal to the sum of the discharges in the constituent channels (taking phase differences into account). The same vertical tide in all parallel channels is assumed. Thus, amplitude and phase of the horizontal tide in each of the constituent channels result from the Lorentz computation.

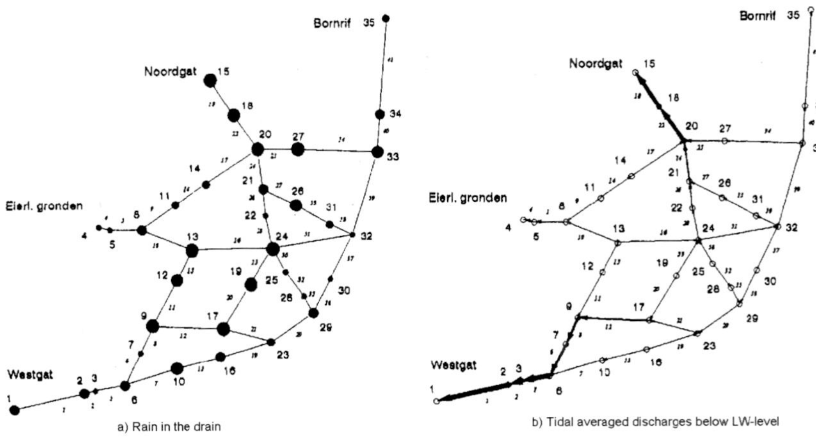

Fig. 7.57 (a) The tide averaged vertical flux of water, thought to be concentrated in the nodal points of the network ("rain in the drain"). The area of the dots indicates the relative quantity of the "rain". (b) Calculated return flux throught the "drains" (channels below LW).

the mass flux and the tidal velocities needed to determine the enhanced bottom friction in the steady flow model.

In the particular case of the Western Wadden Sea, with its high shoals and almost perfect watersheds, one can say for most points beforehand to which inlet the rain will flow. Hence the run-off pattern is primarily a matter of continuity. In other situations, however, other forcing mechanisms may determine the run-off pattern.

The mass flux is only one of the tidal rectification mechanisms. The advection terms in the momentum equations can also give rise to a residual current, via what is usually called the *tidal stress*.[51] Once we use a separate model to compute the residual current, this tidal stress can easily be included, just as a wind stress or any other external forcing.

The idea of separating the oscillatory and the mean part of the current is common practice in wave-driven current modelling. There the mean flow is driven by exactly the same mechanisms: mass flux, effective stress and bed stress enhancement (for instance, see[36]). Non-linear tidal simulation models usually describe the depth-averaged velocity as a function of time and then integrate over the tide to compute the residual current. Thus the mass flux, tidal stress and bed shear stress effects are automatically included, at least if the model is provided with a proper flooding and drying procedure. The practice of depth-averaged flow modelling, however, has also led to the widely spread picture that the residual tidal current should have the same vertical distribution as the instantaneous velocity (usually logarithmic). This is not necessarily the case, due to the phase coupling between water level and velocity (Section 7.4.5).

The residual current model has been verified[26] against a fully non-linear tidal model[fff] of the mouth of the Western Scheldt, a tidal estuary in the south of The Netherlands, also with a pronounced topography. The geometry is much simpler than in the Wadden Sea case, viz. a single channel which bifurcates in the mouth of the estuary. For this geometry, the "rain & drain" approach (while taking tidal stress into account) gives a good approximation of the residual velocities according to a fully non-linear model (Fig. 7.58)

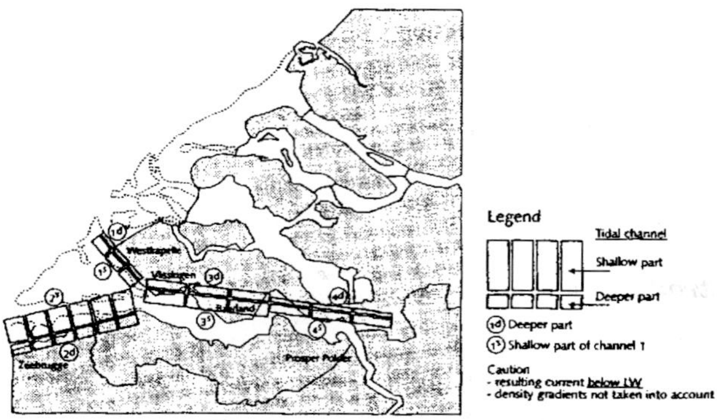

Fig. 7.58a Situation.

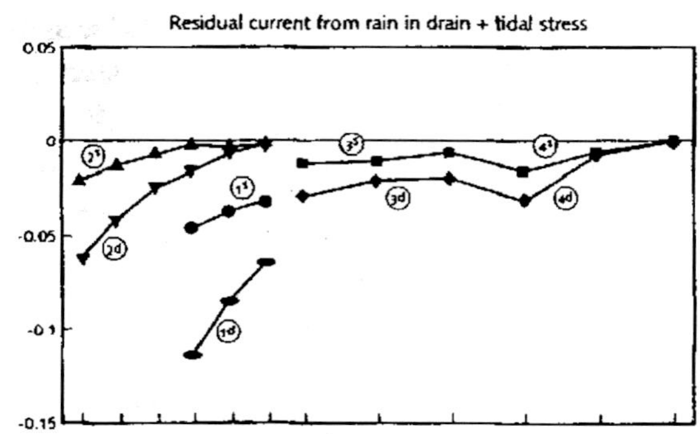

Fig. 7.58b Residual current from rain + tidal stress.

[fff] Used is the DUFLOW model[70] designed for unsteady flow computations in networks of open channel courses. Non-linear friction and inertia terms are implemented. The program allows for composite profiles. The deepest part of the channels should be below LW.

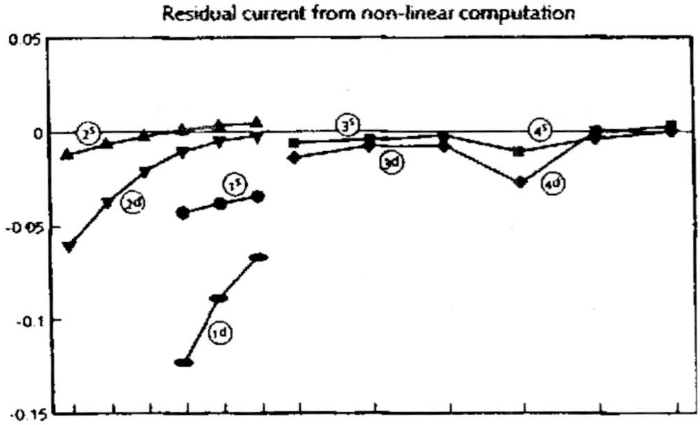

Fig. 7.58c Residual current from non-linear computation.

7.9. Discussion on the Assumptions; Limitations on the Applied Methods in Relation to Other Methods; Acknowledgement

7.9.1. *Introduction*

The model concept which is described in the foregoing should be considered as a first step on the way to long-term morphodynamic modelling of near-resonant multiple-inlet tidal basins. It shows the essence of the procedure to be followed in order to derive a set of morphodynamic evolution equations which encompasses the hydrodynamics without going through a time-consuming simulation process.

The route which was taken herein led to the analysis of the effect of morphological changes on the resonance properties of the basin, and to a linear stability analysis of the morphodynamic system. Another possibility which this concept offers is to consider perturbations to the equilibrium state and describe the evolution from a given initial state to that equilibrium state.

Clearly, the model is still far from complete. The lack of tidal stresses in the residual current model is one gap which has to be filled,[gggg] but a possibly even more important one concerns the effects of overtides, especially those generated by the flooding and drying of the intertidal flats.[71,85] Yet

[gggg]N.B., For the Western Scheldt computations (Fig. 7.34) tidal stress has been taken into account.

the residual current should not be disregarded as a transport agent. Van de Kreeke and Robaczewska,[83] show for the Ems estuary, further eastward in the Wadden Sea, that the contribution of the residual current to the transport even exceeds that due to the M4 overtide. Besides, it must be not too difficult to include the overtides in the model, once it is possible to parameterize their topographical generation. Their inclusion will probably change the outcome of the model, but not the essence of the approach.

The morphological flexibility of the model should probably be extended. Especially if the model is to be utilized for long-term predictions, the intertidal zone has to be made morphologically active, in the vertical (flats level) as well as in the horizontal (flats area vs. channel area). This probably implies that the tidal model runs have to be repeated from time to time, according to the rate of change of the shoals and the channel planform. The frequency of these runs will be much smaller, however, than in the case of straight forward simulation.[25]

It is also clear that the model needs further validation. The strong parallels of the "rain & drain" concept with the well-established practice in wave-driven current modeling gives some confidence, and so does the comparison with the non-linear numerical model for the Western Scheldt, but this is not enough. If the model has to run for the Western Wadden Sea, it also has to be validated for that area. A promising option is a comparison with the residual current maps produced by Ridderinkhof.[62,63] However, in all comparisons the Eulerian kind of definition of residual current in the present paper should be kept in mind.

An interesting extension of the model might be to include fine (cohesive?) sediment. Here the proportion between the transporting agents is quite different, because the sediment is distributed much more evenly over the vertical. Transport of material above the LW-level becomes crucial as well. The near-bed residual current is therefore less important; thus the rain-in-the-drain concept loses its physical meaning. The tidal asymmetry plays a key role.[29] Another approach, like the one of Krol,[46] might be more appropriate. Ways of combining both methods is a challenge for the future.

Particularly for calculation of the effect of silt on morphology, probably however also for the morphology of a sandy bottom, the effect of the salt gradient in the estuary will be of paramount importance. A physical basis for this investigation has been given in Section 7.3.

7.9.2. *Summarizing remarks concerning Sections 7.8 and 7.9*

A model concept is described which has the potential to be the basis of a long-term morphological model for near-resonant multiple-inlet tidal basins. The key element is a set of morphological evolution equations, one for each channel section of the network. These equations are based upon a simplified tidal model in which the oscillatory tidal motion is separated from the residual current. Thus the topography-induced rectification via the residual mass flux above low water can be taken into account via the "rain & drain" approach.

The model was shown to give an indication of the resonance-sensitivity of each of the inlets of the Western Wadden Sea, The Netherlands, to a change of depth in each of the channels in the basin. It was also shown to provide the possibility to assess the morphodynamic stability of the channel system, and to extend it to a long-term prediction model for this type of basins.

Via its connection with the open sea, it provides the possibility to be integrated with a model which describes the morphology of the outer delta and the barrier island coasts.

7.10. Acknowledgements

Concerning this Chapter 7 (Section 7.8 & 7.9) based on papers of de Vriend and Bakker, the authors acknowledge support from Jallah and the importance of the references [42], [22] and [26], which are underlying reports of the present chapter.

References

1. Abamowitz, M. and I. A. Stegun, *Handbook of Mathematical Functions.* Lecture Notes, seventh ed., New York, Dover Publications, Inc., 1970.
2. Aubrey, D. G. and L. Weishar (eds.), *Hydrodynamics and Sediment Dynamics of Tidal Inlets.* Lecture Notes on Coastal and Estuarine Studies, vol. 29, New York, Springer Verlag, 454 pp. 1988.
3. Bakker, W. T., *The dynamics of a coast with a groyne system*, London, Proc. 11th Int. Conf. Coastal Eng., New York, ASCE, pp. 492–517, 1968.
4. Bakker, W. T., *A mathematical theory about sand waves and its application on the Dutch Wadden isle of Vlieland*, Shore Beach, vol. 36, no. 2, 1968.

5. Bakker, W. T., *Effect of tidal resonance on the morphology of Wadden and Estuaries*, Rijkswaterstaat, Netherlands Centre for Coastal Research (internal Report,1993).
6. Bakker, W. T., *Effect of tidal resonance on the morphology of tidal channels*, Proc. of 26th Int. Conf. on Coastal Eng., 1998.
7. Bakker, W. T., Dispersion in the Western Scheldt estuary, Delft Hydraulics, Internal report, 2001.
8. Bakker, W. T., *Vertical circulation in a well-mixed estuary*, Cardiff, Proc. 28th Int. Conf. Coastal Eng., New York, ASCE, pp. 1229–1241, 2002.
9. Bakker, W. T. and M. E. Barber, *Calculation of vertical circulation in tidal estuaries*, Tainan, Proc. of the 5th Int. Conf. on Hydrodnamics, 2002.
10. Bakker, W. T and H. J. De Vriend, Resonance and morphological stability of tidal basins, *Marine Geology* 126, 1995.
11. Bakker, W. T. and D. S. Joustra, *The history of the Dutch coast in the last century*, Washington, Proc. 12th Int. Conf. Coastal Eng., New York, ASCE, pp. 709–728, 1970.
12. Bakker, W. T. and R. Rundberg, Pollendam investigation (in Dutch), Rijkswaterstaat, Study Dept. of Hoorn, Internal rep, 1968.
13. Bakker, W. T. and J. Schaart, Calculation of sand- and silt concentrations in tidal currents, Rijkswaterstaat, Dir. Watermanagement & Coastal Research; Study Dept. at Flushing, Int. Rep. 76-2, 1976.
14. Bakker, W. T., C. van der Kerk and J. H. de Vroeg, *Determination of coastal constants in mathematical line-models*, Loutraki, Proc. 2nd European Workshop on Coastal Zones, 1988.
15. Bakker, W. T. and W. G. M. van Kesteren, *The dynamics of oscillating sheetflow*, Proc. 20th Int. Conf. on Coastal Engng, Taipei, 1986.
16. Beyl, P. R., *Determination of the accuracy of sand transport computations with the method of Bijker-Einstein, found from Western Scheldt data (in Dutch)*, Delft Univ. of Techn., Sect. for Coastal Res. M.Sc. Thesis, 1979.
17. Bijker, E. W., *Some considerations about Scales for coastal models with movable bed*, Ph.D. Thesis Delft University of Technology; Publ. 50 Delft Hydraulics Laboratory, 1967.
18. Bilse, D. P., *A large scale model of the morphological behaviour of the outer delta of tidal inlets*, Delft Hydraulics, Coastal Genesis Report H 1887, 42 pp., 1993.
19. Boon, J. D. and R. J. Byme, On basin hypsometry and the morphodynamic response of coastal inlet systems, *Marine Geo I.*, 40, 27A8, 1981.
20. Bruun, P. (ed.), *Stability of Tidal Inlets.* Amsterdam, Elsevier, Development in Geotechnical Engineering, vol. 23., 510 pp., 1978.
21. Bruun, P. and F. Gerritsen, *Stability of Coastal Inlets.* North Holland, Amsterdam, 123 p., 1960.
22. Bult, G., Wadden and Estuaries: Science and Management (in Dutch), Rijkswaterstaat, RIKZ and CAT Enschede, Intemal Rep., 1994.
23. De Vriend, H. J., Mathematical modelling and large-scale coastal behaviour, *J. Hydraul. Res.* **29**(6), 727, 1991.

24. De Vriend, H. J., W. T. Bakker and D .P. Bilse, A morphological behaviour model for the outer delta of mixed-energy tidal inlets, *Coastal Eng.* **23**(3,4), 305–327, 1994.
25. De Vriend, H. J., M. Capobianco, T. Chesher, H. E. De Swart, B. Latteux and M. J. F. Stive Approaches to longterm modelling of coastal morphology: a review, *Coastal Eng.* **21**(1–3), 225–269, 1993.
26. Dijkstra, A. G. H., Modeling tides and residual currents in the Westem Scheldt. (in Dutch), Rijkswaterstaat, RlKZ, and CAT Enschede, Internal Rep., 1994.
27. Di Silvio, G., *Modelling the morphological evolution of tidal lagoons and their equilibrium configuration*, Ottawa, Proc. XXIIrd IAHR Congress. NRC, pp. 169–175, 1989.
28. Doekes, J., Effect of the closure of the Zuiderzee on the tide in the Wadden Sea (in Dutch), Rijkswaterstaat, Rep. GW1 0-85.001, 13 pp., 1985.
29. Dronkers, J., Tidal asymmetry and estuarine morphology, *Neth. J. Sea Res.*, 1986.
30. Ehlers, J., *The Morphodynamics of the Wadden.* Rotterdam, Balkema, 397 pp., 1988.
31. Escoffier, F. F., The stability of tidal inlets, *Shore & Beach* **8**(4), 114–115, 1940.
32. Escoffier, F. F., *Hydraulics and Stability of Inlets.* U.S. Army Corps of Engineers, 1977.
33. Eysink, W. D., *Morphologic response of tidal basins to changes*, Delft, Proc. 22nd Int. Conf. Coastal Eng., New York, ASCE, pp. 1948–1961, 1990.
34. Eysink, W. D., E. J. Biegel and F. J. M. Hoozemans, Impact of sea level rise on the morphology of the Wadden Sea in the scope of its ecological function, investigations on empirical morphological relations, Delft Hydraulics, lSOS*2 Rep. H 1300, 1992.
35. R. A. Falconer, P. Goodwin and R. G. Marhews (eds.), *Hydraulic and Environmental Modelling of Coastal, Estuarine and River Waters.* Aldershot, Gower Technical, pp. 25–264.
36. Fredsoe, J. and R. Deigaard, *Mechanics of Coastal Sediment Transport.* Singapore, Advanced Series in Ocean Eng., vol. 3., World Scientific, 369 pp., 1992.
37. Gerritsen, F and L. Dekker, *Vergelijkende getijberekening voor de Westerschelde volgens de Methode Lorentz*, Rijkswaterstaat, Adviesdienst Vlissingen notitie WWKZ-84.V259.
38. Hansen, D. V. and M. Rattray, Gravitational circulation in straits and estuaries, *J. Marine Res.* 23, 104–122.
39. Hayes, M. O., *Barrier Island Morphology As Function of Tidal and Wave Regime*, New York, Academic Press, Barrier islands from the Gulf of St, Lawrence to the Gulf of Mexico, pp. 1–27, 1979.
40. Holland, A. M. B. and H. Smit, Fresh water in the Scheldt-estuary: changes in salinity (in Dutch), Rijkswaterstaat, Rep. DGW-93.057.
41. Lppen, A. T., *Estuary and Coastline Hydrodynamics.* New York, McGraw-Hill Book Company Inc., 1966.

42. Jallah, A. N. and W. T. Bakker, Lorentz on PC, Delft University of Technology, Dept. for Hydraulic Eng., Internal Rep., 1994.
43. Jarrett, J. T., *Tidal prism–Inlet area Relationships*, Vicksburg, U.S. Army Corps of Engineers, CERC G.l.T.I. Rept. 3, Ml, 55 pp., 1976.
44. Kamps, S. L., The Dynamics of Wadden and Western Scheldt, (Rijkswaterstaat and CAT ("Hogeschool Enschede") internal report, 1994.
45. Kraus, K. Gingerich and D. L. Kniebel (eds.), *Coastal Sediments '91*, New York, ASCE, pp. 1532–1546, 1991.
46. Krol, M., *The method of averaging in partial differential equations*, University of Utrecht, Ph.D. Thesis, 1990.
47. Leatherman, S. P. (ed.), *Barrier Islands from the Gulf of St. Lawrence to the Gulf of Mexico*. New York, Academic Press, pp. 1–27.
48. List, J. H., B. E. Jaffe and A. H. Sallenger, *Large-scale Coastal Evolution of Louisiana's Barrier Islands*, Coastal Sediments '91, NY, ASCE, pp. 1532–1546, 1991.
49. Lorentz, H. A., *Report State Commission Zuiderzee 1918–1926 (in Dutch)*, The Hague, Algemeene Staatsdrukkerij, 1926.
50. Marino, J. N. and A. J. Mehta, *Sediment Trapping at Florida's East Coast Inlets*. New York, Hydrodynamics and Sediment Dynamics of Tidal Inlets, Lecture Notes on Coastal and Estuarine Studies, vol. 29, Springer Verlag, pp. 284–296, 1988.
51. Nihoul, J. C. J. and F. C. Ronday, The influence of the tidal stress on the residual circulation, *Tellus* 27, 484–489.
52. O'Brien, M. P., Estuary tidal prisms related to entrance areas, *Trans. ASCE* **1**(8), 1931.
53. O'Brien, M. P., Equilibrium flow areas of inlets on sandy coasts, *Proc. ASCE, Waterw. Harbors Coastal Eng.*, 15(WW1), 43–52, 1969.
54. Oertel, G. F., *Processes of Sediment Exchange Between Tidal Inlets, Ebb Deltas and Barrier Islands*. New York, Hydrodynamics and Sediment Dynamics of Tidal Inlets, Lecture Notes on Coastal and Estuarine Studies, vol. 29, Springer Verlag, pp. 297–318, 1988.
55. Oost, A. P., H. De Haas, F. IJnsen, J. M. van den Boogert and P. L. de Boer, The 18.6 yr nodal cycle and its impact on tidal sedimentation, *Sediment. Geol.*, 87, 1–11, 1993.
56. Pieters, T., C. Storm, T. Walhout, T. Ysebaert (ed.), *The Scheldt Estuary, More Than A Navigation Channel*, Rijkswaterstaat, Dept. Zealand & Tidal Waters Dept. Rep. GWWS-91.081, 1991.
57. Pritchard, D. W., The dynamic structure of a coastal plain estuary, *J. Mar. Res.*, 15, 33–42, 1956.
58. Rakhorst, D., *Erosion and sedimentation of the Westem Wadden Sea and the adjacent North Sea: 1926/ 1933-1982 (in Dutch)*, Rijkswaterstaat, North Holland Dept., Rep. 93.ANV.001, 1992.
59. Renger, E., *Two-dimensional stability analysis of tidal basins and tidal flats of larger extent*, New York, Proc. 16th Int. Conf. Coastal Eng., ASCE, pp. 1971–1983, 1978.

60. Renger, E. and H. W. Partenscky, *Sedimentation processes in tidal channels and tidal basins caused by artificial constmctions*, New York, Proc. 17th Int. Conf. Coastal Eng., ASCE, pp. 2481-2494, 1980.
61. Ribberink, 1. and J. H. de Vroeg, Coastal protection Eijerland (Texel): hydraulic and morphologic effect study, Delft Hydraulics, Rep. H 1241, 1992.
62. Ridderinkhof, H., Tidal and residual flows in the westem Dutch Wadden I:numerical model results, *Neth. J. Sea Res.*, 22, 1–22, 1988a.
63. Ridderinkhof, H., Tidal and residual flows in the westem Dutch Wadden II: an analytical model to study a constant flow between connected tidal basins, *Net. J. Sea Res.*, 22, 185–198, 1988b.
64. Rigter, B. P., Reproduction of salinity distribution in tidal rivers, Dekft Hydraulics Lab. Rep. M896-III (in Dutch), 1971.
65. Schelde Informatie Centrum, *Schelde en natuur- Schelde atlas*, www. Scheldenet.nl.
66. Sha, L. P., *Sedimentological studies of the ebb tidal deltas along the West Frisian Islands, The Netherlands*, Utrecht University, Inst. for Geosciences, Ph.D. Thesis, 160 pp., 1990.
67. Sha, L. P., Geological research in the ebb tidal delta of "Het Friesche Zeegat" Rijkswaterstaat, Tidal Waters Division, RGD-Rep. 40010, 1992.
68. Sha, L. P. and J. H. van den Berg, Variation in ebb-tidal delta geometry along the Coast of The Netherlands and the German Bight, *Coastal Res.*, 9, 730–746, 1993.
69. Smit, E. S. P., The influence of tidal inlets on the coast (in Dutch), Delft University of Technology, Section Coastal Eng., Coastal Genesis Rep. 31 pp., May 1987.
70. Spaans, W., N. Booy, N. Praagman, R. Noorman, J. Lander, *Duflow, a macro-computer package for the simulation of one-dimensional unsteady flow in open channel systems*, Delft Univ. of Technology, Dept. of Civil Engng., IHE Delft, RWS, Tidal Waters Dept., 1989.
71. Speer, P. E. and D. G. Aubrey, *A study of non-linear tidal propagation in shallow inlet/estuarine systems, part II: theory*, Est. Coastal Shelf Sci., vol. 21, pp. 207–224, 1985.
72. Steijn, R. C., Some considerations on tidal inlets, Delft Hydraulics, Coastal Genesis Rep. H 840.45, 109 pp., 1991.
73. Steijn, R. C. and T. Louters, *Hydro- and morphodynamics of a mesotidal inlet in the Dutch Wadden Sea*, Brighton, Proc. Oceanology Int. '92 Conf., 1992.
74. Steijn. R. C., T. Louters, A. J. F. van der Spek and H. J. de Vriend, *Numerical model hindcast of the ebb-tidal delta evolution in front of the Deltaworks*, 1989.
75. Stive, M. J. F., *A model for crosshore sediment transport*, Taipei, Proc. 20th Int. Conf. Coastal Eng., New York, ASCE, pp. 1551–1564, 1986.
76. Stive, M. J. F., J. A. Roelvink and H. J. de Vriend, *Large scale coastal evolution concept*, Delft, Proc. 22nd Int. Conf. Coastal Eng., New York, ASCE, pp. 1962–1974, 1990.

77. Terwindt, J. H. and J. A. Battjes, *Research on large-scale coastal behaviour*, Delft, Proc. 22nd Int. Conf. Coastal Eng., New York, ASCE, pp. 1975–1983, 1990.
78. Thijsse, J. Th., *Damped Tides in a Navigation Channel.* Int. Nav, Congress, 1960.
79. Thijsse, J. Th., *Theory of Tides*, Int. Course on Hydraulic Engng, 1965.
80. Van de Kreeke, J., Can multiple inlets be stable? *Est. Coastal Shelf Sci.*, 30, 1990b.
81. Van de Kreeke, J., Stability analysis of a tw~inIet bay system, *Coastal Eng.*, 14, 1990.
82. Van de Kreeke, J. and R.A. Iannuzi, Second-order solution for damped cooscillating tide in narrow channel, *J. Hydr. Eng.*, 1253–1260, Dec. 1998.
83. Van de Kreeke, J. and K. Robaczewska, Tide-induced transport of coarsesediment; application to the Ems estuary, *Neth. J. Sea Res.* **31**(3), 209–220, 1993.
84. Van de Kreeke, J. and Zimmermann, J. T. F., *Gravitational Circulation in Well- and Partially Mixed Estuaries*, The Sea, vol. 9, Ocean Engineering Science, 1989.
85. Van Dongeren, A. R. and H. J. De Vriend, A model of morphological behaviour of tidal basins, *Coastal Eng.* **22**(3,4), 287–310, 1994.
86. Van Overeem, I., R. C. Steijn and G. K. F. M. van Banning, *Simulation of morphodynamics of tidal inlets in the Wadden Sea*, Frankfurt am MaIn, Proc. of the Int. Coastal Congress ICCKid '92, pp. 351–364, 1992.
87. Van Veen, J., *Report on the observations with the "Ocean" in the Vlie tidal inlet (in Dutch)*, Den Haag, Rijkswaterstaat, 61 pp., 1935.
88. Walton, T. L. and W. D. Adams, *Capacity of inlet outer bars to store sand*, Honolulu, Proc. 15th Int. Conf. Coastal Eng., New York, ASCE, pp. 1919–1937, 1976.
89. Wang, Z. B., H. J. de Vriend and T. Louters, *A morphodynamic model for a tidal inlet*, Rotterdam, Balkema, Computer Modeling in Ocean Engineering '91, pp. 235–245, 1991.
90. Wang, Z. B., T. Louters and H. J. De Vriend, *Morphodynamic modelling for a tidal inlet in the Wadden Sea*, Marine Geology, 1994.

Chapter 8

Epilogue: Forum and Future Research

8.1. "There is More Between Water and Earth, Dear Bakker, than is Meant in your Philosophy"

This was the general tendency of the Forum as reaction on the presentation of "Coastal Dynamics".

Rakhorst challenges the floor with the following STATEMENTS:

Statement 1: Holland does not loose sand.

Assuming that the cross shore transport along the Dutch coast at a depth of 15–20 m below MSL is negligible, or minor in shoreward direction, no sand losses occur, directed to the deeper part of the North Sea. Landward transport takes place above water level by wind and furthermore because of the demand for sand of estuarine systems like the Waddenzee. At the Belgian border a net influx of sand is present in the Dutch coastal system; at the German border probably some sand losses occur. The difference between both borders cannot be large. The Dutch coast between Hook of Holland and Groote Keeten near Den Helder does not loose sand (Fig. 8.1). Excluding the outer delta areas and subtracting the supplies the total amount of the 15-year accretion in the outer zone (Fig. 8.1) is 0,8 Mm^3 and in the inner zone it is 7.3 Mm^3.[a] Excluding the "Van Dixhoorn-Triangle" (the latter being the beach zone near Hook of Holland; the narrow strip) the total net erosion is just 2.0 mm^3 in the outer zone, where the inner zone gives an accretion of 0.5 mm^3.

[a]With respect to this subject, refer furthermore to [10], [31] and [38]. (remark Bakker).

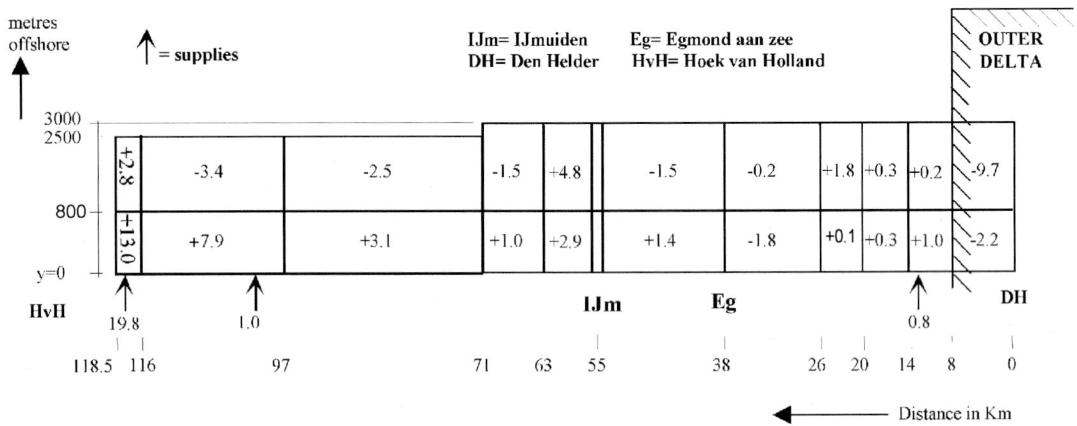

Fig. 8.1 Net erosion/sedimentation (10^6 m^3) of the Holland coast 1970–1985 (1986), not corrected for sand supplies. The main range ($y = 0$) is about 25 m landward of the top of the (most seaward) dune ridge. For situation see Fig. 2.1.

Statement 2: Inner and outer deltas are in equilibrium.

With respect to (sand balances of) estuaries, the existing ideas of asymptotic tendency to equilibrium (apart from the effect of sealevel rise) are over-schematized.

In the first place, the total coastal system (up to the end of the JARKUS ranges; Fig. 2.17 and in the outer deltas above a certain depth) is rather well self-contained. Volumetric computations of de Boer *et al.*[7] do not show much sand feeding or loss from deeper areas. The tidal channels to the former Zuiderzee silted up after the enclosure (1933), where the channel system near Harlingen became better developed (cf. Fig. 8.11). There was a deficit of sand in the Wadden (especially for filling up the Waddensea; cf. channels 6, 14, 18 and 19 to the former Zuiderzee; Fig. 8.11 a,b,e,f) This process of filling up (at the expense of the outer delta) did not take more than half a century. At the moment there is a tendency to a slight reversal: erosion of the tidal basin and extension of the outer delta. The effect could be blurred by effects of the 19-year tidal fluctuation (inclusive higher harmonics) and evidently, by inaccuracy of the measurements.

The statements given above may be substantiated with the following figures.

Studies on erosion and sedimentation in the western Waddenzee show sedimentation of the inner deltas of the Zeegat van Texel and the Zeegat van het Vlie after the closure of the Zuiderzee. Figure 8.2 (from [7]) shows the areas (tidal basins and inlets) considered; refer to Fig. 8.3b for the sedimentation and erosion volumes of those areas. The sedimentation volumes are in the order of the erosion volumes that were found at the outer deltas of both inlets (Figs. 8.3b and 8.2)). The inner and outer deltas of the Eijerlandse Gat and the Zeegat van Ameland show hardly any sedimentation or erosion (Figs. 8.3b and 8.2). There is a tendency that the erosion at the outer deltas follows the sedimentation inside with a delay of a few years. See as well Fig. 8.11.

Thanks to the fact that the erosion of the outer delta and the sedimentation of the inner delta are balanced, no net erosion of the inlet systems (including outer deltas) occurred after the closure of the Zuiderzee. This is in conflict with American empirical formulae on sand contents of outer deltas. Based on these formulae, the outer delta erosion should have been much less. This needs further investigation.[23]

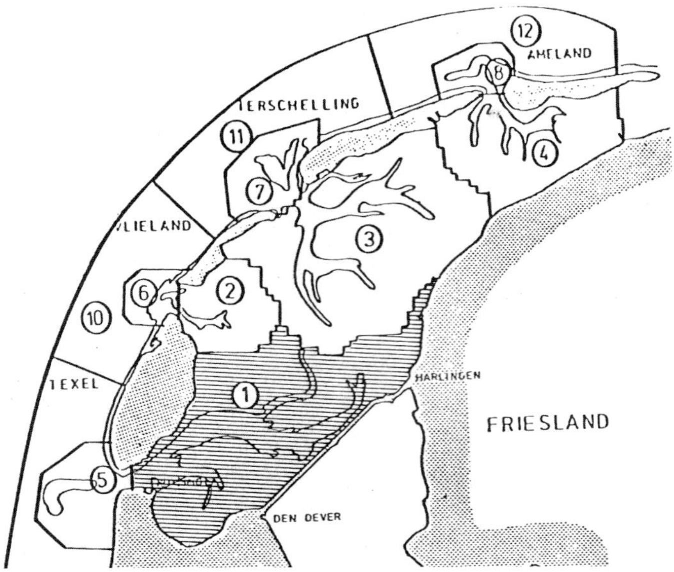

Fig. 8.2 Location flood basins and outer deltas considered in Fig. 8.3b (from de Boer.[7])

The morphological reaction on a sudden event like the closure of a channel, can be assessed by a logarithmic type of equation[b]

$$S_t = S_\infty\{1 - e^{-t\frac{S_1}{S_\infty}}\} \tag{8.1}$$

where S_t is the sedimentation after t years; S_∞, sedimentation at $t = \infty$; S_1, sedimentation first year.

Consideration of the inlet systems "Texel inlet" and "Lauwerszee" indicated the possibility of deriving S_∞ from S_1:

$$S_\infty = 0.323\ S_1^{2.852} \tag{8.2}$$

where S_∞ and S_l should be expressed in mm^3

1. Final sedimentation (S_∞) as result of (only) the Zuiderzee enclosure.
2. Sedimentation as function of time according to Eq. (8.1) $S_\infty = 340 \times 10^6\,\text{m}^3$ and $S_1 = 12.1 \times 10^6\,\text{m}^3$ (in 1932).

[b]Keep the empirical structure of Eqs. (8.1) and (8.2) in mind. Example Eq. (8.1): mathematically at $t = 1$ S_t/S_∞ will allways be larger than $1 - \exp(S_t/S_\infty)$(remark Bakker).

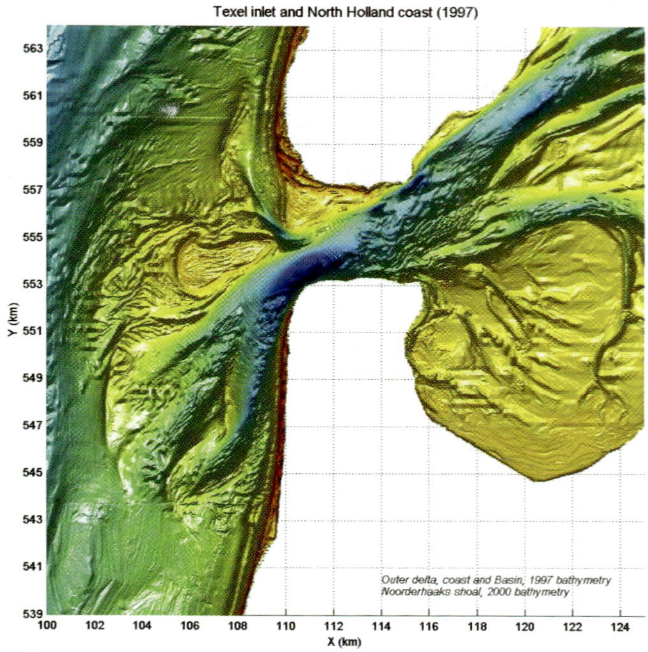

Fig. 8.3a Texel inlet.

3. Final sedimentation (as in "1"), taking autonomous sedimentation (silting up; sea level rise; increase tidal difference) into account.
4. Combination of lines "2" with "3".
5. As "4", taking periodicity of tidal difference ($T = 18.6$ year) into account.
6. As "5", taking phase lag into account.
7. As "6" taking sand mining into account.

* Accretion since 1932, as calculated from measurements (cf. Fig. 8.3b).

The two coordinate systems of this figure have a horizontal shift w.r.t. each other of 1 year and a vertical shift of S_1 (12.1 mm^3). For the first year of sedimentation (1932) is missing Data from [7, 8].

Figure 8.3 shows the sedimentation as a function of time of the inner delta of the Zeegat van Texel. In this curve an 18.6 years cycle may fit, corresponding to the 18.6 years component of the vertical tide with some

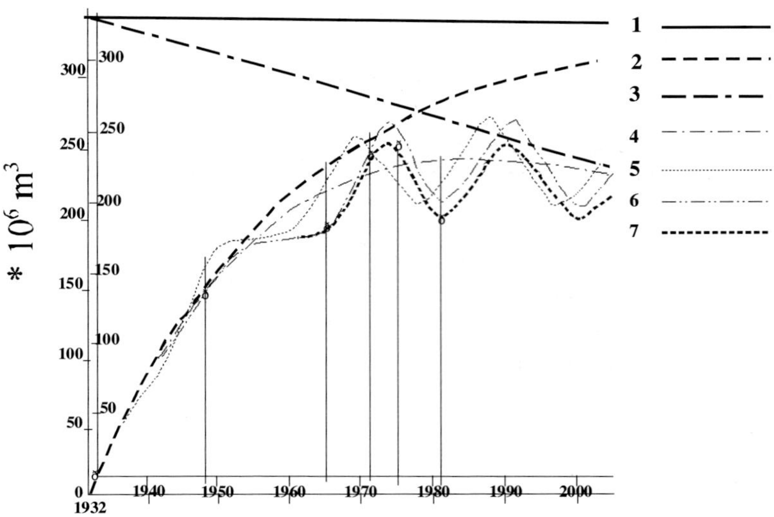

Fig. 8.3b Extent of sedimentation of the inner delta of the Zeegat van Texel (area "1" in Fig. 8.2).

phase shift. On the basis of this figure it can be concluded that the sedimentation in the Waddenzee after the closure of the Zuiderzee in 1932 has come to an end.

Statement 3: Concerning coastal dynamics near breakwaters.

Theoretically, the behavior of the coastline near a breakwater can be simulated using for instance Pelnard–Considère. Simulation also can be performed with a simplified (inertial) model for the (tidal) currents, using two-dimensional potential theory. Accelerations and decelerations result in patterns of sedimentation and erosion (Fig. 8.4). The tide appears to be the main reason for bed changes in a large area around the breakwater. The impact area of wind waves and swell is limited to the coastal sections next to the breakwater.

Figures 8.4 and 8.5 illustrate that large areas at deeper water on both sides of the breakwater are subject to sedimentation, with maximum accretion at some distance from the breakwater. Seawards from the breakwater head a large area is eroding, with maximum erosion in the scour hole in front of the head. Accretion takes place in an area of at least two times the length of the breakwater on both sides of the dam. This can be clearly

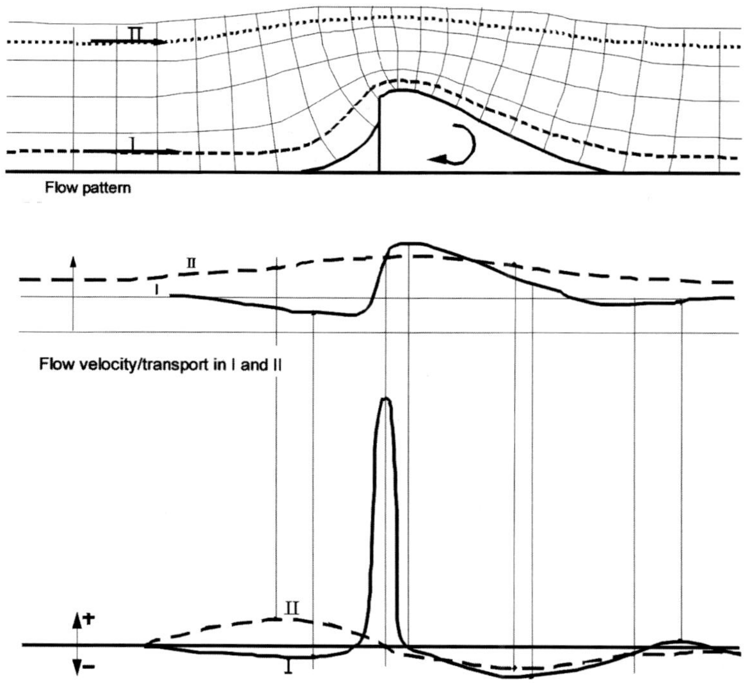

Fig. 8.4 Flow diagram, flow velocity and transport gradient along long dam.

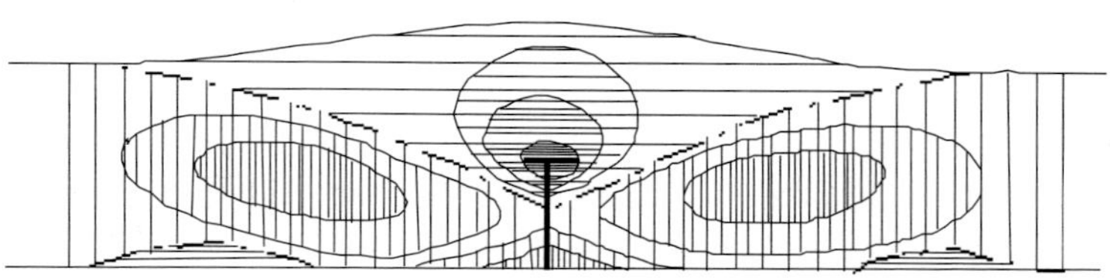

Fig. 8.5 Pattern of sedimentation/erosion by a dam as caused by symmetrical tide.

observed at IJmuiden after extension of the harbor breakwaters (Fig. 8.6). The origin of the accreting volumes is partly the scour area in front of the head and partly slight erosion over a larger area in front of the dam. Together with a cross shore transport towards the coast at deeper water and with a "sand oversaturated" coast as present between Scheveningen and Egmond, this phenomenon results in the so-called steepening of the

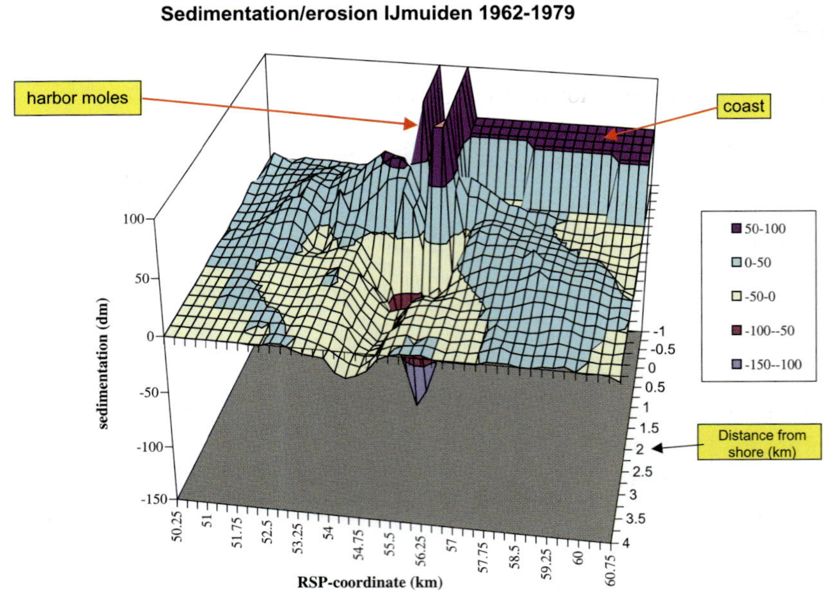

Fig. 8.6a 3D pattern erosion/sedimentation at IJmuiden 1962–1979. The x-axis of the coordinate system is oblique to the coast. Thus the coast on the Southern side of the harbor moles is depicted, where the coast on the Northern side is not.

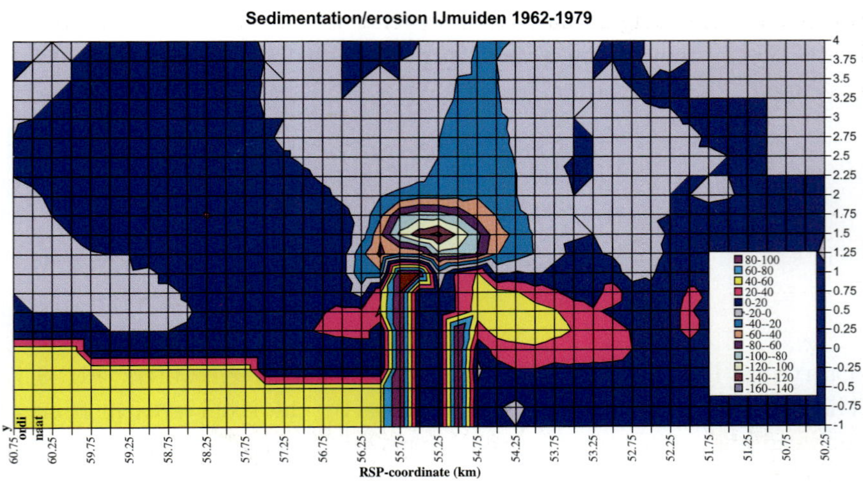

Fig. 8.6b Top view erosion/sedimentation at IJmuiden (1962–1979).

Dutch coast. Steepening of the coast is the result of a process towards a new equilibrium stage.

Statement 4: Erosion: down first, spread later.

Erosion that may occur after a sudden event will initially result in larger depths in a relatively confined area, followed by a more homogeneous erosion over a larger area. In this second phase, the deep initial scour holes are partly filled again. Examples of this phenomenon are scour holes at breakwater heads, channel erosion after closure of the Zuiderzee and scour development after dam breaches.

The first example is given in Fig. 8.7a.d: the "cross-shore dam" (long groin; length ca. 800 m) near Eijerland (North-point of Texel).

This recent example of coastal development, is forecasted by no means (neither by the coastal dynamics theory as displayed in this book, neither by more sophisticated numerical theories). Probably transversal turbulent exchange between sediment-loaded water streaming along the groin and clear water at the lee-side of the groin played a big role. Also in numerical models, up to now this effect is overruled by the effect of littoral current, as soon as waves of some importance occur.

Figure 8.7a,b and c,d shows the initial situation and the one after the construction of the dam (long groin), respectively. The behavior of the coast in this area is determined to a large extent by the behavior of the ebb- and flood channels in the Eijerland inlet (Fig. 8.8a); furthermore a (heavy) revetment, consisting of mattresses prevents scour, which could lead to the loss of the lighthouse (Fig. 8.7b,d). Figure 8.8b,c shows the development of the channel system in the immediate vicinity of the dam.

Fig. 8.7a,b Location Texel Eijerland (1994) copyright: RWS. From [35].

Fig. 8.7c,d Eijerland dam (1996) From [35] copyright: RWS.

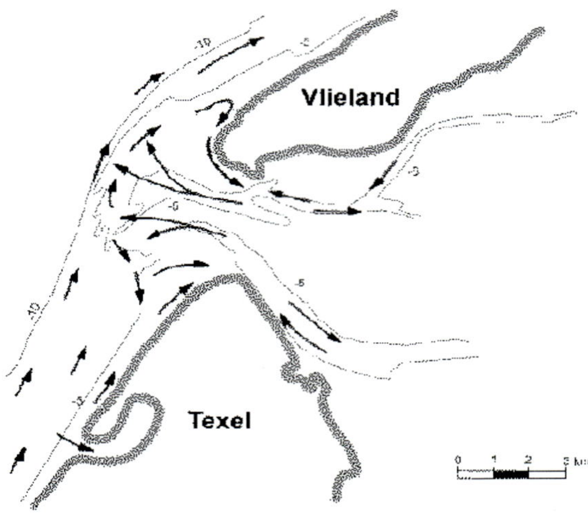

Fig. 8.8a Dominant transport in tidal channels Eijerlandse gat. From [22].

During the design a scour depth at the head of 11–12 m below MSL was predicted. After completion, the initial scour hole became as deep as 18.5 m below MSL. After 3 years the depth was reduced to 11.5 m. Figure 8.9a shows the changes in the cross-section through the scour hole (in the alignment of the dam); the maximum depth in course of time is depicted in Fig. 8.9b. An additional factor in this case is probably the fact

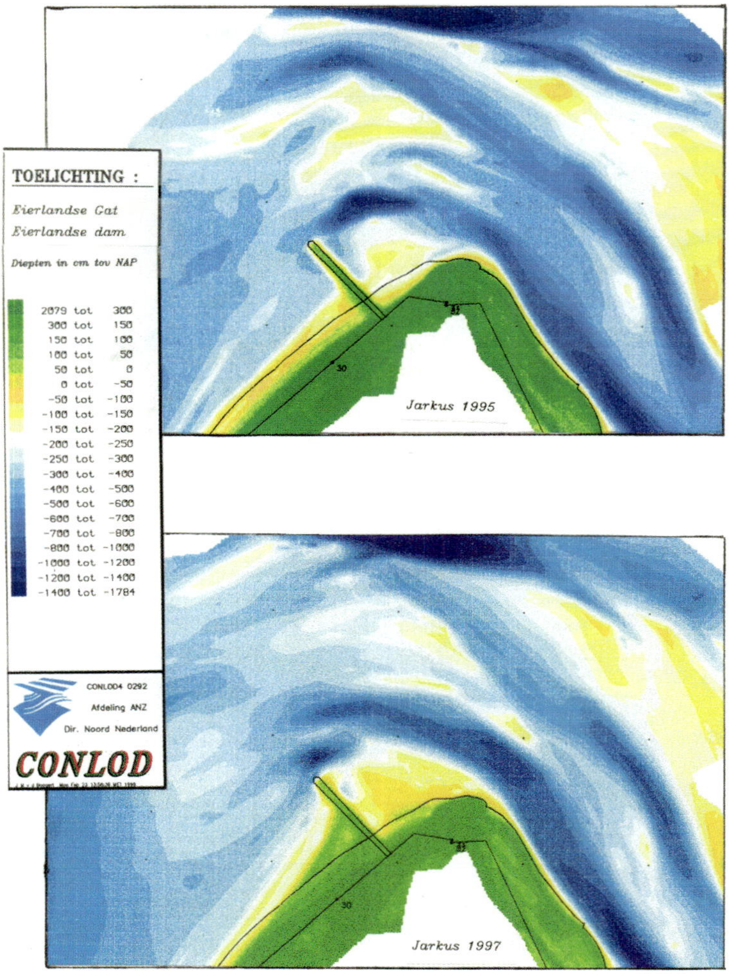

Fig. 8.8b,c Topography Eyerlandse gat; 1995 and 1997. From [22].

that initially both ebb and flood flows were subjected to contraction at the head of the dam, and presently the flood flow only (Fig. 8.8b,c). Also, due to substantial accretion on both sides of the dam, the flow contraction today is much less than in the initial stage. In the design of scour protection works one should account for more extreme conditions in the initial stage compared to the final "equilibrium" state.

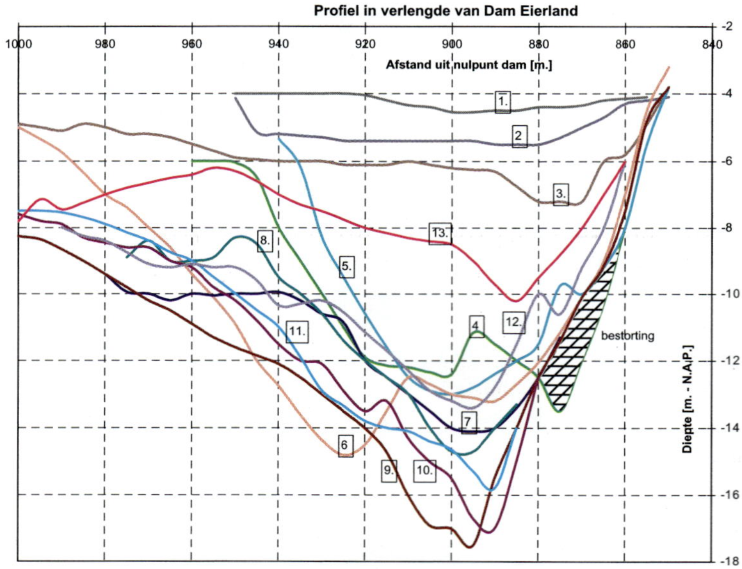

1	2	3	4	5	6	7	8	9	10	11	12	13
1995 May 22	1995 June 29	1995 July 25	1995 Aug 23	1995 Aug 31	1995 Sept 13	1995 Nov 14	1995 Dec 12	1995 Febr 28	1995 Sept 5	1995 Dec 11	1995 Mar 11	1995 Aug 20

Fig. 8.9a Eyerlandse Gat, cross-section scour hole (in alignment dam). Cf. [22].

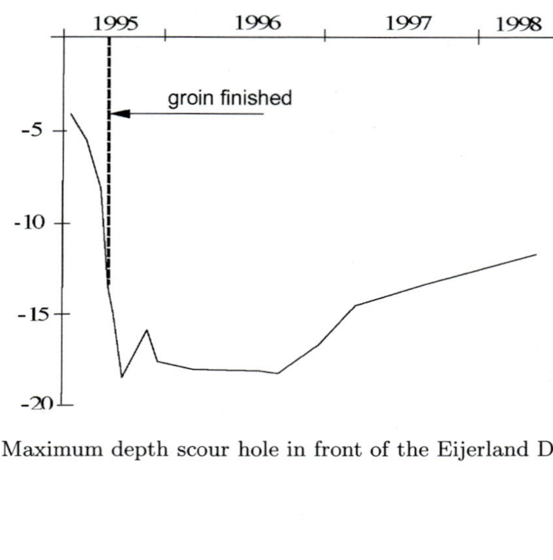

Fig. 8.9b Maximum depth scour hole in front of the Eijerland Dam. Cf. [22].

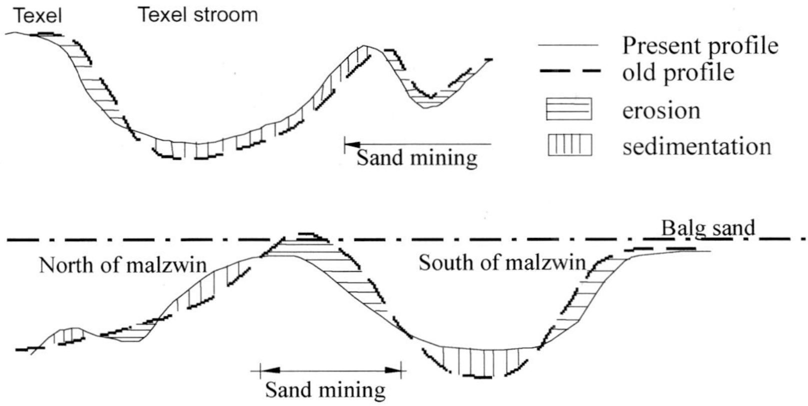

Fig. 8.10 Changes in profiles of "Texelstroom" and "Malzwin". Referring to Fig. 7.3: these are channels 3ab and 2, respectively.

The second example is given in Fig. 8.10,[c] in which reference is made to the development of the Texelstroom and the Malzwin after the closure of the Zuiderzee. Initially, these channels deepened, followed by a process of widening and shallowing. A wide and shallow profile requires a larger cross-section than a narrow and deep profile to accommodate the same tidal volume. This may be the reason that the sedimentation of the Waddenzee was reduced or even stopped in course of time.

Statement 5: Concerning periodicity in coastal dynamics:

Periodicities in hydrodynamic and morphodynamic processes are related to the interactions between the attraction force and motion of the sun, its planets and sub-planets, like the moon. From literature many examples are known of climate cycles due to astronomical periodicities.

The diurnal tide is a well-known cycle in the motion of water. Less known is the 18.6 years periodicity of the vertical tide. At flushing, this periodicity is clearly visible in the development of the tidal differences (Fig. 8.12). As discussed already under "statement 2", the sedimentation curve of the inner delta of the Zeegat van Texel (Fig. 8.3) may be influenced by this 18.6 years cycle.

Apart from the 18.6 years cycle, the annually averaged tidal data give rise to identify periodicities of 4.5 years, 8.85 years, 17.4 years and

[c]For situation refer to Fig. 7.3: these are channel 3ab and 2, respectively.

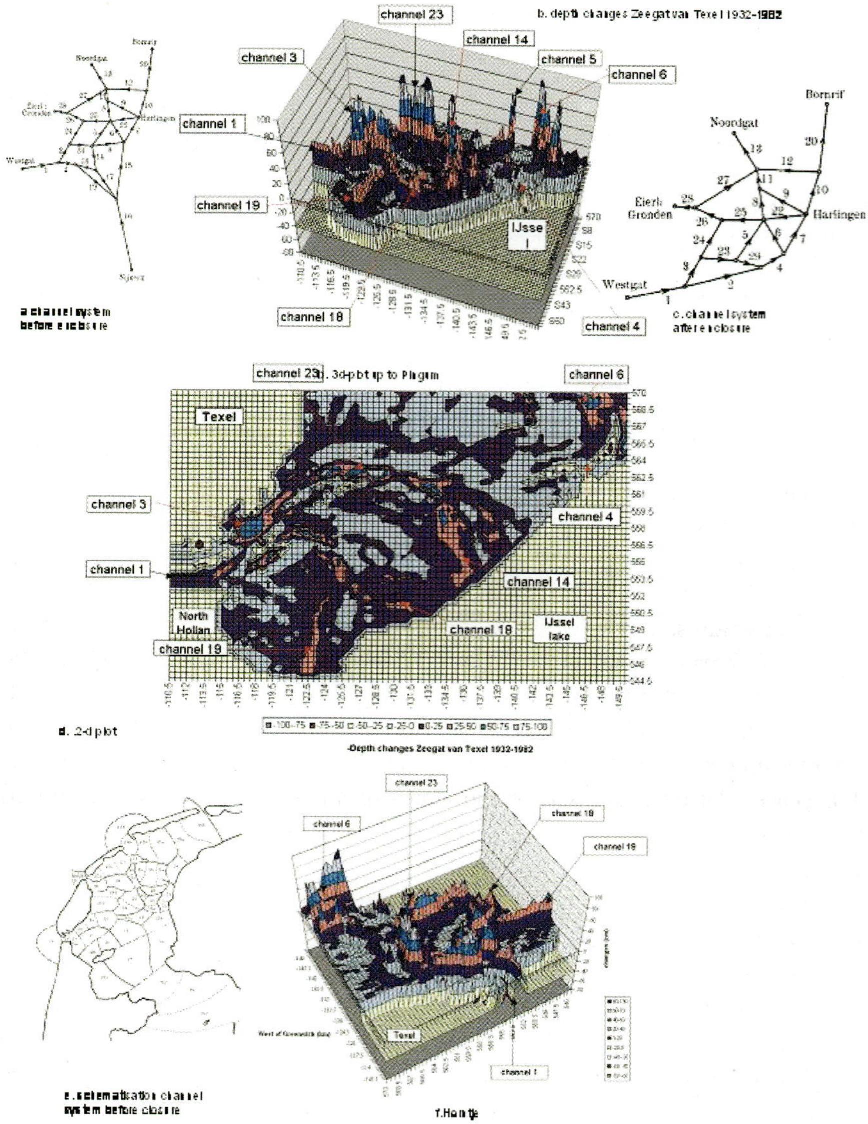

Fig. 8.11 (Rakhorst/Bakker) Erosion/sedimentation in the inner delta of Texel (1932–1982). For situation cf. Fig. 7.54a,b. Figure 7.4 is reproduced in Fig. 8.11c as well. Figure 8.11a shows the Lorentz schematisation of the channel system before the Zuyderzee was enclosed. The schematisation itself (cf. Fig. 7.3) is depicted in Fig. 8.11e. Generally, the erosion/sedimentation graphs (Fig. 8.11b,d,f) demonstrate the siltation of the tidal channels, which became redundant and the more important role of channel 4 in the new situation (erosion).

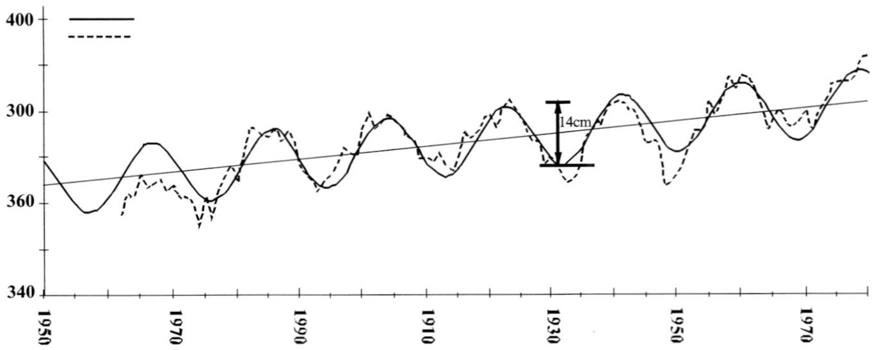

Fig. 8.12 Tidal difference at Vlissingen. From [9].

37.5 years. Probably, there are also periodicities of 60, 84 and a bit more than 180 years.

It is quite difficult to identify these cycles in the registrations of the vertical tide; let alone to identify those in the behavior of the coast.

Large-scale sand waves with a cycle time of 150–200 years are present at the SW corner of Texel, near Terschelling and at Walcheren. Along the Dutch coast sand waves with a period of 75–100 years might be identified. The breaker bars between IJmuiden and the Hondsbossche Zeewering have a cycle time of about 20 years and those in the area south of IJmuiden about 4–5 years. Periodicities with cycle times of 4.5 and 9 years can be found frequently in the behavior of the coastline of northern Holland (Figs. 8.13 and 8.14).

Summarizing, orientation and development of the coast show different periodicities that mostly occur simultaneously, however not always and not everywhere.

Bliek points out, that the relation of Per Bruun & Gerritsen[5] has some elements, which are mostly overlooked. He reports his investigations on ebb/flood dominance in tidal estuaries. His contribution refers to multiple channel estuarine systems, and especially the behaviour of the Western Scheldt estuary (Fig. 7.23).

The Western Scheldt is a macro-tidal estuary, with a semi-diurnal tide. The mean tidal range increases from 3.8 m at the mouth near Flushing to 5.2 m at Antwerp, about 90 km upstream. From there, the tidal range gradually decreases to 2 m at the head of the estuary near the locks of Gent, 155 km upstream.

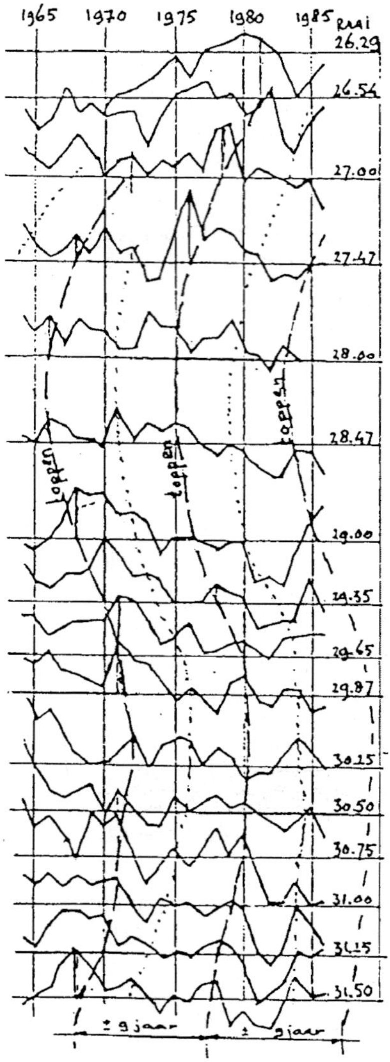

Fig. 8.13 Cross-shore location high-water line; 9 and 4.5 years period, respectively. Cf: Fig. 2.8 concerning range system.

The tidal volume (ebb plus flood volume) varies from 2.2 billion m^3, at Flushing to 1.4 million m^3 at Antwerp. The Western Scheldt is a well-mixed estuary with a relative small river discharge of about 0.25% of the tidal volume at the mouth. The river discharge varies on average between

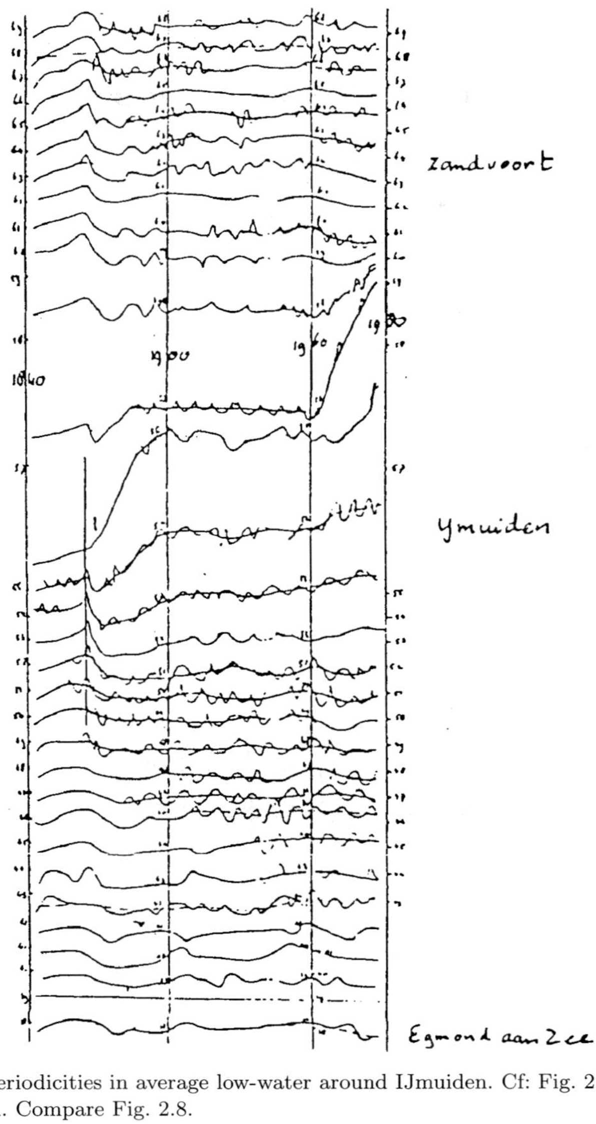

Fig. 8.14 Periodicities in average low-water around IJmuiden. Cf: Fig. 2.11 concerning range system. Compare Fig. 2.8.

$50\,m^3/s$ during summer and $170\,m^3/s$ during winter. Given the strong tidal current and the limited fresh water discharge, the estuary is a tide-dominated system. The estuary covers an area of $300\,km^2$ and consists of well-developed channels and shoals. The estuary is funnel shaped. The width at the mouth is about 5 km with two main channels. At the Belgian–Dutch border the width is reduced to about 1.5 km with one main channel and a shallow secondary channel. The boundaries are almost completely embanked and fringed with mud flats or tidal marshes. The bed material of the Western Scheldt consists of well-sorted fine to medium sands, with an increasing silt content in landward direction up to 5% near the Belgian border.

At 5 locations the ebb and flood dominated main channels cross each other; ebb/flood dominance shifts from left to right bank or from right to left at these "nodes". The distance of these nodes from the gorge of the estuary at Flushing and the total river width near the nodes is given in Table 8.1.

It can be seen that the distance between the nodes increases with increasing river width: between Bath and Valkenisse the distance is 8 km, with an average river width of 2.5 km, and the distance between Baarland and Borssele is 17 km, with an average width of 5 km. In other words, the distance between two succeeding nodes amounts to 3–4 times the total river width.

In the course of time the nodes remain more or less on the same location: the behavior of the channels between the nodes may change significantly however.

The channels in the Western Scheldt are either ebb dominated or flood dominated. The dimensions of the channels depend on the water volume flowing through the channel during the dominant phase of the tide (ebb or flood).

The presence of a relation between tidal volumes and channel cross-section in estuaries is well known in literature.[21,33,37] The main difference is that in this analysis the dominant volume (ebb or flood) and single channels only are considered instead of the total tidal volume and the total cross-section.

Table 8.1 Locations where the ebb and flood dominated main channel crosses each other.

Node	Distance from Gorge (km)	Approx. river width (km)
Bath	63	2.0
Valkenisse	55	3.0
Hansweert	42	4.0
Baarland	28	4.5
Borssele	11	5.5

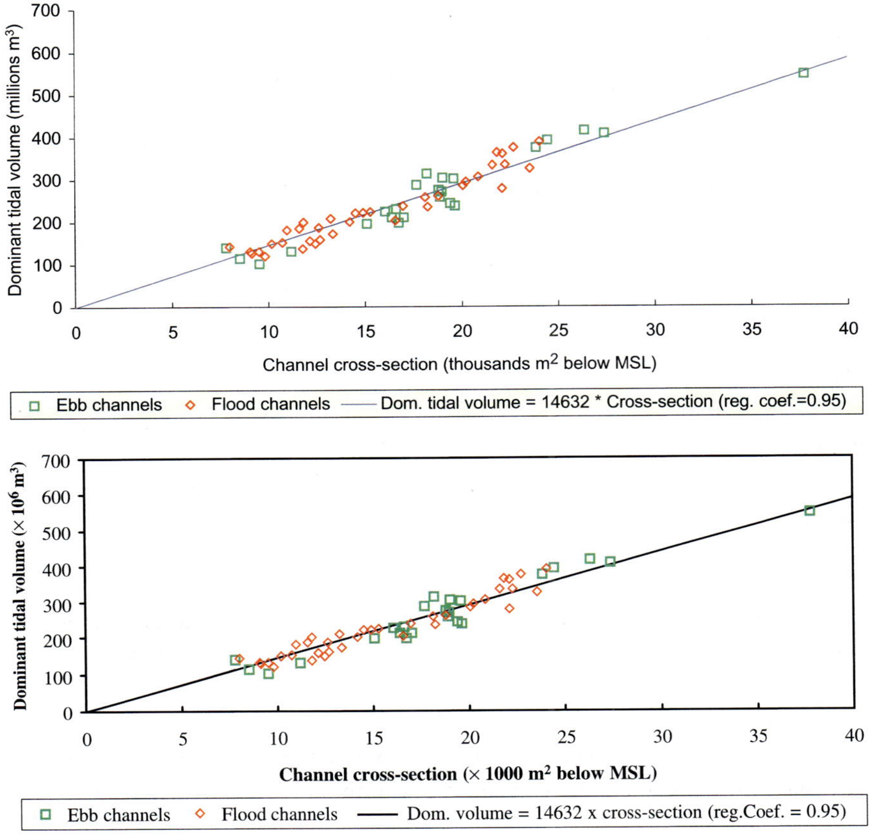

Fig. 8.15 Relation of the dominant tidal volume based on Eq. (8.3).

For the relation between the dominant tidal volume (DV in m^3/tide) and Western Scheldt channel cross-section below mean sea level (A in m^2) a formula of a simple linear shape is found (see Fig. 8.15):

$$DV = 14,632 \times A \tag{8.3}$$

There is no significant difference between ebb channels and flood channels. The relation is only valid for channels with an ebb or flood dominance over the full cross-section. Channels with internally ebb or flood dominated sections in general have a larger profile. The relation is determined by using numerous discharge and profile measurements in the main channels of the estuary from 1932 up to 1991. In this period of 60 years the cross-section and tidal volume of individual channels may have changed substantially. The relation between dominant tidal volume and cross-section has not changed significantly however.

The constant 14,632 cf. Fig. 8.15 has the dimensions [m/tide] and represents the displacement distance of water particles (tidal excursion) during the dominant tidal phase. By taking a duration of a half tidal cycle of 6.2 h, this implies an average flow velocity (in the dominant phase of the tide) of 0.65 m/s. During the non-dominant phase of the tide, the flow velocity will be less, depending on the ratio between dominant and non-dominant tidal volume in that specific channel.

The conclusion that the dimensions of the channels are determined by the dominant tidal phase has some interesting consequences for the channel behavior and the sediment balance.

Apparently, a single channel system, with a flood volume (almost) equal to the ebb volume, is the most effective transporter of water. Both during ebb and diving flood such a channel is used up to its "water transport capacity". A two-channel system with a well-distinguished ebb and flood channels is less effective.

In Fig. 8.16, this is demonstrated for a theoretical case where the ebb channel (channel 1) takes 60% of the total ebb volume and the flood channel

	ebb volume	**flood volume**
channel 1	**60%**	**30%**
channel 2	**40%**	**70%**
total	**100%**	**100%**

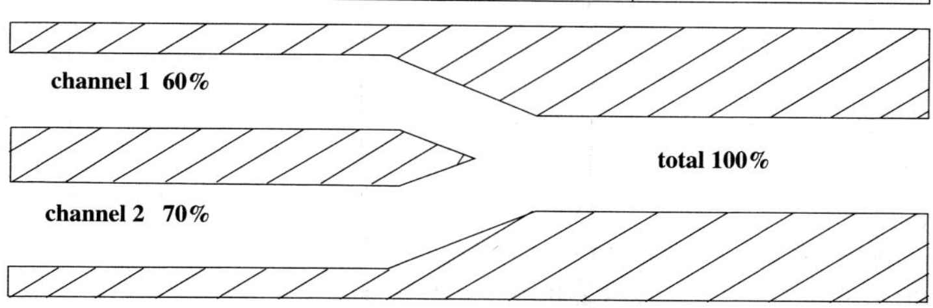

dominance = (60 + 70)/100 = 1.30

Fig. 8.16 Definition of dominance factor of two-channel system.

takes 70% of the total flood volume. In this example, the total cross-section of both channels together is about 130% of the cross-section of a single channel with the same "water transport capacity".

The dominance factor of a two-channel system as defined in Fig. 8.16 is a good indicator for the total cross-section of such a system. Changes in the dominance factor may result in substantial erosion or accretion volumes without a significant impact on the propagation of the tidal wave.

This is illustrated by Van Kleef[36] for the Middelgat area in the Western Scheldt (between the "nodes" of Hansweert and Baarland). He concludes that the relation between the dominance-factor and the total cross-section of the system Middelgat-gat van Ossenisse over the period 1932–1989 is "surprisingly good". The same applies for the total sedimentation in that area.

In the period 1957–1989 a total sedimentation of about 35 million m^3 is found (corresponding to an average layer thickness of approximately 0.60 m).

In the same period, the dominance factor reduced from 1.14 to 1.08.

It can be concluded that a 1% variation of the dominance factor in this area results in an average bed level change of approximately 0.10 m (or a total volume of 5–6 million m^3 of sand), without a significant change of the total tidal volumes. In reality, the tidal volumes even slightly increased in the period 1957–1990.

Remaining question is whether the accretion has initiated the reduction of the dominance factor, or the other way round. The latter is more likely. Channel migration, especially of the Middelgat (which is no longer located in the middle of the estuary but next to the northern border) and a more pronounced development of the Gat van Ossenisse have resulted in a reduction of the Middelgat and a growth of the Gat van Ossenisse. In 1957 the Middelgat still was the most important channel. In 1989 this role was taken over by the Gat van Ossenisse. An interesting detail is that the main fairway to Antwerp was shifted from the Middelgat to the Gat van Ossenisse in the same period.

As a consequence of these developments, also the dominance factor has changed. This resulted in oversized channels (especially the Middelgat) and the conditions for sedimentation were there.

An interesting question is how the dominance concept of multiple channel systems relates to the rain-in-the-drain concept of Bakker and De Vriend.

This question needs some further thinking. The first reaction is that the rain-in-the-drain concept does not contribute to the understanding of the behavior of multiple channel systems.

In the Middelgat case, the ebb-dominated channel (Middelgat) is subjected to serious sedimentation. The reduction of the ebb (and flood) volume in course of time determines the development of this channel. The most important channel, Gat van Ossenisse, is flood dominated, the cross-section being determined by the flood flow. Any rain-in-the-drain contributions to the ebb flow below low water do not affect this channel. Another question is how the rain-in-the-drain concept would affect the equilibrium profile of ebb and flood channels. Would an ebb flow concentration below low water result in a higher sediment transport during ebb and thus in a larger profile of ebb channels than flood channels? This does not follow from the stability analysis of tidal channels given in Fig. 8.15.

The minimum conclusion is that estuaries with more than one channel, like the Western Scheldt, cannot be used for calibration purposes of the rain-in-the-drain concept. Even over a period of 30 years or more the ebb/flood dominances of the channels determine the sedimentation and erosion balance of the estuary. These "internal affairs" over-rule the effects of the rain-in-the-drain concept. In other words, in these estuaries the rain-in-the-drain is a second order phenomenon.

Note that the fresh water discharge of the Scheldt River (50–170 m^3/s) also does not have a significant impact on the estuary development, apart from density effects in the eastern part, where horizontal salt gradients are present (the estuary is well mixed).

By the way, the effects of salt gradients are opposite to the rain-in-the-drain. Especially in the Belgian part of the estuary, near Antwerp, the current profile is more block shaped during flood, while during ebb the currents near the surface are higher.

Steetzel highlighted modern developments concerning *line models,* like the *MOBIC- CUR-* and *DESCAN-model.* Especially near outer deltas, five-layer models could be necessary (Fig. 8.17).

Morphological characteristics of complicated coastal systems can be described using different modelling approaches. One such approach is *process-based modelling* where the physical processes involved are described mathematically, combining a detailed fluid-flow model with a sediment-transport model. By successive iteration the dynamical evolution of an area can be simulated. For the analysis of the predominant processes and circulation patterns, wave, current and sediment transport, process-based

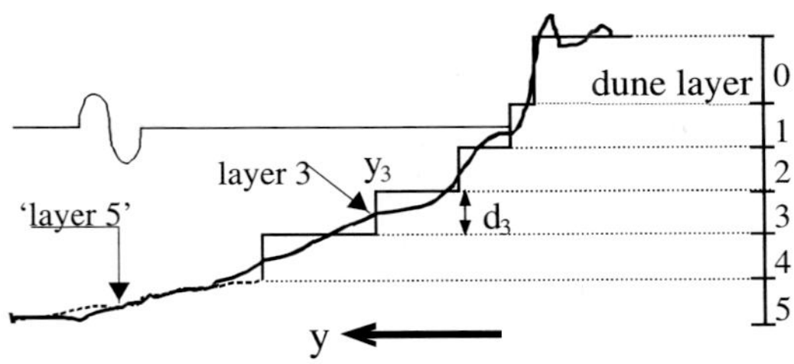

Fig. 8.17 A five-layer model.

models appear to be useful. However, they are less suitable for simulating long time periods, as they require large computation effort and the numerous iterations and accumulation of rounding-off errors may lead to unrealistic results.

For national decisions regarding coastal management, it is important to understand the long-term effects and large-scale implications of both natural processes such as climate-change sea-level rise and major coastal engineering projects such as large-scale land reclamation or the closing of major tidal inlets. To this purpose, within the framework of the national Dutch research program "*COAST*2000*", Steetzel *et al.* developed a pilot-version of a model[29,30] that is basically capable of quantifying the long-term (50–100 years) and large-scale (1–100 km) morphological evolution of the Dutch coast. The concept of this model has been validated and improved further within the framework of two major EU-sponsored projects, viz. the so-called *PACE*- and *SAFE-project* of the Marine Science and Technology Programme (*MAST-III*).

The model developed is based on the multi-layer concept, in which the cross-shore profile is schematised as a number of mutually coupled layers, defined between fixed profile depths, see Fig. 8.17. These layers interact through cross-shore transport. In longshore direction the layers respond to gradients in the longshore transport generated at the profile regions they represent. Though the concept of multi-layer modelling is not new (cf. for instance [11],[d] some recent developments have substantially increased its applicability, such as the extension of the concept for the modelling of the outer delta at tidal inlets is elaborated in more detail by Steetzel.[25]

Earlier versions of this model had the drawback that the interaction between the layers and their response in the longshore direction was

[d]Cf. Section 3.4 in general (remark BAKKER).

determined by a series of pre-defined constants, which had to be determined by the user based on mathematical process-based models or on empirical data. This put considerable restraints on the practical use of the concept. In the present set-up of the model these pre-defined constants have been replaced by formulations to compute cross-shore and longshore sediment transports directly within the model in terms of external conditions such as wave climate, tidal conditions, bathymetry and sediment characteristics. In this way, a very user-friendly behavior-oriented model, the so-called PONTOS-model, has been formulated. As such, the multi-layer concept is now more suitable for the evaluation of changes in a coastal system.

As discussed, new formulations for both the cross-shore interaction and the longshore transports had to be developed and implemented in the model.

The cross-shore interaction is based on the principle of a single wave-based equilibrium profile. Deviations from the equilibrium slope result in a cross-shore exchange of sand between the layers. The rate of adjustment varies with the depth and is amongst others depending on relative water depth and local sediment characteristics. For both the equilibrium slope and the transport rates, formulations have been derived, partly based on empirical formulae and partly based on the results of a series of computations carried out with a process-based model. Relevant input parameters are wave climate and sediment characteristics.[26]

Formulations for both the wave-induced and the tide-induced longshore transport are implemented in the model. These formulations are based on series of computations carried out with a process-based model. Relevant input parameters are wave climate (horizontal and vertical), tide and sediment characteristics. Furthermore, a simple formulation for wave diffraction has been implemented to account for the modified wave field around structures. In a later stage it is envisaged to also include the effect of flow contraction around structures.[28]

In order to test and validate the numerical model, a large number of test cases have been defined and elaborated in detail. Systematic tests have been carried out, for e.g.:

- Cross-shore profile evolution (compound equilibrium profiles for wave and tidal climates, effect of sea level rise, wind climate change);
- Impact of a single groin (oblique wave attack, diffraction effects, combined wave and tidal climates);
- Impact of revetments and dikes (cross-shore profile effects);

- Nourishment of eroding coasts (pre-defined schemes and nourishment requirement assessment using the model's auto-nourishment mode.[6]

The model is set up in such a way that it is generally applicable. A specific application for a part of the Dutch coast has been carried out using the so-called pilot-version of the model. This application focuses on the uninterrupted coast, therefore not including the delta coast in the south and the tidal inlets in the north of the Dutch coast. The objective of this application was to show the ability of the model to simulate the large-scale transport patterns along the Holland coast, taking into account the non-uniform wave and tidal climate.[30]

Roelvink depicted the turf of applicability of the models as treated in this book.

Figure 8.18, from [32], depicts relations between time scale and space scale of models. The models presented could be used for the Meso scale of coastal evolution and the large scale; however the diffusion concept does not hold at the small scale. Reason is, that instantaneous response of hydraulic features to bottom topography is assumed, where in reality there can be retardation: the plunging of waves is retarded with respect to coastal slope and to water depth; currents react with inertia on radiation stress; sediment concentration profiles lag behind the equilibrium concentration profile.

This makes that under some circumstances features can enhance themselves. In the present cross-shore profile models, like the one of Roelvink *et al.*[24] these non-linear effects are incorporated. It might be possible to link various profiles (as simulated by Roelvink) in longshore direction.

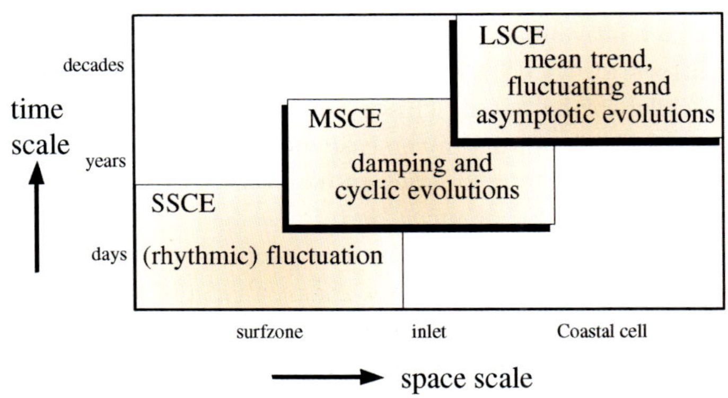

Fig. 8.18 Coastal evolution scale. From [32].

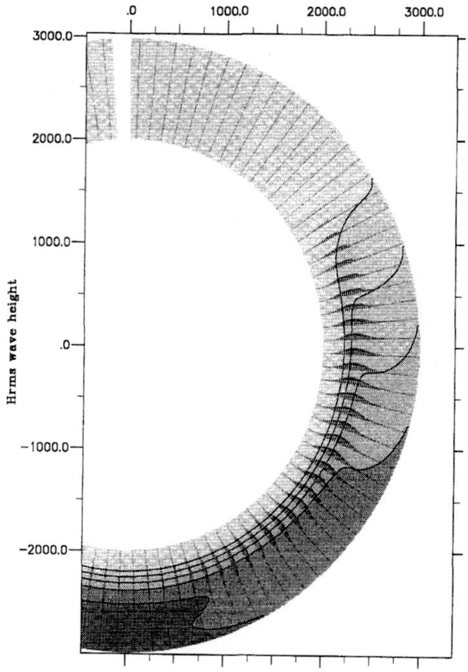

Fig. 8.19 Effect of inertia on the littoral current in case of a relatively small circular island.

In this way a coastal model could be created, not much more intricate than the ones, treated in this book, but able to simulate, for instance, breaker bars.

Figure 8.19 demonstrates the effect of inertia on the littoral current in the case of a relatively small circular island. Large littoral current (and therefore littoral drift) occurs at sites, where waves propagation in deep water is parallel to the beach. This is not CERC-like! On the other hand (as Roelvink told the editor) in the case of a large island where gradients of coastal direction are not so large, result from CERC-formula and more sophisticated theory coincide much better.

Roelse states, that the inaccuracy of measurements can lead to spurious conclusions.

In maps depicting accretion and erosion, "sandwaves" can be observed, with a crest height, having a wavelength of twice the mutual range distance and a crest direction coinciding with the range direction. Reason will be that the sounding results depend on direction of propagation of the measuring vessel. Long-term trends are biased because of different means of monitoring. Easily large inaccuracies of trend of the kind can result from this. From maps of erosion and accretion one even can clearly distinguish sometimes, which area is monitored on the same day and how this changed from year to year. On the other hand, comparison of long-term trends of some area determined independently from an observation time of many years (from maps) by Dutch and Belgian observations respectively, showed only deviations of relatively small order of magnitude.

Roelse presents the actual developments concerning morphological research with respect to the Zealand coast, since 1980. Former evolution has been reported in reference [50] and in Section 2.7.

Many of these modern investigations have been carried out in the scope of the Dutch governmental of chapter 2 research program COAST 2000, which started in 1995.

Earlier, the following research projects were investigated (see, for instance [19])[e]:

(a) The morphological effects of the enclosure of the Brouwershavense Gat[f]; these were hindcasted with the aid of current measurements and tidal computations;
(b) The effect of the reduction of the tidal discharge in the Eastern Scheldt[g] on the tidal channel in the immediate vicinity of the Schouwen coast, linking the outer Delta's North and South of Schouwen (Fig. 2.63).
(c) New empirical relationships between cross-section of channels and tidal discharge were drafted by Gerritsen & de Jong.[12,14] The validity of those relationships in the tidal channels in the embouchure of the former Eastern Scheldt estuary could be demonstrated; it results in narrowing of the tidal channels after the building of the Storm Surge Barrier. No explanation could be given as yet concerning the large-scale sand waves, propagating through the estuary and along the coast.

Refer now to Fig. 2.46 concerning the coastal dynamics of the embouchure of an estuary. This schedule has been used as a basis for the COAST

[e]The attention should be drawn as well to the investigation of Aarninkhoff and v.Kessel,[1] concerning the outer delta of the Eastern Scheldt, Brouwershavense Gat and Haringvliet, published after the date of the "Forum". Way of investigation is in line with the method of the present section and of Section 2.6.

[f]Inlet between Schouwen and Goeree (Fig. 2.64).

[g]Because of the construction of the Eastern Scheldt Barrier, finished in 1984.

2000-program, focusing on two elements:

(1) the large-scale system of shoals and tidal channels; and
(2) the "sand pressure" on the South-Western coast of Walcheren.

The large-scale system of shoals and tidal channels

Consider Fig. 8.20c, showing a map of the sea bottom of the Western Scheldt estuary in 1997.

Topographic and historic investigation reveals the replacement of the central tidal channel, the Deurlo (Fig. 8.20a) orientated in WNW direction by two tidal channels, directed in NW-direction, more or less parallel to the Deurlo. Mind the emergence and the dominance and short-cut of the "Geul van de Rassen", Fig. 8.20a.

Figure 8.20b shows a map of areas of erosion/accretion between 1969 and 1993. Changes are partly natural and partly antropogenic (see below). Parallel areas of accretion/erosion either denote transversal shift of tidal channels, either sandwaves. Also effects of way of measuring (inaccuracy: variance in squat in successive ranges; inaccuracy of determination of water

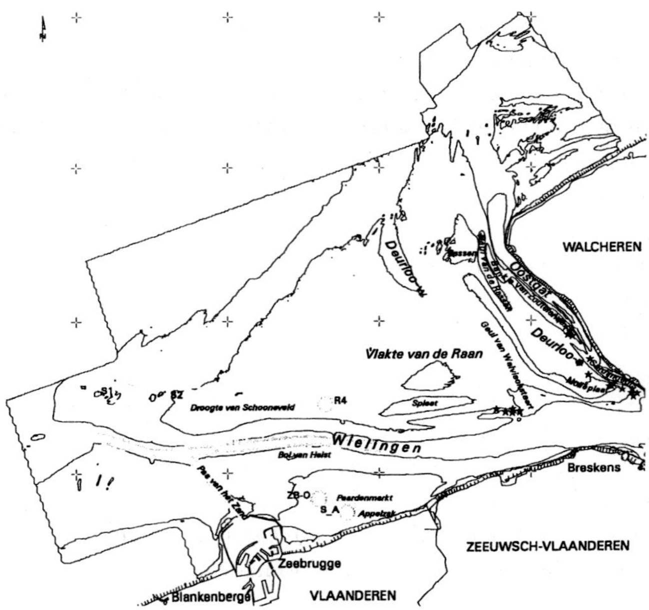

Fig. 8.20a Outer delta Western Scheldt: area under investigation.

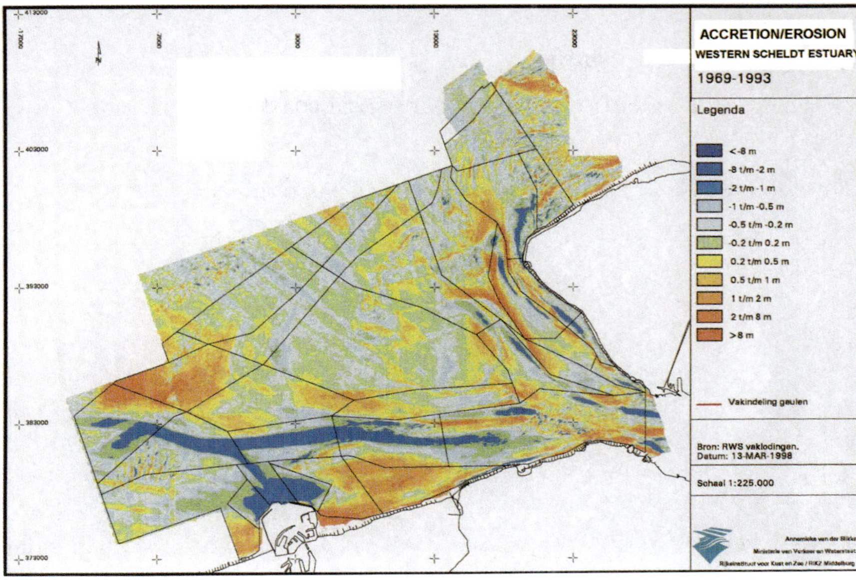

Fig. 8.20b Outer delta Western Scheldt: differences in bottom level 1969–1993.

level) cannot be totally excluded. The total area is subdivided into several sub-areas; the borders of these sub-areas are indicated in Fig. 8.20b,c. Changes in volume in the successive sub-areas have been determined, making use of the successive surveys available. The name assigned to the various sub-areas is given in Fig. 8.20c. Two areas, the scour hole "Pas van 't Zand" (called "Zeebrugge" in Figs. 8.20c and 8.21c) and the "Dumping site Schooneveld" (indicated with relative thick lines on Fig. 8.20b) will be considered in more detail below.

At the Southern side of the Western Scheldt the effect of human interference is dominant:

(a) *Dredging*: The artificial widening of the navigation channel through Scheur[h] and Wielingen (Fig. 8.20c) allows fast propagation of the tidal wave to and through the Western Scheldt. The navigation channel "Pas van 't Zand" (Fig. 8.20a) to Zeebrugge, nearly perpendicular to the direction of the current gives a large catchment area for sand and silt, resulting in the need for much maintenance dredging.

[h] Adjacent to Wielingen, Fig. 8.20a.

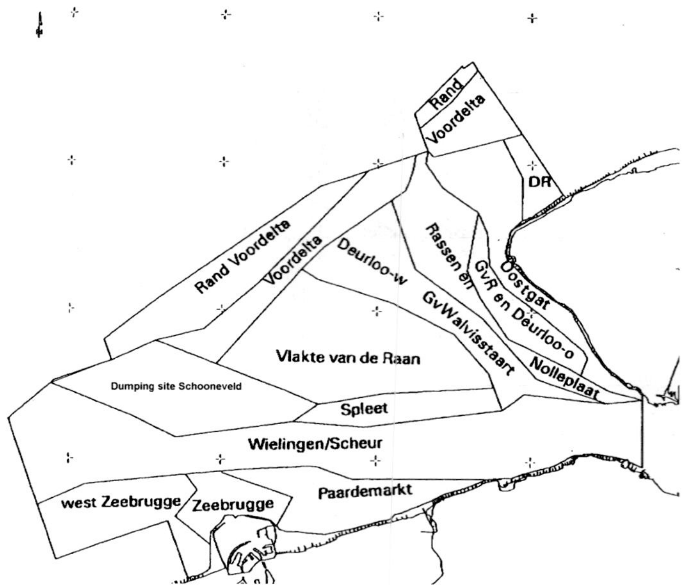

Fig. 8.20c Outer delta Western Scheldt: areas in which sandvolume was determined.

(b) *Extension of harbor moles.* The extension of the harbor moles of Zeebrugge (cf. Figs. 2.74 and 2.75) has a similar effect as the one of IJmuiden (Fig. 8.6a and the schematic Fig. 8.6b). The scour at the seaward ends of the moles causes a broad eroding area ("Scour hole Pas van 't Zand"; Fig. 8.20c), where in the sheltered areas (in the immediate vicinity adjacent to the moles) accretion occurs. The sedimentation at the SW-side and the lee-side (in the immediate vicinity adjacent to the moles) at the NW-side of the harbor point to dominant littoral drift from the South.

(c) *Dumping of dredged material.* At some sites in the Outer Scheldt Estuary dumping takes place of sand and silt originating from the deepening of channels and from maintenance dredging. Of those dumping sites, the one at the Dumping site Schooneveld (Fig. 8.21b,c) is the most important one.

Covering the period 1969–1997 the sediment balance was made for 34 coastal areas, of which the shape was chosen based upon the site of the channels and shoals and based upon the sounding surveys available (Fig. 8.20c). Figure 8.21a–c shows (in the shape of bar graphs) the

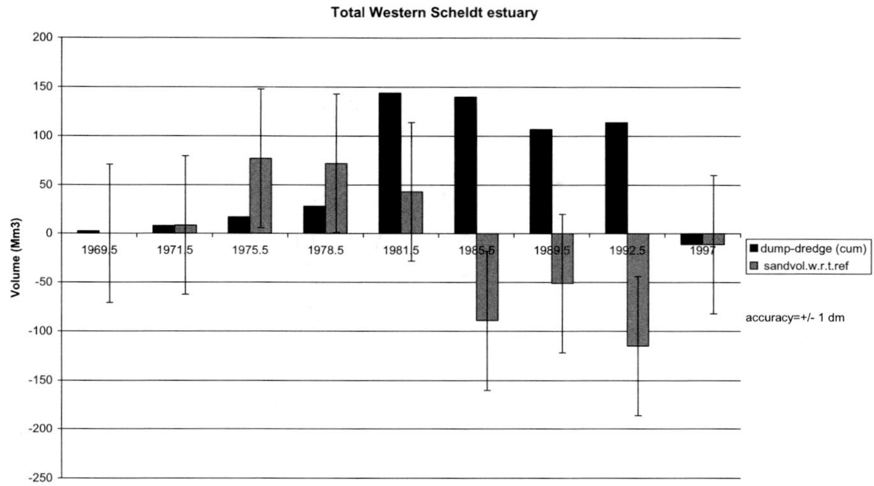

(a)

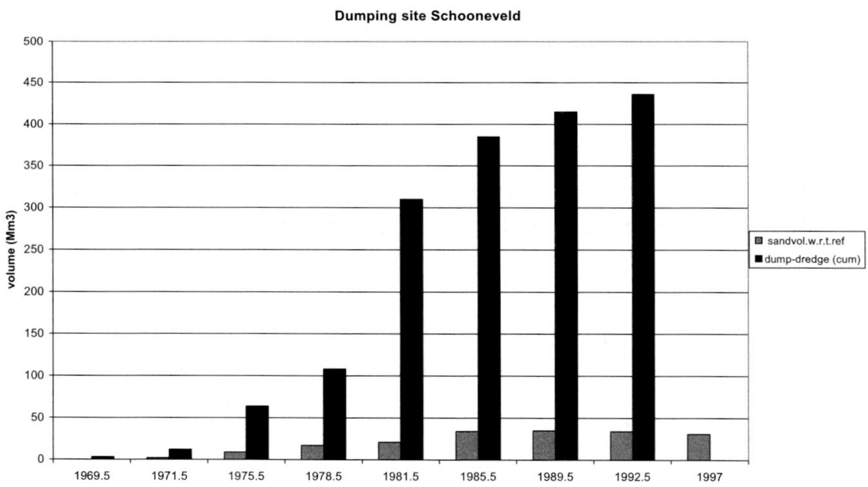

(b)

Fig. 8.21a,b,c Dredging minus dumping compared to increase of sand volume since 1969. (a) Total Western Scheldt; (b) Zeebrugge; (c) dumping site Schooneveld (cf. Fig. 8.20c).

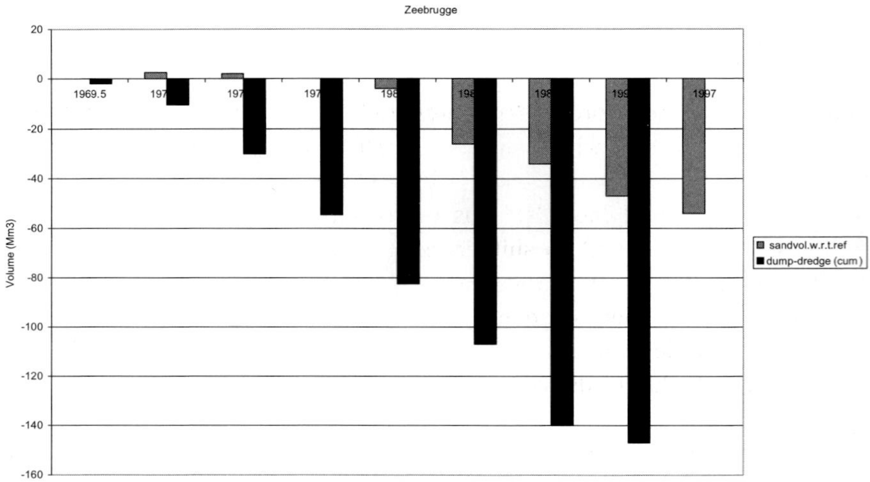

(c)

Fig. 8.21a,b,c (*Continued*)

changes in volume[i] of respectively the total Western Scheldt estuary, of the Scour hole "Pas van 't Zand" ("Zeebrugge") and of the dumping site Schooneveld; furthermore the bar graph shows the human interference (amount of dredging or dumping). In Fig. 8.21a (total Western Scheldt) a measure for the rate of accuracy, i.e. 1 dm accretion, (c.q. erosion over the whole area) is indicated. This shows that the changes are not much larger than this rate of accuracy.[j]

Furthermore, the changes in volume of the dredge and dump areas (Fig. 8.21b,c) clearly show that the human interference is an order of magnitude larger than natural changes (if present). Dredging appears to be rather ineffective. Despite of the scour in front of the harbor moles (caused by those moles) the yearly amount of dredging in the "Pas van 't Zand" (Figs. 8.20 and 8.21b) is 7 times as much as the change in volume of the area monitored. At the dumping site (Fig. 8.21a) the amount of dumping is even 12 times as large as the amount of

[i]The drawn line with dots shows the volume as function of time and the thin line, the linear trend.

[j]One may object, that the accuracy of an average of a lot of stochastic numbers is much better than the accuracy of each of the individual numbers. However, the inaccuracy of the reference level (a systematic error!) forms a very substantial constituent of the total inaccuracy.

accretion monitored. In recent times one even is not able to spot any sedimentation.

Sand pressure. The words "sand pressure" try to express a hypothesis which is implied, and which suggests a net shoreward sand transport in the middle of "Vlakte van de Raan" (Fig. 8.20a). Instead of providing sand nourishment to the beach, it leads to coastal erosion, because it causes the Oostgat (Fig. 8.22a) to shift in coastal direction. The sand supplied causes sedimentation at the seaward bank of the Oostgat (according to this hypothesis). Where on one hand the narrowing of the channel leads to higher velocities and scour, this does not only compensates the sedimentation, but also leads to erosion at the opposing side, i.e. along the coast.

From the current pattern in the area of the shoals no clear evidence for net coastal sand drift can be deduced.

COAST 2000-program contains an investigation of the bottom changes, based upon the scheduled monitoring as usual and a series of additional soundings of detailed areas during two winter seasons (with gales) and furthermore a series of wave registrations, providing data concerning the effect of storms.

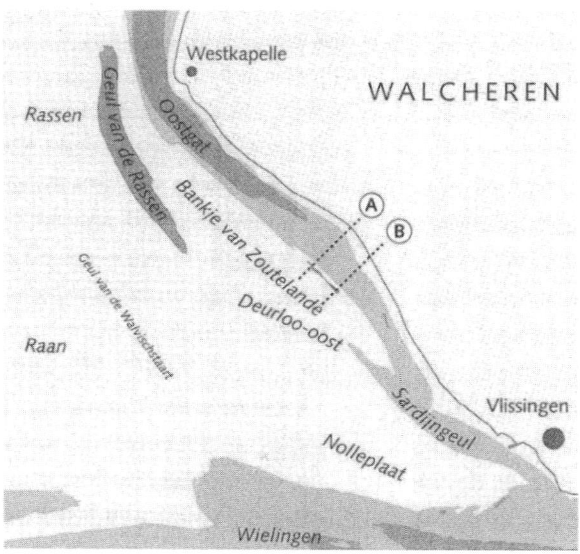

Fig. 8.22a Ranges of which the cross-sections are depicted in Fig. 22b,c.

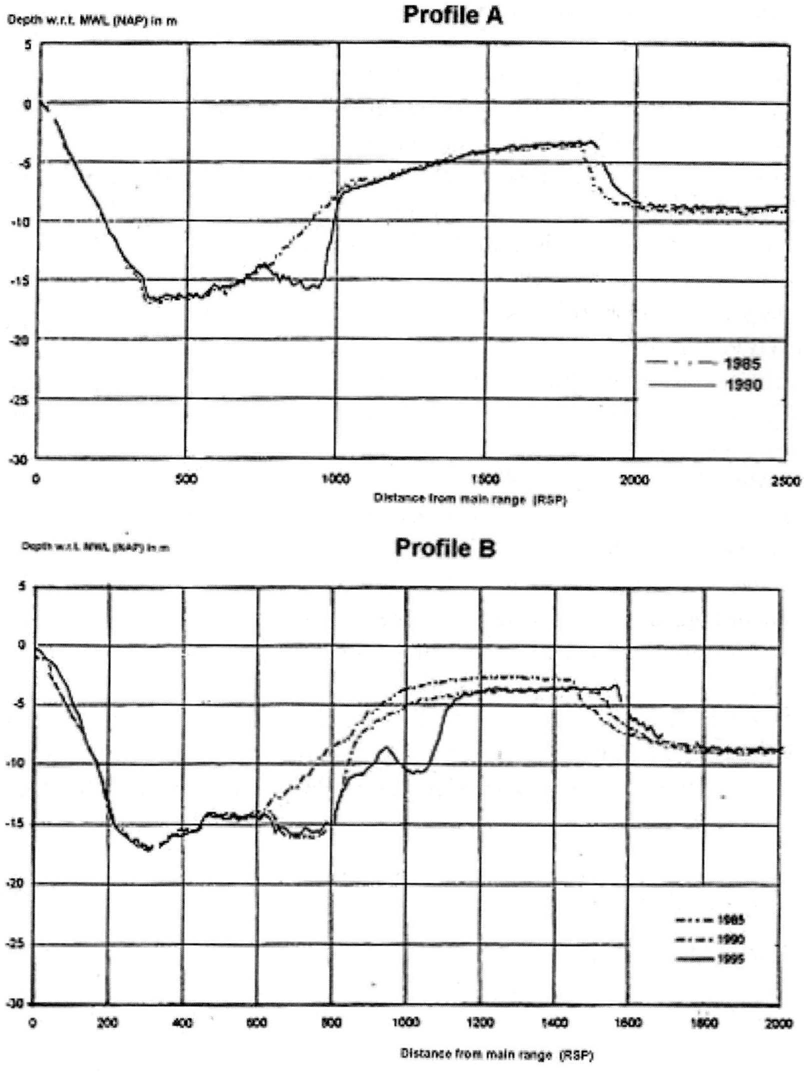

Fig. 8.22b,c Effects of sand mining Bankje van Zoutelande.

Results do not support the hypothesis. Figure 8.22b,c demonstrates that the former sandpits, providing sand for construction of dikes, etc. (years ago), are in no way filled up as yet.

Roelse summarizes in the following way:

(a) Also at the seaward side of the tidal estuary after large-scale human interference the tidal channels apparently show a trend leading to the equilibrium relationships of Gerritsen and de Jong.[12,14]
(b) Up to now, no physical explanation has been found for large-scale migrating sand waves propagating over the shoals.[k]
(c) The system of shoals and tidal channels in the outer delta of the Western Scheldt is rather stable on a scale of centuries. In the central part channels are orientated in the direction NW instead of in direction WNW formerly.
(d) There are serious limitations in the drafting of balances of sediment. These are caused by incomplete or inaccurate information concerning surveys or concerning the data with respect to dredging or dumping of material. Nearly all changes of volume are within the range of accuracy.
(e) Human interference in the Southern part of the outer delta of the Western Scheldt dominates greatly the natural changes.
(f) Considering the volumes of dredging and dumping, in the vicinity of Zeebrugge this system does not seem effective. Quantities of dredging and dumping surpass the resulting deepening, c.q. sedimentation many times.
(g) If one blames the "sand pressure" for the erosion of the SW-side of the island of Walcheren (cf. Fig. 8.20a), this hypothesis cannot be founded as yet. Neither literature review, nor field research proves this theory by now.

Future research

Christel Somers put the question to Bakker:

"What would be your aims, if you had a team of 100 scientists available for coastal dynamics?"

Answer: naturally, the aims would depend on the aims of the sponsor. He anyway would want bang for his bucks. But let the offer be in the same biotope, which created the present book: the one of Dutch coastal engineers.

[k] Those sandwaves are of a different kind compared to (for instance) the Vliehors sandwaves (Subsection 5.3.2). Where the Vliehors sandwaves form a wavy coastline, the present sand waves propagate over the shoals on the outer side of the estuaries. Contrarily to the Vliehors sandwaves, the waves presently considered do not attenuate during propagation.

Anyway, I should want to have security guards for Mr. Fig. 1.1. Workers like him are very vulnerable for management changes. It is quite correct if from time to time question marks are placed near long-term measurement programs and thus I think "popipub"[1] papers (in Delft University terms) are of vital importance for the Dutch coastal engineering of the future. This means papers of the kind, summarized in Chapter 2: phenomenological research, with the physical feeling of Van Veen as ideal and visualized with modern graphical techniques. Results should be presented (at least) in the periodical (5-year) Coastal Management reports of the Rijkswaterstaat. As a tradition, every 5 year a 25-year prediction and every 10 year a 100-year prediction should be given; something in the shape of Fig. 2.8, but extended over the whole coast and over a wider zone.

The team of security guards should contain a hangman for persons who throw away coastal areal photographs of the past because of space problems in storing. In this division of my coastal army the emphasis would be on evaluation of data rather than on handling data.

A technique like Kalman filtering could contribute to provide the extrapolations, where interpolation (probably leading to more accuracy of the extrapolation) could be accomplished by video techniques like the one of Aarninckhof *et al.*[2] and Holman *et al.*[17]

However, for any Dutch coastal area apart from Hook of Holland-Callantsoog any extrapolation in time of coastal shapes would be of doubtful value if the transversal motion of tidal channels would not be incorporated.

Thus research concerning tidal channel rotation (to which in the next paragraph will be referred to) seems of crucial importance.

The team should have an instinct to let itself pay off.

Antennas should be directed on one hand to coastal managers who link vision with feeling for reality, like Waterman[39] and on the other hand the Scientist Army should consist of water wizards who let the waves, tides and density differences do the work instead of the dredgers. In this way they could tell the managers: "Yes, you can". Certainly, present dredgers are mightier than nature; on the other hand, decision making takes so long, that one could gradually mold a coast on a long term basis, learning from signals from nature in the meantime concerning unwanted effects: "Sand diodes". Groskamp[15] converting the random alternate sand motion in a uni-directional one would be one of my favorite topics.

---

[1]Easy-digest publications.

As mentioned, in Chapter 2, coastal changes linked with transversal motion of tidal channels are large compared to the ones occurring along a straight coast; thus research concerning this topic could easily pay off. The teasing D.C.-formula, qua dimension incorrect, is meant to challenge researchers to extend the research of Huys[18] to other areas and to replace D.C. by common sense.

"Borrowing from the sea" instead of "lending to the sea"[m] could be a jackpot as well.

Sand, dumped in this way, that it emerges above the waterline has economic value (tourism, etc.); contrarily to sand under water. If, on the long term, one would fear of coastal erosion, initiated by erosion of "the foundation of the Netherlands" (O.D. −10 to −20 m) it could be wise to anticipate for that by putting additional sand on (or near to) the beach instead of at the deeper part of the shore face[n] Naturally it would erode, but in the meantime it could have provided its benefits.[o]

Chapter 7 (interaction between outer deltas and tidal basins) only emphasizes one aspect of this interaction, often an important aspect (I think) but in no way the only one.

For a realistic simulation, the effect of sand, together with the effect of silt plus the effect of higher harmonics should be considered. In the "rain-in-the-drain" story it is presumed, that fresh water, without sediment flushes in above the low-water line and that it is the local velocity (depth-averaged only over the zone below LW) which determines the sediment transport. For silt transport, naturally those assumptions do not hold. Thens assumptions like the one of Krol (ref. [46] in Chapter 7) might be better. A theory, combining the various aspects (also: higher harmonics) should appear on the list of future research.

It has been assumed, that the 100 coastal engineers I should get to my disposal could be directed to the topics treated in this book. This however covers by no means all the coastal dynamics: As mentioned in ch.., only (some) tools for solving problems in the range of medium- and large scale coastal behavior are given. Thus the microscale (topic of sediment transport under influence of waves and current) was considered out of

[m] Without guarantee! (Wiersma).

[n] Unless one would use the sand for building prismatic shoals, which could be used as sand diodes.[15,34]

[o] Evidently, it also has to be kept in mind, that dumping of sand under water is much cheaper than supply on a beach. The economic optimum (depending on site and circumstances) has to be determined.

scope, as were numerical techniques. Mind also the topics and suggestions, brought forward by the Forum (for instance: the one of Roelvink!).

In order to make coastal research effective, the whole broad scope should be developed simultaneously.

References

1. Aarninkhof, S. G. J. and T. van Kessel, Data Analysis Outer Delta North Zealand (Data Analyse Voordelta; in Dutch). *Large Scale morphological changes 1960–1996*, Delft Hydraulics; Rep. Z 2694, 1999.
2. Aarninkhof, S. G. J. and R. A. Holman, Monitoring the nearshore with video. *Backscatter*, **10**(2), 8–11, 1999.
3. Bakker, W. Th., C. van de Kerk and J. H. de Vroeg, *Determination of coastal constants in mathematical line models, Second European*, workshop on coastal zones Loutraki, Greece, 1988.
4. Bliek, A. J. and C. Storm, *Bars and maintenance dredging in the Western Scheldt*, Proceedings 14th World Dredging Congress Amsterdam, Delft, The Netherlands, pp. 643–656, 1995.
5. Bruun, P. and F. Gerritsen, *Stability of Coastal Inlets* Amsterdam, North Holland Publishing Company, 1958.
6. Capobianco, M., H. Hanson, M. Larson, H. J. Steetzel, *et al.*, *Models for nourishment design and evaluation reviewing their applicability*, Abstract second overall workshop SAFE-project, Sylt, Germany, March, 1998.
7. De Boer, M. *et al.*, *Series: Erosion and Sedimentation in the Wadden Sea.*
8. De Boer, M., G. W. Glim, G. Kool and M. F. Lieshout, Part 1: Erosion and Sedimentation in the inner delta of the Zeegat van Texel 1932–1982 (in Dutch), Rijkswaterstaat, Directorate North Holland, Rep. ANWX 87.H201 (corrected by de Boer: Sept. 1990), 1987.
9. De Boer, M., G. W. Glim, G. Kool and M. F. Lieshout, Part 2: Erosion and Sedimentation in the inner delta of the Eijerlandse Gat; 1934–1982 (in Dutch), Rijkswaterstaat, Directorate North Holland, Rep. ANWX 89.H202, 1989.
10. De Boer, M., G. W. Glim, N. de Graaff, G. Kool and M. F. Lieshout, Part 3: Erosion and Sedimentation in the inner delta of the Zeegat van het Vlie 1933–1983 (in Dutch), Rijkswaterstaat, Directorate North Holland, Rep. ANWX 88.H204, 1988.
11. De Boer, M., G. Kool and M. F. Lieshout, Part 4: Erosion and Sedimentation in the inner delta of the Zeegat van Ameland 1926–1984 (in Dutch), Rijkswaterstaat, Directorate North Holland, Rep. ANVX 91.H202, 1991.
12. De Boer, M., G. W. Glim, N. de Graaff, G. Kool and M. F. Lieshout, Part 5 & 9: Erosion and Sedimentation in the outer delta of the Zeegat van Texel and adjacent coastal areas 1925–1981 (in Dutch), Rijkswaterstaat, Directorate North Holland, Rep. ANWX 88.H201, 1988.
13. De Boer, M., G. W. Glim, G. Kool and M. F. Lieshout, Part 6 & 10: Erosion and Sedimentation in the outer delta of the Eierlandse Gat and

adjacent coastal areas 1926–1983 (in Dutch), Rijkswaterstaat, Directorate North Holland, Rep. ANWX 90.H204, 1990.
14. Kool, G. and M. F. Lieshout, Rijkswaterstaat, Directorate North Holland, Rep. ANVX 91.H201, Supplement Part 6: Erosion and Sedimentation in the outer delta of the Eierlandse Gat 1981–1987 (in Dutch), 1991.
15. De Boer, M., G. W. Glim, G. Kool and M. F. Lieshout, Part 7: Erosion and Sedimentation in the outer delta of the Vlie 1933–1982 (in Dutch), Rijkswaterstaat, Directorate North Holland, Rep. ANWX 86.H210, 1986.
16. ir. Rakhorst H. D., Part 11 (supplement part 7): Erosion and Sedimentation in the outer delta of the Vlie and adjacent coastal areas 1926–1984 (in Dutch), Rijkswaterstaat, Directorate North Holland, Rep. ANV-2000-, 2000.
17. De Boer, M., G. Kool, M. F. Lieshout and D. L. Ulm, Part 8: Erosion and Sedimentation in the outer delta of the Zeegat van Ameland 1926–1982 (in Dutch), Rijkswaterstaat, Directorate North Holland, Rep. ANVX 91.H205, 1991.
18. ir. Rakhorst, H. D., G. Kool and M. F. Lieshout, Part 12 (supplement part 8): Erosion and Sedimentation in the outer delta of the Zeegat van Ameland and adjacent coastal areas 1926–1989 (in Dutch), Rijkswaterstaat, Directorate North Holland, Rep. ANV-92.201, 1993.
19. De Boer, M. and G. C. Visser, Erosion and sedimentation in the Western Waddensea, 1963–1978, Rijkswaterstaat, Study Dept. at Hoorn; Rep. WWKZ-80.H001, 1980.
20. De Ronde, J. G. and J. A. Vogel, sea level rise and hydro meteo scenario's, In: Coastal Protection after 1990; Technical Report 6, Rijkswaterstaat, Coastal Management Report, 1990.
21. De Ruig J., *et al.*, The coast in broader scope (in Dutch), Coastal Management Report 1995, Core report Rijkswaterstaat, Rep. RIKZ 95-005, 1995.
22. De Vroeg, J. H., E. S. P. Smit and W. T. Bakker, *Coastal genesis*, Proc. 21st Int. Conf. On Coastal Engng, Malaga, 1988.
23. Gerritsen, F. and H. de Jong, *Stability parameters of the Western Scheldt Estuary*, Proc. 19th Int. Conf. On Coastal Engng, Houston, pp. 3078–3093, 1984.
24. Gerritsen, F. and H. de Jong, Stability of cross- channel profiles in the Wadden area (in Dutch), Rijkswaterstaat, Study Dept. at Flushing, Rep. WWKZ-84.V016, 1985.
25. Gerritsen, F., H. de Jong and A.Langerak, *Cross-sectional stability of estuary channels in the Netherlands*, Proc. 22nd Int. Conf. On Coastal Engng, Delft, pp. 2922–2935, 1990.
26. Groskamp, B. E. M.-A., Artificial Prismatical Shoals (in Dutch), Report in the scope of a B.Sc. Study; Rijkswaterstaat and Technical College The Hague, Netherlands, 1995.
27. Hanson, H., J. Newe and M. Larson, *Data analysis and one-line model application to the island of Sylt*, Abstract second overall workshop SAFE-project, Sylt, Germany, March, 1998.

28. Holman, R. A., A. H. Sallenger Jr, T. C. Lippmann and J. W. Haines, The application of video image processing to the study of nearshore processes. *Oceanography*, **6**(3).
29. Huys, S. W. E., *Migration of tidal channels in the outer delta (in Dutch)*, University of Utrecht, Earth Sciences, 1992.
30. Kevelam, D. J. and H. Postma, *Morphological Investigation of The Krabbengat (in Dutch)*, Zanen & Verstoep, Civil Engineering Department, 1988.
31. Larson, M., N. C. Kraus and R. A. Wise, *Modelling long-term profile evolution based on equilibrium profile theory*, Abstract second overall workshop PACE-project, London, The United Kingdom, March 1998.
32. O'Brien, M. P., Equilibrium flow areas of inlets on sandy coasts. *Proc. ASCE, J. Waterways, Harbours Coastal Eng. Divi.*, **95**, WWI, 1969.
33. Rakhorst, H. D., *Evaluation of "Seaward Coastal Defense": Texel: Dam Eierland"(in Dutch)*, Rijkswaterstaat, Directorate North Holland, June 30 1999.
34. Rakhorst, H. D., Erosion and sedimentation of the Western part of the Wadden Sea and of the adjacent part of the North Sea; 1926/1933–1982, Rijkswaterstaat, Directorate North Holland, Rep. NH-ANV-2000-20 July 2000.
35. Roelvink, J. A., Th. J. G. P. Meijer, K. Houwman, R. Bakker and R. Spanhoff, *Field Validation and Application of a Coastal Profile Model.* Proc. Coastal Dynamics '95, Gdansk, ASCE, New York, pp. 818–828.
36. Steetzel, H. J., Prediction of development coastline and outer deltas Wadden coast for the period 1990–2040, WL Delft Hydraulics, report H1887 (in Dutch), 1995.
37. Steetzel, H .J., *Using a Modified Multi-layer Model for The Assessment of Large-scale Coastal Evolution.* Abstract second overall workshop PACE-project, London, The United Kingdom, March 1998a.
38. Steetzel, H. J., *Application of A Modified Multi-layer Model for The Assessment of The Evolution of Sylt.* Abstract second overall workshop SAFE-project, Sylt, Germany, March 1998b.
39. Steetzel, H. J. and J. H. de Vroeg, *Beach nourishments & Rock Structures — Behaviour-oriented Modelling of The Combination of Nourishments and Rigid Shore Parallel Structures.* Centre for Civil Engineering Research, Codes and Specifications The Netherlands, 1997.
40. Steetzel, H. J., de Vroeg, J. H. and L. C. van Rijn, Pilot-version of the PonTos-model — Definition and preliminary elaboration of a behaviour-oriented model for the assessment of the large-scale development of the Dutch coast, Joint venture Alkyon Hydraulic Consultancy & Research — WL Delft Hydraulics, Report A066/H3159, 1998.
41. Steetzel, H. J., J. H. de Vroeg, L. C. van Rijn and J. M. T. Stam, *Morphological Modelling using a Modified Multi-layer Approach.* Paper presented at the ICCE in Copenhagen, Denmark, 1998.
42. Stive, M. J. F. and W. D. Eysink, Prediction coastline phase 3: dynamical coastal model (in Dutch), Coastal Management Report 1990 of Ministry of

PublicWorks, Underlying Rep. 3.4 of Techn. Report 5: "Coastal Prediction" Also published as Rep. H 825 of WL|Delft Hydraulics, 1990.

43. Stive, M. J. F., J. A. Roelvink and H. J. de Vriend, *Large-scale coastal evolution concept*, Proc. 22nd Int. Conf. on Coastal Engng, Delft pp. 1962–1975, 1990.
44. Van den Berg, J. H., *Aspects of sediment- and morphodynamics of subtidal deposits of the Oosterschelde.* Rijkswaterstaat Communications No. 43, The Hague, The Netherlands.
45. Van Engelen, H. J., *Sand-diodes for Delfland (in Dutch)*, M.Sc.Thesis Delft University of Technology; Civil Engng; Sect. for Coastal Res, 1995.
46. V. d. Heuvel, Tj. (ed), J. H. M. de Ruig, A. W. Kraak, H. D. Rakhorst, A. Prakken, R. Hillen, P. Roelse, C. Israel and G. H. Snijders, Evaluation of "Seaward coastal defense"(in Dutch), Rijkswaterstaat, RIKZ; Report RIKZ-99.009, 1999.
47. Van Kleef, A. W., *Analysis of the changes in the large scale sediment balance of the Middelgat area in the Western Scheldt*, (internal communication in Dutch), Middelburg, The Netherlands, 1994.
48. van der Kreeke, J. and J. Haring, Equilibrium flow areas in the Rhine-Meuse delta. *Coastal Eng.* 3, 97–111, Delft, The Netherlands, 1979.
49. Van Rijn, L. C., Sand budget and coastal changes of the central Dutch coast between den Helder and Hoek vamn Holland, "Review of research 1990-1995: Dynamics of a central Dutch closed coastal system"; Delft Hydraulics Rep. H2129, 1995.
50. Waterman, R. E., *To An Integral Coastal Management via Building with Nature.* Delft, Waltman, 1990.